# $Q$uantum
# $B$io-Informatics V

Proceedings of Quantum Bio-Informatics 2011

# QP–PQ: Quantum Probability and White Noise Analysis*

Managing Editor: W. Freudenberg
Advisory Board Members: L. Accardi, T. Hida, R. Hudson and
K. R. Parthasarathy

*Q P – P Q*
*Quantum Probability and White Noise Analysis*
**Volume XXX**

# Quantum Bio-Informatics V

## Proceedings of Quantum Bio-Informatics 2011

Tokyo University of Science, Japan          7 – 12 March 2011

### Editors

**Luigi Accardi**
*Università di Roma "Tor Vergata", Italy*

**Wolfgang Freudenberg**
*Brandenburgische Technische Universität Cottbus, Germany*

**Masanori Ohya**
*Tokyo University of Science, Japan*

NEW JERSEY · LONDON · SINGAPORE · BEIJING · SHANGHAI · HONG KONG · TAIPEI · CHENNAI

*Published by*

World Scientific Publishing Co. Pte. Ltd.

5 Toh Tuck Link, Singapore 596224

*USA office:* 27 Warren Street, Suite 401-402, Hackensack, NJ 07601

*UK office:* 57 Shelton Street, Covent Garden, London WC2H 9HE

**British Library Cataloguing-in-Publication Data**
A catalogue record for this book is available from the British Library.

**QP–PQ: Quantum Probability and White Noise Analysis — Vol. 30**
**QUANTUM BIO-INFORMATICS V**
**Proceedings of the Quantum Bio-Informatics 2011**



ISBN 978-981-4460-01-9

Printed in Singapore by Mainland Press Pte Ltd.

# PREFACE

This volume is based on the fifth international conference of quantum bio-informatics held at the QBI Center of Tokyo University of Sciences.

The purpose of the conference is towards new stage making interdisciplinary bridges in mathematics, physics, information and life sciences, in particular, research for new paradigm for information science and life science on the basis of quantum theory.

More than 100 researchers in various fields such as mathematics, physics, information and biology come from all over the world. The conference was held for nearly one week, and we had a lot of fruitful discussion. In this fifth conference, particular attention is come up on quantum entanglement, simulation of bio-systems, brain function, quantum like dynamics and adaptive systems. Most of speakers gave care to the relation between their own topics and the mystery of life.

The papers submitted in this volume are all refereed, whose contents are related to one of the following subjects:

(1) Mathematics of Cryptography and its related topics
(2) Quantum algorithm and computation
(3) Quantum entanglement
(4) Quantum entropy and information dynamics
(5) Quantum dynamics and time operator
(6) Stochastic dynamics and white noise analysis
(7) Brain activity
(8) Quantum like models and PD game
(9) Quantum physics and superconductivity
(10) Quantum tomography and sufficiency
(11) Adaptation in Plants
(12) Alignment of sequences

*Luigi Accardi*
*Wolfgang Freudenberg*
*Masanori Ohya*

# Five years of QBIC

Masanori Ohya

Department of Information Sciences,
Tokyo University of Science, Japan

## 1. Aims of QBIC

The quantum bio-informatics center (QBIC) was founded in 2006 towards new stage making interdisciplinary bridges in philosophy, mathematics, physics, information and life sciences. Our research center (QBIC) tries to find a new paradigm for information science and life science on the basis of quantum mathematics. Several researchers more than 100 on mathematics, physics, information theory and biology who are interested in mathematical study worked together in QBIC during this five years from 2006 to 2010.

To solve the mystery of life is one of the most interesting problems in 21 century. After discovery of DNA, people believes that one key to read the riddle will be hidden in the process how information of life is stored and its change and transmission are made. Concerning the information transmission (communication), quantum information opened a new door and is expected to understand new aspects of existence in-itself.

More concretely, the immensely long DNA, sequence of four bases in the genome, contains information on life, and decoding or changing this sequence is involved in the expression and control of life. In quantum information, meanwhile, we produce various "information" by sequences of two quantum states, and think of ways of processing, communicating and controlling them. It is thought that the problems we can process in time "T" using a conventional computer can be processed in time nearly "log T" using a quantum computer. However, the transmission and processing of information in the living body might be much faster than those of quantum computer and communication.

Seen from this very basic viewpoint, developing the mathematical principles that have been found in quantum information should be useful in constructing mathematical principles for life sciences, which have not been established yet. The mechanism of processing information in life is also expected to be useful for the further growth of quantum information.

## Research project

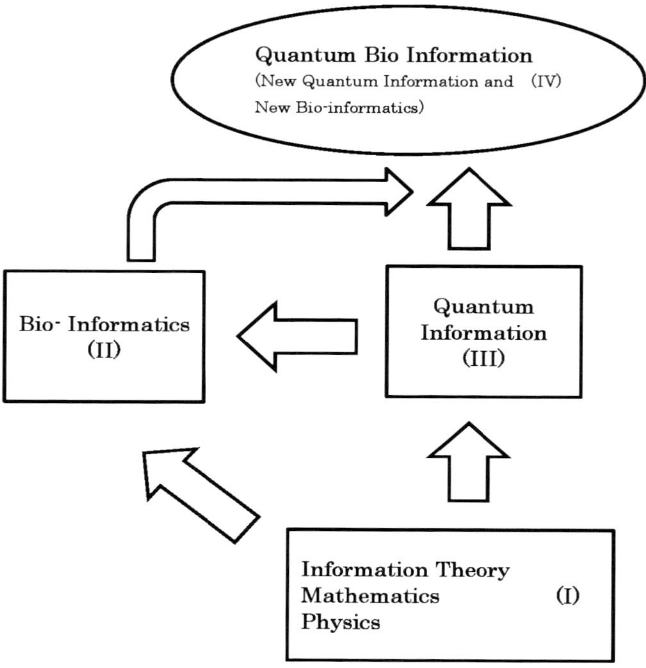



Figure 1.   QBIC Research Project

The way of our research is

(1) to return to the starting point of bio-informatics and quantum information, fields that and to solve these fundamental problems, and

(2) to seriously attempt mutual interaction between the two, with a view to enumerating and solving the many fundamental problems they entail.

In our view, there is no similar research center in the world to return to the basics of bio-information and quantum information and to focus on the correlation between the two with a view to new development of each.

Our way with targets and goal are described in the figure below:

We had more than 200 papers published in this five years. Most of them have first published in the five proceedings of International Conference held in Tokyo University of Science.

I will here review basic results of some achievements in this five years of QBIC.

## 2. Solving the mystery of life

Solving the mystery of life requires several stages (1) Metaphysical, (2) Biological & Physical, (3) Mathematical. The works of the stages (1) and (2) have been done for a long time even in the "new" life science, that is, many philosophical considerations and various experiments have been done, and several (tentative) theories have been made. However it is also true we had not a basic mathematical rule (theory) in the life science so that many researchers could accept it as quantum mechanics.

In order to make such a theory, we have to try to develop fundamentals in various fields (mathematics, physics, information theory) with intention to the goal, i.e., finding the first principle understanding the life itself.

Biological systems are open systems. Biological systems are multi–component and context dependent. Biological subsystems in a biological system are locally interacting each other. Therefore the state of the biological systems depends on its surrounding and the eve of itself. These observations entail that biological systems are adaptive. We have to find a mathematical rule to describe all of those.

However, in order to make our dream realize, we have to develop each field such as mathematics, quantum physics, information, structural biology and bio-informatics so that we can use the fructification to achieve new paradigm as discussed above.

## 3. Some works appeared in conference of QBIC

I itemize some works appeared in the conferences of QBIC during this five years. It is beyond the introduction and my ability to review all works appeared in the five years conferences, so that I only mention some mathematical trials somehow related to life sciences. The fruitful results of various works can been seen in the series of the five proceedings of QBIC conferences.

### 3.1. *Examples of researches in QBIC*

3.1.1. *Concerning <I> of the Figure above*

- White noise, stochastic analysis and some applications to DDS (Hida, Streit, SiSi, Accardi, Volovich, Smolyanov, Fichtner, Inoue, Iriyama, Hara, Ohya); Further developments of Hida calculas and its various applications have been considered, e.g., analysis of drug delivery systems.

- Mathematical physics and noncommutative analysis (Araki, Accardi, Arai, Belavkin, Jamiolkowski, Ojima, Petz, Hiai) have been developed so that we van apply to life sciences, for examples, quantum tomograhy, micro-macro duality.
- Statistics with symmetry (Tomizawa, Miyamoto, Tahata), which enables to analyze several fuctions of human body.
- Supercoductor $\Longrightarrow$ KS model and its development (Kamimura, Sakata, Ushio), which will be related to a realization of qubit.
- Fundamental problems in quantum physics - Bell's inequality, adaptive dynamics, quantum like systems, micro-macro duality, nonequilibrium dynamics (Accardi, Asano, Khrennikov, Volovich, Ojima, Suzuki, Oryu, Ohya)
- Ulitimate secure and fast crypto-algorithm can be found (Acradi, Regoli, Iriyama, Ohya).

## 3.2. *Concerning <II>*

- Protein folding and simulation with brownian molecular dynamics (Yamato, Ando, Takeda, Im).
- Code structure of genes and works of cis-elements (Miyazaki, Sato, Khrennikov, Regoli, Wanke).
- Signal network of envitonmental sensing and adaptation in plants (Kuchitsu).
- Study of biosystems by information measures (R. Belavkin, A. Accardi, Im, Sato, Hara, Ohya), e.g., the most accurate alignment method is founded in QBIC.

## 3.3. *Concerning <III>*

- Quantum teleportation (Fichtner, Freudenberg, Kosakowski, Asano, Tanaka, Ohya), in which new mathematics and its realization are discussed, where the teleportation process can be linear and can use all entangled states.
- Quantum entanglement (Belavkin, Jamiolkowski, Accardi, Matsuoka, Kosakowski, Chruscinski, Majewski, Michalski, Hirota, Ohya), e.g., (1) we can treat even infinite systems; (2) we can treat all correlations including claasical one; (3) .

- Quantum algorithm solving the NPC problem (Volovich, Accardi, Iriyama, Ohya) and its development; e.g., Our SAT alogithm can be used to solve factoring problem of Shor.
- Realization of qubit (Takayanagi, Morinaga).
- Quantum entropy and some applications (Accardi, Belavkin, Petz, Hiai, Araki, Iriyama, Matsuoka, Kosakowski,Watanabe, Suzuki, Ohya), e.g., Entropy production in linear respose dynamics, entropy production in photosythesis.

## 3.4. *Concerning <IV>*

- Generalized Turing machine (Volovich, Iriyama, Ohya) is proposed.
- New description of chaos(Kosakowski, Togawa, Volovich, Inoue, Ohya) $\Longrightarrow$ Adaptive dynamics$\Longrightarrow$ Application of chaos dynamics to the classification of HIV-1 and Influenza A viruses (Sato, Tanabe, Hara)
- Non-Kolmogorov probability and its applications; Adaptive dynamics and lifting are applied to find new probability law (Khrennikov, Accardi, Asano, Basieva, Tanaka, Ohya, Yamato)
- Mathematical model explaining the fuctions of brain are proposed in Fock space (K.-H. Fichtner, L. Fichtner, Freudenberg, Inoue, Ohya).
- Quantum tomography and sufficiency (Jamiolkowski, Petz, Matsuoka, Watanabe, Ohya)
- Quantum algorithm solving the protein folding is studied (Goto, Iriyama, Yamato, Ohya)
- New alignment (MTRAP) of amino acids was proposed in terms of entanglement (Sato, Hara, Ohya)
- Alignment by means of quantum algorithm was made (Iriyama, Sato,Ohya)
- Game theory in non-Kolmogorovian probanility theory has been proposed (Khrennikov, Basieva, Asano, Tanaka, Ohya)

**Remark** New science must be based on new philosophy. Science (theory) without philosophy is fragile. The 21 century is the era not for new technology but for new philosophy crossing our existence.

Finally, I like to ask all of you interested in the QBIC conference: In order to understand the mystery of life and various existence, you dare to have will and intention to use all materials you obtained in several different fields.

## References


1. Accardi, L., Freudenberg, W., Ohya, M., Quantum Bio-Informatics (Quantum Probability and White Noise Analysis, Vol. 21), World Scientific, 2008
2. Accardi, L., Freudenberg, W., Ohya, M., Quantum Bio-Informatics II (Quantum Probability and White Noise Analysis, Vol. 24), World Scientific, , 2009
3. Accardi, L., Freudenberg, W., Ohya, M., Quantum Bio-Informatics III (Quantum Probability and White Noise Analysis, Vol. 26), World Scientific, 2010
4. Accardi, L., Freudenberg, W., Ohya, M., Quantum Bio-Informatics IV (Quantum Probability and White Noise Analysis, Vol. 28), World Scientific, 2011
5. Accardi, L., Freudenberg, W., Ohya, M., Quantum Bio-Informatics V (Quantum Probability and White Noise Analysis), World Scientific, this volume
6. Ohya, M., Volovich, I., Mathematical Foundations of Quantum Information and Computation and Its Applications to Nano- and Bio-systems, Springer, 2011

# CONTENTS

Quantum Bio-Informatics V

pp. 1–13

# COMPLEXITY CONSIDERATIONS QUANTUM COMPUTATION

LUIGI ACCARDI

*Centro Vito Volterra,*
*Università degli Studi di Roma "Tor Vergata", Roma, Italy,*
*E-mail: accardi@volterra.uniroma2.it*

It is usually calimed that quantum computer can outperform classical computer. Is this statement true? We discuss this issue, not in general, but in the context of the most famous algorithm of quantum computation: Shor's algorithm.

## 1. Introduction

Shor's algorithm is supposed to achieve integer factorization faster than classical algorithms. In order to discuss this issue we shortly review Shor's algorithm and the strictly related Simon's period-finding algorithm (see section (2)). Then we argue that, since quantum computer is an analogical machine, the complexity estimates on quantum algorithms should involve the analysis of the concrete implementation of the operations whose use is required by these algorithms. An outline of this analysis is done in section (3).

Finally, in order to compare the performance of Shor's algorithm with some classical probabilistic factorization algorithms, the latter ones are shortly reviewed in section (4).

The essence of the factorization problem can be described as follows:

Given a natural integer $N = pq$, which is the product of two primes $p \neq q$, find $p$ and $q$. If $p$ and $q$ are large and satisfy additional diophantine conditions, the problem is hard and this difficulty has been exploited by a famous cryptographic algorithm.

A classical argument of number theory reduces the factorization problem to the problem of finding the period of the function $a \mapsto y^a \pmod{N}$ (see section (4)).

At the moment there is no classical algorithm that can find the period of the function $a \mapsto y^a \pmod{N}$ (hence the factorization problem) in a num-

2

ber of steps of order $O(\log N)$

It was however known a classical probabilistic algorithm that achieves this goal, not exactly, but with probability of order $O(1/\log N)$.

D. Simon [Sim94] proposed a quantum algorithm that allows to find, using $O(\log N)$ operations and with probability of order $O(1/\log N)$, the period of an arbitrary function $f : \{0, \ldots, N-1\} \to \{0, \ldots, N-1\}$ each value of which can be calculated with an algorithm of complexity of order $O(\log N)$ (with respect to some standard measure of complexity).

P. W. Shor [Sho94a] applied Simon's period finding algorithm to the function $a \mapsto y^a \pmod{N}$ to construct a quantum factorization algorithm which needs a number of steps of the same order of magnitude as the classical probabilistic algorithm and achieves the same result with a probability of the same order of magnitude.

Contrarily to the classical probabilistic algorithm, Simon's (hence Shor's) algorithm is based on additional physical assumptions the experimental verification of some of which is at the moment not available.

The goal of the present note is to point out some of these assumptions.

Some of the considerations in the present notes are contained in the unpublished lectures of the author at the Volterra-CIRM International School "Quantum Computer and Quantum Information", Trento, July 25-31, 2001.


## 2. Simon's period-finding quantum algorithm

Given $N \in \mathbb{N}$ let $f : \{0, 1, \ldots, N-1\} \to \{0, 1, \ldots, N-1\}$ be a periodic function with period $r$, i.e. $r$ is the smallest number in the domain of $f$ such that

$$f(x) = f(x +_N r) \qquad ; \qquad \forall x \in \{0, 1, \ldots, N-1\} \tag{1}$$

where the symbol $+_N$ denotes addition modulo $N$.

Since addition is taken modulo $N$, if (1) is satisfied by $r$, then it is also satisfied by $N - r$. Thus by definition of period, one must have

$$r \leq N - r \Leftrightarrow r \leq N/2$$

Suppose that we know that $f$ is an efficiently computable function, i.e. that, for each $x$, $f(x)$ can be efficiently computed (i.e. in a number of steps which is polynomial in the number of digits of $x$).

If these are the only informations on $f$, the only way to find exactly the period is to carry out an exhaustive search. This requires to calculate $f(x)$ for a set of $x$ of cardinality $N/2$. This algorithm is exponential in the number of bits required to specify $N$, which is of order $\log N$.

In absence of exact results one turns to probabilistic algorithms, either classical or quantum.

As we have seen the performances of the two are essentially the same.

## 2.1. *Ingredients of Simon's quantum period finding algorithm (QPFA)*

The state space of this algorithm is

$$\mathcal{H}^{2^n} \otimes \mathcal{H}^{2^n} \equiv (\mathbb{C}^2)^{\otimes n} \otimes (\mathbb{C}^2)^{\otimes n} \tag{2}$$

where $\mathcal{H} := \mathbb{C}^2$ is the so–called $q$–bit space (the reason why, in (2), one uses two copies of the space $\mathcal{H}^{2^n}$ is explained in Step (3) of the algorithm described in section (2.2)).

In the space $\mathbb{C}^2$ we fix the *computational basis*,

$$|0\rangle := \begin{pmatrix} 0 \\ 1 \end{pmatrix} \quad ; \quad |1\rangle := \begin{pmatrix} 1 \\ 0 \end{pmatrix}$$

which induces the basis (still called computational) in $(\mathbb{C}^2)^{\otimes n}$

$$|\varepsilon_1\rangle \otimes \cdots \otimes |\varepsilon_n\rangle =: |\varepsilon_1, \ldots, \varepsilon_n\rangle \quad ; \quad \varepsilon_j \in \{0, 1\} \tag{3}$$

Identifying the binary string $(\varepsilon_1, \ldots, \varepsilon_n)$ to the binary expansion of a natural integer through the formula

$$x = \sum_{j=1}^{N} \varepsilon_j 2^{j-1} \quad ; \quad x \in \{0, \ldots, N-1 = 2^n - 1\}; \ \varepsilon_j \in \{0, 1\} \tag{4}$$

and extending this notation to the corresponding vectors:

$$|x\rangle = |\varepsilon_1, \ldots, \varepsilon_n\rangle \quad ; \quad x \in \{0, \ldots, N-1\}; \ \varepsilon_j \in \{0, 1\} \tag{5}$$

we will use both the binary and the decimal notation so that the vectors of the form

$$|x\rangle \otimes |y\rangle = |\varepsilon_1, \ldots, \varepsilon_n\rangle \otimes |\eta_1, \ldots, \eta_n\rangle \ ; \ x, y \in \{0, \ldots, N-1\}; \quad \varepsilon_j, \eta_j \in \{0, 1\} \tag{6}$$

define the computational basis for the state space $\mathbb{C}^{2^n} \otimes \mathbb{C}^{2^n}$.

## 2.2. *Steps of Simon's quantum period finding algorithm (QPFA)*

**Step (1).**

The initial state of the quantum system is,

$$|0\rangle_n \otimes |0\rangle_n \in \mathbb{C}^{2^n} \otimes \mathbb{C}^{2^n} \equiv \mathbb{C}^N \otimes \mathbb{C}^N \tag{7}$$

i.e. all $2n$ $q$–bits are in the state $|0\rangle$.

**Step (2).**

Apply to the initial state the unitary operator

$$U_H := H^{\otimes n} \otimes 1$$

where $H$ is the discrete Fourier (or Hadamard) transform on $\mathbb{C}^2$ defined by linear extension of the map:

$$|0\rangle \mapsto \frac{1}{\sqrt{2}}(|0\rangle + |1\rangle) \quad ; \quad |1\rangle \mapsto \frac{1}{\sqrt{2}}(|0\rangle - |1\rangle)$$

and

$$H^{\otimes n} := H \otimes H \otimes \cdots \otimes H \qquad n\text{–times}$$

Since

$$H^{\otimes n}|0\rangle_n = \frac{1}{\sqrt{N}} \sum_{x=0}^{N-1} |x\rangle$$

the action of $U_H$ brings the initial state to

$$\psi_o := U_H|0\rangle_n = \frac{1}{\sqrt{N}} \sum_{x=0}^{N-1} |x\rangle|0\rangle_n \tag{8}$$

**Step (3a).**

Among the unitary extensions of the partial isometry defined by

$$|x\rangle|0\rangle \mapsto |x\rangle|f(x)\rangle \quad ; \quad x \in \{0, \ldots, N-1\} \tag{9}$$

choose one, denoted $U_f$, that can be physically realized.

**Step (3b).**

Realize the physical implementation of $U_f$.

**Step (3c).**

Apply to the state (8) the unitary operator $U_f$. This gives

$$U_f\psi_o =: \psi = \frac{1}{\sqrt{N}} \sum_{x=0}^{N-1} |x\rangle|f(x)\rangle \tag{10}$$

**Step (4a).**

Fix arbitrarily $u \in \{0, \ldots, N-1\}$ and construct the filter defined by the projection

$$P := 1_n \otimes |u\rangle\langle u| \tag{11}$$

**Step (4b).**

Apply the filter (11) to the quantum state described by the vector (10). This amounts to filter all the elements of the ensemble (10) for which $f(x) = u$ and to suppress all the remaining ones.

**Theoretical conclusion from Step 4b**

By by applying the Luders–Zumino formula of the quantum theory of measurement quantum information theorists conclude that the new quantum state of the total system is the one associated to the vector:

$$|\phi\rangle\langle\phi| := \frac{P|\psi\rangle\langle\psi|P}{Tr(P|\psi\rangle\langle\psi|)} = \frac{|P\psi\rangle}{\|P\psi\|}\frac{\langle P\psi|}{\|P\psi\|} \tag{12}$$

where $\psi$ is defined by (10) so that:

$$P\psi = \frac{1}{\sqrt{N}} \sum_{x=0}^{N-1} \langle u|f(x)\rangle|x\rangle|u\rangle = \frac{1}{\sqrt{N}} \sum_{x \in \{0,1,\ldots,N-1\}, f(x)=u} |x\rangle|u\rangle \tag{13}$$

Notice that

$$\|P\psi\|^2 = \frac{|\{x \in \{0,1,\ldots,N-1\}, f(x)=u\}|}{N} = \frac{|f^{-1}(u)|}{N} \tag{14}$$

We will discuss only the case in which $f$ satisfies the following additional conditions:

**Assumption 2.1.** *If $f$ is injective on the interval $[0, r)$.*

**Assumption 2.2.** *$r$ divides $N$ exactly, i.e. independently of $u \in \{0, \ldots, N-1\}$*

$$|f^{-1}(u)| =: M = N/r \in \mathbb{N} \tag{15}$$

In this case from (13), (14) one deduces that

$$\phi = \frac{P\psi}{\| P\psi \|} = \frac{1}{\sqrt{N}} \sum_{x=0}^{N-1} \frac{\langle u|f(x)\rangle}{\| P\psi \|}|x\rangle|u\rangle = \frac{1}{\sqrt{M}} \sum_{j=0}^{M-1} |d_u + jr\rangle|u\rangle \tag{16}$$

where $d_u + jr$, for $j = 0, 1, 2 \ldots M-1$, are all the values of $x$ for which $f(x) = u$ and $d_u < r$.

**Step (5a).**

Construct an apparatus implementing physically the unitary operator $U_{FT} \otimes 1_n$, where $U_{FT}$ is the discrete Fourier transform, given by:

$$U_{FT}|x\rangle = \frac{1}{\sqrt{N}} \sum_{k=0}^{N-1} e^{i2\pi kx/N} |k\rangle$$

**Step (5b).**
Apply to $\phi$ the unitary operator $U_{FT} \otimes 1_n$. This leads to the state

$$\frac{1}{\sqrt{N/r}} \sum_{j=0}^{N/r-1} U_{FT} |d_u + jr\rangle |u\rangle$$

$$= \frac{1}{\sqrt{N}} \frac{1}{\sqrt{N/r}} \sum_{j=0}^{N/r-1} \sum_{k=0}^{N-1} e^{i2\pi k d_u/N} e^{i2\pi kjr/N} |k\rangle |u\rangle$$

$$= \frac{1}{\sqrt{r}} \sum_{k=0}^{N-1} \left( \frac{1}{\sqrt{N/r}} \sum_{j=0}^{N/r-1} e^{i2\pi jrk/N} \right) e^{i2\pi k d_u/N} |k\rangle |u\rangle$$

$$= (U_{FT} \otimes 1_n)\phi \tag{17}$$

Since, if $kr/N$ is not an integer, then

$$\sum_{j=0}^{N/r-1} e^{i2\pi jrk/N} = \frac{e^{i2\pi(N/r)rk/N} - 1}{e^{i2\pi rk/N} - 1} = 0$$

the non zero terms in the $j$–sum are precisely those for which

$$kr/N \in \mathbb{N}$$

i.e. those for which $k$ is a multiple of $M = N/r$. Summing up: at the end of the 5–th step the state of the quantum system is:

$$\xi := (U_{FT} \otimes 1_y)\phi = \frac{1}{\sqrt{r}} \sum_{\{k \in \{0,\dots,N-1\}: k \text{ is a multiple of } M=N/r\}} |k\rangle |u\rangle \tag{18}$$

**Step (6).**
The final step of the algorithm is usually described in the quantum computer literature as follows (see [St97]):
... *The final state of the x register is now measured, and we see that the value obtained must be a multiple of $w/r$* ... (In our notations $w/r = N/r$) In other words, as a result of a measurement, one obtains an integer $k$ satisfying

$$k = \lambda N/r = \lambda M \Leftrightarrow k/N = \lambda/r \tag{19}$$

for some unknown integer $\lambda \in \{0, 1, \ldots, r - 1\}$. Thus, if we make many of these measurements, we have a non zero probability to find a $\lambda$ which is coprime to $r$.

Because of (18), in the relation (19), all these multiples will arise with equal probability $(1/r)$. Therefore one can apply the estimate (29), with $\lambda$ and $r$ replacing $y$ and $N$ respectively, and deduce that

$$P(\{\lambda \in \{2, \ldots, r - 1\} \ : \ \lambda \ \text{is coprime to r}\} \,) \geq \frac{1}{\log r} \qquad (20)$$

If $\lambda$ is coprime to $r$, we reduce the fraction $k/N$ to an irreducible fraction and this gives $\lambda$ and $r$ separately.

If we repeat the measurement of the $|k\rangle$–basis $h = O(\log r) \leq O(\log N)$ times, this will give $h$ possible candidates, $r_1, \ldots, r_h$, for the period and the estimate (20) shows that, with high probability, one of them should be the desired period.

## 3. Complexity considerations on Simon's quantum period finding algorithm (QPFA)

**Step (1).**
The initial state of the quantum system must be physically prepared so that all $2n$ $q$–bits are in the state $|0\rangle$.
An interesting $n$ has an order of a few thousands bits.
**Step (2).**
The unitary operator $U_H := H^{\otimes n} \otimes 1$ must be:
– constructed
– applied to the initial state
**Step (3a).**
One can appeal to a theorem of K.R. Parthasarathy [KRP01a] to conclude that, for any given function $f$, all the unitary extensions of the partial isometry defined by (9) can be physically realized by means of quantum gates, i.e. unitary operators acting only on a single pair of $q$–bits.
However the same theorem gives an upper estimate, on the number of gates to be used, which is exponential in the number of factors. In our case this number is $2n$.
Therefore, in absence of a proof that, among all the unitary extensions of the partial isometry defined by (9), there exists at least one that can be physically realized by a number of quantum gates which is polynomial in $2n$, it makes no sense to speak of the practical realizability of the algorithm.
**Step (3b).**

Even in presence of such a proof the actual physical implementation of the unitary operator might be a formidable task, given the fact that the $q$–bits involved are of order of thousands.

An alternative way could be the discovery of a physically realizable interaction (Hamiltonian) embedding the given unitary in a continuous time evolution. But, even supposing that this can be done, the continuous time evolution will create serious problems due to the extreme non robustness of the algorithm against small perturbations of the unitary operator $U_f$.

**Step (3c).**

Even supposing that the above problems can be solved, the concrete application of the unitary operator $U_f$ to the state (8) is a problem whose solution requires additional costs in terms of time and of experimental work to be done.

**Step (4a).**

The filter defined by the projection (11) must be constructed.

**Step (4b).**

The above comment, on the cost of the concrete realization of Step (3c), also holds for the application of the filter (11) to the quantum state described by the vector (10).

**Theoretical conclusion from Step 4b**

This conclusion heavily depends on the application of the Luders–Zumino formula of quantum measurement theory. This is quite different from the original von Neumann formula and implies that, after an incomplete measurement on a quantum system in a pure state, the system will still remain in a pure state.

Although not logically impossible, such a situation is against physical intuition because an incomplete measurement by definition does not produce maximal information while, in quantum mechanics, a pure state defines a situation of maximal information.

Only some very strong experimental evidence could prove that this natural intuition is wrong.

**Step (5a).**

One must construct an apparatus implementing the discrete Fourier transform on arbitrary quantum states (see above comments to Step (3c)).

**Step (5b).**

One must apply the above apparatus to the quantum state given by (16) (see above comments to Step (3c)).

**Step (6).**

Taken literally, the statement ... *The final state of the x register is now measured ...*, means that the last step of the algorithm consists in the determination of a quantum state.

But it is well known such a determination, in a space of dimension $d$ requires an order of $d$ measurements ($d^2$ in case of a mixture).

In our case $d = 2^{2n}$, i.e. it is exponential in $n$.

One might try a probabilistic approach, choosing at random a $k \in \{0, \ldots, N-1\}$ and evaluating experimentally the transition probability $|\langle k, \xi \rangle|^2$, which will be zero unless $k$ is a multiple of $M = N/r$. But, in the interesting cases, $r$ is of the same order of magnitude of $N$ so that $M = N/r$ is much smaller. This means that on average the number of trials to be done, before a multiple of $M = N/r$ appears, is of order $N$. Since $N = 2^n$, this is again exponential in $n$.

## 4. Classical reduction of the factorization problem to period finding

**Lemma 4.1.** *Let $x \in \mathbb{N}$ and define $x_1 := x + 1$ ; $x_2 := x - 1$ then 2 is the only possible common divisor of $x_1$ and $x_2$. In particular, if $x_1, x_2$ are both odd, then they have no common divisors.*

**Proof** Up to exchange odd indices we can assume that

$$x_1 > x_2$$

Suppose that $n \in \mathbb{N}$ is a common divisor of $x_1, x_2$. Then there are natural integers $x_1^1, x_2^1$ such that

$$x + 1 = nx_1^1 \quad ; \quad x - 1 = nx_2^1$$

$$nx_2^1 + 2 = nx_1^1 \Leftrightarrow x_2^1 + \frac{2}{n} = x_1^1$$

but $2/n \in \mathbb{Z} \Leftrightarrow n = 1, 2$. Thus 2 is the only possible common divisor for $x + 1$ and $x - 1$.

**Lemma 4.2.** *Let $x \in \{2, \ldots, N-2\}$ be any solution of the equation*

$$x^2 = 1 \qquad (mod \ N) \tag{21}$$

*such that $(x \pm 1) \neq 0 \ (mod \ N)$ and define $x_1, x_2 \in \{2, \ldots, N-1\}$ by*

$$x_1 = x + 1 \qquad (mod \ N) ; \qquad x_2 = x - 1 \qquad (mod \ N) \tag{22}$$

*Denote $gcd(x, N)$ the greatest common divisor of $x$ and $\mathbb{N}$.*
*Then the following factorization of $N$:*

$$N = gcd(x_1, N) \cdot gcd(x_2, N) \tag{23}$$

*takes place and is not trivial (i.e. both factors are $\neq 1$).*

**Proof.** In view of (22), (21) is equivalent to

$$x_1 x_2 = (x+1)(x-1) = x^2 + x - x - 1 = 0 \qquad (\mathrm{mod}\ N) \tag{24}$$

which means that the product $x_1 x_2 = (x+1)(x-1)$ is a multiple of $N$. By construction both $x_1$ and $x_2$ can be identified to numbers satisfying

$$1 < x_1, x_2 < N \tag{25}$$

and we know that there exist an integer $\lambda \geq 1$ such that

$$x_1 x_2 = \lambda N$$

For $j = 1, 2$ denote

$$g_j := gcd(x_j, N)$$

Then

$$x_j = g_j y_j$$

where $y_j$ does not divide $N$. In these notations

$$g_1 y_1 g_2 y_2 = \lambda N$$

and, since $y_1 y_2$ does not divide $N$, it must divide $\lambda$. Therefore

$$g_1 g_2 = \frac{\lambda}{y_1 y_2} N =: \lambda' N$$

where $\lambda' := \lambda / y_1 y_2 \in \mathbb{N}$. But $g_1$ and $g_2$ divide $N$ and, being $x_1$ and $x_2$ both odd, they have no common factor. Thus their product divides $N$ so that

$$1 = \lambda' \left( \frac{N}{g_1 g_1} \right)$$

Since both $\lambda'$ and $N/g_1 g_2$ are integers, this identity is possible if and only if

$$\lambda' = N/g_1 g_2 = 1$$

which is the factorization (23). Finally $g_1$ cannot be 1 because otherwise $x_1$ has no common factor with $N$ and therefore the product $x_1 g_2 y_2$ cannot be a multiple of $N$. Since $g_1$ and $g_2$ enter symmetrically in the argument, this factorization (23) is non trivial which is the thesis.

The following Lemma clarifies the connections between the factorization and the period–finding problem.

**Definition 4.1.** Let $V$ be a vector space. a function $V : V \to V$ is called periodic if there exists a vector $r \in V$ such that

$$F(x + r) = F(x) \quad ; \quad \forall x \in V \tag{26}$$

If $V$ is a ring identified to a totally ordered set (e.g. $\{0, 1, \ldots, N\}$ for some $N \in \mathbb{N}$) then the smallest $r$ satisfying (4.1) is called the period of $F$.

**Lemma 4.3.** *If $y \in \mathbb{N}$ is such that the function*

$$F(a) := y^a \ (mod \ N) \quad ; \quad a \in \{0, 1, \ldots, N - 1\} \tag{27}$$

*has an even period $r$, then $y^{r/2}$ is a solution of (21).*

Proof. Under our assumptions $r/2 \in \mathbb{N}$ and

$$(y^{r/2})^2 = y^r = 1 \quad (\mathrm{mod} \ N) \tag{28}$$

### 4.1. *Classical probabilistic factorization algorithms*

**Definition 4.2.** $y \in \mathbb{N}$ is called coprime to $N \in \mathbb{N}$ if $y$ and $N$ have no non trivial common factors.

In this case the minimum $r \in \mathbb{N}$ which satisfies (28) is called the *order of* $y(\mathrm{mod} \ N)$ and denoted $r_{y,N}$.

**Remark.** Otherwise stated, the order of $y(\mathrm{mod} \ N)$ is the period of the function $a \in \{0, 1, \ldots, N - 1\} \mapsto y^a \ (\mathrm{mod} \ N)$. If $y \in \mathbb{N}$ is coprime to $N \in \mathbb{N}$, then $r_{y,N}$ is well defined by Euler theorem and coincides with the period of the function (27).

It is known from number theory that, denoting $P$ the uniform measure on the set $\{0, 1, \ldots, N - 1\}$, i.e.

$$P(x) := 1/N \quad ; \quad x \in \{0, 1, \ldots, N - 1\}$$

as a probability space with , one has:

$$P(\{y \in \{0, 1, \ldots, N - 1\} \ : \ y \text{ is coprime to } N\} \,) \geq \frac{1}{\lg N} \tag{29}$$

This means that the overwhelming majority (more than $N/\log N$) of numbers in $\{0, 1, \ldots, N-1\}$ are coprime with $N$. This fact suggests the following

probabilistic strategy to look for solutions of the factorization problem.
- Pick at random, with uniform distribution, an $y \in \{0, \ldots, N-1\}$.
- By the above discussion the probability that, in $O(\log N)$ independent extractions, $y$ is coprime to $N$ is high.
- If $y$ is coprime to $N$ and $r_{y,N}$ is even, the number $x = y^{r_{y,N}/2}$ is a solution of equation (21).
- If $x$ is not a trivial solution, then by Lemma (4.2) we have a nontrivial factorization of $N$.

Since $y \in \{0, 1, \ldots, N-1\}$ is picked at random, the probability to have such a nontrivial factorization of $N$ is equal to the joint probability of the following three events:

$$[y \text{ is coprime to N}] \cap [r_{y,N} \text{ is even}] \cap [y^{r_{y,N}/2} \neq \pm 1 (\text{mod } N)] \qquad (30)$$

Let us introduce the following assumption.

**Assumption 4.1.** *With respect to the uniform distribution on* $\{0, \ldots, N-1\}$, *the events*

$$[y \text{ is coprime to N}] \text{ and } [r_{y,N} \text{ is even}] \cap [y^{r_{y,N}/2} \neq \pm 1 (\text{mod } N)]$$

*are independent.*

Under the above assumption the probability of the event (30) becomes equal to

$$P([y \text{ is coprime to N}]) \, P\left([r_{y,N} \text{ is even}] \cap [y^{r_{y,N}/2} \neq \pm 1 (\text{mod } N)]\right)$$

and the estimate (29) implies that this is

$$\geq \frac{P\left([y : r_{y,N} \text{ is even and } y^{r_{y,N}/2} \neq \pm 1 (\text{mod } N)] \mid [y \text{ is coprime to N}]\right)}{\lg N} \qquad (31)$$

where $P(\cdot|\cdot)$ denotes conditional probability. This conditional probability is estimated by the following theorem of number theory.

**Theorem 4.1.** *Let $N$ be odd with $k \geq 2$ different primes in its factorization. Then, one has:*

$$P\left([y : r_{y,N} \text{ is even and } y^{r_{y,N}/2} \neq \pm 1 (\text{mod } N)] \mid [y \text{ is coprime to N}]\right)$$

$$\geq 1 - \frac{1}{2^{k-1}} \qquad (32)$$

In particular the probability of the event (30) is estimated by

$$P([y \; : \; r_{y,N} \text{ is even}; \; y^{r/2} \neq \pm 1 (\text{mod } N)] \text{ and } gcd(y, N) = 1)$$

$$\geq \left(1 - \frac{1}{2^{k-1}}\right) \frac{1}{\lg N} \geq \frac{1}{2 \lg N}$$

Once this problem is solved, one picks $y$ at random and calculates $r_{y,N}$ according to Theorem (4.1).

In $O(\log N)$ trials, the probability that the pair $(y, r_{y,N})$ satisfies (30) in greater than $1/2 \lg N$.

Given a pair $(y, r_{y,N})$, satisfying (30), one solves the factorization problem using Lemma (4.2).

## References


1. Ekert A., Jozsa R.: Quantum computation and Shor's factoring algorithm, Rev. Mod. Phys. 68 (1996) 733
2. Parthasarathy K.R.: A remark on the unitary group of a tensor product of $n$ finite dimensional Hilbert spaces, Preprint Volterra n. 479 (2001)
3. Shor P. W.: "Algorithms for quantum computation: Discrete logarithms and factoring." In: Proceedings of the 35th IEEE Annual Symposium on Foundations of Computer Science, S. Goldwasser (ed.) IEEE Computer Society Press, New York (1994) 124-134
4. Shor P. W.: Polynomial-time algorithms for prime factorization and discrete logarithms on a quantum computer, in: Proc. 35th Annual Symp. on Foundations of Computer Science, Santa Fe, IEEE Computer Society Press (1994); revised version 1995a preprint quant-ph/9508027
5. Simon D.: On the power of quantum computation, in Proc. 35th Annual Symposium on Foundations of Computer Science IEEE Computer Society Press, Los Alamitos (1994) 124-134
6. Andrew Steane: Quantum computing quant-ph/9708022

Quantum Bio-Informatics V
© 2013 World Scientific Publishing Co. Pte. Ltd.
pp. 15–24

# QUANTUM MARKOV CHAINS AND ISING MODEL ON CAYLEY TREE

LUIGI ACCARDI

*Centro Interdisciplinare Vito Volterra*
*II Università di Roma "Tor Vergata"*
*Via Columbia 2, 00133 Roma, Italy*
*E-email: accardi@volterra.uniroma2.it*

FARRUKH MUKHAMEDOV[*]    MANSOOR SABUROV[‡]

*Department of Computational & Theoretical Sciences*
*Faculty of Science, International Islamic University Malaysia*
*P.O. Box, 141, 25710, Kuantan*
*Pahang, Malaysia*
[*] *E-mail: far75m@yandex.ru; farrukh_m@iiu.edu.my*
[‡] *E-mail: msaburov@gmail.com*

In the present paper we study forward Quantum Markov Chains (QMC) associated with Ising model on Cayley tree of order $k$. Using the tree structure of graphs, we give a construction of quantum Markov chains on a Cayley tree. By means of such constructions we prove the existence of a phase transition for the Ising model on a Cayley tree of order $k$ in QMC scheme. By the phase transition we mean the existence of two not quasi equivalent QMC for the given family of interaction operators.

*Keywords*: quantum Markov Chain; Cayley tree; Isingh model; phase transition

## 1. Introduction

One of the basic open problems in quantum probability is the construction of a theory of quantum Markov fields, that is quantum process with multi-dimensional index set. This program concerns the generalization of the theory of Markov fields (see [10,14]) to noncommutative setting, naturally arising in quantum statistical mechanics and quantum filed theory.

The quantum analogues of Markov chains were first constructed in [1], where the notion of quantum Markov chain on infinite tensor product algebras was introduced. Nowadays, quantum Markov chains have become a standard computational tool in solid state physics, and several natural applications have emerged in quantum statistical mechanics and quantum information theory. The reader is referred to [12,16,18] and the references cited therein, for recent developments of the theory and the applications.

First attempts to construct a quantum analogue of classical Markov fields have been done in [15,2]. Investigations of phase transitions of spin models on hierarchical lattices showed that there are exact calculations of various physical quantities (see for example, [8]). On the other hand, the study of exactly solved models deserves some general interest in statistical mechanics [8]. Therefore, it is natural to investigate quantum Markov fields over hierarchical lattices. For example, a Cayley tree is the simplest hierarchical lattice with non-amenable graph structure. First attempts to investigate quantum Markov chains over such trees was done in [7], such studies were related to investigate thermodynamic limit of valence-bond-solid models on a Cayley tree [11]. Therefore, in [6] we have introduced a hierarchy of notions of Markovianity for states on discrete infinite tensor products of $C^*$–algebras and for each of these notions we constructed some explicit examples. We showed that the construction of [3] can be generalized to trees. Note that a noncommutative extension of classical Markov fields, associated with Ising and Potts models on a Cayley tree, were investigated in [17]. In the classical case, Markov fields on trees are also considered in [19,20].

If a tree is not one-dimensional lattice, then it is expected (from a physical point of view) the existence of a phase transition for quantum Markov chains constructed over such a tree. In [4] we have provided a construction of forward QMC, such states are different from backward QMC. In that construction, a QMC is defined as a weak limit of finite volume states with boundary conditions. Such a QMC depends on the boundary conditions. For by means of the provided construction we proved uniqueness QMC, associated with $XY$-model on a Cayley tree of order two. In [5] it was established the existence of a phase transition for $XY$-model on the Cayley tree of order three. Combining results of the papers [4,5] one can state that by the increasing the dimension of the Cayley tree we are getting the phase transition for $XY$-model on the Cayley tree in QMC scheme.

Unlike $XY$-model on the Cayley tree, our goal, in this paper, is to establish the existence of a phase transition for the Ising model on the Cayley tree of any order. Using the construction defined in [4,5] we show the existence of a phase transition for the Ising model on a Cayley tree of order

$k$ in QMC scheme.

## 2. Preliminaries

Let $\Gamma^k = (L, E)$ be a semi-infinite Cayley tree of order $k \geq 1$ with the root $x^0$ (i.e. each vertex of $\Gamma^k$ has exactly $k + 1$ edges, except for the root $x^0$, which has $k$ edges). Here $L$ is the set of vertices and $E$ is the set of edges. The vertices $x$ and $y$ are called *nearest neighbors* and they are denoted by $l = < x, y >$ if there exists an edge connecting them. A collection of the pairs $< x, x_1 >, \ldots, < x_{d-1}, y >$ is called a *path* from the point $x$ to the point $y$. The distance $d(x, y), x, y \in V$, on the Cayley tree, is the length of the shortest path from $x$ to $y$.

Recall a coordinate structure in $\Gamma^k$: every vertex $x$ (except for $x^0$) of $\Gamma^k$ has coordinates $(i_1, \ldots, i_n)$, here $i_m \in \{1, \ldots, k\}$, $1 \leq m \leq n$ and for the vertex $x^0$ we put $(0)$. Namely, the symbol $(0)$ constitutes level 0, and the sites $(i_1, \ldots, i_n)$ form level $n$ ( i.e. $d(x^0, x) = n$) of the lattice.

Let us set

$$W_n = \{x \in L : d(x, x_0) = n\}, \qquad \Lambda_n = \bigcup_{k=0}^{n} W_k, \qquad \Lambda_n^c = \bigcup_{k=n}^{\infty} W_k$$

For $x \in \Gamma_+^k$, $x = (i_1, \ldots, i_n)$ denote

$$S(x) = \{(x, i) :\ 1 \leq i \leq k\},$$

here $(x, i)$ means that $(i_1, \ldots, i_n, i)$. This set is called a set of *direct successors* of $x$.

The algebra of observables $\mathcal{B}_x$ for any single site $x \in L$ will be taken as the algebra $M_d$ of the complex $d \times d$ matrices. The algebra of observables localized in the finite volume $\Lambda \subset L$ is then given by $\mathcal{B}_\Lambda = \bigotimes_{x \in \Lambda} \mathcal{B}_x$. As usual if $\Lambda^1 \subset \Lambda^2 \subset L$, then $\mathcal{B}_{\Lambda^1}$ is identified as a subalgebra of $\mathcal{B}_{\Lambda^2}$ by tensoring with unit matrices on the sites $x \in \Lambda^2 \setminus \Lambda^1$. Note that, in the sequel, by $\mathcal{B}_{\Lambda,+}$ we denote the positive part of $\mathcal{B}_\Lambda$. The full algebra $\mathcal{B}_L$ of the tree is obtained in the usual manner by an inductive limit

$$\mathcal{B}_L = \overline{\bigcup_{\Lambda_n} \mathcal{B}_{\Lambda_n}}.$$

In what follows, by $\mathcal{S}(\mathcal{B}_\Lambda)$ we will denote the set of all states defined on the algebra $\mathcal{B}_\Lambda$.

Consider a triplet $\mathcal{C} \subset \mathcal{B} \subset \mathcal{A}$ of unital $C^*$-algebras. Recall that a *quasi-conditional expectation* with respect to the given triplet is a completely

18

positive (CP) identity preserving linear map $\mathcal{E} : \mathcal{A} \to \mathcal{B}$ such that $\mathcal{E}(ca) = c\mathcal{E}(a)$, for all $a \in \mathcal{A}$, $c \in \mathcal{C}$.

A state $\varphi$ on $\mathcal{B}_L$ is called a *forward quantum d-Markov chain (QMC)*, associated to $\{\Lambda_n\}$, on $\mathcal{B}_L$ if for each $\Lambda_n$, there exist a quasi-conditional expectation $\mathcal{E}_{\Lambda_n^c}$ with respect to the triplet $\mathcal{B}_{\Lambda_{n+1}^c} \subseteq \mathcal{B}_{\Lambda_n^c} \subseteq \mathcal{B}_{\Lambda_{n-1}^c}$ and a state $\hat{\varphi}_{\Lambda_n^c} \in \mathcal{S}(\mathcal{B}_{\Lambda_n^c})$ such that for any $n \in \mathbb{N}$ one has

$$\hat{\varphi}_{\Lambda_n^c}|_{\mathcal{B}_{\Lambda_{n+1}\setminus\Lambda_n}} = \hat{\varphi}_{\Lambda_{n+1}^c} \circ \mathcal{E}_{\Lambda_{n+1}^c}|_{\mathcal{B}_{\Lambda_{n+1}\setminus\Lambda_n}} \tag{1}$$

and

$$\varphi = \lim_{n\to\infty} \hat{\varphi}_{\Lambda_n^c} \circ \mathcal{E}_{\Lambda_n^c} \circ \mathcal{E}_{\Lambda_{n-1}^c} \circ \cdots \circ \mathcal{E}_{\Lambda_1^c} \tag{2}$$

in the weak-* topology.

Note that (1) is an analogue of the DRL equation from classical statistical mechanics [10,14], and QMC state is thus the counterpart of the infinite-volume Gibbs measure.

**Remark 2.1.** We point out that in [6] a forward QMC was called a generalized quantum Markov state, and the existence of the limit (2) under the condition (1) was proved there as well.


## 3. Construction of QMC on the Cayley tree

In this section, we recall a construction of forward quantum $d$-Markov chain (see [4]).

Let us rewrite the elements of $W_n$ in the following order, i.e.

$$\overrightarrow{W_n} := \left(x_{W_n}^{(1)}, x_{W_n}^{(2)}, \cdots, x_{W_n}^{(|W_n|)}\right), \quad \overleftarrow{W_n} := \left(x_{W_n}^{(|W_n|)}, x_{W_n}^{(|W_n|-1)}, \cdots, x_{W_n}^{(1)}\right).$$

Note that $|W_n| = k^n$. Vertices $x_{W_n}^{(1)}, x_{W_n}^{(2)}, \cdots, x_{W_n}^{(|W_n|)}$ of $W_n$ can be represented in terms of the coordinate system as follows

$$x_{W_n}^{(i)} = (1, 1, \cdots, 1, i), \quad i = 1, \cdots, k$$
$$x_{W_n}^{(k+i)} = (1, 1, \cdots, 2, i), \quad i = 1, \cdots, k$$
$$\vdots$$
$$x_{W_n}^{(|W_n|-k+i)} = (k, k, , \cdots, k, i), \quad i = 1, \cdots, k$$

Analogously, for a given vertex $x$, we shall use the following notation for the set of direct successors of $x$:

$$\overrightarrow{S(x)} := ((x, 1), (x, 2), \cdots (x, k)), \quad \overleftarrow{S(x)} := ((x, k), (x, k-1), \cdots (x, 1)).$$

In what follows, for the sake of simplicity, we will use notation $i \in \overrightarrow{S(x)}$ (resp. $i \in \overleftarrow{S(x)}$ instead of $(x,i) \in \overrightarrow{S(x)}$ (resp. $(x,i) \in \overleftarrow{S(x)}$).

Assume that for each edge $< x,y > \in E$ of the tree an operator $K_{<x,y>} \in \mathcal{B}_{\{x,y\}}$ is assigned. We would like to define a state on $\mathcal{B}_{\Lambda_n}$ with boundary conditions $w_0 \in \mathcal{B}_{(0),+}$ and $\mathbf{h} = \{h_x \in \mathcal{B}_{x,+}\}_{x \in L}$.

Let us denote

$$K_{[m-1,m]} := \prod_{x \in \overrightarrow{W}_{m-1}} \prod_{y \in \overrightarrow{S(x)}} K_{<x,y>},$$

$$\mathbf{h}_n^{1/2} := \prod_{x \in \overrightarrow{W}_n} h_x^{1/2}, \qquad \mathbf{h}_n := \mathbf{h}_n^{1/2}(\mathbf{h}_n^{1/2})^*,$$

$$K_n := w_0^{1/2} K_{[0,1]} K_{[1,2]} \cdots K_{[n-1,n]} \mathbf{h}_n^{1/2},$$

$$\mathcal{W}_{n]} := K_n K_n^*,$$

It is clear that $\mathcal{W}_{n]}$ is positive.

In what follows, by $\mathrm{Tr}_\Lambda : \mathcal{B}_L \to \mathcal{B}_\Lambda$ we mean normalized partial trace (i.e. $\mathrm{Tr}_\Lambda(\mathbb{1}_L) = \mathbb{1}_\Lambda$, here $\mathbb{1}_\Lambda = \bigotimes_{y \in \Lambda} \mathbb{1}$), for any $\Lambda \subseteq_{\mathrm{fin}} L$. For the sake of shortness we put $\mathrm{Tr}_{n]} := \mathrm{Tr}_{\Lambda_n}$.

Let us define a positive functional $\varphi_{w_0,\mathbf{h}}^{(n,f)}$ on $\mathcal{B}_{\Lambda_n}$ by

$$\varphi_{w_0,\mathbf{h}}^{(n,f)}(a) = \mathrm{Tr}(\mathcal{W}_{n+1]}(a \otimes \mathbb{1}_{W_{n+1}})), \tag{3}$$

for every $a \in \mathcal{B}_{\Lambda_n}$. Note that here, $\mathrm{Tr}$ is a normalized trace on $\mathcal{B}_L$ (i.e. $\mathrm{Tr}(\mathbb{1}_L) = 1$).

To get an infinite-volume state $\varphi^{(f)}$ on $\mathcal{B}_L$ such that $\varphi^{(f)} \lceil_{\mathcal{B}_{\Lambda_n}} = \varphi_{w_0,\mathbf{h}}^{(n,f)}$, we need to impose some constrains to the boundary conditions $\{w_0, \mathbf{h}\}$ so that the functionals $\{\varphi_{w_0,\mathbf{h}}^{(n,f)}\}$ satisfy the compatibility condition, i.e.

$$\varphi_{w_0,\mathbf{h}}^{(n+1,f)} \lceil_{\mathcal{B}_{\Lambda_n}} = \varphi_{w_0,\mathbf{h}}^{(n,f)}. \tag{4}$$

**Theorem 3.1.** *(4) Let the boundary conditions $w_0 \in \mathcal{B}_{(0),+}$ and $\mathbf{h} = \{h_x \in \mathcal{B}_{x,+}\}_{x \in L}$ satisfy the following conditions:*

$$\mathrm{Tr}(w_0 h_0) = 1 \tag{5}$$

$$\mathrm{Tr}_{x]}\left[ \prod_{y \in \overrightarrow{S(x)}} K_{<x,y>} \prod_{y \in \overrightarrow{S(x)}} h^{(y)} \prod_{y \in \overleftarrow{S(x)}} K_{<x,y>}^* \right] = h^{(x)}, \tag{6}$$

*for every $x \in L$. Then the functionals $\{\varphi_{w_0,\mathbf{h}}^{(n,f)}\}$ satisfy the compatibility condition (4). Moreover, there is a unique forward quantum d-Markov chain $\varphi_{w_0,\mathbf{h}}^{(b)}$ on $\mathcal{B}_L$ such that $\varphi_{w_0,\mathbf{h}}^{(f)} = w - \lim_{n \to \infty} \varphi_{w_0,\mathbf{h}}^{(n,f)}$.*

20

We say that there exists a *phase transition* for a family of operators $\{K_{<x,y>}\}$ if (5), (6) have at least two $(u_0, \{h_x\}_{x \in L})$ and $(v_0, \{s_x\}_{x \in L})$ solutions such that the corresponding quantum $d$-Markov chains $\varphi_{u_0,\mathbf{h}}$ and $\varphi_{v_0,\mathbf{s}}$ are not quasi equivalent. Otherwise, we say there is no phase transition (see [5]).

## 4. QMC associated with Ising model and main results

In this section, we define the model and formulate the main results of the paper. In what follows we consider a semi-infinite Cayley tree $\Gamma^k = (L, E)$ of order $k$. Our starting $C^*$-algebra is the same $\mathcal{B}_L$ but with $\mathcal{B}_x = M_2(\mathbb{C})$ for $x \in L$. By $\sigma_x^{(u)}, \sigma_y^{(u)}, \sigma_z^{(u)}$ we denote the Pauli spin operators at site $u \in L$ given by

$$\mathbf{1}^{(u)} = \begin{pmatrix} 1 & 0 \\ 0 & 1 \end{pmatrix}, \quad \sigma_x^{(u)} = \begin{pmatrix} 0 & 1 \\ 1 & 0 \end{pmatrix}, \quad \sigma_y^{(u)} = \begin{pmatrix} 0 & -i \\ i & 0 \end{pmatrix}, \quad \sigma_z^{(u)} = \begin{pmatrix} 1 & 0 \\ 0 & -1 \end{pmatrix}. \tag{7}$$

For every edge $<u, v> \in E$ put

$$K_{<u,v>} = \exp\{\beta H_{<u,v>}\}, \quad \beta > 0, \tag{8}$$

where

$$H_{<u,v>} = \frac{1}{2}\left(\mathbf{1}^{(u)}\mathbf{1}^{(v)} + \sigma_z^{(u)}\sigma_z^{(v)}\right). \tag{9}$$

Such kind of Hamiltonian is called *quantum **Ising** model* per edge $<x, y>$.

Now taking into account the following equalities

$$H_{<u,v>}^m = H_{<u,v>} = \frac{1}{2}\left(\mathbf{1}^{(u)}\mathbf{1}^{(v)} + \sigma_z^{(u)}\sigma_z^{(v)}\right),$$

one finds

$$K_{<u,v>} = \mathbf{1}^{(u)}\mathbf{1}^{(v)} + (\exp\beta - 1)H_{<u,v>}. \tag{10}$$

It follows from (9) that

$$K_{<u,v>} = K_0\mathbf{1}^{(u)}\mathbf{1}^{(v)} + K_3\sigma_z^{(u)}\sigma_z^{(v)}. \tag{11}$$

where, $K_0 = \frac{\exp\beta + 1}{2}$, $K_3 = \frac{\exp\beta - 1}{2}$.

The main result of the present paper concerns the existence of the phase transition for the model (8). Namely, we have the following result.

**Theorem 4.1.** *Let $\{K_{<x,y>}\}$ be given by (8) on the Cayley tree of order $k \geq 2$ and $\theta = \exp(2\beta)$, $\beta > 0$. Then the following assertions hold true:*

*(i) If $\theta \leq \frac{k+1}{k-1}$ then there is a unique forward quantum d-Markov chain.*

*(ii) If $\theta > \frac{k+1}{k-1}$ then there is a phase transition for a given model, i.e. there are two distinct forward quantum d-Markov chains.*

## 5. Diagonalizability of forward QMC

In this section we shall reduce equations (6) to some dynamical system. Our goal is to describe all solutions $\mathbf{h} = \{h_x\}$ of that equation.

Furthermore, we shall assume that $h_u = h_v$ for every $u, v \in W_n$, $n \in \mathbb{N}$. Hence, we denote $h_x^{(n)} := h^{(x)}$ and $h_y^{(n+1)} := h^{(y)}$ if $x \in W_n$ and $y \in W_{n+1}$. Now from (8),(9) one can see that $K_{<u,u>} = K^*_{<u,v>}$, therefore, equation (6) can be reduced to the dynamics of the following mapping $f : \mathbb{R}^2_+ \to \mathbb{R}^2_+$ given by

$$\begin{cases} x' = \dfrac{2\theta \sqrt[k]{x} - 2\sqrt[k]{y}}{\theta^2 - 1} \\[2mm] y' = \dfrac{-2\sqrt[k]{x} + 2\theta \sqrt[k]{y}}{\theta^2 - 1} \end{cases} \tag{12}$$

where $\beta > 0$, $\theta = \exp 2\beta$, and $(x', y') = f(x, y)$. Then one has that $\theta > 1$.

**Remark 5.1.** The dynamical system $f : \mathbb{R}^2_+ \to \mathbb{R}^2_+$ given by (12) is well-defined if and only if $x, y \geq 0$ and $\frac{1}{\theta^k} y \leq x \leq \theta^k y$. In what follows, we will only consider the case of which $x > 0$ and $y > 0$.

Investigating the dynamical system (12), one can prove the absence of periodic points for any $\theta > 1$. Moreover, we can find the existence of the fixed points of the mentioned dynamical system. Note that each fixed point of the dynamical system (12) defines boundary conditions which are solutions of (5), (6). Namely, we have that if $\theta \leq \frac{k+1}{k-1}$ then there are boundary conditions $(w_0(\alpha_0), \{h_x(\alpha_0)\})$ of the model (10)

$$w_0(\alpha_0) = \frac{1}{\alpha_0}\sigma_0, \quad h_x(\alpha_0) = \alpha_0\sigma_0^{(x)} \tag{13}$$

and if $\theta > \frac{k+1}{k-1}$ then apart previous one, there are two extra boundary conditions $(w_0(\beta), \{h_x(\beta)\})$ and $(w_0(\gamma), \{h_x(\gamma)\})$ of the model (10)

$$w_0(\beta) = \frac{1}{\beta_0}\sigma_0, \quad h_x(\beta) = \beta_0\sigma_0^{(x)} + \beta_3\sigma_3^{(x)} \tag{14}$$

$$w_0(\gamma) = \frac{1}{\gamma_0}\sigma_0, \quad h_x(\gamma) = \gamma_0\sigma_0^{(x)} + \gamma_3\sigma_3^{(x)} \tag{15}$$

here $\alpha_0 = \sqrt[k-1]{\Theta^k}$, and $\beta = (\beta_0, \beta_3)$, $\gamma = (\gamma_0, \gamma_3)$ are vectors with

$$\beta_0 = \frac{A_2 \sqrt[k-1]{B_2} + B_2 \sqrt[k-1]{B_2}}{2}, \quad \beta_3 = \frac{A_2 \sqrt[k-1]{B_2} - B_2 \sqrt[k-1]{B_2}}{2},$$

$$\gamma_0 = \frac{A_3 \sqrt[k-1]{B_3} + B_3 \sqrt[k-1]{B_3}}{2}, \quad \gamma_3 = \frac{A_3 \sqrt[k-1]{B_3} - B_3 \sqrt[k-1]{B_3}}{2}.$$

Note that thesis boundary conditions (13), (14), (15) define (see Theorem 3.1) the forward QMC. Hence, the existence of the boundary conditions imply the existence of forward QMC for the model (10) for any $\theta > 1$.

We are going to prove diagonalizability of the forward QMC corresponding to any boundary conditions which are solutions of (5) and (6).

The diagonal subalgebra $M_2^d(\mathbb{C})$ of the algebra $M_2(\mathbb{C})$ is defined as follows

$$M_2^d(\mathbb{C}) = \{a \in M_2(\mathbb{C}) : a = a_0 \mathbf{1} + a_3 \sigma_z\}.$$

Since the elements $\mathbf{1}, \sigma_x, \sigma_y, \sigma_z$ are basis in the algebra $M_2(\mathbb{C})$ then every element $a \in M_2(\mathbb{C})$ can be written in the following form $a = a_0 \mathbf{1} + a_1 \sigma_x + a_2 \sigma_y + a_3 \sigma_z$. Then for any $a \in M_2(\mathbb{C})$ an element $a_d = a_0 \mathbf{1} + a_3 \sigma_z$ is called *its diagonal part* and an element $a_{xy} = a_1 \sigma_x + a_2 \sigma_y$ is called *its $xy-$part*. It is clear that a linear span of the elements $\mathbf{1}, \sigma_z$ is a commutative diagonal subalgebra $M_2^d(\mathbb{C})$. In these notions, any element $a \in M_2(\mathbb{C})$ can be written as $a = a_d + a_{xy}$.

Let us consider a conditional expectation $E : M_2(\mathbb{C}) \to M_2^d(\mathbb{C})$ as follows $E(a) = e_{11} a e_{11} + e_{22} a e_{22}$, where $e_{11}, e_{22}$ are two minimal projectors of the algebra $M_2(\mathbb{C})$. It is clear that $E(a) = a^d$.

The diagonal subalgebra $\mathcal{B}_L^d$ of the full algebra $\mathcal{B}_L$ is defined by an inductive limit

$$\mathcal{B}_L^d = \overline{\bigcup_{\Lambda_n} \mathcal{B}_{\Lambda_n}^d}^{\|\cdot\|}, \quad \mathcal{B}_{\Lambda_n}^d = \bigotimes_{u \in \Lambda_n} \mathcal{B}_u^d, \quad \mathcal{B}_u^d = M_2^d(\mathbb{C}), \ \forall \, u \in \Lambda.$$

We will define a conditional expectation $\mathcal{E} : \mathcal{B}_L \to \mathcal{B}_L^d$ on the algebra $\mathcal{B}_L$ as follows. The value of this expectation at any linear generator $a_{\Lambda_n} = \bigotimes_{x \in \Lambda_n} a_x$ of the algebra $\mathcal{B}_{\Lambda_n}$ defines

$$\mathcal{E}(a_{\Lambda_n}) = \bigotimes_{x \in \Lambda_n} E(a_x) \tag{16}$$

and the linear extension of (16) defines it on the whole algebra $\mathcal{B}_{\Lambda_n}$. By inductive limit we define the expectation $\mathcal{E}$ on the full algebra $\mathcal{B}_L$. For any

$a_\Lambda \in \mathcal{B}_\Lambda$ the value $\mathcal{E}(a_\Lambda)$ is called *a diagonal part* of $a_\Lambda$ and denote by $a_\Lambda^d = \mathcal{E}(a_\Lambda)$.

**Theorem 5.1.** *Let* $\varphi_{w_0,\mathbf{h}}^{(f)}$ *be a forward QMC of the model* (10) *with boundary conditions which are solutions of* (5) *and* (6). *Let* $\mathcal{E} : \mathcal{B}_L \to \mathcal{B}_L^d$ *be a conditional expectation given* (16) *and* $\theta > 1$. *Then for any* $a \in \mathcal{B}_\Lambda$

$$\varphi_{w_0,\mathbf{h}}^{(f)}(a) = \varphi_{w_0,\mathbf{h}}^{(f)}(\mathcal{E}(a)).$$

Note that in [13] it has been proved that any quantum Markov state over one dimensional lattice is diagonalizable.

## Acknowledgement

The authors would like to thank Professor Noboru Watanabe and Professor Masanori Ohya for their kind invitation to ICQBIC 2011. The second named author (F.M.) acknowledges the Junior Associate scheme of the Abdus Salam International Centre for Theoretical Physics, Trieste, Italy.

## References


1. Accardi L., On the noncommutative Markov property, *Func. Anal. Appl.*, **9** (1975) 1–8.
2. Accardi L., Fidaleo F., Quantum Markov fields, *Inf. Dim. Analysis, Quantum Probab. Related Topics* **6** (2003) 123–138.
3. Accardi L., Frigerio A., Markovian cocycles, *Proc. Royal Irish Acad.* **83A** (1983) 251-263.
4. Accardi L., Mukhamedov, F. Saburov M. On Quantum Markov Chains on Cayley tree I: uniqueness of the associated chain with $XY$-model on the Cayley tree of order two. *Inf. Dim. Analysis, Quantum Probab. Related Topics* (accepted) `arXiv:1004.3623`.
5. Accardi L., Mukhamedov, F. Saburov M. On Quantum Markov Chains on Cayley tree II: Phase transitions for the associated chain with $XY$-model on the Cayley tree of order three. *Ann. Henri Poincare* (in press) `arXiv:1004.3623`.
6. Accardi L., Ohno, H., Mukhamedov, F., Quantum Markov fields on graphs, *Inf. Dim. Analysis, Quantum Probab. Related Topics* **13**(2010), 165–189.
7. Affleck L, Kennedy E., Lieb E.H., Tasaki H., Valence bond ground states in isortopic quantum antiferromagnets, *Commun. Math. Phys.* **115** (1988), 477–528.
8. Baxter R. J. *Exactly Solved Models in Statistical Mechanics*, London/New York: Academic, 1982.
9. Bratteli O., Robinson D.W., Operator algebras and quantum statistical mechanics. 1. Texts amd Monographs in Physics. Springer-Verlag, New York, 1987.

10. Dobrushin R.L., Description of Gibbsian Random Fields by means of conditional probabilities, *Probability Theory and Applications* **13**(1968) 201–229

11. Fannes M., Nachtergaele B. Werner R. F., Ground states of VBS models on Cayley trees, *J. Stat. Phys.* **66** (1992) 939–973.

12. Fannes M., Nachtergaele B. Werner R. F., Finitely correlated states on quantum spin chains, *Commun. Math. Phys.* **144** (1992) 443–490.

13. Fidaleo F., Mukhamedov F., Diagonalizability of non homogeneous quantum Markov states and associated von Neumann algebras, *Probab. Math. Stat.* **24** (2004), 401–418.

14. Georgi H.-O. *Gibbs measures and phase transitions*, de Gruyter Studies in Mathematics vol. 9, Walter de Gruyter, Berlin, 1988.

15. Liebscher V., Markovianity of quantum random fields, Proceedings Burg Conference 15–20 March 2001, W. Freudenberg (ed.), World Scientific, QP–PQ Series 15 (2003) 151–159.

16. Matsui, T., A characterization of pure finitely correlated states, *Infin. Dimens. Anal. Quantum Probab. Relat. Top.* **1** (1998) 647–661.

17. Mukhamedov F.M. On factor associated with the unordered phase of $\lambda$-model on a Cayley tree. *Rep. Math. Phys.* **53**(2004) 1–18.

18. Ohya M., Petz D., Quantum entropy and its use, Springer, Berlin, Heidelberg, New York, 1993.

19. Preston C., Gibbs States on Countable Sets, Cambridge University Press, London, 1974.

20. Spataru A., Construction of a Markov Field on an infinite Tree, *Advances in Math* **81**(1990), 105–116.

Quantum Bio-Informatics V
© 2013 World Scientific Publishing Co. Pte. Ltd.
pp. 25–35

# MATHEMATICAL ASPECTS OF CONSERVED QUANTITIES IN A GENERAL CLASS OF QUANTUM SYSTEMS

ASAO ARAI

*Department of Mathematics,*
*Hokkaido University*
*Sapporo 060-0810, Japan*
*E-mail: arai@math.sci.hokudai.ac.jp*

We present some results of mathematically rigorous analysis on conserved quantities in a general class of quantum systems whose Hamiltonian H is of the form H=Q2 with Q being a self-adjoint operator. The class includes supersymmetric quantum systems.

*Keywords* : Conserved quantity; Dirac operator; Hamiltonian; Self-adjoint operator; Spectrum; Strong commutativity; Strong anti-commutativity; Supersymmetry.

## 1. Introduction

In a quantum system whose Hamiltonian is given by $H$, a self-adjoint operator on a Hilbert space $\mathcal{H}$, the time development of the quantum system is governed by the strongly continuous one-parameter unitary group $\{e^{-itH/\hbar}\}_{t\in\mathbb{R}}$ generated by $H$, where $i$ is the imaginary unit, $t$ is a time parameter and $\hbar > 0$ is a parameter denoting physically the Planck constant divided by $2\pi$. For a linear operator $A$ on $\mathcal{H}$, its time development is defined by

$$A_H(t) := e^{itH/\hbar} A e^{-itH/\hbar}, \quad t \in \mathbb{R}, \tag{1.1}$$

called the Heisenberg operator of $A$ with respect to $H$. We denote the domain of $A$ by $D(A)$. An operator $A$ is called a *conserved quantity* of $H$ if, for all $t \in \mathbb{R}$, the *operator equality*

$$A_H(t) = A \tag{1.2}$$

holds (i.e., $D(A(t)) = D(A)$ and $A(t)\psi = A\psi, \forall \psi \in D(A), \forall t \in \mathbb{R}$). Note that, in this definition, $A$ is not necessarily self-adjoint. We say that a conserved quantity $A$ of $H$ is symmetric (resp. self-adjoint) if $A$ is a symmetric (resp. self-adjoint) operator.

In the physics literature, one often defines a conserved quantity of the Hamiltonian $H$ as a linear operator $A$ which commutes with $H$: $[A, H] := AH - HA = 0$ with $D([A, H]) := D(AH) \cap D(HA)$. But $[A, H] = 0$ does not necessarily imply the *operator equality* (1.2), while the operator equality (1.2) implies $[A, H] = 0$. The definition of a conserved quantity given here is proper in the sense that its time development is independent of the time $t \in \mathbb{R}$.

As is well known, conserved quantities play important roles in quantum mechanics. The existence of a *non-trivial* conserved quantity in a quantum system makes it possible to reduce the Hamiltonian of the quantum system to a smaller closed subspace so that one may analyze properties of the Hamiltonian easier. Usually conserved quantities are individually discussed in each concrete quantum system. A general and unified treatment of conserved quantities in quantum systems seems to be missing in the literature. Taking this situation into account, it may be interesting and important to investigate structures of conserved quantities in a general class of quantum systems. This is one of the motivations for the present work. In this paper we present some results of mathematical aspects on conserved quantities in a general class of quantum systems. For results without proof, see [4] or [5].

In order that the reader can easily understand an essential nature of the problem we are concerned with in the present paper, we begin with an interesting and non-trivial example in a relativistic quantum mechanics.

## 2. A Free Relativistic Quantum Particle Revisited

Consider a quantum system of a free relativistic quantum particle with mass $m > 0$ whose Hamiltonian is given by the free Dirac operator with mass $m > 0$:

$$Q_m := \sum_{j=1}^{3} c\alpha_j p_j + mc^2\beta$$

acting in $L^2(\mathbb{R}^3; \mathbb{C}^4) = \oplus^4 L^2(\mathbb{R}^3)$, where $c > 0$ is the speed of light in the vacuum, $\alpha_j, \beta$ are $4 \times 4$ Hermitian matrices satisfying anti-commutation relations

$$\{\alpha_j, \alpha_k\} := \alpha_j\alpha_k + \alpha_k\alpha_j = 2\delta_{jk}, \quad \{\alpha_j, \beta\} = 0, \quad \beta^2 = 1 \quad (j, k = 1, 2, 3),$$

and $p_j := -i\hbar D_{x_j}$ with $D_{x_j}$ being the generalized partial differential operators in the $j$th variable $x_j$ in $\mathbf{x} = (x_1, x_2, x_3) \in \mathbb{R}^3$. The following facts are well known [9] :

(i) $Q_m$ is self-adjoint with domain $D(Q_m) = \cap_{j=1}^{3} D(p_j)$.

(ii) Let $\sigma(Q_m)$ denote the spectrum of $Q_m$. Then $\sigma(Q_m) = (-\infty, -mc^2] \cup [mc^2, \infty)$.

(iii) $Q_m$ is bijective and $Q_m^{-1}$ is bounded with operator norm $\|Q_m^{-1}\| = 1/mc^2$.

The Heisenberg operator of the position operator $x_j$ with respect to $Q_m$ ($j = 1, 2, 3$) is defined by

$$x_j(t) := e^{itQ_m/\hbar} x_j e^{-itQ_m/\hbar}, \quad t \in \mathbb{R}. \tag{2.1}$$

**Theorem 2.1.** [9]

(i) *(Invariance of domain $D(x_j)$)*

$$e^{-itQ_m/\hbar} D(x_j) = D(x_j), \quad \forall t \in \mathbb{R}, j = 1, 2, 3.$$

(ii) *(Explicit form of $x_j(t)$) Let*

$$v_j := c^2 p_j Q_m^{-1}, \quad u_j := c\alpha_j - c^2 p_j Q_m^{-1}$$

*and*

$$z_j(t) := \frac{\hbar}{2i} Q_m^{-1} (e^{2itQ_m/\hbar} - 1) u_j, \quad j = 1, 2, 3.$$

*Then*

$$x_j(t) = x_j + v_j t + z_j(t) \quad on \ D(x_j), j = 1, 2, 3. \tag{2.2}$$

The part $x_j + v_j t$ on the right hand side of (2.2) may be interpreted as a uniform linear motion along the $j$th direction with velocity $v_j$. Hence $\mathbf{v} := (v_1, v_2, v_3)$ is called the *classical velocity*, which may be a quantum version of a velocity of a classical relativistic particle. On the other hand, the part $z_j(t)$ on the right hand side of (2.2), which is a bounded operator and describes an oscillating motion, is interpreted as a quantum effect, called the *Zitterbewegung*. Indeed one can easily prove the following facts:

$$\lim_{\hbar \to 0} \|z_j(t)\| = 0,$$

so that $\lim_{\hbar \to 0} \|x_j(t) - (x_j + tv_j)\| = 0$. This means that, in the classical limit $\hbar \to 0$, $x_j(t)$ tends asymptotically to the uniform linear motion (which is a free motion in the usual sense).

The $j$th component of the standard velocity operator of the particle under consideration is defined by the *strong* derivative of $x_j(t)$ in $t$:

$$w_j(t)\psi := \text{s-}\frac{\mathrm{d}x_j(t)\psi}{\mathrm{d}t}, \quad \psi \in D(x_j), \quad j = 1, 2, 3.$$

One can show that

$$w_j(t) = e^{itQ_m/\hbar} c\alpha_j e^{-itQ_m/\hbar} \quad \text{on } D(x_j).$$

On the other hand, it follows from (2.2) that, for all $t \in \mathbb{R}$ and $j = 1, 2, 3$,

$$w_j(t) = v_j + e^{2itQ_m/\hbar} u_j \quad \text{on } D(x_j).$$

Thus, for all $t \in \mathbb{R}$ and $j = 1, 2, 3$,

$$e^{itQ_m/\hbar} c\alpha_j e^{-itQ_m/\hbar} = v_j + e^{2itQ_m/\hbar} u_j, \tag{2.3}$$

where this equation originally holds on $D(x_j)$, but it can be extended to the operator equality, since the operators on the both sides are bounded. In particular, we have

$$\alpha_j = \hat{v}_j + \hat{u}_j \tag{2.4}$$

with

$$\hat{v}_j := \frac{v_j}{c}, \quad \hat{u}_j := \frac{u_j}{c}. \tag{2.5}$$

This gives a decomposition of $\alpha_j$, which is remarkable in the following sense:

    (i) $[\alpha_j, Q_m] \neq 0$, $\{\alpha_j, Q_m\} \neq 0$.
    (ii) (commutativity) $[Q_m, \hat{v}_j] \subset 0$.
    (iii) (anti-commutativity) $\{Q_m, \hat{u}_j\} \subset 0$.
    (iv) (anti-commutativity) $\{\hat{v}_j, \hat{u}_j\} = 0$,

where, for two linear operators $A$ and $B$, "$A \subset B$" means that $B$ is an extension of $A$, i.e., $D(A) \subset D(B)$ and $A\psi = B\psi$ for all $\psi \in D(A)$.

Based on (ii) and (iii), we call $\hat{v}_j$ (resp. $\hat{u}_j$) the *commutative* (*anti-commutative*) part of $\alpha_j$ with respect to $Q_m$

The following theorem shows that the commutative (resp. anti-commutative) part of $\alpha_j$ can be recovered as a weak limit (in time) of the Heisenberg operator (resp. a "deformed" Heisenberg operator) of $\alpha_j$ with respect to $Q_m$:

**Theorem 2.2.** *For $j = 1, 2, 3$,*

$$\text{w-}\lim_{t \to \pm\infty} e^{itQ_m/\hbar} \alpha_j e^{-itQ_m/\hbar} = \hat{v}_j, \tag{2.6}$$

$$\text{w-}\lim_{t \to \pm\infty} e^{itQ_m/\hbar} \alpha_j e^{itQ_m/\hbar} = \hat{u}_j, \tag{2.7}$$

*where* w-lim *means weak limit.*

**Remark 2.1.** The operator $e^{itQ_m/\hbar}\alpha_j e^{itQ_m/\hbar}$ on the left hand side of (2.7) is *not* the Heisenberg operator of $\alpha_j$ with respect to $Q_m$ (the operator coming to the right of $\alpha_j$ is $e^{+itQ_m/\hbar}$, not $e^{-itQ_m/\hbar}$) and suggests a new class of time-dependent quantities to be investigated more generally.

**Proof.** We denote by $\langle \cdot, \cdot \rangle$ the inner product of $L^2(\mathbb{R}^3; \mathbb{C}^4)$. For all $\psi, \phi \in L^2(\mathbb{R}^3; \mathbb{C}^4)$, we have by (2.3)

$$\left\langle \psi, e^{itQ_m/\hbar}\alpha_j e^{-itQ_m/\hbar}\phi \right\rangle = \langle \psi, \hat{v}_j \phi \rangle + \left\langle \psi, e^{2itQ_m/\hbar}\hat{u}_j \phi \right\rangle$$

It is known that $Q_m$ is purely absolutely continuous. Hence the second term on the right hand side converges to 0 as $t \to \pm\infty$. Thus (2.6) holds.

One can show (see (ii) and (iii) described above) that, for all $t \in \mathbb{R}$,

$$e^{itQ_m/\hbar}\hat{v}_j = \hat{v}_j e^{itQ_m/\hbar}, \quad e^{itQ_m/\hbar}\hat{u}_j = \hat{u}_j^{-itQ_m/\hbar}.$$

Hence, by (2.4),

$$e^{itQ_m/\hbar}\alpha_j e^{itQ_m/\hbar} = \hat{v}_j e^{2itQ_m/\hbar} + \hat{u}_j.$$

Thus, in the same way as in the preceding case, we obtain (2.7). ∎

We note that

$$H_m := Q_m^2 = c^2 p^2 + m^2 c^4,$$

where $p^2 := \sum_{j=1}^{3} p_j^2 = -\hbar^2 \Delta$ ($\Delta$ is the three-dimensional generalized Laplacian) and

$$e^{itH_m/\hbar}\alpha_j e^{-itH_m/\hbar} = \alpha_j, \quad \forall t \in \mathbb{R}.$$

This means that $\alpha_j$ is a self-adjoint conserved quantity of $H_m$.

In the same way as in the case of $\alpha_j$, we see that $\beta$ is a self-adjoint conserved quantity of $H_m$ with the following decomposition:

$$\beta = \beta_+ + \beta_-, \tag{2.8}$$

where $\beta_+ := mc^2 Q_m^{-1}, \beta_- := \beta - mc^2 Q_m^{-1}$. It is easy to see that $[Q_m, \beta_+] = 0$ on $D(Q_m)$; $\{Q_m, \beta_-\} = 0$ on $D(Q_m)$; $\{\beta_+, \beta_-\} = 0$. Hence the decomposition (2.8) has a structure similar to that of the decomposition (2.4) of $\alpha_j$.

It should be noted also that $H_m$ is a supersymmetric Hamiltonian and $Q_m$ is a supercharge [9] .

Thus it would be natural to ask to what extent the structure revealed here on conserved quantities $\alpha_j$ and $\beta$ of $H_m$ can be taken over to a conserved quantity in a general quantum system.

## 3. Uniqueness Theorem on a Decomposition of a Linear Operator and Some Consequences

In the question raised in the preceding section, we want to treat not only bounded conserved quantities but also unbounded ones. For unbounded operators, however, the usual concepts of commutativity and anti-commutativity are almost useless. But it has been known that there are stronger concepts of commutativity and anti-commutativity useful in treating unbounded operators. We first recall the definition of them:

**Definition 3.1.** Let $S$ be a self-adjoint operator on a complex Hilbert space $\mathcal{H}$ and $B$ be a linear operator on $\mathcal{H}$.

    (i) $B$ *strongly commutes with* $S$ if $e^{itS}B \subset Be^{itS}$, $\forall t \in \mathbb{R}$.

    (ii) $B$ *strongly anti-commutes with* $S$ if $e^{itS}B \subset Be^{-itS}$, $\forall t \in \mathbb{R}$.

**Remark 3.1.** In the case where $B$ is self-adjoint, (i) $B$ strongly commutes with $S$ if and only if the spectral measure $E_B(\cdot)$ of $B$ and that of $S$ commute: for all Borel sets $J, K \subset \mathbb{R}$, $E_S(J)E_B(K) = E_B(K)E_S(J)$ [7] ; (ii) $B$ is strongly anti-commutes with $S$ if and only if $E_S(-J)B \subset BE_S(J)$ for all Borel sets $J \subset \mathbb{R}$, where $-J := \{-x \in \mathbb{R} | x \in \mathbb{R}\}$ [10].

In what follows, $Q$ is a self-adjoint operator on a complex Hilbert space $\mathcal{H}$. Let

$$H_Q := Q^2. \tag{3.1}$$

**Theorem 3.1.** (***Uniqueness of a decomposition***) *Let* $Q$ *be injective and* $A$ *be a closed linear operator on* $\mathcal{H}$. *Suppose that there exists a pair* $(A_+, A_-)$ *of closed linear operators* $A_\pm$ *(on* $\mathcal{H}$*) satisfying the following conditions (i)–(iv) :*

    *(i)* $A = A_+ + A_-$.

    *(ii)* $A_+$ *strongly commutes with* $Q$.

    *(iii)* $A_-$ *strongly anti-commutes with* $Q$.

    *(iv)* *There exists a subspace* $\mathcal{D} \subset D(A_+) \cap D(A_-)$ *which is a common core of* $A_\pm$.

*Then the pair* $(A_+, A_-)$ *is unique. Moreover,* $A$ *and* $A_\pm$ *are conserved quantities of* $H_Q$.

**Remark 3.2.** In applications, one may have the case where $A$ and $A_\pm$ are self-adjoint, and $A_+$ and $A_-$ strongly anticommute. In this case, one can apply the theory of strongly anti-commuting self-adjoint operators developed by Vasilesc [10], Pedersen [6], Samoilenko [8] and the present author [1–3]. For example, one can construct a Hilbert space representation of $\mathsf{su}(2, \mathbb{C})$ (the Lie algebra of the two-dimensional special unitary group $SU(2)$) from $A_\pm$ [1].

Theorem 3.1 has some consequences. Let $\mathcal{H}_{\mathrm{ac}}(Q)$ be the absolutely continuous subspace of $Q$:

$$\mathcal{H}_{\mathrm{ac}}(Q) := \{\psi \in \mathcal{H} | \text{the measure } \|E_Q(\cdot)\psi\|^2 \text{ on } \mathbb{R} \text{ is absolutely}$$
$$\text{continuous with respect to the Lebesgue measure on } \mathbb{R}\}.$$

**Theorem 3.2.** *Assume that $Q$ is injective. Let $A$ be a closed linear operator on $\mathcal{H}$. Suppose that there exists a pair $(A_+, A_-)$ of densely defined closed linear operators $A_\pm$ satisfying (i)–(iii) in Theorem 3.1. Then the following (i) and (ii) hold:*

*(i) For all $\psi \in D(A) \cap \mathcal{H}_{\mathrm{ac}}(Q)$,*

$$\text{w-}\lim_{t \to \pm\infty} e^{itQ} A e^{-itQ} \psi = A_+ \psi.$$

*(ii) For all $\psi \in D(A) \cap \mathcal{H}_{\mathrm{ac}}(Q)$,*

$$\text{w-}\lim_{t \to \pm\infty} e^{itQ} A e^{itQ} \psi = A_- \psi.$$

**Proof.** (i) For a densely defined linear operator $L$ on a Hilbert space, we denote its adjoint by $L^*$. Since $A_\pm$ are densely defined closed linear operators, $D(A_\pm^*)$ are dense. By Theorem 3.1 (i)–(iii), we have operator equality

$$e^{itQ} A e^{-itQ} = A_+ + A_- e^{-2itQ}, \quad \forall t \in \mathbb{R}.$$

Hence, for all $\phi \in D(A_-^*)$ and $\psi \in D(A) \cap \mathcal{H}_{\mathrm{ac}}(Q)$,

$$\langle \phi, e^{itQ} A e^{-itQ} \psi \rangle = \langle \phi, A_+ \psi \rangle + \langle A_-^* \phi, e^{-2itQ} \psi \rangle.$$

By the assumption that $\psi \in \mathcal{H}_{\mathrm{ac}}(Q)$, $\lim_{t \to \pm\infty} \langle A_-^* \phi, e^{-2itQ} \psi \rangle = 0$. Hence

$$\lim_{t \to \pm\infty} \langle \phi, e^{itQ} A e^{-itQ} \psi \rangle = \langle \phi, A_+ \psi \rangle.$$

By a limiting argument, one can extend this result to all $\phi \in \mathcal{H}$.

(ii) As in the preceding case, we have operator equality

$$e^{itQ} A e^{itQ} = A_+ e^{2itQ} + A_-, \quad \forall t \in \mathbb{R}.$$

Hence one can proceed as in part (i) to obtain the desired result. $\blacksquare$

We next describe a supersymmetric structure associated with a decomposition of $A$ in Theorem 3.1.

Assume that $Q$ is injective. Let $A$ be a self-adjoint operator on $\mathcal{H}$. Suppose that there exists a pair $(A_+, A_-)$ of self-adjoint operators $A_\pm$ satisfying (i)–(iii) in Theorem 3.1 and $A_-$ is injective. Let

$$A_- = U|A_-|$$

be the polar decomposition of $A_-$. Then $U$ is unitary and self-adjoint and

$$UQ = -QU.$$

Hence $(\mathcal{H}, H_Q, Q, U)$ is a supersymmetric quantum mechanics [9] . This implies that, for example, the spectrum $\sigma(Q)$ of $Q$ is symmetric with respect to the origin in $\mathbb{R}$.

## 4. Existence Theorems

In this section we state results on the existence of a decomposition of a linear operator stated in Theorem 3.1.

Throughout this section, we assume the following:

**Hypothesis (Q)** : $Q$ is an injective self-adjoint operator on $\mathcal{H}$.

Let $A$ be a closed linear operator on $\mathcal{H}$. Then we can define the following operators:

$$A_\pm := \frac{1}{2}(A \pm QAQ^{-1})$$

with $D(A_\pm) = \mathcal{D}_{A,Q} := D(A) \cap D(QAQ^{-1})$.

It is obvious that

$$A_+ + A_- \subset A. \tag{4.1}$$

**Remark 4.1.** (i) If $QAQ^{-1} = A$ on $\mathcal{D}_{A,Q}$, then $A_+ \subset A$ and $A_- \subset 0$. (ii) If $QAQ^{-1} = -A$ on $\mathcal{D}_{A,Q}$, then $A_+ \subset 0$ and $A_- \subset A$. Hence, operator relation (4.1) is "non-trivial" only if

$$QAQ^{-1} \neq A \text{ and } QAQ^{-1} \neq -A \text{ on } \mathcal{D}_{A,Q}.$$

Roughly speaking, this is the case where $[Q, A] \neq 0$ and $\{Q, A\} \neq 0$.

### 4.1. *Bounded Conserved Quantities*

For definiteness, we first consider the case of bounded conserved quantities. We denote by $\mathfrak{B}(\mathcal{H})$ the set of bounded linear operators $B$ on $\mathcal{H}$ with $D(B) = \mathcal{H}$.

**Lemma 4.1.** *Let $A \in \mathfrak{B}(\mathcal{H})$ and $QAQ^{-1}$ be a densely defined bounded linear operator on $\mathcal{H}$. Then $A_{\pm}$ are closable and*

$$\bar{A}_{\pm} = \frac{1}{2}(A \pm \overline{QAQ^{-1}}), \tag{4.2}$$

*where, for a closable linear operator $C$, we denote its closure by $\bar{C}$.*

**Theorem 4.1.** *Let $A \in \mathfrak{B}(\mathcal{H})$ be a conserved quantity of $H_Q$ defined by (3.1). Suppose that $QAQ^{-1}$ be densely defined and bounded. Then*

$$A = \bar{A}_+ + \bar{A}_-. \tag{4.3}$$

*Moreover, $\bar{A}_+$ strongly commutes with $Q$ and $\bar{A}_-$ strongly anti-commutes with $Q$.*

**Corollary 4.1.** *Under the same assumption as in Theorem 4.1,*

$$\text{w-}\lim_{t \to \pm\infty} e^{itQ} A e^{-itQ} \psi = \bar{A}_+ \psi, \quad \forall \psi \in \mathcal{H}_{\mathrm{ac}}(Q),$$

$$\text{w-}\lim_{t \to \pm\infty} e^{itQ} A e^{itQ} \psi = \bar{A}_- \psi, \quad \forall \psi \in \mathcal{H}_{\mathrm{ac}}(Q),$$

For a symmetric operator $S$ and a unit vector $\psi \in D(S)$, we define

$$(\Delta S)_\psi := \|(S - \langle \psi, S\psi \rangle)\psi\|,$$

the uncertainty of $S$ in the vector state $\psi$.

**Corollary 4.2.** (***Uncertainty relation***) *Under the same assumption as in Theorem 4.1, for all unit vectors $\psi \in D(Q)$,*

$$(\Delta Q)_\psi (\Delta A)_\psi \geq \left| \langle Q\psi, \bar{A}_- \psi \rangle \right|.$$

**Corollary 4.3.** (***Spectral symmetry***) *Suppose that the same assumption as in Theorem 4.1 holds and that $A$ is self-adjoint with property*

$$\{\psi \in \mathcal{H} | A\psi = \overline{QAQ^{-1}}\psi\} = \{0\}.$$

*Then the spectrum $\sigma(Q)$ is symmetric with respect to the origin in $\mathbb{R}$.*

**Theorem 4.2.** *Let $A \in \mathfrak{B}(\mathcal{H})$ such that $D((QAQ^{-1})^2)$ is dense and $QAQ^{-1}$ is bounded. Suppose that $A^2$ strongly commutes with $Q$. Then $\bar{A}_+$ and $\bar{A}_-$ anti-commute:*

$$\{\bar{A}_+, \bar{A}_-\} = 0.$$

## 4.2. *Unbounded Conserved Quantities*

**Theorem 4.3.** *Let $A$ be a self-adjoint conserved quantity of $H_Q$. Suppose that $D(Q) \subset \mathcal{D}_{A,Q} = D(A) \cap D(QAQ^{-1})$. Then:*

    (i) *The operator $A_+|D(Q)$ is essentially self-adjoint and the closure*

$$\hat{A}_+ := \overline{A_+|D(Q)}$$

*strongly commutes with $Q$.*

(ii) *The operator $A_-|D(Q)$ is essentially self-adjoint and the closure*

$$\hat{A}_- := \overline{A_-|D(Q)}$$

*strongly anticommutes with $Q$.*

(iii) *Suppose that there exists a constant $c \geq 0$ such that*

$$\|QAQ^{-1}\psi\|^2 \leq \|A\psi\|^2 + c\|\psi\|^2, \quad \forall \psi \in D(Q).$$

    *Then*

$$A = \hat{A}_+ + \hat{A}_-.$$

**Corollary 4.4.** *Under the same assumption as in Theorem 4.3, for all $\psi \in D(A) \cap \mathcal{H}_{\mathrm{ac}}(Q)$*

$$\text{w-}\lim_{t \to \pm\infty} e^{itQ} A e^{-itQ} \psi = \hat{A}_+ \psi,$$

$$\text{w-}\lim_{t \to \pm\infty} e^{itQ} A e^{itQ} \psi = \hat{A}_- \psi.$$

**Corollary 4.5.** *(**Uncertainty relation**) Under the same assumption as in Theorem 4.3, for all unit vectors $\psi \in D(Q) \cap D(A)$,*

$$(\Delta Q)_\psi (\Delta A)_\psi \geq \left| \left\langle Q\psi, \hat{A}_- \psi \right\rangle \right|,$$

**Corollary 4.6.** *(**Spectral symmetry**) In addition to the same assumption as in Theorem 4.3, suppose that $\hat{A}_-$ is injective. Then $\sigma(Q)$ is symmetric with respect to the origin in $\mathbb{R}$.*

**Theorem 4.4.** *Let the same assumption as in Theorem 4.3 be satisfied. Suppose that*

$$\|QAQ^{-1}\psi\| = \|A\psi\|, \quad \psi \in D(Q).$$

*Then $\hat{A}_+$ and $\hat{A}_-$ strongly anti-commute.*

## Acknowledgments

The author would like to thank Professor M. Ohya for inviting him to present a lecture at the International Conference QBIC'11 (Noda Campus of TUS, March 7-14, 2011). This work was supported by the Grant-In-Aid No.21540206 for Scientific Research from Japan Society for the Promotion of Science (JSPS).

## References

1. A. Arai, Commutation properties of anticommuting self-adjoint operators, spin representation and Dirac operators, Integr. Equat. Oper. Th. **16** (1993), 38–63
2. A. Arai, Characterization of anticommutativity of self-adjoint operators in connection with Clifford algebra and applications, Integr. Equat. Oper. Th. **17**(1993), 451–463.
3. A. Arai, Analysis on anticommuting self-adjoint operators, Adv. Stud. Pure Math. **23** (1994), 1–15.
4. A. Arai, Operator Theory on a Commutation Relation, mp_arc 06-36. Unpublished.
5. A. Arai, Conserved quantities in a general class of quantum systems and applications to supersymmetric quantum mechanics, in preparation.
6. S. Pedersen, Anticommuting self-adjoint operators, J. Funct. Anal. **89** (1990), 428–443.
7. M. Reed and B. Simon, Methods of Modern Mathematical Physics Vol.I, Academic Press, New York, 1972.
8. Yu.S. Samoilenko, Spectral Theory of Families of Self-Adjoint Operators, Kluwer Academic Publishers, Dordrecht, 1991.
9. B. Thaller, The Dirac Equation, Springer-Verlag, Berlin, Heidelberg, 1992.
10. F.-H. Vasilescu, Anticommuting self-adjoint operators, Rev. Roum. Math. Pures et Appl. **28** (1983), 77–91.

Quantum Bio-Informatics V
© 2013 World Scientific Publishing Co. Pte. Ltd.
pp. 37–48

# OSCILLATIONS AND ROLLING FOR DUFFING'S EQUATION

I.YA. AREF'EVA, E.V. PISKOVSKIY AND I.V. VOLOVICH[*]

*Steklov Mathematical Institute of the Russian Academy of Sciences,
Gubkina Str. 8, 119991 Moscow, Russia;
National Research Nuclear University "MEPhI",
Kashirskoe Highway 31, 115409 Moscow, Russia*
$^{*}E-mail:volovich@mi.ras.ru$

The Duffing equation has been used to model nonlinear dynamics not only in mechanics and electronics but also in biology and in neurology for the brain process modeling. Van der Pol's method is often used in nonlinear dynamics to improve perturbation theory results when describing small oscillations. However, in some other problems of nonlinear dynamics particularly in case of Duffing–Higgs equation in field theory, for the Einsten-Friedmann equations in cosmology and for relaxation processes in neurology not only small oscillations regime is of interest but also the regime of slow rolling. In the present work a method for approximate solution to nonlinear dynamics equations in the rolling regime is developed. It is shown that in order to improve perturbation theory in the rolling regime it turns out to be effective to use an expansion in hyperbolic functions instead of trigonometric functions as it is done in van der Pol's method in case of small oscillations. In particular the Duffing equation in the rolling regime is investigated using solution expressed in terms of elliptic functions. Accuracy of obtained approximation is estimated. The Duffing equation with dissipation is also considered.

## 1. Introduction

The Duffing equation is one of the central equations used to describe the nonlinear dynamics in mechanics, electronics, ecology, secure communications and also in biology for brain process modeling [1,2]. The Duffing equation has the form

$$\ddot{q} + \alpha\dot{q} + \beta q + \epsilon q^3 = \gamma\cos(\nu t), \qquad (1)$$

where $q = q(t)$ is a real-valued function and $\alpha, \beta, \epsilon, \gamma, \nu$ are constants. There are various methods in nonlinear dynamics which are typically used to investigate the behaviour of the system in the neighborhood of the free oscillations see [3,4,5,6]. However, in case of the Higgs equation in field theory,

38

the Einstein-Friedmann equation in cosmology and the relaxation processes in neurology and also in some other problems of nonlinear dynamics not only regime of small oscillations is of interest but also the slow rolling regime.

In small oscillations regime the perturbations in the vicinity of solutions to linear equation of harmonic oscillator are investigated:

$$\ddot{q} + \omega^2 q = 0, \tag{2}$$

where $\omega > 0$. In the rolling (or climbing) regime the perturbations of solutions to the following equation are studied:

$$\ddot{q} - \mu^2 q = 0, \tag{3}$$

where $\mu > 0$.

In the present work the slow rolling regime is considered for the Duffing equation of the special form which describes the motion of the classical particle in the double well potential (Higgs equation, see [7]) with dissipation:

$$\ddot{q} - \mu^2 q + \alpha\dot{q} + \epsilon q^3 = 0. \tag{4}$$

An intriguing analogy is observed between the results obtained and the van der Pol method formulae. The equations (2) and (3) are connected via change $\omega = i\mu$; however to estimate the accuracy of the formulae obtained, the van der Pol's method cannot be directly used and we use different method – a method of expansion of elliptic functions in hyperbolic ones.

The solution to the Duffing equation (4) for the case $\alpha = 0$ is known to be expressed in terms of elliptic functions, and formulae obtained by means of the van der Pol method used to describe small oscillations are in correspondence with expansion of elliptic function in trigonometric series. In the present work the rolling regime for the Duffing equation is investigated using the representation of the solution in terms of elliptic functions. The Duffing equation with dissipation is considered as well. It is shown that in order to improve perturbation theory in the rolling regime the expansion in terms of hyperbolic functions but not in trigonometric functions as in the case of oscillations turns out to be more effective.

It is well-known that simple equations of nonlinear dynamics allow exact solutions in terms of elliptic functions. Correspondence between these exact solutions and approximate van der Pol method is considered in [4], where an equation for anharmonic oscillator with real frequency is investigated.

The simple oscillation equation considered in nonlinear dynamics has the following form:

$$\ddot{q} + \omega^2 q = \epsilon f(q, \dot{q}), \tag{5}$$

where $q = q(t)$ – a real-valued function of time, frequency $\omega$ – a positive number, $\epsilon$ – small parameter.

In applications the following equation is often considered

$$\ddot{q} - \mu^2 q = \epsilon f(q, \dot{q}), \tag{6}$$

where $\mu^2 > 0$. It comes from (5) via change $\omega = i\mu$. Such equation can be called an equation of rolling.

For instance, this form is of the Higgs equation (see [7]), the Friedman equation of this form is used to describe dynamics of tachyon field in cosmology.

By analogy with the Krylov-Bogoliubov expansion method, used to solve equations describing oscillations (5), we will look for general solution to the equation describing rolling (6) in the form of expansion

$$q = a \sinh \psi + \epsilon u_1(a, \psi) + \epsilon^2 u_2(a, \psi) + ... \tag{7}$$

Here $a$ and $\psi$ are functions of time which should be determined. We will show that solutions to some equations of the type (6) can be represented in the form

$$q(t) = A \sinh(\mu(A)t) + \epsilon a_1 \sinh(2\mu(A)t) + \epsilon^2 a_2 \sinh(3\mu(A)t) + .... \tag{8}$$

Here in order to simplify notation the case of $q(0) = 0$ is considered, $A$– arbitrary constant,

$$\mu(A) = \mu(A, \epsilon) = \mu + \epsilon \mu_1(A) + .... \tag{9}$$

The series (8) converges when the time variable $t$ takes certain values, see below.

In the next section the Duffing–Higgs equation and the anharmonic oscillator are considered. In Sect.3 solutions to the Duffing–Higgs equation are presented. Then in Sect.4 a hyperbolic analogue of the van der Pol method is constructed. Finally in Sect.5 an approximate solution of the nonlocal nonlinear equation is presented.

## 2.  Higgs equation and anharmonic oscillator

In this section exact and approximate solutions to some nonlinear equations are considered. The solution to the equation for anharmonic oscillator (the Duffing equation)

$$\ddot{q}(t) + \omega^2 q(t) = -\epsilon q^3(t), \quad \omega > 0, \ \epsilon > 0$$

has the representation

$$q(t) = a\,\mathrm{cn}(\Omega t + b, k),$$

where $\mathrm{cn}(u, k)$ – elliptic cosine [8], $a$ - amplitude and $b$ - phase are arbitrary constants, $\Omega$ and $k$ - frequency and modulus of elliptic function are obtained from parameters of the equation $\omega, \epsilon$ and depend on amplitude $A$:

$$\Omega = \sqrt{\omega^2 + \epsilon a^2}, \qquad k = \sqrt{\frac{\epsilon a^2}{2(\omega^2 + \epsilon^2)}}.$$

The function $\mathrm{cn}(\Omega t + b, k)$ can be expanded in trigonometric functions

$$\mathrm{cn}(\Omega t + b, k) = \frac{2\pi}{k\mathbf{K}} \sum_{n=1}^{\infty} \frac{\mathsf{q}^{n-1/2}}{1 + \mathsf{q}^{2n-1}} \cos\left((2n-1)\frac{\pi(\Omega t + b)}{2\mathbf{K}}\right), \qquad (10)$$

where

$$\mathsf{q} = e^{-\pi\frac{\mathbf{K}'}{\mathbf{K}}}, \qquad \mathsf{q}' - \pi\frac{\mathbf{K}}{\mathbf{K}'}. \qquad (11)$$

Here $\mathbf{K} = \mathbf{K}(k)$ and $\mathbf{K}' = \mathbf{K}'(k)$ – complete elliptic integrals of the first kind (10). In particular, with $b = -\mathbf{K}$ we obtain:

$$\mathrm{cn}(\Omega t - \mathbf{K}, k) = \frac{2\pi}{k\mathbf{K}} \sum_{n=1}^{\infty} \frac{\mathsf{q}^{n-1/2}}{1 + \mathsf{q}^{2n-1}} \sin\left((2n-1)\frac{\pi\Omega t}{2\mathbf{K}}\right) \qquad (12)$$

From this representation one can obtain asymptotic expansion with small $\epsilon$, that is obtained by means of the Krylov-Bogoliubov averaging method [4], and meets initial condition $q(0) = 0$,

$$q(t) = A\sin\left(\omega(1 + \epsilon\frac{3A^2}{8\omega^2})t\right) - \epsilon\frac{A^3}{32\omega^2}\sin\left(3\omega(1 + \epsilon\frac{3A^2}{8\omega^2})t\right) + \dots \qquad (13)$$

Here $A$ is an arbitrary real parameter.

The Higgs equation

$$\ddot{q}(t) - \mu^2 q(t) = -\epsilon q^3(t), \quad \mu > 0, \ \epsilon > 0 \qquad (14)$$

is considered in the next Section, and its solution is given in terms of elliptic functions. The choice of elliptic functions depends on the sign of the solution energy, the latter is defined by initial conditions. In particular, with $E > 0$ the solution has the following form

$$q(t) = a\,\mathrm{cn}(\Omega t + b, k), \tag{15}$$

where $a$, $b$ are arbitrary constants, and $\Omega$ and $k$ are defined with the following relations:

$$\Omega = \sqrt{-\mu^2 + \epsilon a^2}, \qquad k = \sqrt{\frac{\epsilon a^2}{2(-\mu^2 + \epsilon a^2)}}. \tag{16}$$

This solution is periodical with period $T = 2\mathbf{K}/\Omega$. Here $\mathbf{K} = \mathbf{K}(k)$. With $b = -\mathbf{K}$ the solution (15) corresponds to initial conditions $q(0) = 0$. With $t$ such that $|t| < T/2$, the solution can be expanded in hyperbolic series:

$$\mathrm{cn}(\Omega t - \mathbf{K}, k) = \frac{2\pi}{k\mathbf{K}'} \sum_{n=1}^{\infty} (-1)^n \frac{\mathsf{q}'^{n-1/2}}{1 + \mathsf{q}'^{2n-1}} \sinh((2n-1)\frac{\pi\Omega t}{2\mathbf{K}'}) \tag{17}$$

Here $\mathsf{q}' = \mathsf{q}(k)$. From the latter with small $\epsilon$ we deduce asymptotic expansion. This series coincides with the series that is obtained by extension of the Krylov-Bogoliubov formulae for anharmonic oscillator in the case of the Higgs equation:

$$q(t) = A \sinh\left(\mu(1 + \frac{3A^2}{8\mu^2}\epsilon)t\right) - \frac{\epsilon A^3}{32\mu^2} \sinh\left(3\mu(1 + \frac{3A^2}{8\mu^2}\epsilon)t\right) + \dots \tag{18}$$

For the first time the formula (18) was obtained in [9] where also a non-local equation [10] was considered and also the equation with damping. This expansion can be formally obtained from the averaging method result (13) by means of change $\omega \to i\mu$, $A \to -iA$.

## 3. Solution to the Higgs equation

The Higgs equation

$$\ddot{q} - \mu^2 q = -\epsilon\, q^3, \quad \epsilon > 0, \tag{19}$$

with the initial conditions

$$q(0) = q_0, \quad \dot{q}(0) = v_0 \tag{20}$$

has solution belonging to one of the following three types. Type of the solution is defined by energy

$$E = \frac{1}{2}v_0^2 - \frac{1}{2}\mu^2 q_0^2 + \frac{1}{4}\epsilon q_0^4 \tag{21}$$

42

If $E > 0$ the solution has the following representation

$$q(t) = a\,\mathrm{cn}(\Omega t + b, k) \tag{22}$$

$$a^2 = \frac{\mu^2}{\epsilon}\left(1 + \sqrt{1 + \frac{4\epsilon E}{\mu^4}}\right) \tag{23}$$

$$\Omega^2 = \mu^2\sqrt{1 + \frac{4\epsilon E}{\mu^4}} \tag{24}$$

$$k^2 = \frac{1}{2} + \frac{1}{2}\frac{1}{\sqrt{1 + \frac{4\epsilon E}{\mu^4}}}, \tag{25}$$

$b$ is defined from the condition $q(0) = a\,\mathrm{cn}(b, k)$.

We will discuss the case with initial data of the special form $q(0) = 0$. In this case $b = -\mathbf{K}$ and

$$q(t) = a\,\mathrm{cn}\left(\Omega t - \mathbf{K}, k\right) \tag{26}$$

The energy $E$ is expressed by means of parameter $k'$, $k^2 + k'^2 = 1$, by

$$E = \frac{\mu^4 k'^2 \left(1 - k'^2\right)}{\epsilon \left(2 k'^2 - 1\right)^2} \tag{27}$$

## 4. Approximate solutions

In this section exact solutions approximations will be presented and their approximation accuracy will be estimated in the limit of coupling constant tends to zero.

The function $\mathrm{cn}(u - \mathbf{K}, k)$ with $|u| < \mathbf{K}$ and small $k'$ has the following asymptotic expansion

$$\mathrm{cn}(u - \mathbf{K}, k) = \frac{\pi}{k\mathbf{K}'}\left\{\frac{\sinh(u')}{\cosh(\rho')} - \frac{\sinh(3u')}{\cosh(3\rho')} + \frac{\sinh(5u')}{\cosh(5\rho')} + \ldots\right\} \tag{28}$$

where

$$\rho' = \frac{\pi\mathbf{K}}{2\mathbf{K}'}, \quad u' = \frac{\pi u}{2\mathbf{K}'} \tag{29}$$

The series (28) converges when $|\Re u| < \mathbf{K}$ and $-1 < k < 1$.

One can prove the following

**Proposition 1.** *For all $\delta$ from $0 < \delta < \sqrt{2}$ there exists such $k_0 = k_0(\delta)$, that for all*

$$k_0 \le k < 1 \tag{30}$$

*and*

$$0 < t \leq t_0, \tag{31}$$

*holds*

$$\left| \mathrm{cn}(\Omega t - \mathbf{K}, k) - \frac{\pi}{k\mathbf{K}'} \frac{\sinh(\dfrac{\pi\Omega t}{2\mathbf{K}'})}{\cosh(\dfrac{\pi\mathbf{K}}{2\mathbf{K}'})} \right| < \delta. \tag{32}$$

$k_0$ *is defined by*

$$\frac{4\left(1 - k_0^2\right)^{3/2}}{k_0} = \delta \tag{33}$$

*and $t_0 > 0$ is given by*

$$t_0 = \frac{1}{3\Omega} \ln\left(\frac{k\delta}{4k'^3}\right), \tag{34}$$

*and the condition $t_0 > 0$, i.e. $\frac{k\delta}{4k'^3} > 1$, is satisfied for all $1 > k > k_0$ by choice $k_0$ according to (33).*

The proof of Proposition 1 is based on

$$\exp\left(-\frac{\pi\mathbf{K}}{2\mathbf{K}'}\right) - \sqrt{1 - k^2} \leq 0, \tag{35}$$

$$\frac{\pi}{2\mathbf{K}'} \leq 1 \tag{36}$$

and is presented in [11].

Let us consider the approximation of the exact solution

$$q(t) = a\,\mathrm{cn}\left(\Omega t - \mathbf{K}, k\right) \tag{37}$$

by means of the first term of the following expansion (28)

$$q(t) = a\,\mathrm{cn}(\Omega t - \mathbf{K}, k) = a\,\frac{\pi}{k\mathbf{K}'} \frac{\sinh(\dfrac{\pi\Omega t}{2\mathbf{K}'})}{\cosh(\dfrac{\pi\mathbf{K}}{2\mathbf{K}'})} + \ldots \tag{38}$$

We have

**Proposition 2** *For all $\delta_s$ from*

$$0 < \delta_s < \frac{2\mu}{\sqrt{\epsilon}} \tag{39}$$

*with*

$$t < \frac{1}{3\Omega} \ln(\frac{\mu^4}{4\epsilon E} \delta_s) \tag{40}$$

*and*

$$\epsilon \frac{4\,E}{\mu^4} < \delta_s \tag{41}$$

*it holds*

$$\left| q(t) - a\frac{\pi}{k\mathbf{K}'} \frac{\sinh(\dfrac{\pi\Omega t}{2\mathbf{K}'})}{\cosh(\dfrac{\pi\mathbf{K}}{2\mathbf{K}'})} \right| < \delta_s, \tag{42}$$

*where a is given by (23), $\Omega$ is given by (24), and k - by (25).*

Then, we can prove

**Proposition 3.** *The first term of (38) can be presented in the form (18)*

$$q_1(t) \equiv a\frac{\pi}{k\mathbf{K}'} \frac{\sinh(\dfrac{\pi\Omega t}{2\mathbf{K}'})}{\cosh(\dfrac{\pi\mathbf{K}}{2\mathbf{K}'})} = A\sinh\left(\mu(1+\frac{3A^2}{8\mu^2}\epsilon)t\right) + ..., \tag{43}$$

*with accuracy*

$$|q_1(t) - q_{1,appr}(t)| < A\cosh\left(\mu(1+\frac{3A^2}{8\mu^2}\epsilon)t\right) \cdot \left(\frac{15}{64}\frac{E^2}{\mu^8}\epsilon^2 + O\left(\epsilon^3\right)\right)\mu t \tag{44}$$

*here*

$$q_{1,appr}(t) \equiv A\sinh\left(\mu(1+\frac{3A^2}{8\mu^2}\epsilon)\,t\right) \tag{45}$$

*and*

$$A = a\frac{\pi}{k\mathbf{K}'}\frac{1}{\cosh(\dfrac{\pi\mathbf{K}}{2\mathbf{K}'})} = \frac{\sqrt{2E}}{\mu}\left(1 - \frac{9E}{16\mu^4}\epsilon + \frac{271E^2}{256\mu^8}\epsilon^2 + \mathcal{O}(\epsilon^3)\right) \tag{46}$$

The proof of this proposition is given in [11].

As a final result we have the following

**Proposition 4** *Let $q(t)$ be the solution to the Higgs equation (14) with initial conditions $q(0) = 0, \dot{q}(0) = A\mu$, where $A$ is an arbitrary constant. Then there exists such quantity*

$$\mu(A) = \mu(1 + \frac{3A^2}{8\mu^2}\epsilon) + O(\epsilon^2),$$

*that for all $\delta > 0$ there exist constants $C_1 > 0, C_2 > 0$ and $\epsilon_0 > 0$ such that for all $0 \le \epsilon < \epsilon_0$ the following inequality holds*

$$|q(t) - A\sinh(\mu(A)t)| < \delta, \tag{47}$$

*for all*

$$0 \le t < C_1 \log \frac{C_2}{\epsilon} \tag{48}$$

## 5. Nonlocal Nonlinear Equation

We are interested in getting approximation solutions for nonlocal version of the Higgs equation, see eq. (55) below. The linear approximation for this equation is

$$e^{\tau \partial_t^2}_{D,0} \equiv \frac{1}{2\sqrt{\tau\pi}} \int_0^\infty dt' \left[ -\frac{(t-t'^2)}{4\tau} - e^{-\frac{(t+t'^2)}{4\tau}} \right] \Phi(t'^2 \Phi(t), \quad t > 0. \tag{49}$$

Solutions to Eq. (49) are given by a linear combination

$$\Phi_D(t) = \sum_n \left( \frac{1}{2} B_n (e^{\alpha_n t} - e^{-\alpha_n t}) + \frac{1}{2} B_n^* (e^{\alpha_n^* t} - e^{-\alpha_n^* t}) \right), \tag{50}$$

where $\alpha_n$ are solutions to the equation

$$e^{\tau \alpha_n^2} = m^2, \tag{51}$$

$$\alpha_n = \pm\sqrt{\frac{\ln m^2 + 2\pi i n}{\tau}}, n = 0, \pm 1, \pm 2, \dots \tag{52}$$

For the $\alpha_0$ mode,

$$\Phi(t) = B_0 \sinh \alpha_0 t$$

we have

$$e^{\tau \partial_t^2}_{D,0} \sinh rt = e^{\tau r^2} \sinh rt, \quad t > 0, \quad e^{\tau r^2} = m^2. \tag{53}$$

Indeed, the action of the operator $e^{\tau \partial_t^2}_0$ on the function $e^{rt}$ is given by the formula

$$e^{\tau \partial_t^2}_{D,0} e^{rt} = \frac{1}{2} e^{rt + r^2 \tau} \left[ 1 + \operatorname{erf}\left( r\sqrt{\tau} + \frac{t}{2\sqrt{\tau}} \right) \right]$$
$$- \frac{1}{2} e^{-rt + r^2 \tau} \left[ 1 + \operatorname{erf}\left( r\sqrt{\tau} - \frac{t}{2\sqrt{\tau}} \right) \right]$$

since

$$\frac{1}{2\sqrt{\tau\pi}} \int_0^\infty dt' \left[ -\frac{(t-t'^2}{4\tau} - e^{-\frac{(t+t'^2}{4\tau}} \right] e^{rt'}$$

$$= \frac{1}{2} e^{rt+r^2\tau} \left[ 1 + \mathrm{erf}\left( r\sqrt{\tau} + \frac{t}{2\sqrt{\tau}} \right) \right]$$

$$- \frac{1}{2} e^{-rt+r^2\tau} \left[ 1 + \mathrm{erf}\left( r\sqrt{\tau} - \frac{t}{2\sqrt{\tau}} \right) \right] \tag{54}$$

Hence, we get (53).

Nonlocal Higgs equation has the form

$$\frac{1}{2\sqrt{\tau\pi}} \int_{-\infty}^\infty dt' \left( \frac{\xi^2((t-t'^2-2\tau)}{4\tau^2} - \mu^2 \right) e^{-\frac{(t-t'^2}{4\tau}} \Phi(t'^3(t)). \tag{55}$$

This equation is related with superstring field theory and has many cosmological applications [10,9]. Equation (55) for $\xi = 0$ an an odd function $\Phi(-t) = -\Phi(t)$ has the form

$$\frac{1}{2\sqrt{\tau\pi}} \int_0^\infty dt' \left( e^{-\frac{(t-t'^2}{4\tau}} - e^{-\frac{(t+t'^2}{4\tau}} \right) \Phi(t'^3(t)). \tag{56}$$

It is known [12] that this equation has a solution interpolating between two vacua $\Phi = \pm\frac{\mu}{\sqrt{\epsilon}}$.

It is also known [13] that for $\xi > \xi_{cr}$ equation (56) has oscillation solutions with period $T_\xi$ and within one period a good approximation for these solutions is

$$\Phi(t) = a\sinh\left(\Omega t\right) - \frac{\epsilon a^3}{32\mu^2} e^{-9\lambda\Omega^2} \sinh\left(3\Omega t\right) + ... \tag{57}$$

where $\Omega$ is a root of

$$\Omega^2 - \frac{\mu^2}{\xi^2} - \frac{3}{4}\frac{\epsilon}{\xi^2} a^2 e^{-\lambda\Omega^2} = 0. \tag{58}$$

Approximation (57) is valid for small $\lambda$ and small $t$, $t < T_\xi/4$. Solutions to (58) are given by the Lambert W function satisfying a relation $W(x)e^{W(x)} = x$,

$$\Omega = \pm\sqrt{\frac{\mu^2}{\xi^2} + \frac{1}{\lambda}\mathrm{W}\left( \frac{3\lambda\epsilon a^2}{4\xi^2} e^{-\frac{\mu^2}{\xi^2}\lambda} \right)} \tag{59}$$

Expanding the right hand site of (59) on $\lambda$ we get

$$\Omega = \pm \left( \sqrt{\frac{\mu^2}{\xi^2} + \frac{3}{4}\frac{\epsilon}{\xi^2}a^2} - \frac{3}{8}\frac{\epsilon}{\xi^2}a^2\lambda\sqrt{\frac{\mu^2}{\xi^2} + \frac{3}{4}\frac{\epsilon}{\xi^2}a^2} + \mathcal{O}(\lambda^2) \right) \qquad (60)$$

Note, that if $\lambda = 0$ one gets an approximate solution to the local equation

$$(\xi^2\partial^2 - \mu^2)q(t) = -\epsilon q^3(t) \qquad (61)$$

and the approximate solution

$$q(t) = A\sinh\left(\frac{\mu}{\xi^2}(t + \frac{3A^2}{8\mu^2}\epsilon t)\right) - \frac{\epsilon A^3}{32\mu^2}\sinh\left(\frac{3\mu}{\xi^2}(t + \frac{3A^2}{8\mu^2}\epsilon t)\right) + ... \qquad (62)$$

## 6. Conclusion

In this paper the rolling regime for the Duffing–Higgs equation was considered and a new method to get approximate solutions has been presented. In cosmology a nonlocal version of the equation (6) is considered in [10]. In certain circumstances one can reduce it to a local problem with modified parameters of initial problem, while the asymptotic expansion for the equation has useful applications in investigation of nonlocal problems [9].

Let us note also that the equation (6) along with the equation (5) constitute an important example in functional mechanics [14,15].

## Acknowledgments

The present work is partially supported by the following grants RFFI 11-01-00894-a(I.A.) and RFBR 11-01-00828-a, NSch-2928.2012.1, 11-01-12114-ofim-2011 (E.P. and I.V).

## References

1. The Duffing equation: Nonlinear oscillators and their behaviour, Eds.I.Kovacic and M.J.Brennan, Wiley, 2011.
2. E.C. Zeeman, Duffing's equation in brain modelling, Bulletin of the Institute of Mathematics and Its Applications, 12(7):207-214, (1976).
3. V.I. Arnold, V.V. Kozlov, A.I. Neishtadt, *Mathematical aspects of classical and celestial mechanics*, Springer, 2006.
4. N.N. Bogoliubov, Y.A. Mitropolski, *Asymptotic Methods in the Theory of Non-Linear Oscillations*, New York: Gordon and Breach, 1961.
5. V. V. Kozlov, *Symmetries, topology and resonances in Hamiltonian mechanics*, Ergebnisse der Mathematik und ihrer Grenzgebiete (3) [Results in Mathematics and Related Areas (3)], 31, Springer-Verlag, Berlin, 1996.

6. M. Ohya and I. Volovich, *Mathematical Foundations of Quantum Information. and Computation and Its Applications to Nano- and Bio-systems*, Springer, 2011.

7. S. Weinberg, The quantum theory of fields, Cambridge University Press, 1996.

8. N.I. Akhiezer, *Elements of the theory of elliptic functions*, AMS, Providence, RI, 1990.

9. I.Ya. Aref'eva and I.V.Volovich, *Cosmological Daemon*, JHEP, 1108 (2011) 102, arXiv:1103.0273.

10. I.Ya. Aref'eva, *Nonlocal String Tachyon as a Model for Cosmological Dark Energy*, 2006 AIP Conf. Proc. **826**, pp. 301–311, arXiv: astro-ph/0410443

11. I.Ya. Aref'eva, E.V. Piskovskiy, and I.V.Volovich, *Rolling in nonlinear dynamics and and elliptic functions*, submitted to Theor. and Math. Phys.

12. V.S. Vladimirov, Ya.I. Volovich, Theor.Math.Phys. **138** (2004) 297-309, arXiv: math-ph/0306018.

13. Ya. I. Volovich, Numerical study of nonlinear equations with infinite number of derivatives, J. Phys. A36 (2003) 8685, arXiv: math-ph/0301028

14. I.V. Volovich, Bogolyubov equations and functional mechanics, Theor. Mathem. Phys. 164:3 (2010) 354-362.

15. E.V. Piskovskiy, I.V. Volovich, *On the Correspondence between Newtonian and Functional Mechanics*,Proceedings volume in QP-PQ by World Scientific (ICQBIC2010).

Quantum Bio-Informatics V
© 2013 World Scientific Publishing Co. Pte. Ltd.
pp. 49–56

# GENERAL FORMALISM OF DECISION MAKING BASED ON THEORY OF OPEN QUANTUM SYSTEMS

M. ASANO, M. OHYA

*Department of Information Sciences, Tokyo University of Science*
*Yamasaki 2641, Noda-shi, Chiba, 278-8510 Japan*

I. BASIEVA, A. KHRENNIKOV*

*International Center for Mathematical Modelling*
*in Physics and Cognitive Sciences*
*Linnaeus University, S-35195, Växjö, Sweden*
*$^*E-mail : andrei.khrennikov@lnu.se*

We present the general formalism of decision making which is based on the theory of open quantum systems. A person (decision maker), say Alice, is considered as a quantum-like system, i.e., a system which information processing follows the laws of quantum information theory. To make decision, Alice interacts with a huge mental bath. Depending on context of decision making this bath can include her social environment, mass media (TV, newspapers, INTERNET), and memory. Dynamics of an ensemble of such Alices is described by Gorini-Kossakowski-Sudarshan-Lindblad (GKSL) equation. We speculate that in the processes of evolution biosystems (especially human beings) designed such "mental Hamiltonians" and GKSL-operators that any solution of the corresponding GKSL-equation stabilizes to a diagonal density operator (In the basis of decision making.) This limiting density operator describes population in which all superpositions of possible decisions has already been resolved. In principle, this approach can be used for the prediction of the distribution of possible decisions in human populations.

## 1. Introduction

Recently the mathematical formalism of quantum theory started to be used to describe the process of information by cognitive systems[1-5] and especially for decision making[6-14]. We call such models *quantum-like* (QL) to distinguish them from quantum physical models. The basic assumption of all QL models is that the mental state of a decision maker, say Alice, can be described by a vector in a complex Hilbert space representing in general nontrivial superposition of basis vectors corresponding to possible decisions. According to Bohr experimental information and, in particular, probabilities are *classical.* Therefore even in our QL-models statistics of decisions has to be described by classical probability. We supposed[12−14] that, to make a decision, Alice interacts with a mental bath. Depending on context of decision making this bath can include her social environment, mass media (TV, newspapers, INTERNET), and memory. Dynamics of an ensemble of such Alices is described by Gorini-Kossakowski-Sudarshan-Lindblad (GKSL) equation. We seculate that in the processes of evolution biosystems (especially human beings) designed such "mental Hamiltonians" and GKSL-operators that any solution of the corresponding GKSL-equation stabilizes to a diagonal density operator (In the basis of decision making.) This limiting density operator describes population in which all superpositions of possible decisions has already been resolved. In principle, this approach can be used for the prediction of the distribution of possible decisions in human populations.

In papers[12−14] the aforementioned approach was elaborated for a special problem of game theory, Prisoner's Dilemma. In this simple model Alice interacts with a mental bath based on her (working) memory related to possible strategies of her coplayer, say Bob, and her reflections on his possible strategies. Although such a coplayer is an important personage of the story about the Prisoner's Dilemma, our approach[12−14], in fact, does not rely to real actions of Bob. Alice processes her own reflections on possible actions of Bob (constrained by her own actions). Therefore, in principle, Bob can be excluded from the model and we can apply our scheme of decision making to Alice who treats an arbitrary problem. We shall proceed in this way.

The only restriction to the problem under considerations is that it can be represented in the form of a vector of questions ("subproblems") $A = (A_1, ..., A_n)$. Each question has two answers $A_j = \pm 1$, (yes/no). Questions are *compatible,* i.e., Alice can come to a vector of joint answers

$$(\alpha_1...\alpha_n), \tag{1}$$

where $\alpha_j = \pm 1$. However, these questions are in general *nonseparable.* Alice can have such states of mind in which questions cannot be split from each other. These are *entangled states of mind:*

$$|\psi\rangle = \sum c_\alpha |\alpha\rangle, \tag{2}$$

where $\alpha = \alpha_1...\alpha_n$. In decision making entanglement can be interpreted in the following way. Alice cannot process questions separately, i.e., to make decisions for each of them and them simply form the vector of answers (1).

## 2. Problem solving as decoherence

In this paper we describe dynamics of the process of decision making in the aforementioned class of problems. It is well known that in physics the quantum state dynamics is described by Schrödinger's equation. This dynamics is unitary (roughly speaking it is combined of a family of rotations; in principle, this family can be infinite). Busemeyer et al.[10] used this equation to model the dynamics of the process of decision making in games of the Prisoner's Dilemma type. However, a possibility to describe dynamics of Alice by Schrödinger's equation is questionable. This equation describes dynamics of an isolated system, i.e., a system which does not interact with the environment. In the modern complex society, Alice definitely cannot be considered as an isolated social system. She is in the permanent contact with the social environment of increasing complexity including mass media (especially TV) and recently INTERNET. The influence of the environment induces random fluctuations of opinions and choices in Alice's mind.

We are interested in the "unstable" part of population involved in the process of decision making for some problem; those who have no concrete opinions and who will make their choices through a long chain of opinion-fluctuations. If Alice were an isolated social system, then the only possibility to describe transition from the mental state of superposition of choices to the state corresponding to the concrete choice was to use the projection postulate of quantum mechanics (the von Neumann postulate). This state reduction process, from superposition to one of its components, is called *the state collapse.* It is imagined as instantaneous (the jump-type) transition from one state to another. The state collapse might be used to describe the situation in which Alice makes her choice spontaneously. This type of behavior cannot be completely excluded from consideration, but such cases are not statistically significant. Moreover, by quantum ideology the state collapse occurs when an isolated system driven by Schrödinger's equation interacts practically instantaneously with a measurement device. This corresponds to the following decision making context: Alice who has been totally isolated from the mental bath related to the problem is suddenly asked to make her choice. We are interested in Alices who is not isolated from the mental environment.

Therefore let us take more seriously a role which the social environment plays in the process of decision making. We apply to social science the theory of open quantum systems, i.e., systems which interact with a large thermostat (bath). Since a bath is a huge physical systems with millions of variables (cf. with complexity of the "social bath" around an American citizen during the election campaign), it is in general impossible to provide the exact mathematical description of the dynamics of a quantum system interacting with a bath. Physicists proceed under a few assumptions providing a possibility to describe this dynamics at least approximately. In quantum physics interaction of a quantum system with a bath is described by a quantum version of the master equation for Markovean dynamics. Quantum Markovian dynamics given by the *Gorini-Kossakowski-Sudarshan-Lindblad* (GKSL) equation[15] is the most popular approximation of quantum dynamics in the presence of interaction with a bath.

We remind shortly the origin of the GKSL-dynamics. The starting point is that the state of a composite system, a quantum system $s$ combined with a bath, is a pure quantum state, complex vector $\Psi$, which evolution is described by Schrödinger's equation. This is an evolution in a Hilbert space of the huge dimension (since a bath has so many degrees of freedom). The existence of the Schrödinger dynamics in the huge Hilbert space has a merely theoretical value. Observers are interested in the dynamics of the state $\phi_s$ of the quantum system $s$. The next fundamental assumption in derivation of the GKSL-equation is the Markovness of the evolution, the absence of long term memory effects. It is assumed that interaction with the bath destroys such effects. Thus, the GKSL-evolution is the Markovian evolution. Finally, we point to the condition of the factorizability of the initial state of a composite system (a quantum system coupled with a bath), $\Psi = \phi_s \otimes \phi_{\text{bath}}$, where $\otimes$ is the sign of the tensor product. Physically factorization is equivalent to the absence of correlations (at the beginning of evolution; later they are induced by the interaction term of Hamiltonian – the generator of evolution). One of distinguishing features of the evolution under the mentioned assumptions is the existence of one or a few *equilibrium points.* The state of the quantum system $s$ stabilizes to one of such points in the process of evolution; a pure initial state, a complex vector $\psi_s$, is transformed into a mixed state, a density matrix $\rho_s(t)$ (classical state without superposition effects).

In contrast to the GKSL-evolution, the Schrödinger evolution does not induce stabilization; any solution different from an eigenvector of Hamiltonian will oscillate for ever. Another property of the Schrödinger dynamics is that it always transfers a pure state into a pure state, i.e., a vector into a vector: quantumness if it was originally present in a state (in the form of superposition) cannot disappear in the process of continuous dynamical evolution. Transition from quantum indeterminism to classical determinism can happen only as the result of the collapse of the quantum state. We remark that Busemeyer at al.[10] have used the Schrödinger equation. They were confronted with problem of selection of the moment of time corresponding to decision-collapse[10].

On one hand, in our model of decision making we would like to escape the usage of the state collapse. On the other hand, to make a decision, Alice has to make transition from quantum to classical representation of her preferences. The GKSL-evolution provides such a possibility (and without "quantum jumps"). Alice's mental state smoothly evolves (fluctuating, but with dumping of fluctuations) to the final classical decision state. As was mentioned in introduction, first time this idea has been explored in the QL-model for Prisoner's Dilemma[12].

We remark that in each application of this QL-model one should check carefully whether *social counterparts of physical conditions of applicability of quantum Markovian dynamics* can be found.

## 3. Dynamics of decision making

The quantum dynamical equation has the form $ih\frac{\partial \psi}{\partial t}(t) = H\psi(t)$, where $H$ is the operator of energy, Hamiltonian, and $h$ is the Planck constant. This equation has been used for the description of mental dynamics[16] (in particular, the dynamics of expectations of the financial market). The question of the mental interpretation of an analog of the Planck constant $h$ has been discussed. This is a complicated problem. In general a positive constant in front of the time derivative can be interpreted as the time scale parameter[16]. Since the usage of the symbol $h$ may be a source of misunderstanding (especially for physically educated readers), we shall use a new scaling parameter, say $\tau$. It determines the time scale of updating of the mental state of Alice. We rewrite the dynamical equation as

$$i\tau\frac{\partial \psi}{\partial t}(t) = H\psi(t), \tag{3}$$

The most general Hamiltonian $H$ in the space of mental states has the form

$$H = H_{\text{stab}} + H_{\text{flip}}, \tag{4}$$

where $H_{\text{stab}}$ is the part of Hamiltonian responsible for stability of distribution of opinions about various possible selections of decisions. It is given by

$$H_{\text{stab}} = \sum_{\alpha} \lambda_{\alpha,\alpha} |\alpha\rangle\langle\alpha| \tag{5}$$

where $\alpha = \alpha_1...\alpha_n$. And $H_{\text{flip}}$ is the part of Hamiltonian responsible for flipping from one selection of the strategy-vector to another. It is given by

$$H_{\text{flip}} = \sum_{\alpha \neq \beta} \lambda_{\alpha,\beta} |\alpha\rangle\langle\beta| \tag{6}$$

In the absence of the $H_{\text{flip}}$-component, the probabilistic structure of superposition is preserved. Only phases of choices evolve in the rotation-like way.

In the presence of the flipping component $H_{\text{flip}}$ the distribution of probabilities of choices of various strategies changes in the process of evolution. Such flipping from one strategy to another makes the state dynamics really quantum. (In fact, for political and social technologies the most important is the flipping part of the Hamiltonian.)

In physics the dynamics of a system in a bath is described by the quantum analog of the master equation, the GKSL-equation, see section 2. We write this equation by using the time scaling constant $\tau$, instead of the Planck constant:

$$i\tau \frac{\partial \rho}{\partial t}(t) = [H, \rho(t)] + L(\rho(t)), \tag{7}$$

where $L$ is a linear operator acting in the space of operators on the complex Hilbert space. The general form of $L$ was found by Gorini, Kossakowski, Sudarshan, and Lindblad . At the moment we are not interested in (a rather complex) structure of $L$. For our applications, it is sufficient to know that it can be expressed through matrix multiplication for a family of matrices. Under natural selection of matrices (mental Hamiltonian $H$ and the components of the operator $L$) any solution of this equation stabilizes to a diagonal density matrix

$$\rho_{\text{decision}} = \text{diag}(\text{p}_\alpha). \tag{8}$$

This matrix describes the distribution of firmly established decisions.

Consider two populations, say $\mathcal{E}_1$ and $\mathcal{E}_2$. Suppose that in our QL model the first one is described by a pure state $|\psi\rangle$, see (2), and the second one by the density matrix (8). Moreover, suppose that the complex amplitudes given by the coefficients in the expansion (2) produce the same probabilities as the density matrix, i.e.,

$$p_\alpha = |c_\alpha|^2. \tag{9}$$

One may ask: "What is the difference?" At the first sight there is no difference at all, since we obtain the same probability distribution of preferences. However, the distributions of mental state in ensembles $\mathcal{E}_1$ and $\mathcal{E}_2$

are totally different. All people in $\mathcal{E}_1$ are in the same state of indeterminisity (superposition) $\psi$. They are in doubts. They are ready to change they opinions (to create a new superposition of opinions). In particular, $\mathcal{E}_1$ is a proper population for political manipulations. Opposite to $\mathcal{E}_1$, the population $\mathcal{E}_2$ consists of people who have already resolved they doubts, whose mental states have already been reduced to states of the form $\alpha = \alpha_1...\alpha_n$, definite choices.

**Conclusion.** *We presented the general QL-model of decision making (problem solving). It is based on application of theory of open quantum systems to cognitive and social sciences.*

# References


1. A. Khrennikov, Quantum-like brain: Interference of minds, *BioSystems* **84**, 225-241 (2006).
2. L. Accardi, A. Khrennikov, M. Ohya, The problem of quantum-like representation in economy, cognitive science, and genetics, in: L. Accardi, W. Freudenberg, M. Ohya (Eds.), Quantum Bio-Informatics II: From Quantum Information to Bio-Informatics, WSP, Singapore, 1-8 (2008).
3. K.-H. Fichtner, L. Fichtner, W. Freudenberg, and M. Ohya, On a quantum model of the recognition process, *QP-PQ: Quantum Prob. White Noise Analysis* **21**, 64-84 (2008).
4. A. Khrennikov, *Ubiquitous quantum structure: from psychology to finances,* Springer, Berlin-Heidelberg-New York, 2010.
5. E. Conte, A. Khrennikov, O. Todarello, A. Federici, J. P. Zbilut, Mental states follow quantum mechanics during perception and cognition of ambiguous figures, *Open Systems and Information Dynamics* **16**, 1-17 (2009).
6. J. R. Busemeyer, Z. Wang, and J. T. Townsend, Quantum dynamics of human decision making, *J. Math. Psychol.* **50**, 220-241 (2006).
7. L. Accardi, A. Khrennikov, M. Ohya, Quantum Markov model for data from Shafir-Tversky experiments in cognitive psychology, *Open Systems and Information Dynamics* **16**, 371-385 (2009).
8. A. Khrennikov and E. Haven, Quantum mechanics and violations of the sure-thing principle: the use of probability interference and other concepts, *J. Math. Psychology* **53**, 378-388 (2009).
9. A. Khrennikov, Quantum-like model of cognitive decision making and information processing, *Biosystems* **95**, 179-187 (2009).
10. E. M. Pothos, J. R. Busemeyer, A quantum probability explanation for violation of rational decision theory, *Proc. Royal. Soc.* B **276**, 2171-2178 (2009).
11. T. Cheon and I. Tsutsui, Classical and quantum contents of solvable game theory on Hilbert space, *Phys. Lett.* A **348**, 147-152 (2006).
12. M. Asano, A. Khrennikov, M. Ohya, Quantum-like model for decision making process in two players game. A Non-Kolmogorovian model, *Found. Phys.* **41**, 538–548 (2010).

13. M. Asano, M. Ohya, Y. Togawa and A. Khrennikov, I. Basieva, Quantum-like model for Prisoner's Dilemma Games, Quantum Bioinforamtics-4; *QP-PQ: Quantum Probability and White Noise Analysis* **28** (2010).

14. M. Asano, M. Ohya, Y. Togawa and A. Khrennikov, I. Basieva, Quantum-like Representation of Bayesian Inference. ADVANCES IN QUANTUM THEORY: American Institute of Physics **1327**, 57-62 (2011).

15. R. S. Ingarden, A. Kossakowski, and M. Ohya, *Information dynamics and open systems: classical and quantum approach,* Kluwer, Dordrecht, 1997.

16. A. Khrennikov, *Information dynamics in cognitive, psychological, social, and anomalous phenomena,* Ser.: Fundamental Theories of Physics, Kluwer, Dordreht, 2004.

Quantum Bio-Informatics V

© 2013 World Scientific Publishing Co. Pte. Ltd.

pp. 57–67

# QUANTUM-LIKE REPRESENTATION OF NON-BAYESIAN INFERENCE

M. ASANO[†], I. BASIEVA[††], A. KHRENNIKOV[††], M. OHYA[†] AND Y. TANAKA[†]

[†] *Department of Information Sciences, Faculty of Science and Technology, Tokyo University of Science, Noda City, Chiba 278, Japan*

[††]*International Center for Mathematical Modelling in Physics and Cognitive Sciences, Linnaeus University, Växjö, S-35195 Sweden*

This research is related to the problem of "irrational decision making or inference" that have been discussed in cognitive psychology. There are some experimental studies, and these statistical data cannot be described by classical probability theory [1,2,3,4,5]. The process of decision making generating these data cannot be reduced to the classical Bayesian inference. For this problem, a number of quantum-like coginitive models of decision making was proposed [6,7,8,9,10,11,12,13,14,15,16,17,18,19,20]. Our previous work [21] represented in a natural way the classical Bayesian inference in the frame work of quantum mechanics. By using this representation, in this paper, we try to discuss the non-Bayesian (irrational) inference that is biased by effects like the quantum interference. Further, we describe "psychological factor" disturbing "rationality" as an "environment" correlating with the "main system" of usual Bayesian inference.

## 1. Bayesian inference

Bayesian inference is a method of statistical inference in which Bayes' theorem is used to calculate how the degree of belief in a proposition changes due to evidence. To explain its process, let us consider the event system denoted by $S_1 = \{A, B\}$ where the events $A$ and $B$ are mutually exclusive. A decision-making entity, which we call Alice, has a belief for the occurrence of the event $A$ (or $B$). In the philosophy of Bayesian probability, the degree of belief is represented by probability, $P(A)$ ( or $P(B)$) which is called *prior probabilities*. Here, another event system $S_2 = \{C, D\}$ is introduced, and Alice knows the conditional probabilities $P(C|A)$, $P(C|B)$, $P(D|A)$ and $P(D|B)$ for the events $C$ and $D$. In such situation, Alice can *update* her beliefs for the events $A$ and $B$ when she sees the occurrence of the event $C$ or $D$. The updated belief is calculated by Bayes' theorem: If Alice sees the occurrence of $C$, she can update $P(A)$ to

$$P(A|C) = \frac{P(C|A)P(A)}{P(C|A)P(A) + P(C|B)P(B)}.$$

If Alice sees the occurrence of $D$, she can update $P(A)$ to

$$P(A|D) = \frac{P(D|A)P(A)}{P(D|A)P(A) + P(D|B)P(B)}.$$

These conditional probabilities are called the *posterior probabilities.*


## 2. Prediction State Vector

In our previous work, we represented the process of Bayesian inference by using the following state vector which is defined on Hilbert space $\mathcal{H}_1 \otimes \mathcal{H}_2 = \mathbb{C}^2 \otimes \mathbb{C}^2$;

$$\begin{aligned}
|\Phi\rangle = \sqrt{P(A')}\,|A'\rangle \otimes (\sqrt{P(C'|A')}\,|C'\rangle + \sqrt{P(D'|A')}\,|D'\rangle) \\
+ \sqrt{P(B')}\,|B'\rangle \otimes (\sqrt{P(C'|B')}\,|C'\rangle + \sqrt{P(D'|B')}\,|D'\rangle). \quad (1)
\end{aligned}$$

This vector is called the *prediction state vector.* The set of vectors $\{|A'\rangle, |B'\rangle\}$ is a set of orthogonal basis on $\mathcal{H}_1$, and $\{|C'\rangle, |D'\rangle\}$ is a set of orthogonal basis on $\mathcal{H}_2$. The vector $|A'\rangle$ ($|B'\rangle$) represents Alice's intention to judge *"the event $A(B)$ occurs in the system $S_1$."* , and $|C'\rangle$ ($|D'\rangle$) is Alice's intention to judge *"the event $C(D)$ occurs in the system $S_2$".* The form of $|\Phi\rangle$ indicates that these intentions coexist in psychological meaning. The value of $\sqrt{P(A')}$ and $\sqrt{P(B')}$ specifies the degrees of intentions $A'$ and $B'$. In the form of $|\Phi\rangle$, Alice feels *causality* between $S_1$ and $S_2$: The events in $S_1$ are *causes* and the events in $S_2$ are *results.* Then, the degree of intention $C'$ ($D'$) is given by $\sqrt{P(C'|A')}$ or $\sqrt{P(C'|B')}$ ($\sqrt{P(D'|A')}$ or $\sqrt{P(D'|B')}$), which is determined depending on the intention $A'$ or $B'$. The square of $\sqrt{P(A')}$ is equivalent to a prior probability $P(A)$ in the Bayesian inference. If Alice knows the conditional probabilities $P(C|A)$ and $P(C|B)$, she will give the degree of intention $C'$ as $\sqrt{P(C'|A')} = \sqrt{P(C|A)}$ or $\sqrt{P(C'|B')} = \sqrt{P(C|B)}$.

When Alice sees the occurrence of the event $C$ in $S_2$, for example, she can judge "the event $C$ occurs" and then the intention $D'$ is *vanished* instantaneously. In our formalism, such vanish is represented as the *reduction of the prediction state* $|\Phi\rangle\langle\Phi|$ by the projection operator $M_{C'} = I \otimes |C'\rangle\langle C'|$:

$$M_{C'}|\Phi\rangle = \sqrt{P(C'|A')P(A')}\,|A'\rangle \otimes |C'\rangle + \sqrt{P(C'|B')P(B')}\,|B'\rangle \otimes |C'\rangle. \quad (2)$$

The projected state vector $M_{C'}|\Phi\rangle$ is normalized as

$$\frac{\sqrt{P(C'|A')P(A')}}{\sqrt{P(C'|A')P(A') + P(C'|B')P(B')}}\,|A'\rangle \otimes |C'\rangle$$
$$+\frac{\sqrt{P(C'|B')P(B')}}{\sqrt{P(C'|A')P(A') + P(C'|B')P(B')}}\,|B'\rangle \otimes |C'\rangle. \qquad (3)$$

In this form, one can interpret that the degrees $\sqrt{P(A')}$ and $\sqrt{P(B')}$ are updated to the degrees which squares are

$$\frac{P(C'|A')P(A')}{P(C'|A')P(A') + P(C'|B')P(B')}$$

and

$$\frac{P(C'|B')P(B')}{P(C'|A')P(A') + P(C'|B')P(B')}.$$

These values correspond to the posterior probabilities $P(A'|C')$ and $P(B'|C')$ in the Bayesian inference, if $P(C'|A') = P(C|A)$ and $P(C'|B') = P(C|B)$.

## 3. Quantum-like Bayesian Updating

In the previous section, we represented the Bayesian updating in the term of reduction of the prediction state. As seen in Eq. (2), the projection $|C'\rangle\langle C'|$ or $|D'\rangle\langle D'|$ provides such reduction. In this section, we consider different reduction by projection $|\phi\rangle\langle\phi|$ or $|\phi_\perp\rangle\langle\phi_\perp|$ which is given as

$$|\phi\rangle = \sqrt{\delta}\,|C'\rangle + \sqrt{1-\delta}\mathrm{e}^{\mathrm{i}\theta}\,|D'\rangle, \quad |\phi_\perp\rangle = \sqrt{1-\delta}\,|C'\rangle - \sqrt{\delta}\mathrm{e}^{\mathrm{i}\theta}\,|D'\rangle. \qquad (4)$$

Here, $\delta$ and $\theta$ are real numbers with $0 \le \delta \le 1$ and $0 \le \theta \le 2\pi$. The vectors $|\phi\rangle$ and $|\phi_\perp\rangle$ are orthogonal; $\langle\phi|\phi_\perp\rangle = \langle\phi_\perp|\phi\rangle = 0$. For example, by $|\phi\rangle\langle\phi|$, the prediction state vector $|\Phi\rangle$ of Eq. (1) is projected to

$$(I \otimes |\phi\rangle\langle\phi|)|\Phi\rangle = (\sqrt{P(A',C')}\sqrt{\delta} + \sqrt{P(A,'D')}\sqrt{1-\delta}\mathrm{e}^{-\mathrm{i}\theta})\,|A',\phi\rangle$$
$$+(\sqrt{P(B',C')}\sqrt{\delta} + \sqrt{P(B',D')}\sqrt{1-\delta}\mathrm{e}^{-\mathrm{i}\theta})\,|B',\phi\rangle, \qquad (5)$$

From this result, we define the following posterior probabilities for event $A'$ and $B'$.

$$P(A'|\phi) = \frac{1}{T}(P(A',C')\delta + P(A',D')(1-\delta)$$
$$+ 2\sqrt{P(A',C')P(A',D')}\sqrt{\delta(1-\delta)}\cos\theta),$$
$$P(B'|\phi) = \frac{1}{T}(P(B',C')\delta + P(B',D')(1-\delta)$$
$$+ 2\sqrt{P(B',C')P(B',D')}\sqrt{\delta(1-\delta)}\cos\theta), \qquad (6)$$

where $T$ is the normalization factor; $T = \mathrm{tr}((I \otimes |\phi\rangle \langle\phi|) |\Phi\rangle \langle\Phi|)$. Similarly, from the projection $|\phi_\perp\rangle \langle\phi_\perp|$, we define

$$
\begin{aligned}
P(A'|\phi_\perp) &= \frac{1}{K}(P(A',C')(1-\delta) + P(A',D')\delta \\
&\quad - 2\sqrt{P(A',C')P(A',D')}\sqrt{\delta(1-\delta)}\cos\theta), \\
P(B'|\phi_\perp) &= \frac{1}{K}(P(B',C')(1-\delta) + P(B',D')\delta \\
&\quad - 2\sqrt{P(B',C')P(B',D')}\sqrt{\delta(1-\delta)}\cos\theta),
\end{aligned}
\tag{7}
$$

where $K = \mathrm{tr}((I \otimes |\phi_\perp\rangle \langle\phi_\perp|) |\Phi\rangle \langle\Phi|)$. In general, the above posterior probabilities cannot be discussed in the usual Bayesian inference based on classical probability theory. One can see it in the term of quantum interference, for example $2\sqrt{P(A',C')P(A',D')}\sqrt{\delta(1-\delta)}\cos\theta$ in Eq. (6), which is a specific effect in quantum mechanics. Hereafter, we call the updating by $|\phi\rangle \langle\phi|$ ( or $|\phi_\perp\rangle \langle\phi_\perp|$), *the quantum-like Bayesian updating*.

## 3.1. *Biased Posterior Probability*

It is important to discuss how Alice's inference is explained by the framework of the quantum-like updating. In order to do this, firstly, we rewrite the posterior probabilities of Eq. (6) and Eq. (7) to the following forms;

$$
\begin{aligned}
P(A'|\phi) &= \frac{P(A')}{T}(\cos^2\frac{\theta}{2}\cos^2(\lambda_A - \omega) + \sin^2\frac{\theta}{2}\cos^2(\lambda_A + \omega)), \\
P(B'|\phi) &= \frac{P(B')}{T}(\cos^2\frac{\theta}{2}\cos^2(\lambda_B - \omega) + \sin^2\frac{\theta}{2}\cos^2(\lambda_B + \omega)), \\
P(A'|\phi_\perp) &= \frac{P(A')}{K}(\cos^2\frac{\theta}{2}\sin^2(\lambda_A - \omega) + \sin^2\frac{\theta}{2}\sin^2(\lambda_A + \omega)), \\
P(B'|\phi_\perp) &= \frac{P(B')}{K}(\cos^2\frac{\theta}{2}\sin^2(\lambda_B - \omega) + \sin^2\frac{\theta}{2}\sin^2(\lambda_B + \omega)).
\end{aligned}
\tag{8}
$$

Here, we define

$$
\begin{aligned}
\sqrt{P(C'|A')} &= \cos\lambda_A, \quad \sqrt{P(D'|A')} = \sin\lambda_A, \\
\sqrt{P(C'|B')} &= \cos\lambda_B, \quad \sqrt{P(D'|B')} = \sin\lambda_B, \\
\sqrt{\delta} &= \cos\omega, \quad \sqrt{1-\delta} = \sin\omega,
\end{aligned}
\tag{9}
$$

with $0 \leq \lambda_A, \lambda_B, \omega \leq \frac{\pi}{2}$. In Eq. (8), $\cos^2(\lambda_A \mp \omega)$, $\sin^2(\lambda_A \mp \omega)$, $\cos^2(\lambda_B \mp \omega)$ and $\sin^2(\lambda_B \mp \omega)$ have the phases shifted from $\lambda_{A,B}$ by $\omega$. We call them *biased conditional probabilities* and denote them by $P_{\mp\omega}(C'|A')$, $P_{\mp\omega}(D'|A')$, $P_{\mp\omega}(C'|B')$ and $P_{\mp\omega}(D'|B')$, respectively. The posterior probabilities of

Eq. (8) are mixtures of these biased conditional probabilities:

$$P(\cdot|\phi) = \frac{P(\cdot)}{T}\left[\cos^2\frac{\theta}{2}P_{-\omega}(C'^2\frac{\theta}{2}P_{+\omega}(C'|\cdot)\right],$$

$$P(\cdot|\phi_\perp) = \frac{P(\cdot)}{K}\left[\cos^2\frac{\theta}{2}P_{-\omega}(D'^2\frac{\theta}{2}P_{+\omega}(D'|\cdot)\right]. \tag{10}$$

We call these probabilities *biased posterior probabilities.*


### 3.2. *Reliability of Knowledge*

The quantum-like updating is useful to discuss the following situation:

Alice is informed of the occurrence of the event $C$ or $D$ from an informational source, but she cares *reliability of its knowledge* and does *not* believe the knowledge perfectly.

In this situation, even if Alice gets the knowledge of the occurrence of $C$, she cannot completely dispel the intention $D'$ that judges the occurrence of $D$. We represents the fluctuation between intentions $C'$ and $D'$, which is remained after getting the knowledge, by the vector $|\phi\rangle = \sqrt{\delta}\,|C'\rangle + e^{i\theta}\sqrt{1-\delta}\,|D'\rangle$. Here, we call $\sqrt{\delta}$ *reliability of knowledge* and denotes the degree of remained intention $D'$ by $\sqrt{1-\delta}$. Similarly, the vector $|\phi_\perp\rangle = \sqrt{1-\delta}\,|C'\rangle - e^{i\theta}\sqrt{\delta}\,|D'\rangle$ implies the fluctuation in the case that Alice gets the knowledge of the occurrence of $D$. The biased posterior probabilities $P(\cdot|\phi)$ and $P(\cdot|\phi_\perp)$ in Eq. (10) can be interpreted as the updated beliefs for the events $A$ and $B$ by the knowledges which Alice does not believe perfectly.


### 3.3. *Irrationality*

It should be noted that the biased posterior probabilities in Eq. (10) are specified by another parameter $\theta$, which cannot be discussed in the classical probability theory. According to the sense of Bayesian probability, $\delta$ is the probability of Alice's belief for the event $C$ $(D)$ after getting the knowledge of $C$ $(D)$, and then the posterior probabilities $P(A\,or\,B|\text{knowledge-}C)$ will be simply estimated as

$$P(\cdot|\text{knowledge-}C) = \frac{P(\cdot)}{T}\left[\delta P(C|\cdot) + (1-\delta)P(D|\cdot)\right]. \tag{11}$$

On the other hands, our biased posterior probabilities $P(\cdot|\phi)$ are described as a mixture of the biased conditional probabilities $P_{\mp\omega}(C'|\cdot)$ whose distribution is given by $\cos^2\frac{\theta}{2}$ and $\sin^2\frac{\theta}{2}$. Here, we consider the case of $\theta = \frac{\pi}{2}$, that is, the case of the *fair* mixing with $\cos^2 \frac{\theta}{2} = \sin^2 \frac{\theta}{2} = \frac{1}{2}$. Then, $P(\cdot|\phi)$ coincide with the classical solution $P(\cdot|\text{knowledge-}C)$, which is *rational* from the view point of classical probability theory. In this meaning, Alice at $\theta \neq \frac{\pi}{2}$ estimates *irrational* posterior probabilities. The difference between $\cos^2 \theta/2$ and $\sin^2 \theta/2$ will specify this type of irrationality.

## 4. Bayesian Updating Biased by Psychological Factor

As discussed in the previous section, the biased posterior probabilities are specified by the parameters $\sqrt{\delta} = \cos \omega$ and $\theta$. The parameter $\omega$ (or $\delta$) is related with the reliability of knowledge, and $\theta$ is related with Alice's irrationality. Alice will hold some sort of *psychological factor* specified by the reliability and the irrationality, and it disturbs the usual Bayesian updating process.

In this section, we represent the psychological factor as a state vector on an *environment system* that correlates with a main system for the usual Bayesian updating. By this representation, we re-formalize the process of biased (quantum-like) Bayesian updating.

### 4.1. *System of Psychological Factor*

To describe the usual Bayesian updating, we introduced the prediction state vector as Eq. (4):

$$|\Phi\rangle = \sqrt{P(A')}\,|A'\rangle \otimes (\sqrt{P(C'|A')}\,|C'\rangle + \sqrt{P(D'|A')}\,|D'\rangle)$$
$$+ \sqrt{P(B')}\,|B'\rangle \otimes (\sqrt{P(C'|B')}\,|C'\rangle + \sqrt{P(D'|B')}\,|D'\rangle).$$

For such main system, we define an environment system such as

$$|\Phi\rangle \otimes |E\rangle \equiv |\Phi, E\rangle \in \mathcal{H}_1 \otimes \mathcal{H}_2 \otimes \mathcal{H}_3 \tag{12}$$

The state vector $|E\rangle$ in the environment system implies the psychological factor which Alice holds. In the compound system $|\Phi, E\rangle$, Alice makes a correlated state by a unitary operation $V$ on $\mathcal{H}_1 \otimes \mathcal{H}_2 \otimes \mathcal{H}_3$;

$$V\,|\Phi, E\rangle \equiv |\Theta\rangle \tag{13}$$

In such the correlated state, entangled state in terms of quantum mechanics, we can discuss the affections to the usual Bayesian process by the psychological factor.

### 4.2. *Bias Operator*

Here, we introduce the operator $V$ defined by

$$V \equiv I \otimes U_{-\omega} \otimes |E_{-\omega}\rangle \langle E_{-\omega}| + I \otimes U_{+\omega} \otimes |E_{+\omega}\rangle \langle E_{+\omega}| . \tag{14}$$

We call this *bias operator*. The vectors $|E_{\mp\omega}\rangle$ are orthogonal basis on $\mathcal{H}_3$, and the operators $U_{\mp\omega}$ on $\mathcal{H}_2$ are the unitary operators defined by the matrices

$$\begin{pmatrix} \cos\omega & \pm\sin\omega \\ \mp\sin\omega & \cos\omega \end{pmatrix} \tag{15}$$

with the basis $|C'\rangle$ and $|D'\rangle$. One can check that these operators generate the biased conditional probabilities;

$$\begin{aligned}
(I \otimes U_{\mp\omega}) |\Phi\rangle &= \sqrt{P(A')} |A'\rangle \otimes (\sqrt{P_{\mp\omega}(C'|A')} |C'\rangle + \sqrt{P_{\mp\omega}(D'|A')} |D'\rangle) \\
&\quad + \sqrt{P(B')} |B'\rangle \otimes (\sqrt{P_{\mp\omega}(C'|B')} |C'\rangle + \sqrt{P_{\mp\omega}(D'|B')} |D'\rangle) \\
&\equiv |\Phi_{\mp\omega}\rangle
\end{aligned} \tag{16}$$

As seen in the structure of $V$ in Eq. (14), the basis $|E_{\mp\omega}\rangle$ are psychological factors making Alice to bias $P(C'|\cdot)$ toward $P_{\mp\omega}(C'|\cdot)$ and $P(D'|\cdot)$ toward $P_{\mp\omega}(D'|\cdot)$. Let us consider that the vector $|E\rangle$ is expanded as

$$|E\rangle = \cos\frac{\theta}{2} |E_{-\omega}\rangle + \sin\frac{\theta}{2} |E_{+\omega}\rangle . \tag{17}$$

The coefficients $\cos\frac{\theta}{2}$ and $\sin\frac{\theta}{2}$ mean the degrees of the psychological factors $E_{\mp\omega}$. Then, the correlated state $|\Theta\rangle$ is described by

$$|\Theta\rangle = \cos\frac{\theta}{2} |\Phi_{-\omega}\rangle \otimes |E_{-\omega}\rangle + \sin\frac{\theta}{2} |\Phi_{+\omega}\rangle \otimes |E_{+\omega}\rangle . \tag{18}$$

### 4.3. *Redefinition of Biased Posterior Probability*

In this section, we derive the biased posterior probabilities of Eq. (10) from the correlated state $|\Theta\rangle \langle \Theta|$. Firstly, we define the state

$$\mathrm{tr}_{\mathcal{H}_3}(|\Theta\rangle \langle \Theta|) = \cos^2\frac{\theta}{2} |\Phi_{-\omega}\rangle \langle \Phi_{-\omega}| + \sin^2\frac{\theta}{2} |\Phi_{+\omega}\rangle \langle \Phi_{+\omega}| \equiv \tilde{\rho}, \tag{19}$$

and call this *biased prediction state*. The affections to the usual Bayesian inference by the reliability of the knowledge and the irrationality are all reflected on this $\tilde{\rho}$. When Alice obtains the knowledge-$C$, the biased prediction state $\tilde{\rho}$ is updated by the projection $M_{C'} = I \otimes |C'\rangle \langle C'|$ such as

$$\frac{M_{C'} \tilde{\rho} M_{C'}}{\mathrm{tr}(M_{C'} \tilde{\rho})} \equiv \tilde{\rho}_{C'} . \tag{20}$$

Similarly, by the knowledge-$D$, $\tilde{\rho}$ is updated as

$$\frac{M_{D'}\tilde{\rho}M_{D'}}{\mathrm{tr}(M_{D'}\tilde{\rho})} \equiv \tilde{\rho}_{D'}. \tag{21}$$

By using these updated states, we define the following probabilities;

$$\mathrm{tr}(M_{A'}\tilde{\rho}_{C'}) \equiv \tilde{P}(A'|C'), \ \ \mathrm{tr}(M_{B'}\tilde{\rho}_{C'}) \equiv \tilde{P}(B'|C'),$$
$$\mathrm{tr}(M_{A'}\tilde{\rho}_{D'}) \equiv \tilde{P}(A'|D'), \ \ \mathrm{tr}(M_{B'}\tilde{\rho}_{D'}) \equiv \tilde{P}(B'|D'), \tag{22}$$

where $M_A = |A'\rangle\langle A'| \otimes I$ and $M_B = |B'\rangle\langle B'| \otimes I$. One can easily check that these probabilities are equivalent to the biased posterior probabilities defined in Eq. (10);

$$\tilde{P}(\cdot|C') = P(\cdot|\phi), \ \ \tilde{P}(\cdot|D') = P(\cdot|\phi_\perp). \tag{23}$$

### 4.4. *Biased Posterior Probability by PVM Operators*

In the subsection 4.1, we introduced the environment system of psychological factor $|E\rangle$ and the operator $V$ which makes a correlation between the main system of Bayesian updating and the environment. Generally, the form of $V$ specifies psychological affections to the usual updating, and the form of Eq. (14) is just an example. In this subsection, instead of this $V$, we introduce another $V$ as

$$V \equiv |A'\rangle\langle A'| \otimes \left(U_{-\omega_{A'}} \otimes |E_{-\omega_{A'}}\rangle\langle E_{-\omega_{A'}}| + U_{+\omega_{A'}} \otimes |E_{+\omega_{A'}}\rangle\langle E_{+\omega_{A'}}|\right)$$
$$+ |B'\rangle\langle B'| \otimes \left(U_{-\omega_{B'}} \otimes |E_{-\omega_{B'}}\rangle\langle E_{-\omega_{B'}}| + U_{+\omega_{B'}} \otimes |E_{+\omega_{B'}}\rangle\langle E_{+\omega_{B'}}|\right), \tag{24}$$

where the form of $U_{\mp\omega_{A',B'}}$ is same with that of Eq. (15), but $\omega_A \neq \omega_B$ in general. The set of $\{|E_{\mp\omega_A}\rangle\}$ and the set of $\{|E_{\mp\omega_B}\rangle\}$ are sets of orthogonal basis on $\mathcal{H}_3$. We assume that the psychological factor $|E\rangle$ is expanded as

$$\begin{aligned}
|E\rangle &= \cos\frac{\theta_{A'}}{2}|E_{-\omega_{A'}}\rangle + \sin\frac{\theta_{A'}}{2}|E_{+\omega_{A'}}\rangle \\
&= \cos\frac{\theta_{B'}}{2}|E_{-\omega_{B'}}\rangle + \sin\frac{\theta_{B'}}{2}|E_{+\omega_{B'}}\rangle.
\end{aligned} \tag{25}$$

Then, from Eqs. (20),(21) and (22), one can derive the posterior probabilities as

$$
\begin{aligned}
\tilde{P}(A'|C') &= \frac{P(A')}{T}\left[\cos^2\frac{\theta_{A'}}{2}P_{-\omega_{A'}}(C'|A'^2\frac{\theta_{A'}}{2}P_{+\omega_{A'}}(C'|A')\right], \\
\tilde{P}(B'|C') &= \frac{P(B')}{T}\left[\cos^2\frac{\theta_{B'}}{2}P_{-\omega_{B'}}(C'|B'^2\frac{\theta_{B'}}{2}P_{+\omega_{B'}}(C'|B')\right], \\
\tilde{P}(A'|D') &= \frac{P(A')}{K}\left[\cos^2\frac{\theta_{A'}}{2}P_{-\omega_{A'}}(D'|A'^2\frac{\theta_{A'}}{2}P_{+\omega_{A'}}(D'|A')\right], \\
\tilde{P}(B'|D') &= \frac{P(B')}{K}\left[\cos^2\frac{\theta_{B'}}{2}P_{-\omega_{B'}}(D'|B'^2\frac{\theta_{B'}}{2}P_{+\omega_{B'}}(D'|B')\right].
\end{aligned}
\tag{26}
$$

Here, we show that the posterior probabilities can be derived in the sense of biased posterior probability discussed in Sec. 3. Let us consider the following PVM (Projective-Valued Measure) operators on $\mathcal{H}_1 \otimes \mathcal{H}_2$;

$$
\begin{aligned}
M_1 &= |A'\rangle\langle A'| \otimes |\phi_A\rangle\langle\phi_A| + |B'\rangle\langle B'| \otimes |\phi_B\rangle\langle\phi_B|, \\
M_2 &= |A'\rangle\langle A'| \otimes |\phi_{A\perp}\rangle\langle\phi_{A\perp}| + |B'\rangle\langle B'| \otimes |\phi_{B\perp}\rangle\langle\phi_{B\perp}|,
\end{aligned}
\tag{27}
$$

where

$$
\begin{aligned}
|\phi_x\rangle &= \cos\omega_{x'}|C'\rangle + \sin\omega_{x'}\mathrm{e}^{\mathrm{i}\theta_{x'}}|D'\rangle, \\
|\phi_{x'\perp}\rangle &= \sin\omega_{x'}|C'\rangle - \cos\omega_{x'}\mathrm{e}^{\mathrm{i}\theta_{x'}}|D'\rangle.
\end{aligned}
\tag{28}
$$

These forms imply that Alice has two kinds of reliability of knowledge and the irrationality, which are determined with depending on the intention $|A'\rangle$ and $|B'\rangle$. One can easily check that the reduction of prediction state by the PVM $M_1$ or $M_2$ provides the bias posterior probability which coincides with the result in Eq. (26);

$$
\begin{aligned}
\frac{M_1|\Phi\rangle}{\sqrt{\mathrm{tr}(M_1|\Phi\rangle\langle\Phi|)}} &= \sqrt{\tilde{P}(A'|C')}|A',\phi_A\rangle + \sqrt{\tilde{P}(B'|C')}|B',\phi_B\rangle, \\
\frac{M_2|\Phi\rangle}{\sqrt{\mathrm{tr}(M_2|\Phi\rangle\langle\Phi|)}} &= \sqrt{\tilde{P}(A'|D')}|A',\phi_{A\perp}\rangle + \sqrt{\tilde{P}(B'|D')}|B',\phi_{B\perp}\rangle.
\end{aligned}
\tag{29}
$$

## References


1. Conte, E., Khrennikov, A., Todarello, O., Federici, A., Zbilut, J. P., Mental states follow quantum mechanics during perception and cognition of ambiguous figures, *Open Systems and Information Dynamics*, **16** (2009) 1-17.

2. Conte, E., Todarello, O., Federici, A. , Vitiello, F., Lopane, M., Khrennikov, A. and Zbilut, J. P., Some remarks on an experiment suggesting quantum-like behavior of cognitive entities and formulation of an abstract quantum mechanical formalism to describe cognitive entity and its dynamics, *Chaos, Solitons and Fractals,* **31** (5) (2007) 1076-1088.

3. Croson, R. The Disjunction Effect and Reason-Based Choice in Games, *Organizational Behavior and Human Decision Processes* **80** (1999) 118-133.

4. Shafir, E. and Tversky, A., Thinking through uncertainty: nonconsequential reasoning and choice. *Cognitive Psychology* **24** (1992) 449-474.

5. Tversky, A., Shafir, E., The disjunction effect in choice under uncertainty, Psychological Science **3** (1992) 305-309.

6. Accardi, L., Khrennikov, A. and Ohya, M., Quantum Markov model for data from Shafir-Tversky experiments in cognitive psychology. *Open Systems and Information Dynamics,* **16** (2009) 371-385.

7. Asano, M., Khrennikov, A. and Ohya, M., Quantum-like model for decision making process in two players game, A Non-Kolmogorovian model, Found. Phys. **41** (2010) 538-548.

8. Busemeyer, J. B., Wang, Z. and Townsend, J. T., Quantum dynamics of human decision making, *J. Math. Psychology* **50** (2006) 220-241.

9. Busemeyer, J R., Wang, Z. and Lambert-Mogiliansky, A., Empirical Comparison of Markov and quantum models of decision making, *J. Math. Psychology* **53** (2009) 423-433.

10. Busemeyer, J. R., Trueblood, J., Comparison of quantum and bayesian models of inference. In P. Bruza, D. Sofge, W. Lawless, K. van Rijsbergen, M. Klusch (Eds.), Quantum interaction-Springer (2009) 29-43.

11. Busemeyer, J. R., Pothos, E. M., Franco, R. and Trueblood, J. S., A quantum theoretical explanation for probability judgment errors, Psychological Review **118** (2) (2011) 193-218.

12. Cheon T. and Takahashi, T., Classical and quantum contents of solvable game theory on Hilbert space, Phys. Lett. A **348** (2006) 147-152.

13. Cheon T. and Takahashi, T., Interference and inequality in quantum decision theory, *Phys. Lett.* A **375** (2010) 100-104.

14. Danilov, V.I. and Lambert-Mogiliansky, A., Measurable systems and behavioral sciences, Mathematical Social Sciences **55**(3)(2008) 315-340,

15. Danilov, V.I. and Lambert-Mogiliansky, A., Expected utility theory under non-classical uncertainty, Theory and Decision **68**(1)(2010) 25-47.

16. Franco, R. The conjunction fallacy and interference effects, J. Math. Psychol. **53** (2009) 415-422.

17. Khrennikov, A., Quantum-like brain: Interference of minds, *BioSystems* **84** (2006) 225-241.

18. Khrennikov A., Haven E., Quantum mechanics and violations of the sure-thing principle: the use of probability interference and other concepts, *Journal of Mathematical Psychology* **53** (2009) 378-388.

19. Lambert-Mogiliansky, A., Zamir, S., and Zwirn, H., Type Indeterminancy: A model of the KT (Kahneman-Tversky)-man. J. Math. Psychol. **53** (5) (2009) 349-361.

20. Pothos, E. M., and Busemeyer, J. R., A quantum probability explanation for violation of rational decision theory, *Proc. Royal. Soc.* B **276** (2009) 2171-2178.

21. M.Asano, M. Ohya, Y.Tanaka, A. Khrennikov and I. Basieva, Quantum-like Representation of Bayesian Updating, American Institute of physics: Proceedings of the International Conference on Advances in Quantum Theory **1327** (2011) 57-62

Quantum Bio-Informatics V
© 2013 World Scientific Publishing Co. Pte. Ltd.
pp. 69–84

# A MATHEMATICAL TREATMENT OF JOINT AND CONDITIONAL PROBABILITY

MASANORI ASANO, MASANORI OHYA, YOSHIHARU TANAKA*

*Department of Information Sciences, Tokyo University of Science
2641 Yamazaki, Noda city, Chiba, Japan
*E-mail: tanaka@is.noda.tus.ac.jp*

ICHIRO YAMATO

*Department of Biological Science and Technology, Tokyo University of Science
2641 Yamazaki, Noda city, Chiba, Japan
E-mail: iyamato@rs.noda.tus.ac.jp*

IRINA BASIEVA, ANDREI KHRENNIKOV+

*International Center for Mathematical Modeling in Physics and Cognitive
Sciences Linnaeus University, S-35195, Växjö, Sweden
+ E-mail: andrei.khrennikov@lnu.se*

There exist several phenomena breaking the classical probability laws. Such systems are adaptive (context dependent) systems. In this paper, we present a new mathematical formula to compute the probability in those systems by using the concepts of the adaptive dynamics and lifting theory.

Keywords: quantum probability, lifting, adaptive dynamics, cognitive science

## 1. Introduction

There exist several phenomena breaking the classical probability laws such as quantum interference, quantum-like interference in cognitive science[3], the game of prisoner's dilemma (PD game)[6,7], the lactose-glucose interference in *Escherichia coli* (*E. coli*) growth[10]. It is important to notice that these phenomena are context dependent, so that they are adaptive to the context of the surroundings. In such systems, the conditional probability can not be defined in usual mathematical framework. These phenomena will require us a change of classical probability law, e.g. total probability law[9].

It is well known that in quantum systems the conditional probability does not exist (see the section 2) in the sense of classical systems, so that the naive total probability law should be reconsidered[16]. Same situation is occurred even in non-quantum systems. One of our trials is to create a new mathematical model which will describe in the unified framework both "traditional quantum phenomena" and recently found quantum-like phenomena outside of physics. In this paper we review the concepts of the adaptive dynamics[19] and lifting theory[5] to make a mathematical framework for the study of these contextual dependent systems.

## 2. Conditional probability and joint probability in quantum systems

The conditional probability and the joint probability do not generally exist in quantum system, which is an essential difference from classical system. We will review these facts for the sequel uses[24].

Let $\mathcal{H}$, $\mathcal{K}$ be the Hilbert spaces describing the system of interest, $\mathcal{S}(\mathcal{H})$ be the set of all states or probability measures on $\mathcal{H}$, $\mathcal{O}(\mathcal{H})$ be the set of all observables or events on $\mathcal{H}$ and $\mathcal{P}(\mathcal{H}) \subset \mathcal{O}(\mathcal{H})$ be the set of projections in $\mathcal{O}(\mathcal{H})$.

In classical probability, the joint probability for two events $A$ and $B$ is

$$\mu(A \cap B)$$

and the conditional probability is defined by

$$\frac{\mu(A \cap B)}{\mu(B)}.$$

In quantum probability, if the von Neumann-Lüder projection rule is correct, after a measurement of $F \in \mathcal{P}(\mathcal{H})$, a state $\rho$ is considered to be

$$\rho_F = \frac{F \rho F}{\mathrm{tr}\rho F}.$$

When we observe an event $E \in \mathcal{P}(\mathcal{H})$, the expectation value becomes

$$\mathrm{tr}\rho_F E = \frac{\mathrm{tr}F\rho FE}{\mathrm{tr}\rho F} = \frac{\mathrm{tr}\rho FEF}{\mathrm{tr}\rho F}. \tag{1}$$

This expectation value can be a candidate of the *conditional probability in QP (quantum probability)*.

There is another candidate for the conditional probability in QP, which is a direct generalization of CP (classical probability). This alternative expression of joint probability and the conditional probability in QP are expressed as

$$\varphi(E \wedge F)$$

and

$$\frac{\varphi(E \wedge F)}{\varphi(F)}, \tag{2}$$

where $\varphi$ is a state (a measure) and $\wedge$ is the meet of two events (projections) corresponding to $\cap$ in CP, and for the state describing by a density operator, we have

$$\varphi(\cdot) = \mathrm{tr}\rho(\cdot).$$

We ask when the above two expressions (1) and (2) in QP are equivalent. From the next proposition, $\varphi(\cdot \wedge F)/\varphi(F)$ is not a probability measure (state) on $\mathcal{P}(\mathcal{H})$.

**Proposition 2.1.** *(1) When $E$ commutes with $F$, the above two expressions are equivalent, namely,*

$$\frac{\varphi(FEF)}{\varphi(F)} = \frac{\varphi(E \wedge F)}{\varphi(F)}.$$

*(2) When $EF \neq FE$, $\frac{\varphi(\cdot \wedge F)}{\varphi(F)}$ is not a probability on $\mathcal{P}_{\mathcal{H}}$, so that the above two expressions are not equivalent.*

**Proof.** (1) $EF = FE$ implies $E \wedge F = EF$ and $FEF = EFF = EF^2 = EF$, so that

$$\frac{\varphi(E \wedge F)}{\varphi(F)} = \frac{\varphi(FEF)}{\varphi(F)} = \frac{\varphi(EF)}{\varphi(F)}.$$

(2) Put $K_\varphi(E \mid F) \equiv \frac{\varphi(E \wedge F)}{\varphi(F)}$ and put $z \in \mathrm{linsp}\{x, y\}$, $z \neq x, y$ for any $x, y \in \mathcal{H}$. Take the projections $P_x = |x\rangle\langle x|$, $P_y = |y\rangle\langle y|$, $P_z = |z\rangle\langle z|$ such that $(P_x \vee P_y) \wedge P_z = P_z$ and $P_x \wedge P_z = 0 = P_y \wedge P_z$. Then

$$K_\varphi((P_x \vee P_y) \wedge P_z \mid F) = K_\varphi(P_z \mid F) \neq 0,$$

$$K_\varphi(P_x \wedge P_z \mid F) + K_\varphi(P_y \wedge P_z \mid F) = 0.$$

Therefore

$$K_\varphi((P_x \vee P_y) \wedge P_z \mid F) \neq K_\varphi(P_x \wedge P_z \mid F) + K_\varphi(P_y \wedge P_z \mid F)$$

so that $K_\varphi\left(\cdot \mid F\right)$ is not a probability measure on $\mathcal{P}_\mathcal{H}$. $\qquad\square$

In CP, the joint distribution for two random variables $f$ and $g$ is expressed as

$$\mu_{f,g}\left(\Delta_1, \Delta_2\right) = \mu\left(f^{-1}\left(\Delta_1\right) \cap g^{-1}\left(\Delta_2\right)\right)$$

for any Borel sets $\Delta_1, \Delta_2 \in B\left(\mathbb{R}\right)$. The corresponding quantum expression is either

$$\varphi_{A,B}\left(\Delta_1, \Delta_2\right) = \varphi\left(E_A\left(\Delta_1\right) \wedge E_B\left(\Delta_2\right)\right)$$

or

$$\varphi(E_A\left(\Delta_1\right) \cdot E_B\left(\Delta_2\right))$$

for two observables $A$, $B$ and their spectral measures $E_A(\cdot)$, $E_B(\cdot)$ such that

$$A = \int aE_A\left(da\right), \quad B = \int bE_B\left(da\right).$$

It is easily checked that neither one of the above expressions satisfies neither the condition of probability measure nor the marginal condition unless $AB = BA$, so that they can not be the joint quantum probability in the classical sense.

Let us explain the above situation, as an example, in a physical measurement process. When an observable $A$ has a discrete decomposition like

$$A = \sum_k a_k F_k, \ F_i \perp F_j \ (i \neq j),$$

the probability obtaining $a_k$ by measurement in a state $\rho$ is

$$p_k = \mathrm{tr}\rho F_k$$

and the state $\rho$ is changed to a (conditional) state $\rho_k$ such that

$$\rho_k = \frac{F_k \rho F_k}{\mathrm{tr}\rho F_k} \equiv P_\rho\left(\cdot | F_k\right).$$

After the measurement of $A$, we will measure a similar type observable $B$ (i.e., $B = \sum_j b_j E_j$, $(E_i \perp E_j \ (i \neq j))$) and the probability obtaining $b_j$ after we have obtained the above $a_k$ for the measurement of $A$ is given by

$$\begin{aligned}
p_{jk} &= \left(\mathrm{tr}\rho F_k\right)\left(\mathrm{tr}\rho_k E_j\right) \\
&= \mathrm{tr}\rho F_k E_j F_k \\
&= P_\rho\left(E_j | F_k\right) \mathrm{tr}\rho F_k.
\end{aligned} \tag{3}$$

This $p_{jk}$ satisfies

$$\sum_{j,k} p_{jk} = 1,$$

$$\sum_j p_{jk} = \mathrm{tr}\rho F_k = p_k, \tag{4}$$

but not

$$\sum_k p_{jk} = \mathrm{tr}\rho E_j$$

unless $E_j F_k = F_k E_j$ $(\forall j, k)$ so that $p_{jk}$ is not considered as a joint quantum probability distribution. More intuitive expression breaking the usual classical probability law is the following:

$$p_{jk} = P(B = b_j \mid A = a_k)P(A = a_k) \ \cdots \text{and}$$

$$P(B = b_j) \neq \sum_k P(B = b_j \mid A = a_k)P(A = a_k)$$

*Therefore we conclude in quantum system that the above two candidates can not satisfy the properties of both conditional and joint probabilities in the sense of classical system.* The above discussion shows that the order of the measurement of two observables $A$ and $B$ is essential and it gives us a different expectation value, hence the state change.

## 3. Liftng and joint probability

In order to partially solve the difficulty of the nonexistence of joint quantum distribution, the notion of compound state [22] satisfying the marginal condition is useful. We discuss a bit general notion named "lifting"[4] and present several examples of the lifting in order to discuss a new scheme of probability containing both classical and quantum.

**Definition 3.1.** Let $\mathcal{A}_1, \mathcal{A}_2$ be C*-algebras and let $\mathcal{A}_1 \otimes \mathcal{A}_2$ be a fixed C*-tensor product of $\mathcal{A}_1$ and $\mathcal{A}_2$. A *lifting* from $\mathcal{A}_1$ to $\mathcal{A}_1 \otimes \mathcal{A}_2$ is a weak *-continuous map

$$\mathcal{E}^* : \mathcal{S}(\mathcal{A}_1) \to \mathcal{S}(\mathcal{A}_1 \otimes \mathcal{A}_2)$$

If $\mathcal{E}^*$ is affine and its dual is a completely positive map, we call it a linear lifting; if it maps pure states into pure states, we call it pure.

74

The algebras $\mathcal{A}_1$, $\mathcal{A}_2$ can be considered as two systems of interest, for instance, $\mathcal{A}_1$ is an objective system for a study and $\mathcal{A}_2$ is the subjective system or the surrounding of $\mathcal{A}_1$.

Note that to every lifting from $\mathcal{A}_1$ to $\mathcal{A}_1 \otimes \mathcal{A}_2$ we can associate two channels: one from $\mathcal{A}_1$ to $\mathcal{A}_1$, defined by

$$\Lambda^* \rho_1(A_1) \equiv (\mathcal{E}^* \rho_1)(A_1 \otimes 1) \quad ; \quad \forall A_1 \in \mathcal{A}_1$$

another from $\mathcal{A}_1$ to $\mathcal{A}_2$, defined by

$$\Lambda^* \rho_1(A_2) \equiv (\mathcal{E}^* \rho_1)(1 \otimes A_2) \quad ; \quad \forall A_2 \in \mathcal{A}_2$$

In general, a state $\varphi \in \mathcal{S}(\mathcal{A}_1 \otimes \mathcal{A}_2)$ such that

$$\varphi \mid_{\mathcal{A}_1 \otimes 1} = \rho_1 \quad ; \quad \varphi \mid_{1 \otimes \mathcal{A}_2} = \rho_2$$

is called a compound state of the states $\rho_1 \in \mathfrak{S}(\mathcal{A}_1)$ and $\rho_2 \in \mathfrak{S}(\mathcal{A}_2)$. Remark here that the above compound state is nothing but the joint probability in CP.

The following problem is important in several applications: Given a state $\rho_1 \in \mathcal{S}(\mathcal{A}_1)$ and a channel $\Lambda^* : \mathcal{S}(\mathcal{A}_1) \to \mathcal{S}(\mathcal{A}_2)$, find a standard lifting $\mathcal{E}^* : \mathcal{S}(\mathcal{A}_1) \to \mathcal{S}(\mathcal{A}_1 \otimes \mathcal{A}_2)$ such that $\mathcal{E}^* \rho_1$ is a compound state of $\rho_1$ and $\Lambda^* \rho_1$. Several particular solutions of this problem have been proposed by Ohya, Ceccini and Petz, however an explicit description of all the possible solutions to this problem is still missing, which might be related to find a new scheme of probability theory.

*However it is not sure that one can resolve the difficulty of quantum probability if one can solve this problem. The compound state corresponds to the joint probability in classical systems, but there is still ambiguity to define the conditional state in quantum systems. As pointed out in Introduction, the usual conditional probability meets an inadequacy to interpret a certain phenomenon, in which it is important not to manage to set the conditional state by mimicking the classical one but to make a mathematical rule to set new treatment of probabilistic aspects of such a phenomenon.*

**Definition 3.2.** A lifting from $\mathcal{A}_1$ to $\mathcal{A}_1 \otimes \mathcal{A}_2$ is called nondemolition for a state $\rho_1 \in \mathcal{S}(\mathcal{A}_1)$ if $\rho_1$ is invariant for $\Lambda^*$ i.e., if for all $a_1 \in \mathcal{A}_1$

$$(\mathcal{E}^* \rho_1)(a_1 \otimes 1) = \rho_1(a_1)$$

The idea of this definition being that the interaction with system 2 does not alter the state of system 1.

**Definition 3.3.** A transition expectation from $\mathcal{A}_1 \otimes \mathcal{A}_2$ to $\mathcal{A}_1$ is a completely positive linear map $\mathcal{E} : \mathcal{A}_1 \otimes \mathcal{A}_2 \rightarrow \mathcal{A}_1$ satisfying

$$\mathcal{E}(1_{\mathcal{A}_1} \otimes 1_{\mathcal{A}_2}) = 1_{\mathcal{A}_1}.$$

Let an initial state (resp. input signal) is changed (resp. transmitted) to the final state resp. output state) due to a dynamics $\Lambda^*$ (resp. channel). Here $\mathcal{A}_1$ (resp. $\mathcal{A}_2$) is interpreted as the algebra of observables of the input (resp. output) system and $\mathcal{E}^*$ describes the interaction between the input and the output. If $\rho_1 \in \mathcal{S}(\mathcal{A}_1)$ is the initial state, then the state $\rho_2 = \Lambda^* \rho_1 \in \mathcal{S}(\mathcal{A}_2)$ is the output state.

In several important applications, the state $\rho_1$ of the system before the interaction (preparation, input signal) is not known and one would like to know this state knowing only $\Lambda^* \rho_1 \in \mathcal{S}(\mathcal{A}_2)$, i.e., the state of the apparatus after the interaction (output signal). From a mathematical point of view this problem is not well posed, since the map $\Lambda^*$ is usually not invertible. The best one can do in such cases is to acquire a control on the description of those input states which have the same image under $\Lambda^*$ and then choose among them according to some statistical criterion.

Let us show some important examples of liftings and channels below

## Example 3.1. : Isometric lifting.

Let $V : \mathcal{H}_1 \rightarrow \mathcal{H}_1 \otimes \mathcal{H}_2$ be an isometry

$$V^* V = 1_{\mathcal{H}_1}.$$

Then the map

$$\mathcal{E} : x \in \mathbf{B}(\mathcal{H}_1) \otimes \mathbf{B}(\mathcal{H}_2) \rightarrow V^* x V \in \mathbf{B}(\mathcal{H}_1)$$

is a transition expectation in the sense of Accardi, and the associated lifting maps a density matrix $w_1$ in $\mathcal{H}_1$ into

$$\mathcal{E}^* w_1 = V w_1 V^*$$

in $\mathcal{H}_1 \otimes \mathcal{H}_2$. Liftings of this type are called isometric. Every isometric lifting is a pure lifting. In this case the channel $\Lambda^* : \mathcal{H}_1 \rightarrow \mathcal{H}_1$ is given by $\text{tr}_{H_2} \mathcal{E}^*$.

It is the particular isometric lifting characterized by the properties.

$$\mathcal{H}_1 = \mathcal{H}_2 =: \Gamma(\mathbb{C}) \text{ (Fock space over } \mathbb{C}) = L^2(\mathbb{R})$$

$$V : \Gamma(\mathbb{C}) \rightarrow \Gamma(\mathbb{C}) \otimes \Gamma(\mathbb{C})$$

is characterized by the expression

$$V \left| \theta \right\rangle = \left| \alpha\theta \right\rangle \otimes \left| \beta\theta \right\rangle$$

where $\left| \theta \right\rangle$ is the normalized coherent vector parametrized by $\theta \in \mathbb{C}$ and $\alpha, \beta \in \mathbb{C}$ are such that

$$|\alpha|^2 + |\beta|^2 = 1$$

Notice that this liftings maps coherent states into products of coherent states. So it maps the simplex of the so called classical states (i.e., the convex combinations of coherent vectors) into itself. Restricted to these states it is of convex product type explained below, but it is not of convex product type on the set of all states. Denoting, for $\theta \in \mathbb{C}$, $\omega_\theta$ the coherent state on $\mathbf{B}(\Gamma(\mathbb{C}))$, namely,

$$\omega_\theta(b) = \left\langle \theta, b\theta \right\rangle \;\; ; \;\; b \in \mathbf{B}(\Gamma(\mathbb{C}))$$

then for any $b \in \mathbf{B}(\Gamma(\mathbb{C}))$

$$(\mathcal{E}^*\omega_\theta)(b \otimes 1) = \omega_{\alpha\theta}(b),$$

so that this lifting is not nondemolition. These equations mean that, by the effect of the interaction, a coherent signal (beam) $\left| \theta \right\rangle$ splits into 2 signals (beams) still coherent, but of lower intensity, but the total intensity (energy) is preserved by the transformation.

Finally we mention two important beam splitting which are used to discuss quantum gates and quantum teleportation.

(1) Superposed beam splitting:

$$V_s \left| \theta \right\rangle \equiv \frac{1}{\sqrt{2}} (\left| \alpha\theta \right\rangle \otimes \left| \beta\theta \right\rangle - i \left| \beta\theta \right\rangle \otimes \left| \alpha\theta \right\rangle)$$

(2) Beam splitting with two inputs and two output: Let $\left| \theta \right\rangle$ and $\left| \gamma \right\rangle$ be two input coherent vectors. Then

$$V_d \left( \left| \theta \right\rangle \otimes \left| \gamma \right\rangle \right) \equiv \left| \alpha\theta + \beta\gamma \right\rangle \otimes \left| -\bar{\beta}\theta + \bar{\alpha}\gamma \right\rangle,$$

where $|\alpha|^2 + |\beta|^2 = 1$. These extend linearly to isometry, and their isometric liftings are neither of convex product type nor nondemolition type.

**Example 3.2.** *Quantum measurement:* If a measuring apparatus is prepared by an positive operator valued measure $\{Q_n\}$ then the state $\rho$ changes to a state $\Lambda^*\rho$ after this measurement, $\rho \rightarrow \Lambda^*\rho = \sum_n Q_n\rho Q_n.$

**Example 3.3.** *Reduction (Open system dynamics):* If a system $\Sigma_1$ interacts with an external system $\Sigma_2$ described by another Hilbert space $\mathcal{K}$ and the initial states of $\Sigma_1$ and $\Sigma_2$ are $\rho_1$ and $\rho_2$, respectively, then the combined state $\theta_t$ of $\Sigma_1$ and $\Sigma_2$ at time $t$ after the interaction between two systems is given by

$$\theta_t \equiv U_t(\rho_1 \otimes \rho_2)U_t^*,$$

where $U_t = \exp(-itH)$ with the total Hamiltonian $H$ of $\Sigma_1$ and $\Sigma_2$. A channel is obtained by taking the partial trace w.r.t. $\mathcal{K}$ such as

$$\rho_1 \to \Lambda^*\rho_1 \equiv \mathrm{tr}_\mathcal{K}\theta_t.$$

**Example 3.4. : The compound lifting.**

Let $\Lambda^* : \mathcal{S}(\mathcal{A}_1) \to \mathcal{S}(\mathcal{A}_2)$ be a channel. For any $\rho_1 \in \mathcal{S}(\mathcal{A}_1)$ in the closed convex hull of the extremal states, fix a decomposition of $\rho_1$ as a convex combination of extremal states in $\mathcal{S}(\mathcal{A}_1)$

$$\rho_1 = \int_{\mathcal{S}(\mathcal{A}_1)} \omega_1 d\mu$$

where $\mu$ is a Borel measure on $\mathcal{S}(\mathcal{A}_1)$ with support in the extremal states, and define

$$\mathcal{E}^*\rho_1 \equiv \int_{\mathcal{S}(\mathcal{A}_1)} \omega_1 \otimes \Lambda^*\omega_1 d\mu$$

Then $\mathcal{E}^* : \mathcal{S}(\mathcal{A}_1) \to \mathcal{S}(\mathcal{A}_1 \otimes \mathcal{A}_2)$ is a lifting, nonlinear even if $\Lambda^*$ is linear, and it is a nondemolition type.

The most general lifting, mapping $\mathcal{S}(\mathcal{A}_1)$ into the closed convex hull of the extremal product states on $\mathcal{A}_1 \otimes \mathcal{A}_2$ is essentially of this type. This nonlinear nondemolition lifting was first discussed by Ohya to define the compound state and the mutual entropy for quantum information communication [22,23]. The above is a bit more general because we shall weaken the condition that $\mu$ is concentrated on the extremal states used in [22].

Therefore once a channel is given, by which a lifting of convex product type can be constructed. For example, the von Neumann quantum measurement process is written, in the terminology of lifting, as follows: Having measured a compact observable $A = \sum_n a_n P_n$ (spectral decomposition with $\sum_n P_n = I$) in a state $\rho$, the state after this measurement will be

$$\Lambda^*\rho = \sum_n P_n\rho P_n$$

and a lifting $\mathcal{E}^*$, of convex product type, associated to this channel $\Lambda^*$ and to a fixed decomposition of $\rho$ as $\rho = \sum_n \mu_n \rho_n$ $(\rho_n \in \mathcal{S}(\mathcal{A}_1))$ is given by :

$$\mathcal{E}^* \rho = \sum_n \mu_n \rho_n \otimes \Lambda^* \rho_n.$$

**Example 3.5.** Amplifier channel: To recover the loss in the course of a quantum communication, we need to amplify the signal (photon). In quantum optics, a linear amplifier is usually expressed by means of annihilation operators $a$ and $b$ on $\mathcal{H}$ and $\mathcal{K}$, respectively :

$$c = \sqrt{G} a \otimes I + \sqrt{G-1} I \otimes b^*$$

where $G(\geq 1)$ is a constant and $c$ satisfies CCR (i.e., $[c,\, c^*] = I$) on $\mathcal{H} \otimes \mathcal{K}$. This expression however is not convenient to compute several information quantities, like entropy. The lifting expression of the amplifier is good for such uses and it is given as follows: Let $c = \mu a \otimes I + \nu I \otimes b^*$ with $|\mu|^2 - |\nu|^2 = 1$ and $|\gamma\rangle$ be the eigenvector of $c$ : $c|\gamma\rangle = \gamma |\gamma\rangle$. For two coherent vectors $|\theta\rangle$ on $\mathcal{H}$ and $|\theta'\rangle$ on $\mathcal{K}$ , $|\gamma\rangle$ can be written by the squeezing expression : $|\gamma\rangle = |\theta \otimes \theta' \,;\, \mu, \nu\rangle$ and the lifting is defined by an isometry

$$V_{\theta'} |\theta\rangle = |\theta \otimes \theta' \,;\, \mu, \nu\rangle$$

such that

$$\mathcal{E}^* \rho = V_{\theta'} \rho V_{\theta'}^* \quad \rho \in \mathcal{S}(\mathcal{H}).$$

The channel of the amplifier is

$$\Lambda^* \rho = \mathrm{tr}_\mathcal{K} \mathcal{E}^* \rho.$$

Finally we note that a channel is determined by a lifting and conversely a lifting is constructed by a channel.


## 4. Adaptive Dynamics

The idea of the adaptive dynamics has implicitly appeared in the study of compound dynamics, chaos and the SAT algorithm. The name of the adaptive dynamics was deliberately used in [19]. The AD has two aspects, one of which is the "observable-adaptive" and another is the "state-adaptive".

*The observable-adaptive dynamics is a dynamics characterized as follows: (1) Measurement depends on how to see an observable to be measured. (2) The interaction between two systems depends on how a fixed observable exists, that is, the interaction is related to some aspects of observables to be measured or prepared.*

*The state-adaptive dynamics is a dynamics characterized as follows: (1)Measurement depends on how the state to be used exists, as same as the observable. (2)The correlation between two systems interaction depends on how the state of at least one of the systems at one instant exists e.g., the interaction Hamiltonian depends on the state at that.*

The idea of observable-adaptivity comes from studying chaos. We claimed that any observation will be unrelated or even contradicted to mathematical universalities such as taking limits, sup, inf, etc. Observation of chaos is a result due to taking suitable scales of, for example, time, distance or domain, and it will not be possible in the limiting cases. Examples of the observable-adaptivity are used to understand chaos [20,18] and examine the violation of Bell's inequality, namely the chameleon dynamics of Accardi [1]. The idea of the state-adaptivity is implicitly started in constructing a compound state for quantum communication [21,22,23,5] Examples of the state-adaptivity are seen in an algorithm solving NP complete problem, i.e., a pending problem for more than 30 years asking whether there exists an algorithm solving a NP complete problem in polynomial time, as discussed [25,24,4].

We will discuss in the section 5 how we can apply the adaptive dynamics to a bio-system or a psycho-system. The concept of the adaptivity is naturally existed in such systems. Our formulation here contains some treatments shown in the book [24] to understand the evolution of HIV-1, the brain function and the irrational behavior of prisoners.

## 5. New views of probability both in classical and quantum systems

In this section we discuss how to use the concept of lifting to explain phenomena breaking the usual probability law.

Let $\mathcal{A}, \mathcal{B}$ be C*-algebras describing the systems for a study, more specifically, let $\mathcal{A}, \mathcal{B}$ be the sets of all observales in Hilbert spaces $\mathcal{H}, \mathcal{K}$; $\mathcal{A} = \mathcal{O}(\mathcal{H})$, $\mathcal{B} = \mathcal{O}(\mathcal{K})$. Let $\mathcal{E}^*$ be a lifting from $\mathcal{S}(\mathcal{H})$ to $\mathcal{S}(\mathcal{H} \otimes \mathcal{K})$, so that its dual map $\mathcal{E}$ is a mapping from $\mathcal{A} \otimes \mathcal{B}$ to $\mathcal{A}$. There are several liftings for various different cases to be considered: (1) If $\mathcal{K}$ is $\mathbb{C}$, then the lifting $\mathcal{E}^*$ is nothing but a channel from $\mathcal{S}(\mathcal{H})$ to $\mathcal{S}(\mathcal{H})$. (2) If $\mathcal{H}$ is $\mathbb{C}$, then the lifting $\mathcal{E}^*$ is a channel from $\mathcal{S}(\mathcal{H})$ to $\mathcal{S}(\mathcal{K})$. Further $\mathcal{K}$ or $\mathcal{H}$ can be decomposed as $\mathcal{K} = \otimes_i \mathcal{K}_i$ (resp. $\oplus_i \mathcal{K}_i$), and so for $\mathcal{H}$, so that $\mathcal{B}$ can be $\otimes_i \mathcal{B}_i$ (resp. $\oplus_i \mathcal{B}_i$) and so for $\mathcal{A}$.

*The adaptive dynamics is considered that the dynamics of a state or* *an observable after an instant (say the time $t_0$) attached to a system of interest is affected by the existence of some other observable and state at that instant.* Let $\rho \in \mathcal{S}(\mathcal{H})$ and $A \in \mathcal{A}$ be a state and an observable before $t_0$, and let $\sigma \in \mathcal{S}(\mathcal{H} \otimes \mathcal{K})$ and $Q \in \mathcal{A} \otimes \mathcal{B}$ be a state and an observable to give an effect to the state $\rho$ and the observable $A$.In many cases, the effect to the state is dual to that to the observable, so that we will discuss the effect to the state only. This effect is described by a lifting $\mathcal{E}^*_{\sigma Q}$,so that the state $\rho$ becomes $\mathcal{E}^*_{\sigma Q}\rho$ first, then it will be $\mathrm{tr}_{\mathcal{K}}\mathcal{E}^*_{\sigma Q}\rho \equiv \rho_{\sigma Q}$. The adaptive dynamics is the whole process such as

$$Adaptive\ Dynamics:\ \ \rho \Rightarrow \mathcal{E}^*_{\sigma Q}\rho \Rightarrow \rho_{\sigma Q} = \mathrm{tr}_{\mathcal{K}}\mathcal{E}^*_{\sigma Q}\rho$$

That is, what we need is how to construct the lifting for each problem to be studied, that is, we properly construct the lifting $\mathcal{E}^*_{\sigma Q}$ by choosing $\sigma$ and $Q$ properly.

The state change discussed in Section 2 is a naive example of the adaptive dynamics, in which the lifting is given as follows: $Q = A = \sum_k a_k F_k \in \mathcal{A}$, $F_i \perp F_j$ $(i \neq j)$ and $\mathcal{E}^*_{\sigma Q} \equiv \{\mathcal{E}^*_{F_k A}\}$ (this case the state $\sigma$ is not needed) such that for any $B \in \mathcal{A}$

$$P(B \mid A = a_k) = \mathrm{tr}\, B\mathcal{E}^*_{F_k A}\rho = \mathrm{tr}\, BF_k\rho F_k / \mathrm{tr}\, F_k\rho.$$

In this case the lifting is a channel from $\mathcal{S}(\mathcal{H})$ to $\mathcal{S}(\mathcal{H})$, the case (1) above. It is true that we do not know whether this "projection rule" can describe almost all probabilistic phenomena in nature or not.

Let us go back to the discussion using lifting $\mathcal{E}^*_{\sigma Q}$ above. The expectation value of another observable $B \in \mathcal{A}$ or $\mathcal{A} \otimes \mathcal{B}$ in the adaptive state $\rho_{\sigma Q}$ is

$$\mathrm{tr}\rho_{\sigma Q}B = \mathrm{tr}_{\mathcal{H}}\mathrm{tr}_{\mathcal{K}}B\mathcal{E}^*_{\sigma Q}\rho.$$

Now suppose that there are two quantum event systems $A = \{a_k \in \mathbb{R}, F_k \in \mathcal{A}\}$ and $B = \{b_j \in \mathbb{R}, E_j \in \mathcal{A}\}$, where we do not assume $F_k$, $E_j$ are projections, but they satisfy the conditions $\sum_k F_k = I$, $\sum_j E_j = I$ as POVM (positive operator valued measure) corresponding to the partition of a probability space in classical system. Then the "joint-like" probability obtaining $a_k$ and $b_j$ might be given by the *formula*

$$P(a_k, b_j) = \mathrm{tr}\, E_j \boxdot F_k \mathcal{E}^*_{\sigma Q}\rho, \tag{5}$$

where $\boxdot$ is a certain operation (relation) between $A$ and $B$, more generally one can take a certain operator function $f(E_j, F_k)$ instead of $E_j \boxdot F_k$. If $\sigma, Q$

are independent from any $F_k$, $E_j$ and the operation $\boxdot$ is the usual tensor product $\otimes$ so that $A$ and $B$ can be considered in two independent systems or to be commutative, then the above "joint-like" probability becomes the joint probability. However if not such a case, e.g., $Q$ is related to $A$ and $B$, the situation will be more subtle. Therefore the problem is how to set the operation $\boxdot$ and how to construct the lifting $\mathcal{E}^*_{\sigma Q}$ in order to describe the particular problems associated to systems of interest. We in the sequel discuss this problem in the contextual dependent systems like bio-systems and psyco-systems mentioned in Introduction. That is, we discuss how to apply the formula 5 to the following three problems breaking the usual probability law.

Let us consider a simple and intuitive example: When one takes sugar S and chocolate C and he is asked whether it is sweet (1) or not so (2). Then the simple classical probability law may not be satisfied, that is,

$$P(C = 1) \neq P(C = 1|S = 1)P(S = 1) + P(C = 1|S = 2)P(S = 2)$$

because the LHS $P(C = 1)$ will be very close to 1 but the RHS will be less than $\frac{1}{2}$. After taking very sweet sugar, he will taste the chocolate is not so sweet. Taking sugar changes his taste, i.e., the situation of the tongue changes. The conditional probability should be defined on the basis of such a change, so that it is observable-adaptive quantity. The $P(C = *|S = *)$ should be written as $Padap$ $(C = *|S = *)$ and its proper mathematical description (definition) should be given, that is, we will give a mathematical formula to compute the LHS and the RHS above.

Here, we give *the initial state of tongue* by

$$\rho \equiv |x_0\rangle \langle x_0|, \quad |x_0\rangle = \frac{1}{\sqrt{2}}(|e_1\rangle + |e_2\rangle)$$

with $e_1 = (1,0)^T$ and $e_2 = (0,1)^T$. When one takes "sugar", the operator corresponding to taking "sugar" will be given as

$$S = \begin{pmatrix} \lambda_1 & 0 \\ 0 & \lambda_2 \end{pmatrix},$$

where $|\lambda_1|^2 + |\lambda_2|^2 = 1$. When he takes sugar, he will taste that it is sweet with the probability $|\lambda_1|^2$ and non-sweet with the probability $|\lambda_2|^2$. Therefore $|\lambda_1|^2$ should be much higher than $|\lambda_2|^2$. We express the situation of the tongue changes as

$$\rho \rightarrow \rho_S = \Lambda_S^*(\rho) \equiv \frac{S^* \rho S}{tr\,|S|^2 \rho}.$$

We can use the above two expressions $\rho_S$,which give us the same result for the computation of the probability. For some time duration, the tongue becomes dull to sweetness, so the tongue state can be written by means of a certain "exchanging" operator $X = \begin{pmatrix} 0 & 1 \\ 1 & 0 \end{pmatrix}$ such that

$$\rho_S^a = X \rho_S X,$$

where $"a"$ means the adaptive change. Then similarly as sugar, when one takes a chocolate, the state will be $\rho_{S \to C}^a$ given by

$$\rho_{S \to C}^a = \Lambda_C^*(\rho_S^a) \equiv \frac{C^* \rho_S^a C}{tr \, |C|^2 \, \rho_S^a},$$

where $C$ will be given as

$$C = \begin{pmatrix} \mu_1 & 0 \\ 0 & \mu_2 \end{pmatrix}$$

with $|\mu_1|^2 + |\mu_2|^2 = 1.$ Common experience tells us that $|\lambda_1|^2 \geq |\mu_1|^2 \geq |\mu_2|^2 \geq |\lambda_2|^2$ and the first two are much larger than the last two.

As shown above, the adaptive set $\{\sigma, Q\}$ is the set $\{S \, (= \sigma_S), X, C\}$, we introduce the following *nonlinear demolition* lifting:

$$\mathcal{E}_{\sigma Q}^*(\rho)(= \mathcal{E}_{S \ (=\sigma_S) X C}^*(\rho)) \equiv \rho_S \otimes \rho_{S \to C}^a = \Lambda_S^*(\rho) \otimes \Lambda_C^*(X \Lambda_S^*(\rho) X),$$

which implies the joint probabilities $P(S = j, C = k) \ (j, k = 1, 2)$ as

$$P(S = j, C = k) = tr E_j \otimes E_k \mathcal{E}_{\sigma Q}^*(\rho).$$

The probability that one tastes sweetness of the chocolate after tasting sugar is

$$P(C = 1, S = 1) + P(C = 1, S = 2) = \frac{|\lambda_2|^2 \, |\mu_1|^2}{|\lambda_2|^2 \, |\mu_1|^2 + |\lambda_1|^2 \, |\mu_2|^2}.$$

Note that this probability is much less than

$$P(C = 1) = tr E_1 \Lambda_C^*(\rho) = |\mu_1|^2,$$

which is the probability of sweetness tasted by the neutral tongue $\rho$. In this sense, the usual probability law

$$P(C = 1) = P(S = 1, C = 1) + P(S = 2, C = 1)$$

is not satisfied.

We have other applications[8,9,6,7] of our new formula (5) from several different fields such as biology, psychology, game theory.

# References


1. Accardi, L., *Urne e camaleonti: Dialogo sulla realta, le leggi del caso e la teoria quantistica*, Il Saggiatore, 1997 (English edition, World Scientific, 2002; japanese edition, Makino, 2002, russian edition, Regular and Chaotic dynamics, 2002)
2. Accardi, L., Khrennikov, A., Ohya, M., The problem of quantum-like representation in economy, cognitive science, and genetics, *Quantum Bio-Informatics II (QP-PQ:Quantum Prob. White Noise Analysis Vol.21)*, World Scientific, 1-8, 2008
3. Accardi, L., Khrennikov, A., Ohya, M., Quantum Markov Model for Data from Shafir-Tversky Experiments in Cognitive Psychology, *Open Systems and Information Dynamics,* Vol.16, 371-85, 2009
4. Accardi, L., Ohya, M., A Stochastic Limit Approach to the SAT Problem, Open Systems and Information dynamics, Vol.11, 1-16, 2004
5. Accardi, L., Ohya, M., Compound Channels, Transition Expectations, and Liftings, Appl. Math. Optim., Vol.39, 33-59, 1999
6. Asano, M., Ohya, M., Khrennikov, A., Quantum-Like Model for Decision Making Process in Two Players Game, Found. of Phys., Vol.41(3), 538-548, 2010
7. Asano, M., Ohya, M., Tanaka, Y., Khrennikov, A., Basieva, I., On application of Gorini-Kossakowski-Sudarshan-Lindblad equation in cognitive psychology, Open Systems and Information Dynamics, Vol.18(1), 55-69, 2011
8. Asano, A., Basieva. I., Khrennikov, A., Tanaka, Y., Yamato, I., Quantum-like model for the adaptive dynamics of the genetic regulation of E. coli's metabolism of glucose/lactose, Systems and Synthetic Biology, to be published.
9. Asano, M., Ohya, M., Tanaka, Y., Khrennikov, A., Basieva, I., Quantum-like Representation of Bayesian Updating, Proc. of the Int'l Conference on Advances in Quantum TheoryAmerican Institute of Phys., Vol.1327, 57-62, 2011
10. Basieva, I., Khrennikov, A., Ohya, M., Yamato, I., Quantum-like interference effect in gene expression: glucose-lactose destructive interference, Systems and Synthetic Biology, Vol.5, 59-68, 2011
11. Cheon, T., Takahashi, T., Interference and inequality in quantum decision theory, Phys. Lett. A, Vol.375, 100-104, 2010
12. Fichtner, K.-H., Fichtner, L., Freudenberg, W., Ohya, M., On a quantum model of the recognition process, *Quantum Bio-Informatics (QP-PQ:Quantum Prob. White Noise Analysis Vol.21)*, World Scientific, 64-84, 2008
13. Fichtner, K.-H., Fichtner, L., Freudenberg, W., Ohya, M., "Quantum models of the recognition process - on a convergence theorem", Open Systems and Information Dynamics, Vol.17, 161-187, 2010
14. Inoue, K., Ohya, M., Sato, K., Application of chaos degree to some dynamical systems, Chaos, Soliton and Fractals, Vol.11, 1377-1385, 2000
15. Inoue, K., Ohya, M., Volovich, I.V., Semiclassical properties and chaos degree

for the quantum baker's map, J. of Math. Phys., Vol.43, 734, 2002

16. Khrennikov, A., *Ubiquitous quantum structure: from psychology to finance*, Springer, Heidelberg-Berlin-New York, 2010

17. Khrennikov, A., *Contextual approach to quantum formalism (Fundamental Theories of Physics)*, Springer, Heidelberg-Berlin-New York, 2009

18. Kossakowski, A., Ohya, M., Togawa, Y., How can we observe and describe chaos?, Open System and Information Dynamics, Vol.10(3), 221-233, 2003

19. Ohya, M., Adaptive Dynamics and its Applications to Chaos and NPC Problem, *Quantum Bio-Informatics II (QP-PQ:Quantum Prob. White Noise Analysis Vol.21)*, World Scientific, 181-216 (2008)

20. Ohya, M., Complexities and their applications to characterization of chaos, Int'l J. of Theoretical Phys., Vol.37(1), 495-505, 1998

21. Ohya, M., Note on quantum proability, L.Nuovo Cimento, Vol.38(11), 203-206, 1983

22. Ohya, M., On compound state and mutual information in quantum information theory, IEEE Trans.Information Theory, Vol.29, 770–777, 1983

23. Ohya, M., Some aspects of quantum information theory and their applications to irreversible processes, Rep. on Math. Phys., Vol.27, 19-47, 1989

24. Ohya, M., Volovich, I.V., *Mathematical Foundations of Quantum Information and Computation and Its Applications to Nano- and Bio-systems*, Springer, 2011

25. Ohya, M., Volovich, I.V., New quantum algorithm for studying NP-complete problems, Rep. Math. Phys., Vol.52(1), 25-33, 2003

Quantum Bio-Informatics V
© 2013 World Scientific Publishing Co. Pte. Ltd.
pp. 85–94

# ENTANGLED STATES PREPARATION IN CLUSTERS OF THREE RESONANTLY INTERACTING FLUORESCENT PARTICLES

IRINA BASIEVA

*Linnaeus University, 351 95 Växjö*
*Prokhorov General Physics Institute, Vavilova str. 38, Moscow, 119991, Russia*

Quantum dynamics of the states of three resonantly interacting two-level fluorescent particles is considered. It is shown that resonant laser excitation may lead to state inversion and to entanglement of energy states of the particles. The exact analytical solution of the problem in the case of zero decoherence is given. It describes complex dynamics in the case of strong laser radiation as well as easier and more perspective for practical implementation case of weak laser field. For the latter situation, the approximate solution obtained by the perturbation theory application is also presented and tested against the exact theoretical solution and numerical simulation.

## 1. Introduction

Coherent quantum optical phenomena are of fundamental importance for quantum information processing. There is a vast literature devoted to search and study of the properties of paired, tripled, quadrupled, etc. quantum objects that have extremely good isolation from the environment, allowing for pertaining coherence during quantum manipulations. The particles in a cluster may have strong coherent interaction with each other and be driven to a particular quantum state by laser radiation, e.g. to entangled states [1, 2] which are basic for quantum information processing, communication, or quantum data protection [3, 4, 5, 6, 7, 8]. In solid state, coherence is taken away by the phonons (vibrations of the crystal lattice), so for the quantum experiments the condition of weak electron-phonon interaction is of the highest priority. Rare-earth ions in crystals have uniquely weak electron-phonon interaction because internal optical 4f shell electrons are safely isolated by the outer $5s^2$ and $5p^6$ electrons.

Pair and tetramer clusters of neodymium in fluoride crystals can be considered as an example of the suitable quantum particles. They have strong coherent quadrupole-quadrupole interaction at the $^4I_{9/2} - {}^4G_{5/2}$ transition [9]. Physical and chemical properties and thermodynamics of clusterization are studied in detail in [10, 11, 12]. The clusters may have high concentration, small spectral broadening [ 13 ] and small decoherence (as attested by accumulated photon echo experiments [ 14 , 15 , 16 ]). The aforementioned properties allow to plan and perform coherent experiments of entangling states in clusters of rare-earth ions in crystals at temperatures of about $2 - 30^{\circ}$K [A1].

The clusters of three resonantly interacting fluorescent particles can be engineered, e.g. as quantum dots. In the present paper the analytical solution for quantum states dynamics of such three-particle clusters under monochromatic laser radiation is given.

## 2.  Theoretical model

We model three equidistant fluorescent particles as two-level ions having energy gap $\hbar\omega_0$ each. Each ion resonantly interacts with the others which we characterize by the parameter $V$. We consider the evolution of the energy states of the system under the influence of a linearly polarized electromagnetic field of the zero phase (the phase is actually insignificant for our results) and of the central frequency $\hbar\omega_0$, the same as isolated ion would have. When referring to specific intensities, time periods, ion-ion interaction parameter $V$ values, we use the numbers valid for optical control of neodymium ions in fluorides.

We use one-half pseudo-spin formalism (spin projection $s_z = -1/2$ corresponds to the ground, and $s_z = +1/2$ to the excited state of an ion), expressing the Hamiltonian through the system's total spin operator and its projections.

We have studied the equation for density matrix of the four states of trimer in the rotating frame. The transitions between the states characterized by the different total momentum values, are prohibited in our framework. Here, we consider only the transitions between the so-called "optically bright" states having total momentum $J=3/2$, the same as in the ground level of the system, and use the value of $J_z$ – projection of momentum to axis z  to identify the certain energy state.

Without laser radiation, the states can be presented  analogously to the Dicke's work [17], in the basis of eigenstates of operators $J^2$ and  $J_z$. Ion-ion interaction $V$ leads to the splitting of the intermediate levels of the cluster of the three ions. Levels shift by itself can be considered as a proof for entanglement of a kind – when shifted intermediate levels (one-exciton level or biexciton level) are occupied, it means that there are one or two excitons in the system as a whole, the excitation cannot be assigned to a particular ion, so it is not localized.

The state $|-3/2\rangle$ ($|J=3/2, J_z=-3/2\rangle$) corresponds to the ground (or vacuum) level (we also speak about the level population, described by the diagonal element of density matrix $\rho_{-3/2-3/2}$). Zero energy value is assigned to this state.

The state $|-1/2\rangle$ ($|J=3/2, J_z=-1/2\rangle$) corresponds to one-exciton level ($\rho_{-1/2-1/2}$) and has the energy $\hbar\omega_0-2\hbar V$.

The state $|1/2\rangle$ ($|J=3/2, J_z=1/2\rangle$) corresponds to two-exciton level ($\rho_{1/2\,1/2}$) and, noteworthy, its energy $2\hbar\omega_0-2\hbar V$ is $\hbar\omega_0$ distant from the previous one-exciton level, meaning that one photon from laser radiation at the central frequency $\hbar\omega_0$ may lead to the transition between these two intermediate levels.

And, finally, the fourth state $|3/2\rangle$ ($|J=3/2, J_z=3/2\rangle$) corresponds to three-exciton level ($\rho_{3/2\,3/2}$) and has the energy of exactly $3\hbar\omega_0$.

Non-diagonal elements of density matrix describe the coherence between the levels, $e.g.$ $\rho_{-3/2;-1/2}$ is considered as a measure of coherence between the vacuum level $|-3/2\rangle$ and one-exciton level $|-1/2\rangle$.

We consider states dynamics under the laser radiation of the amplitude $E$ and of the form $E\cos(\hbar\omega_0 t)$. We introduce the parameter of ion-radiation interaction $A=dE$, which characterizes how strong ion's dipole momentum interacts with the laser field. We study the master equation for density matrix, not taking the decoherence into account:

$$i\hbar\dot{\rho} = [H,\rho].$$

(1)

It is so happens, that in our model the transitions are possible only between the neighboring energy states.

Now, we use the rotating wave frame approximation allowing us to obtain the Hamiltonian in the form [18, 19],:

$$H = \begin{pmatrix} 2.25V & A\sqrt{3}/2 & 0 & 0 \\ A\sqrt{3}/2 & 0.25V & A & 0 \\ 0 & A & 0.25V & A\sqrt{3}/2 \\ 0 & 0 & A\sqrt{3}/2 & 2.25V \end{pmatrix}.$$

(2)

From now on, we will work in the frame rotating with the central frequency $\hbar\omega_0$ and the solutions are also presented in this frame.

## 3. Results and discussion

### 3.1. *Exact analytical solution*

We have found the eigensystem of Hamiltonian (2):

$$E_1 = \hbar(5V/4 - A/2 + \Delta_1) \tag{3}$$

$$\psi_1 = \frac{1}{\sqrt{2}}\left\{\sin\alpha\left(|3/2\rangle - |-3/2\rangle\right) + \cos\alpha\left(|1/2\rangle - |-1/2\rangle\right)\right\} \tag{4}$$

$$E_2 = \hbar(5V/4 - A/2 - \Delta_1) \tag{5}$$

$$\psi_2 = \frac{1}{\sqrt{2}}\left\{\cos\alpha\left(|3/2\rangle - |-3/2\rangle\right) + \sin\alpha\left(|-1/2\rangle - |1/2\rangle\right)\right\} \tag{6}$$

$$E_3 = \hbar(5V/4 + A/2 + \Delta_2) \tag{7}$$

$$\psi_3 = \frac{1}{\sqrt{2}}\left\{\sin\beta\left(|3/2\rangle + |-3/2\rangle\right) + \cos\beta\left(|1/2\rangle + |-1/2\rangle\right)\right\} \tag{8}$$

$$E_4 = \hbar(5V/4 + A/2 - \Delta_2) \tag{9}$$

$$\psi_4 = \frac{1}{\sqrt{2}}\left\{\cos\beta\left(|3/2\rangle + |-3/2\rangle\right) - \sin\beta\left(|1/2\rangle + |-1/2\rangle\right)\right\} \tag{10}$$

where:

$$\Delta_1 \equiv \sqrt{A^2 + AV + V^2} \tag{11}$$

$$\sin\alpha = \frac{1}{\sqrt{2}}\left[1 + \frac{V + A/2}{\Delta_1}\right]^{1/2} \tag{12}$$

$$\cos\alpha = \frac{1}{\sqrt{2}}\left[1 - \frac{V + A/2}{\Delta_1}\right]^{1/2} \tag{13}$$

$$\Delta_2 \equiv \sqrt{A^2 - AV + V^2} \tag{14}$$

$$\sin\beta = \frac{1}{\sqrt{2}}\left[1 + \frac{V - A/2}{\Delta_2}\right]^{1/2} \tag{15}$$

$$\cos\beta = \frac{1}{\sqrt{2}}\left[1 - \frac{V - A/2}{\Delta_2}\right]^{1/2} \tag{16}$$

Let us also denote the differences between eigenvalues as: $\omega_{ij} = \left(E_i - E_j\right)/\hbar$.

We have obtained the exact analytical solution in the form:

$$\rho(t) = \frac{1}{4} R\rho(0) R^+ \tag{17}$$

$$R_{3/2;3/2} = R_{-3/2;-3/2} =$$
$$e^{-iE_1 t}\sin^2\alpha + e^{-iE_2 t}\cos^2\alpha + e^{-iE_3 t}\sin^2\beta + e^{-iE_4 t}\cos^2\beta \tag{18}$$

$$R_{3/2;1/2} = R_{1/2;3/2} = R_{-3/2;-1/2} = R_{-1/2;-3/2} =$$
$$\left[\left(e^{-iE_1t} - e^{-iE_2t}\right)\sin 2\alpha + \left(e^{-iE_3t} - e^{-iE_4t}\right)\sin 2\beta\right]/2 \tag{19}$$

$$R_{3/2;-1/2} = R_{3/2;-1/2} = R_{1/2;-3/2} = R_{-3/2;1/2} = \tag{20}$$
$$\left[\left(e^{-iE_2t} - e^{-iE_1t}\right)\sin 2\alpha + \left(e^{-iE_3t} - e^{-iE_4t}\right)\sin 2\beta\right]/2$$

$$R_{3/2;-3/2} = R_{-3/2;3/2} =$$
$$-e^{-iE_1t}\sin^2\alpha - e^{-iE_2t}\cos^2\alpha + e^{-iE_3t}\sin^2\beta + e^{-iE_4t}\cos^2\beta \tag{21}$$

$$R_{1/2;1/2} = R_{-1/2;-1/2} =$$
$$e^{-iE_1t}\cos^2\alpha + e^{-iE_2t}\sin^2\alpha + e^{-iE_3t}\cos^2\beta + e^{-iE_4t}\sin^2\beta \tag{22}$$

$$R_{1/2;-1/2} = R_{-1/2;1/2} =$$
$$-e^{-iE_1t}\cos^2\alpha - e^{-iE_2t}\sin^2\alpha + e^{-iE_3t}\cos^2\beta + e^{-iE_4t}\sin^2\beta \tag{23}$$

If we start with a pure state $\rho(0)=|\psi(0)><\psi(0)|$, then the density matrix evolution is simplified to $\rho(t)=|\psi(t)><\psi(t)|$, where $|\psi(0)>=R|\psi(0)>$. For example, starting with the ground state $\rho(0)=|-3/2><-3/2|$, we have $|\psi(t)>=R|-3/2>=R_{3/2;-3/2}|3/2>+ R_{1/2;-3/2}|1/2> +R_{-1/2;-3/2}|-1/2> +R_{-3/2;-3/2}|-3/2>$ and the density matrix evolution is:

$$\rho(t) = \frac{1}{4}\begin{pmatrix} \left|R_{3/2;-3/2}\right|^2 & R_{3/2;-3/2}\overline{R}_{3/2;-1/2} & R_{3/2;-3/2}\overline{R}_{3/2;1/2} & R_{3/2;-3/2}\overline{R}_{3/2;3/2} \\ R_{3/2;-1/2}\overline{R}_{3/2;-3/2} & \left|R_{3/2;-1/2}\right|^2 & R_{3/2;-1/2}\overline{R}_{3/2;1/2} & R_{3/2;-1/2}\overline{R}_{3/2;3/2} \\ R_{3/2;1/2}\overline{R}_{3/2;-3/2} & R_{3/2;1/2}\overline{R}_{3/2;-1/2} & \left|R_{3/2;1/2}\right|^2 & R_{3/2;1/2}\overline{R}_{3/2;3/2} \\ R_{3/2;3/2}\overline{R}_{3/2;-3/2} & R_{3/2;3/2}\overline{R}_{3/2;-1/2} & R_{3/2;3/2}\overline{R}_{3/2;1/2} & \left|R_{-3/2;-3/2}\right|^2 \end{pmatrix}. \tag{24}$$

Let us explicitly write down the terms responsible for the dynamics and entanglement of the ground and maximally excited three-exciton states:

$$\tilde{\rho}_{3/2;3/2} = \frac{1}{2}\begin{pmatrix} 1+\sin^2\alpha\left\{\cos^2\alpha\left[\cos(\omega_{12}t)-1\right]-\sin^2\beta\cos(\omega_{31}t)-\cos^2\beta\cos(\omega_{14}t)\right\} - \\ -\cos^2\alpha\left\{\sin^2\beta\cos(\omega_{32}t)+\cos^2\beta\cos(\omega_{42}t)\right\}+\sin^2\beta\cos^2\beta\left[\cos(\omega_{34}t)-1\right] \end{pmatrix}$$

$$\tilde{\rho}_{3/2;-3/2} = \tilde{\rho}^*_{-3/2;3/2} = \frac{1}{8}\begin{pmatrix} \sin^2 2\alpha\left[1-\cos(\omega_{12}t)\right]+\sin^2 2\beta\left[\cos(\omega_{34}t)-1\right]- \\ 4i\left\{\cos^2\alpha\cos^2\beta\sin(\omega_{42}t)-\sin^2\alpha\cos^2\beta\sin(\omega_{14}t)+\right. \\ \left.\cos^2\alpha\sin^2\beta\sin(\omega_{32}t)+\sin^2\alpha\sin^2\beta\sin(\omega_{31}t)\right\} \end{pmatrix} \tag{25}$$

$$\tilde{\rho}_{-3/2;-3/2} = \frac{1}{2}\begin{pmatrix} 1+\sin^2\alpha\left\{\cos^2\alpha\left[\cos(\omega_{12}t)-1\right]+\sin^2\beta\cos(\omega_{31}t)+\cos^2\beta\cos(\omega_{14}t)\right\} - \\ +\cos^2\alpha\left\{\sin^2\beta\cos(\omega_{32}t)+\cos^2\beta\cos(\omega_{42}t)\right\}+\sin^2\beta\cos^2\beta\left[\cos(\omega_{34}t)-1\right] \end{pmatrix}$$

### 3.2. *Approximate solution for the case of weak laser field*

When the interaction of the ion with the laser field is weaker than the ion-ion interaction ($A<<V$), we can use a perturbation theory approximation, obtaining a much simpler expression for the dynamics of the ground state:

$$\rho_{ground} = \left(1 + \cos \omega' t\right)/2 ,\qquad(26)$$

the excited state:

$$\rho_{excited} = \left(1 - \cos \omega' t\right)/2 ,\qquad(27)$$

and non-diagonal term, responsible for their entanglement:

$$\rho_{ent} = -\sin \omega' t / 2 \qquad(28)$$

If the initial state of the system is the vacuum state $|\text{-}3/2\rangle$, it will entangle with the three-exciton state $|3/2\rangle$, see Fig. 1. The oscillation frequency for Eqs. (26) – (28) is $\omega' = 3A^3/(8V^2)$. The transition from the ground state to three-exciton state is a three-photon process. The laser photon frequency failed twice to match the energy gaps between the ground state and one-exciton state, and between one-exciton state and biexciton state. The frequency of the oscillations is proportional to the square of the ratio $A/V$ (the small parameter used to obtained the approximate expressions) because the process is twice not resonant.

Approximate solution (26) – (28) is shown in Fig.1 by dash lines, proving that even the value $A/V=0.375$ is small enough for our approximations be valid for the first half of the oscillations period at least.

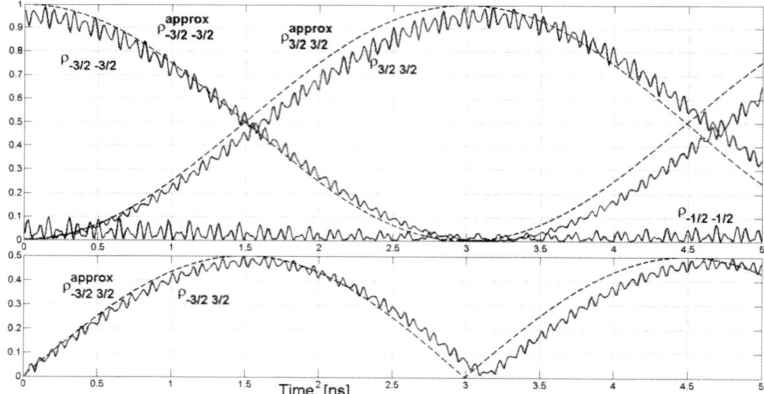



Figure 1. Entanglement of the levels $\rho_{-3/2;-3/2}$ and $\rho_{3/2;3/2}$ starting from the ground level $\rho_{-3/2;-3/2}$, laser intensity $I=1.6\cdot10^8 W/cm^2$ ($A=0.375V$).

If the initial state of the system is the one-exciton state $|\text{-}1/2\rangle$, then it can be driven by the central frequency laser radiation to the biexciton state $|1/2\rangle$. Fig. 2 demonstrates the levels population dynamics and entanglement. Exact numerical solution is shown by dots and fully coincides with the solid lines corresponding to approximate analytical solution (26) – (28) where the frequency $\omega'=2A$ (we have the period somewhat smaller than 6ns). Exact matching is due to the fact that the approximation condition of ion-ion interaction $A$ being smaller than ion-laser radiation interaction $V$ is fulfilled much better here *(A=V/100)* than in the case shown in Fig.1 *(A=0,375V)*.

This transition from one-exciton state to biexciton state is resonant, so the oscillations frequency does not contain the ratio $A/V$. Naturally, it is much easier to implement, and we can see from comparing Fig. 1 and Fig.2, that to reach maximal entanglement after the pulse of the same duration (1.5ns in Figs.), laser intensities almost three orders higher are needed in the case of three-photon transition comparing to the resonant, one-photon, transition.

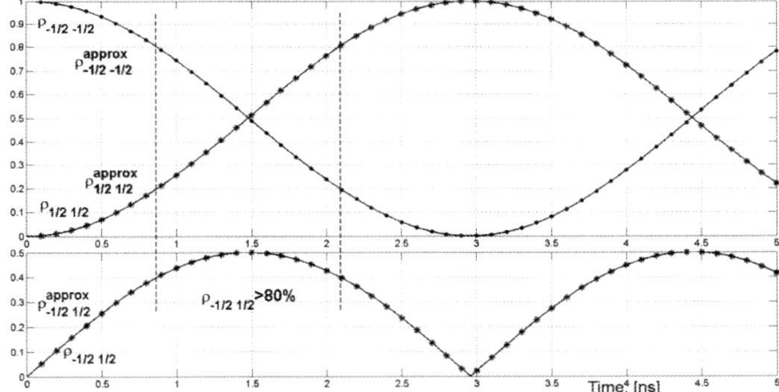


Figure 2. Entanglement of the one-exciton level $\rho_{-1/2;-1/2}$ and biexciton level $\rho_{1/2;1/2}$ starting from the one-exciton level $\rho_{-1/2;-1/2}$, $A=V/100$ (laser intensity I=1.1·10⁵W/cm²).

Maximal (100%) entanglement occurs at the one-fourth or three-fourths of the period (1.5 ns and 4.5 ns), when populations of the lower and the upper levels coincide and other levels populations disappear. Wide range of the entanglement (±30% around the moment of maximal entanglement) shows that we have high fidelity entanglement >80% even at the moments when levels populations are 20% and 80%.

### 3.3. *States dynamics under strong laser field*

The following Figs. 3 and 4 demonstrate populations evolution under the strong laser field $A>V$. Distinct symbols (dots, asterisks, triangles, diamonds) correspond to the numerical simulation, and solid lines correspond to exact analytical formulas (18) – (24).

In the case when coherent interaction of the ions with the laser radiation $A$ is of the same order or greater than ion-ion interaction $V$, the oscillations are fast

92

and relatively complicated. However, here moments of maximal entanglement can be also pointed out, see Fig. 3, t=0.245ns. Besides, Fig. 3 proves that analytical formulas fit the numerical simulation precisely, so that one can use expressions (18) – (24) for prediction of level populations behavior without special simulation.

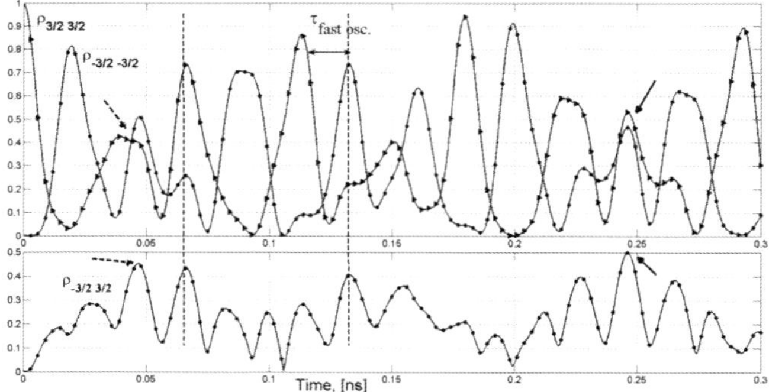


Figure 3. Entanglement of the ground level $\rho_{-3/2\ -3/2}$ and three-exciton level $\rho_{3/2\ 3/2}$ starting from the ground state when ion-laser radiation interaction is greater than ion-ion interaction $A=3V$.

In Fig. 3 we observe two kinds of oscillations – fast oscillations (with the period of about 0.02ns) can be associated with the strength of the ion-radiation interaction, and the slower ones (0.1ns), defined by the relation of the interactions. Increasing the laser radiation leads to sharpening the peaks of maximal entanglement and noticeable entanglement (up to 80%) can be reached extremely fast (at about 0.04ns, see the dashed arrows in Fig.3).

While the entanglement can be obtained after a short pulse, the difficulty here may be in the precise hitting the maximal entanglement point, because it oscillates sharply from the maximum to approximately 25%. In the case of weaker intensities, see Figs. 1 and 2, the entanglement changes smoothly.

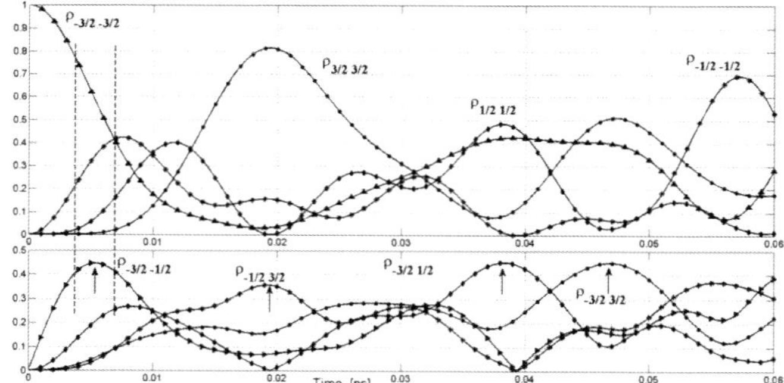


Figure 4. Entanglement of the ground level $\rho_{-3/2\ -3/2}$ and three-exciton level $\rho_{3/2\ 3/2}$ and other levels population dynamics starting from the ground state $\rho_{-3/2\ -3/2}$. Ion-laser radiation interaction is stronger than ion-ion interaction $A=3V$.

Fig. 4 demonstrates the complex level population evolution together with the entanglement between all levels. The parameters used are the same as at the previous Fig. 3, yet we magnify the shorter time stage to make the Figure more readable.

We can see a stepwise entanglement of the level pairs – at first, high entanglement occurs between the vacuum $|-3/2\rangle$ and one-exciton $|-1/2\rangle$ states (at $t\approx0.005$ns), then vacuum state becomes entangled with the biexciton state $|1/2\rangle$ at $t\approx0.04$ns, and, finally, vacuum state gets entangled with the three-exciton $|3/2\rangle$ state at $t\approx0.047$ns.

### 3.4.  *Conclusion*

We derived an exact analytical solution for the dynamics of the energy states of cluster of three resonantly interacting fluorescent particles under laser radiation with the central frequency. Moments of maximal entanglement of vacuum and three-exciton state or of one-exciton and two-exciton state can be easily obtained from the analytical solution. The solution contains a number of periodic functions and for the case of relatively weak laser radiation can be considerably simplified to include just one periodic function. The case of strong laser radiation demonstrates much more complex beatings, however, good entanglement is also possible in this case and can be described using the exact analytical solution.

## References


1. D. M. Greenberger, M. A. Horne, A. Shimony, and A. Zeilinger, *Am. J. Phys.* **58**, 1131 (1990).
2. M. Nielsen, I. Chuang, Quantum Computation and Quantum Communication, Cambridge University Press, Cambridge, 2000.
3. C. H. Bennet and D. P. DiVincenzo, Nature (London) **404**, 247 (2000).
4. *Quantum computation and quantum information*, ed. by C. Macchiavello, G. M. Palma and A. Zeilinger (World Scientific, Singapour, 2001).
5. P. Bori, W. Langbein, S. Schneider et al., *Phys. Rev. Lett.* **87**, 157401 (2001).
6. D. Birkedal, K. Leosson, and J. M. Hvam, *Phys. Rev. Lett.* **87**, 227401 (2001).
7. T. T. Basiev, A. Ya. Karasik, V. V. Fedorov, and K. W. Ver Steeg, *JETP* **86**, 156 (1998).
8. G. Chen, T. H. Stievater, E. T. Batteh, et al., *Phys. Rev. Lett.* **88**, 117901 (2002).

9. T.T. Basiev, V.V. Fedorov, A.Ya. Karasik, K.K. Pukhov, *J. Luminesc.* **81**, 189 (1999).

10. V.V. Osiko, Yu.K. Voron'ko, A.A. Sobol, Crystals 10, Springer, Berlin, 1984.

11. V. V. Osiko, *Solid state physics* **7** (5), 1294 (1965).

12. V. V. Osiko, Yu. K. Voronko, A. M. Prokhorov, I. A. Tscherbakov, *JETP* **60** (3), 943 (1971).

13. V.V. Fedorov, W. Beck, T.T. Basiev, A. Ya. Karasik, C. Flytzanis, *J. Chem. Phys.* **257**, 275 (2000).

14. K.Ver Steeg, R.Reeves, T.T. Basiev, and A.Ya.Karasik, *J. Luminesc.* **60/61**, 742 (1994).

15. K.W.Ver Steeg, R.J.Reeves, T.T. Basiev, A.Ya.Karasik, and R.C.Powell, *Phys. Rev. B* **51** (9), 6085 (1995).

16. T. T. Basiev, A.Ya. Karasik, V.V. Fedorov, *JETP* **113** (1), 278 (1998).

17. R. H. Dicke, *Phys. Rev.* **93**, 99 (1954).

18. L. Quiroga and N. F. Johnson, *Phys. Rev. Lett.* **83**, 2270 (1999).

19. E. Merzbacher, in Quantum Mechanics (John Wiley & Sons, New York) 1998.

Quantum Bio-Informatics V
© 2013 World Scientific Publishing Co. Pte. Ltd.
pp. 95–115

# MINIMUM OF INFORMATION DISTANCE CRITERION FOR OPTIMAL CONTROL OF MUTATION RATE IN EVOLUTIONARY SYSTEMS

ROMAN V. BELAVKIN

*School of Engineering and Information Sciences*
*Middlesex University, London NW4 4BT, UK*
*E-mail: R.Belavkin@mdx.ac.uk*
*E-mail: wspc@wspc.com*

Evolutionary dynamics studies changes in populations of species, which occur due to various processes such as replication and mutation. Here we consider this dynamics as an example of Markov evolution on a simplex of probability measures describing the populations, and then define optimality of this evolution with respect to constraints on information distance between these measures. We show how this convex programming problem is related to a variational problem of optimizing Markov transition kernel subject to a constraint on Shannon's mutual information. This relation is represented by the Pythagorean theorem in information geometry considered on the simplex of joint probability measures. We discuss the application of this variational approach to optimization of a stochastic search in metric spaces, and in particular to optimization of mutation rate parameter during the search for optimal DNA sequences in evolutionary systems.

Keywords: Replicator-mutator equation; Markov evolution; Mutual information; Information geometry; Hamming space.

## 1. Introduction

Some of the most important processes driving evolution of biological organisms are replication and mutation. Organisms that are well-adapted to their environment tend to have high replication rates (fitness), but changes in the environment may require the development of new features, which the organisms may acquire by mutation. The random nature of mutation, however, means that finding the right solution can be extremely difficult, and most mutations have neutral or deleterious effects.

Perhaps, the earliest mathematical work on the role of mutation in adaptation is due to Ronald Fisher, who studied this phenomenon by considering mutation as a random motion between points of Euclidean space representing different species[1]. The geometry of Euclidean space lead to a simple and very influential conclusion that adaptation is more likely to occur by small mutations. This theory, however, was developed before the discovery of DNA, which means that random motion between DNA sequences may be a better model of mutation. It has been shown recently that geometry of a finite space, such as a Hamming space of DNA sequences, implies different relation between the size mutation and adaptation[2,3]. In particular, mutation size maximizing the probability of adaptation varies as a function of fitness of the organism. In fact, the advantages of using variable mutation rates have been known for some time in the fields of evolutionary computation and operations research[4,5]. Our aim is to develop mutation rate control functions optimizing the evolution.

Kinetic equations that are used to model evolutionary dynamics of biological organisms is an important example of (generally non-linear) Markov evolution of probability measures of populations at different generations (time moments). We shall consider evolution as a stochastic search for optimal sequences, and we shall optimize the corresponding controlled Markov process with respect to an information-theoretic criterion. The possibility of relating probabilistic interpretation of information with 'hereditary information' and adaptation based on the mechanisms of mutation was predicted by Kolmogorov[6], and this work defines this relation explicitly.

In the next section, we overview some basic ideas of evolutionary dynamics and Fisher's geometric model of adaptation. Information-theoretic variational problems for optimization of controlled Markov process will be defined in Section 3. Section 4 presents general expressions for stochastic search in metric spaces, and then specific expressions for evolutionary dynamics in Hamming spaces of DNA sequences. The paper concludes by discussing and comparing three control functions of the mutation rate.


## 2. Replicator-Mutator Dynamics

Let $\Omega$ be the set of different species of organisms – the *quasispecies*[11]. The *fitness function* $f : \Omega \to \mathbb{R}$ defines their replication rates. If $\mu(E)$ is a measure of the overall abundance of quasispecies in $E \subseteq \Omega$ and $P(E) = \mu(E)/\mu(\Omega)$ is their frequency (relative abundance), then the evolutions $\mu(t) := \mu_t(E)$ and $p(t) := P_t(E)$ of these measures are described by exponential laws:

$$\mu(t) = e^{tf}\,\mu(0)\,, \qquad p(t) = e^{tf-\Psi(t)}\,p(0) \tag{1}$$

where $\Psi(t) = \ln \int e^{tf(\omega)} \, dP_t(\omega)$ is the cumulant generating function. Its first derivative $\Psi'(t)$, in particular, is the expected value $\mathbb{E}\{f\}(t) = \int f(\omega) \, dP_t(\omega)$ (i.e. the first cumulant). Differential equations describing changes of $\mu$ and $p$ in continuous time are obtained by taking the derivatives at $t = 0$:

$$\mu'(t) = f \, \mu(t) \,, \qquad p'(t) = [f - \mathbb{E}\{f\}(t)] \, p(t) \tag{2}$$

These are so-called *replicator* equations, and one can see that fitness function defines generators of these evolutions, the so-called *selection pressure*. Note that generally $f$ may also change with time, such as when the selection pressure depends on the distribution $p(t)$ for interacting species. With constant selection, the dynamics is quite simple: The abundance of species with $f(\omega) > \mathbb{E}\{f\}$ exponentially increases compared to species with $f(\omega) \leq \mathbb{E}\{f\}$. The fixed point of this evolution is the Dirac measure $\delta_\top(E)$, where $\top \in \Omega$ is any 'top' species with $f(\top) = \sup f(\omega)$, sometimes referred to as the *wild type*.

Observe that the replicator dynamics preserves the supports $\mathrm{supp}(p) := \{\omega : p(\omega) \neq 0\}$ of the measures for all $t \in [0, \infty)$, which means it cannot introduce new species. Mutation is the process $\Omega \ni a \mapsto b \in \Omega$ of random change from one species to another after (or during) the replication process, and it plays the role of innovation process increasing the diversity of quasispecies. It can be described by Markov transition probabilities $P(b \mid a)$, and the products $P(b \mid a) \cdot f(a)$ define the selection-mutation rates. The selection-mutation matrix $A = [P(b \mid a) \cdot f(a)]$ becomes the generator of the replicator-mutator evolution:

$$\mu'(t) = A \, \mu(t) \,, \qquad p'(t) = [A - \mathbb{E}\{f\}(t)] \, p(t) \tag{3}$$

If $P(b \mid \top) > 0$, then mutation introduces genetic drift from the top element $\top \in \Omega$ slowing down the evolution. On the other hand, the support of measures $p(t)$ may not include $\top$ with the highest replication rate, and therefore mutation can speed up the evolution by introducing $\top$ into the population. This shows that mutation can have different effects. Generally, mutation $a \mapsto b$ can be deleterious, if $f(a) > f(b)$, neutral, if $f(a) = f(b)$, or advantageous, if $f(a) < f(b)$. The problem of maximizing advantageous mutations is fundamental both for theory of evolution and for applications of evolutionary algorithms.

If the set $\Omega$ is equipped with a metric $d : \Omega \times \Omega \to [0, \infty)$, then radius $r = d(a, b)$ characterizes the 'size' of mutation $a \mapsto b$. Fisher [1] represented species by points in Euclidean space $\mathbb{R}^l$ of $l \in \mathbb{N}$ traits, and therefore

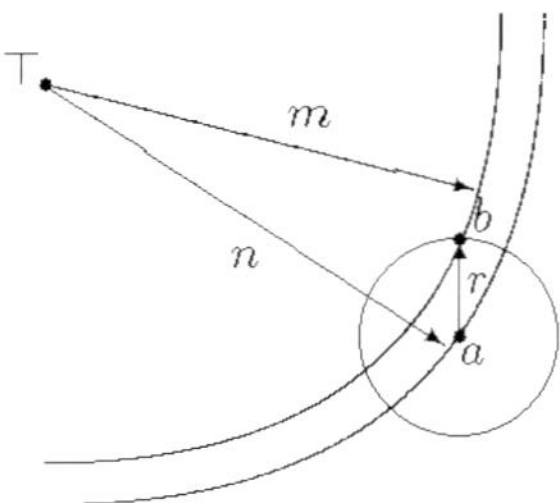

Figure 1. Mutation of $a$ to $b$ by radius $d(a,b) = r$. The distances $d(\top, a) = n$ and $d(\top, b) = m$ to an optimal element $\top$ define fitness of $a$ and $b$. The intersection of spheres $S(a, r)$ and $S(\top, m)$ define probability $P(m \mid n, r)$.

equipped $\Omega$ with Euclidean metric $d_E(a, b) = \|a - b\|_2$. He also identified fitness $f(\omega)$ with the negative distance $d_E(\top, \omega)$ to a single top element. Figure 1 depicts mutation of point $a \in S(\top, n) := \{\omega : d_E(\top, \omega) = n\}$ into $b \in S(\top, m)$. Adaptation corresponds to condition $m < n$. Fisher used the geometry of Euclidean space to show that probability of advantageous mutation (adaptation) decreases exponentially as mutation size increases, and therefore mutations of small radii are more likely to be advantageous. Moreover, this preference for small mutations does not depend on the distance $d_E(\top, a)$ of the parent from the optimum.

In this paper, we continue our previous work[2,3] on a generalization of Fisher's ideas to general metric spaces, and in particular to the space of DNA sequences equipped with the Hamming or Levenshtain metrics. One fundamental difference is that these spaces are finite and have very different geometries from Euclidean space. For this reason, probability of adaptation in these spaces can be maximized by mutations of different radii $r = d(a, b)$ depending on the distance $d(\top, a)$ between the parent sequence and the optimum. This variable preference for different mutation sizes leads naturally to the problem of optimal control of mutation.

## 3. Optimal control of Markov evolution

Let $p(x_{t+1} \mid x_t) := dP(f(b) = x_{t+1} \mid f(a) = x_t)/dx$ be the probability density of transition between two values of fitness associated with mutation probability $P(b \mid a)$. If the operator $T(\cdot) = \int_{x_t} p(x_{t+1} \mid x_t)(\cdot)\, dx$ does not depend on time, then $T^t$ defines a linear transformation of initial distribution $p(0) := dP_0(x)$ of fitness values into distribution $p(t) := dP_t(x)$ of fitness values after $t \geq 0$ generations:

$$p(1) = Tp(0) = \int_{x_0} p(x_1 \mid x_0)\, p(x_0)\, dx \quad \Rightarrow \quad p(t) = T^t p(0)$$

Adaptation is an increase of expected fitness $\mathbb{E}\{f\}(t) \geq \mathbb{E}\{f\}(0)$.

Let $\mu$ be a control parameter that changes the probability $p_\mu(x_{t+1} \mid x_t)$. In particular, $\mu$ can be the *mutation rate* parameter controlling expected mutation radius. The collection of pairs $(x, \mu)$ corresponds to a control function $\mu(x)$ parametrizing Markov operators $T_{\mu(x)}$. In this case, the replicator-mutator evolution can be controlled and optimized. For example, one can maximize expected fitness $\mathbb{E}_{\mu(x)}\{f\}(t)$ at $t = \lambda$ or cumulative expected fitness on $t \in [0, \lambda]$. Exact and approximate solutions have been obtained for some specific cases of the first problem [2,3]. Unfortunately, deriving closed-form solutions for the general case is not feasible. However, an interesting approach is to consider the evolution in terms of changes in information distance $I(p, q)$ between probability measures at different time moments.

## 3.1. *Minimum information distance criterion*

Let $C_c(\Omega)$ be the space of continuous real functions $f : \Omega \to \mathbb{R}$ with compact support, equipped with the supremum norm $\| \cdot \|_\infty$. Its dual is the space $\mathcal{M}(\Omega)$ of real Radon measures, equipped with the norm $\| \cdot \|_1$. Positive measures $\mu$ are positive linear functionals on $C_c(\Omega)$, and probability measures $P$ are elements of the simplex:

$$\mathcal{P}(\Omega) := \{\mu \in \mathcal{M}(\Omega) : \mu \geq 0\,,\ \|\mu\|_1 = 1\}$$

If $\Omega$ is separable and completely metrizable, then all probability measures on $\Omega$ are Radon (i.e. they are points of $\mathcal{P}(\Omega)$). The replicator-mutator evolution (3) corresponds to some parametrized trajectory $t \mapsto p(t) \in \mathcal{P}(\Omega)$. Thus, optimization of linear transformations $p(t) = T^t_{\mu(x)} p(0)$ over control functions $\mu(x)$ can be considered along the problem of optimization of the trajectories (evolutions) they generate. This variational problem can be formulated by defining appropriate functionals on $\mathcal{P}(\Omega)$.

The simplex $\mathcal{P}(\Omega)$ can be equipped with a topology, defined by some information distance $I : \mathcal{P} \times \mathcal{P} \to [0, \infty]$. For example, one can consider topology of the *total variation* metric $I_V(p, q) = \|p - q\|_1$, induced by the standard $\| \cdot \|_1$-norm topology of $\mathcal{M}(\Omega)$. Another example, which

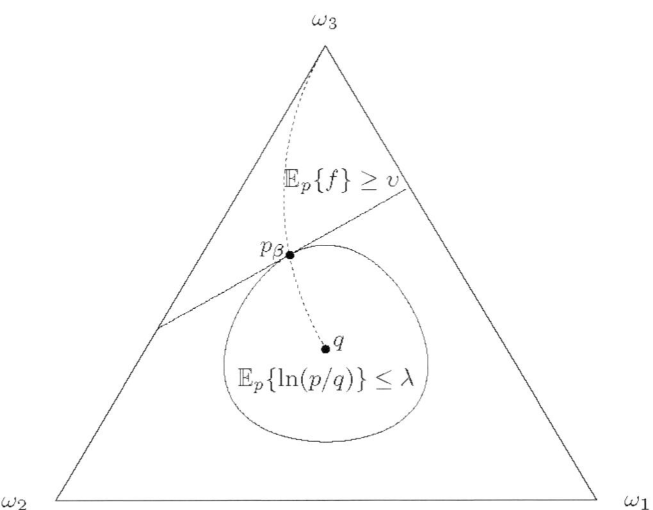



Figure 2. 2-Simplex $\mathcal{P}(\Omega)$ of probability measures over set $\Omega = \{\omega_1, \omega_2, \omega_3\}$. Level sets represent expected fitness $\mathbb{E}_p\{f\} = v$ and the Kullback-Leibler divergence $\mathbb{E}_p\{\ln(p/q)\} = \lambda$ (i.e. information distance). Probability measure $p_\beta$ is the solution to variational problems (4) and (5). Dashed curve represents optimal evolution (10).

is important in information theory and physics, is the Kullback-Leibler divergence[12]:

$$I_{KL}(p, q) := \mathbb{E}_p\{\ln(p/q)\}$$

Although it is not a metric (e.g. $I_{KL}(p, q) \neq I_{KL}(q, p)$ for some $p$ and $q$), $I_{KL}$ is unique in the sense of additivity: $I_{KL}(p_1 p_2, q_1 q_2) = I_{KL}(p_1, q_1) + I_{KL}(p_2, q_2)$ for all $p_1 p_2$, $q_1 q_2 \in \mathcal{P}(\Omega)$. Thus, we consider trajectories, each point $p(t)$ of which is a solution to the following problem:

$$\text{maximize} \quad \mathbb{E}_p\{f\} \quad \text{subject to} \quad \mathbb{E}_p\{\ln(p/q)\} \leq \lambda \tag{4}$$

Here, the maximization is over all probability measures $p \in \mathcal{P}(\Omega)$. This linear programming problem is equivalent to the following convex programming problem:

$$\text{minimize} \quad \mathbb{E}_p\{\ln(p/q)\} \quad \text{subject to} \quad \mathbb{E}_p\{f\} \geq v \tag{5}$$

If measure $p = p(t)$ is obtained from $q = p(0)$ as a linear transformation $p = T^t q$, then minimization of time $t$ (the power of Markov operator $T^t$)

is related to minimization of information distance $\mathbb{E}_p\{\ln(p/q)\}$ of $p$ from $q$. The constraint $\mathbb{E}_p\{f\} \geq v$ in problem (5) defines the infimum of information distance $\lambda = \mathbb{E}_p\{\ln(p/q)\}$. On the other hand, to achieve divergence $\lambda$ in the least number $t$ of transformations $q \mapsto Tq$, one has to maximize information distance $\mathbb{E}_{Tq}\{\ln(Tq/q)\}$ of each transformation. As will be shown next, this corresponds to minimization of mutual information between the input and output of a channel defined by operator $T$.

### 3.2. *Information distance and mutual information*

Let $\mathcal{P}(\Omega^2)$ be the set of all probability measures on the Cartesian product $\Omega^2 = \Omega \times \Omega$. Elements of $\mathcal{P}(\Omega^2)$ are joint probability measures $P(A_i \times B_j)$, $A_i \times B_j \subseteq \Omega^2$, which uniquely define the marginal measures $Q(A_i) = \int_B dP(A_i \times B)$, $P(B_j) = \int_A dP(A \times B_j)$ and conditional probabilities via factorization:

$$P(A_i \times B_j) = Q(A_i)\, P(B_j \mid A_i) = P(A_i \mid B_j)\, P(B_j)$$

Markov transition kernel $dP(b \mid a)$ defines a channel from $A_i$ to $B_j$. The mutual information between the input $a \in A_i$ and output $b \in B_j$ was defined by Shannon [8] as information divergence of joint measure $P(A_i \times B_j)$ from the product of its marginals $Q(A_i)\, P(B_j)$, or as the expectation of the information divergence of conditional probability $P(B_j \mid a)$ from the marginal $P(B_j)$, taken with respect to a fixed marginal $Q(A_i)$:

$$I_S\{a, b\} := \int_A dQ(a) \int_B \ln\left[\frac{dP(b \mid a)}{dP(b)}\right] dP(b \mid a) \tag{6}$$

**Proposition 3.1.** *Let $T(\cdot) = \int_A dP(b \mid a)\,(\cdot)$ be a Markov morphism $T : \mathcal{P}(\Omega) \to \mathcal{P}(\Omega)$ of the simplex $\mathcal{P}(\Omega)$. Let $q := dQ(a)$ be the input probability measure, $p := dP(b)$ be the output probability measure $p = Tq$, and let $w := dP(a, b) = dP(b \mid a)\, dQ(a)$ be the corresponding joint measure. Then the mutual information of the channel corresponding to $T$ is the difference*

$$I_S\{a, b\} = \gamma - \lambda \tag{7}$$

*where $\gamma$ is information distance $\mathbb{E}_w\{\ln(w/q \cdot q)\}$ in $\mathcal{P}(\Omega^2)$, and $\lambda$ is information distance $\mathbb{E}_p\{\ln(p/q)\}$ in $\mathcal{P}(\Omega)$.*

**Proof.** Indeed, multiplying $dP(b \mid a)$ and $dP(b)$ by $dQ(b)$ in equation (6) we have

$$\int_{A \times B} \ln \left[ \frac{dP(a,b)}{dQ(a)\, dP(b)} \right] dP(a,b)$$

$$= \int_{A \times B} \ln \left[ \frac{dP(a,b)\, dQ(b)}{dQ(a)\, dP(b)\, dQ(b)} \right] dP(a,b)$$

$$= \int_{A \times B} \ln \left[ \frac{dP(a,b)}{dQ(a)\, dQ(b)} \right] dP(a,b) - \int_{A \times B} \ln \left[ \frac{dP(b)}{dQ(b)} \right] dP(a,b)$$

Denoting $w := dP(a,b)$, $p := dP(b)$, $q := dQ(a)$, $q \cdot p := dQ(a)\, dP(b)$ and $q \cdot q = dQ(a)\, dQ(b)$ this can be written as

$$\underbrace{\mathbb{E}_w \left\{ \ln \left( \frac{w}{q \cdot p} \right) \right\}}_{I_S \{a,b\}} = \underbrace{\mathbb{E}_w \left\{ \ln \left( \frac{w}{q \cdot q} \right) \right\}}_{\gamma} - \underbrace{\mathbb{E}_p \left\{ \ln \left( \frac{p}{q} \right) \right\}}_{\lambda}$$

as required. $\qquad \square$

**Remark 3.1.** Equation (7) is a special case of the Pythagorean theorem in information geometry[13], which states that for all $w$, $p$ and $q \in \mathcal{P}(\Omega)$ such that $\mathbb{E}_w \{\ln(p/q)\} = \mathbb{E}_p \{\ln(p/q)\}$ the following triangle equality holds:

$$I_{KL}(w,q) = I_{KL}(w,p) + I_{KL}(p,q)$$

Thus, the Kullback-Leibler information divergence has been compared to a squared distance[14]. Clearly, the hypotenuse of the triangle is marked by the pair of measures $(w,q)$, while catheti by $(w,p)$ and $(p,q)$. Proposition 3.1 considers joint measures in $\mathcal{P}(\Omega^2)$. The hypotenuse is marked by the pair of joint measures $(w, q \cdot q)$, and catheti by $(w, q \cdot p)$ and $(q \cdot p, q \cdot q)$. The product measures $q \cdot q$ and $q \cdot p$ belong to submanifold $\mathcal{P}_0(\Omega^2) \subset \mathcal{P}(\Omega^2)$ describing independent random variables. It is easy to show that $I_{KL}(q \cdot p, q \cdot q) = I_{KL}(p,q)$. Thus, the evolution $p(t) \in \mathcal{P}(\Omega)$ corresponds to evolution of product measures $q \cdot p(t) \in \mathcal{P}(\Omega^2)$, $p(0) = q$. The cathetus $(q \cdot p, q \cdot q)$ belongs to submanifold $\mathcal{P}_0(\Omega^2)$. Mutual information is represented by the cathetus $(w, q \cdot p)$, which is 'orthogonal' to submanifold $\mathcal{P}_0(\Omega^2)$. Joint measure $w$ is not arbitrary, but such that $p$ and $q$ are its marginals.

Proposition 3.1 shows that an increase of information distance between $p = Tq$ and $q$ corresponds to an increase of divergence of joint measure $w$ from $q \cdot q$ or to a decrease of mutual information. Therefore, minimization of the number $t$ of times operator $T$ should be applied to transform $q = p(0)$ into $p(t) = T^t p(0)$ corresponds to a minimization of mutual information of $T$, because this maximizes information divergence of $p(s+1)$ from $p(s)$

in each transformation $p(s + 1) = Tp(s)$, $s \in [0, t)$. The corresponding variational problem is formulated by constraining mutual information from above.

Given a utility or fitness function $f_2 : \Omega \times \Omega \to \mathbb{R}$ (e.g. $f_2(a, b) = f(b) - f(a)$, where $f$ is a utility of fitness function on $\Omega$), we consider transformations $T(\cdot) = \int dP(b \mid a)(\cdot)$ that are solutions to the following problem:

$$\text{maximize} \quad \mathbb{E}_w\{f_2\} \quad \text{subject to} \quad I_S\{a, b\} \leq \gamma \qquad (8)$$

Here, the maximization is over all joint probability measures $w \in \mathcal{P}(\Omega^2)$ with a fixed marginal (input) measure $q = dQ(a)$. This is equivalent to maximization over all Markov transition kernels $dP(b \mid a)$. The corresponding convex programming problem is as follows:

$$\text{minimize} \quad I_S\{a, b\} \quad \text{subject to} \quad \mathbb{E}_w\{f_2\} \geq v \qquad (9)$$

Problems (8) and (9) appear similar to problems (4) and (5), but there is a subtle difference: In the former, variation of Markov transition kernels $dP(b \mid a)$ also changes the output measure $dP(b) = \int_A dP(b \mid a) \, dQ(a)$.

### 3.3. *General solutions of variational problems*

Convex programming problems can be solved using the standard method of Lagrange multipliers. The following is the Lagrange function for problem (5):

$$K(p, \beta) = \mathbb{E}_p\{\ln(p/q)\} + \beta[v - \mathbb{E}_p\{f\}]$$

where $\beta$ is the Lagrange multiplier corresponding to $v$. Because it is convex, the necessary and sufficient conditions for its infimum are obtained by taking its gradient to zero:

$$\nabla_p K(p, \beta) = \ln(p/q) + 1 - \beta f = 0, \quad \nabla_\beta K(p, \beta) = v - \mathbb{E}_p\{f\} = 0$$

Taking into account the normalization condition $\mathbb{E}_p\{1\} = 1$ the optimal solutions are defined by equations:

$$p(\beta) = e^{\beta f - \Psi(\beta)} \, p(0), \quad p(0) = q, \quad \mathbb{E}_{p(\beta)}\{f\} = v \qquad (10)$$

One can see that the first equation defines trajectory $p(\beta)$ of solutions corresponding to the replicator dynamics with constant selection (2):

$$p'(\beta) = [f - \mathbb{E}\{f\}(\beta)] \, p(\beta)$$

104

The inverse of the second equation defines Lagrange multiplier from the value of the constraint: $\beta = (\mathbb{E}\{f\})^{-1}(v)$. Because $\mathbb{E}\{f\}(\beta) = \Psi'(\beta)$, it is clear that $\beta = (\Psi^*)'(v)$, where $\Psi^*(v) = \sup[\beta v - \Psi(\beta)]$ is the Legendre-Fenchel dual of $\Psi(\beta)$. The value $\lambda = \Psi^*(v)$ is the value of information divergence $I_{KL}(p,q)$ achieved in problems (4) and (5).

Technically, the solution of problem (9) is identical to that of problem (5), and the optimality conditions for joint measures are as follows:

$$w(\beta) = e^{\beta\, f_2 - \Psi(\beta)} q \cdot p(\beta), \quad p(0) = q, \qquad \mathbb{E}_{w(\beta)}\{f_2\} = v \qquad (11)$$

The first equation implies that optimal Markov transition kernels have the following form:

$$dP_\beta(b \mid a) = e^{\beta\, f_2(a,b) - \Psi(\beta,a)}\, dP(b)$$

where the normalizing factor depends on $a \in A$:

$$e^{\Psi(\beta,a)} = \int_B e^{\beta\, f_2(a,b)}\, dP(b)$$

Similarly, normalization factor $e^{\Psi(\beta,b)}$ in $dP(a \mid b)$ depends on $b \in B$. These dependencies, however, can be removed if the products $e^{-\Psi(\beta,a)}\, dQ(a)$ and $e^{-\Psi(\beta,b)}\, dP(b)$ do not depend on $a \in A$ and $b \in B$ respectively. Indeed, observe that

$$dQ(a) = \int_B dP(a \mid b)\, dP(b) = \int_B e^{\beta\, f_2(a,b) - \Psi(\beta,b)}\, dQ(a)\, dP(b)$$

so that

$$\int_B e^{\beta\, f_2(a,b) - \Psi(\beta,b)}\, dP(b) = 1$$

If the Radon-Nikodym derivative $e^{-\Psi(\beta,b)}\, dP(b)/d\mu(b)$ is constant, then $e^{\Psi(\beta,b)} = [dP(b)/d\mu(b)] \int_B e^{\beta\, f_2(a,b)}\, d\mu(b)$, and

$$dP_\beta(a \mid b) = \frac{e^{\beta\, f_2(a,b)}\, dQ(a)\, d\mu(b)}{dP(b) \int_B e^{\beta\, f_2(a,b)}\, d\mu(b)} = \frac{dP_\beta(b \mid a)\, dQ(a)}{dP(b)}$$

Therefore, transition kernel $dP_\beta(b \mid a)$ takes the following simple form

$$dP_\beta(b \mid a) = e^{\beta\, f_2(a,b) - \Psi_0(\beta)}\, d\mu(b), \qquad \Psi_0(\beta) = \ln \int_B e^{\beta\, f_2(a,b)}\, d\mu(b) \qquad (12)$$

In particular, one can show that these simplified expressions can be used in the case when $\Omega = (\Omega, +)$ is a locally compact group with invariant measure $\mu$, and the utility function is translation invariant: $f_2(a+c, b+c) = f_2(a,b)$. An important example is when $\Omega$ is a vector space, and $f_2(a,b)$ depends only on the difference $a - b$ [15].

## 4. Optimal Search of DNA Sequences

Fisher's geometric model of adaptation in Euclidean space can be generalized to any topological space, in which convergence of some sequence $\{\omega_n\}$ to an optimal point $\top$ corresponds to increasing values $f(\omega_n)$ of some objective function (e.g. fitness values). We shall first consider the case of a metric space, and then specialize the results to a Hamming space.

### 4.1. *Stochastic search in metric spaces*

Let $(\Omega, d)$ be a metric space. A search in $\Omega$ is the problem of constructing a sequence $\{\omega_n\}$ converging to some target point $\top \in \Omega$, and it corresponds to an optimization problem of maximizing some objective function $f : \Omega \to \mathbb{R}$ such that $\sup f(\omega) = f(\top)$. If the objective function is sufficiently smooth in the neighbourhood of $\top$, then $f(\omega)$ can be replaced in this neighbourhood by negative distance $-d(\top, \omega)$. The difficulty is that the objective function may be not smooth (i.e. rugged), its values may be unknown or there can be some cost associated with its computation.

A stochastic search is the problem of constructing a sequence $\{p(t)\}$ of probability measures on $\Omega$, and then evaluating the objective function $f$ on a random sample of points $\omega$ from $p(t)$. The expected values $\mathbb{E}\{f\}(t)$ converge to $\sup f(\omega)$ if the sequence $\{p(t)\}$ converges to the Dirac measure $\delta_\top(d\omega)$.

Transformations from $p(t) := dP_t(\omega)$ to $p(t+1) := dP_{t+1}(\omega)$ can be realized by a Markov process $dP(b \mid a)$ on $\Omega$ such that $p(t+1) = Tp(t) = \int_A dP(b \mid a) \, dP_t(a)$. This corresponds to 'moving' points $a$ in the sample from $p(t)$ to some new points $b$ in the sample from $p(t+1)$. We shall refer to this random process as *mutation* of points $a$ to points $b$, and the distance $d(a, b) = r$ will be called the *mutation radius*. This process is shown schematically on Figure 1, and it is a generalization of Fisher's geometric theory of adaptation[1] from Euclidean space $\mathbb{R}^l$ of $l$ traits to a general metric space $(\Omega, d)$. Thus, adaptation corresponds to mutation of point $a$ in the sphere $S(\top, n) := \{\omega : d(\top, \omega) = n\}$ of radius $n$ to point $b \in S(\top, m)$ such that $m < n$.

. The problem of optimization if this Markov process can be formulated using the following considerations:

(1) The transition kernels should maximize the expected value of the difference $f(b) - f(a)$. Note that if $f(\omega)$ is sufficiently smooth, then it can be replaced by $-d(\top, \omega)$, and the transition kernel should maximize the expected difference $d(\top, a) - d(\top, b)$.

(2) The transition kernel must satisfy a constraint on mutual information $I_S\{a,b\}$. Indeed, suppose that the value of information divergence of $Tp(t)$ from $p(t)$ is $\lambda > 0$. Then it is clear from the relation $I_S\{a,b\} = \gamma - \lambda$ of Proposition 7 that $I_S\{a,b\} \leq \gamma - \lambda < \sup I_S\{a,b\}$. In fact, the supremum $I_S\{a,b\}$ is achieved if and only if $dP(b \mid a)$ corresponds to an injective mapping between $a$ and $b$, so that $Tp(t) = p(t)$ and $\lambda = 0$.

Thus, the optimization problem is a special case of variational problem (8):

$$\text{maximize} \quad \mathbb{E}_w\{d(\top, a) - d(\top, b)\} \quad \text{subject to} \quad I_S\{a,b\} \leq \gamma - \lambda \quad (13)$$

Its solutions are exponential transition kernels $dP_\beta(b \mid a) = \exp\{\beta\,[d(\top, a) - d(\top, b)] - \Psi(\beta, a)\}$. Moreover, the factors $\exp\{\beta\,d(\top, a)\}$ can be removed, so that the kernels take the following form:

$$dP_\beta(b \mid a) = e^{-\beta\,d(\top, b) - \Psi(\beta)}\,dP(b)\,, \qquad e^{\Psi(\beta)} = \int_B e^{-\beta\,d(\top, b)}\,dP(b) \quad (14)$$

Note that $a$ does not appear in the equations above explicitly, but the Lagrange multiplier $\beta$ depends on random mutual information $i(a,b) = \ln[dP(b \mid a)/dP(b)]$ between $a$ and $b$. Indeed, generally $dP(b \mid a) = \exp\{i(a,b)\}\,dP(b)$, and therefore $i(a,b) = -\beta\,d(\top, b) - \Psi(\beta)$. If $\beta$ is related to the expected value of $i(a,b)$ with respect to $dP(b \mid a)$, then $\beta$ depends only on $a$. Let us now analyse this dependence.

Because the sequence $\{p(t)\}$ of measures $p(t) = T^t p(0)$ is supposed to converge to $\delta_\top(d\omega)$ (i.e. $\delta_\top = T\delta_\top$ is a fixed point of $T$), the point $\top \in \Omega$ must be an absorbing state $dP_\beta(\top \mid \top) = 1$ of this Markov process. This implies $\beta \to \infty$ as $d(\top, a) \to 0$, which corresponds to supremum of random mutual information $i(a,b)$ and $T$ mapping $\top$ to $\top$. Note that the identity function $\mathrm{id} : \Omega \to \Omega$ corresponds to supremum of (expected) mutual information $I_S\{a,b\} = \mathbb{E}_w\{i(a,b)\}$, while $I_S\{a,b\} < \sup I_S\{a,b\}$ imply that $\beta < \infty$, and the optimal operator $T$ is non-deterministic. Indeed, it follows directly from the exponential form of optimal joint measures (11) that $dP_\beta(b \mid a) > 0$ for all $dP(b) > 0$ when $\beta < \infty$. In fact, this non-deterministic property of optimal Markov transition kernels follows from mutual absolute continuity of joint measures $w(\beta)$ and the product of marginals $w(0) = q \cdot q$, and it characterizes solutions to more general variational problems with constraints on any closed functional $I : \mathcal{P} \times \mathcal{P} \to \mathbb{R} \cup \{\infty\}$, which has a strictly convex dual[16] (i.e. not just only Kullback-Leibler divergence).

The constraint on mutual information implies that the mutation radius $d(a,b) > 0$ for $a \neq \top$ or $d(\top,a) > 0$. In fact, $d(a,b) \geq |d(\top,a) - d(\top,b)|$, which follows from the triangle inequality. However, $d(a,b) \to 0$ as $d(\top,a) \to 0$ or $\beta(a) \to \infty$. Denoting by $\gamma(a) - \lambda(a)$ the constraint on the expectation of $i(a,b)$ taken with respect to $dP(b \mid a)$, we can now reformulate the variational problem as follows:

$$\text{minimize} \quad \mathbb{E}\{d(a,b) \mid a\} \quad \text{subject to} \quad \mathbb{E}\{i(a,b) \mid a\} \leq \gamma(a) - \lambda(a) \quad (15)$$

Optimal transition kernels are $dP_\beta(b \mid a) = \exp\{-\beta\, d(a,b) - \Psi(\beta,a)\}\, dP(b)$. As was pointed out in Section 3.3, if $\Omega$ is a locally compact group with invariant measure $\mu$, and the metric is translation invariant $d(a,b) = d(a + c, b + c)$, then the transition kernels take a more simple form:

$$dP_\beta(b \mid a) = e^{-\beta\, d(a,b) - \Psi_0(\beta)}\, d\mu(b), \qquad e^{\Psi_0(\beta)} = \int_B e^{-\beta\, d(a,b)}\, d\mu(b) \quad (16)$$

where $\beta$ depends on the constraint $\gamma(a) - \lambda(a)$.

In some applications, the information $\mathbb{E}\{i(a,b) \mid a\}$ is the same for all points in the sphere $S(\top,n) := \{a : d(\top,a) = n\}$. In particular, this is the case when the information (feedback) is communicated by the value $f(a)$ of an objective function, which for smooth $f$ can be identified with $-d(\top,a) = -n$. Then the probability of mutation by radius $r = d(a,b)$ is:

$$dP(r \mid n) = \nu(r)\, e^{-\beta\, r - \Psi_0(\beta)}\, dr, \qquad e^{\Psi_0(\beta)} = \int_{r=0}^{\infty} \nu(r)\, e^{-\beta\, r}\, dr \quad (17)$$

where $\nu(r)$ is the Radon-Nikodym derivative of the Hausdorff measure of the sphere $S(a,r)$ of radius $r$ taken with respect to Lebesgue measure $dr$.

One can see from Figure 1 that the probability $dP(m \mid n, r)$ that after mutating by radius $r = d(a,b)$ point $a$ in the sphere $S(\top,n)$ is 'moved' to point $b$ in the sphere $S(\top,m)$ is defined by the measure of the intersection of the spheres $S(\top,m)$ and $S(a,r)$ with $d(\top,a) = n$. Denoting the Radon-Nikodym derivative of this measure with respect to $dr$ by $\nu(m \cap r \mid n)$, this probability is:

$$dP(m \mid n, r) = \frac{\nu(m \cap r \mid n)}{\nu(r)} \quad (18)$$

The probability $dP_\beta(m \mid n) = \int_{r=0}^{\infty} dP(m \mid n, r)\, dP_\beta(r \mid n)$ of a transition from sphere $S(\top,n)$ to sphere $S(\top,m)$ is

$$dP_\beta(m \mid n) = \int_{r=0}^{\infty} \nu(m \cap r \mid n)\, e^{-\beta\, r - \Psi_0(\beta)}\, dr \quad (19)$$

This transition probability allows one to compute expectations $\mathbb{E}\{f\}(t) = \mathbb{E}\{-d(\top, \omega)\}(t)$ of the objective function during the Markov evolution $p(t) = T^t_{\beta(a)}p(0)$ and consequently optimize parameters $\beta$ of the transition probabilities $P_\beta(b \mid a)$ for each $a$.

### 4.2. *Stochastic search in the Hamming space*

Let us denote by $\mathcal{H}^l_\alpha := \{1, \ldots, \alpha\}^l$ the set of all sequences of letters from a finite alphabet $\{1, \ldots, \alpha\}$ and length $l$. The alphabet can be equipped with operations of addition and multiplication such that $\mathcal{H}^l_\alpha$ becomes a linear algebra over Galois field $GF(\alpha)$, and it is an example of a finite vector space (there are $\alpha^l$ points in $\mathcal{H}^l_\alpha$). The space $\mathcal{H}^l_\alpha$ can be equipped with the Hamming metric $d_H(a, b) := |\{i : a_i \neq b_i\}|$ counting the number of different letters. Hamming metric can also be defined as $d_H(a, b) := \|a - b\|_H$, where $\|\cdot\|_H : \mathcal{H}^l_\alpha \to \mathbb{N}_0$ is the Hamming weight counting the number of letters in a sequence not equal to the additive unit of the field $GF(\alpha)$. Thus, addition of sequences results in a substitution of some letters, which corresponds to simple point-mutation, and the Hamming distance counts the number of substitutions.

Let us consider the problem of stochastic search in a Hamming space $\mathcal{H}^l_\alpha$ and the corresponding variational problems (13) and (15). Because Hamming distance $d_H(a, b) = \|a - b\|_H$ is translation invariant $d_H(a, b) = d_H(a + c, b + c)$, we can use equation (16) for the optimal transition kernel:

$$P_\beta(b \mid a) = e^{-\beta\, d_H(a,b) - \Psi_0(\beta)}, \quad e^{\Psi_0(\beta)} = \sum_{b \in \mathcal{H}^l_\alpha} e^{-\beta\, d_H(a,b)} \tag{20}$$

where the Lagrange multiplier $\beta$ depends on $a$.

**Proposition 4.1.** *Optimal transitions (20) between sequences in Hamming space $\mathcal{H}^l_\alpha$ correspond to independent mutation of each letter in a sequence with probability $\mu = \upsilon/l$. The asymptotic of this process as $l \to \infty$ corresponds to Poisson process with mutation rate $\upsilon$.*

**Proof.** The number of sequences in a sphere $S(a, r) := \{b : d_H(a, b) = r\}$ in Hamming space $\mathcal{H}^l_\alpha$ is $\nu(r) = (\alpha - 1)^r \binom{l}{r}$, and therefore

$$e^{\Psi_0(\beta)} = \sum_{r=0}^{l} (\alpha - 1)^r \binom{l}{r} e^{-\beta r} = [1 + (\alpha - 1)e^{-\beta}]^l$$

Thus, for a parent sequence that is $n = d_H(\top, a)$ letters away from $\top$, the probability of mutation by radius $r$ is:

$$P_\beta(r \mid n) = (\alpha - 1)^r \binom{l}{r} \frac{e^{-\beta r}}{[1 + (\alpha - 1)e^{-\beta}]^l}$$

The expected mutation radius is

$$\mathbb{E}\{r\} = \frac{d}{d\beta} \Psi_0(\beta) = \frac{l}{1 + e^\beta/(\alpha - 1)}$$

Variational problem (15) of minimizing expected mutation radius $\mathbb{E}\{r\}$ subject to a constraint to mutual information is equivalent to the problem of minimizing mutual information subject to constraint $\mathbb{E}\{r\} \leq \upsilon$ on the expected mutation radius. Thus, using the above expression for $\mathbb{E}\{r\}$, we obtain the following expression for parameter $\beta$:

$$\beta = \ln\left(\frac{l - \upsilon}{\upsilon}\right) + \ln(\alpha - 1)$$

It is easy to see that $P_\beta(r \mid n)$ corresponds to binomial distribution with probability of success $\mu = \upsilon/l$. Indeed, changing parametrization from $\beta$ to $\upsilon$, the probability $P_\beta(r \mid b)$ can be written as follows

$$P_\upsilon(r \mid n) = \binom{l}{r}\left(\frac{\upsilon}{l - \upsilon}\right)^r \left(1 + \frac{\upsilon}{l - \upsilon}\right)^{-l} = \binom{l}{r}\left(\frac{\upsilon}{l}\right)^r \left(1 - \frac{\upsilon}{l}\right)^{l-r}$$

As is well-known, the limit of distribution $P_\upsilon(r \mid n)$ as $l \to \infty$ is Poisson distribution:

$$\lim_{l \to \infty}\left[P_\upsilon(r \mid n) = \frac{l!}{(l - \upsilon)^r(l - r)!}\frac{\upsilon^r}{r!}\left(1 - \frac{\upsilon}{l}\right)^l\right] = \frac{\upsilon^r}{r!}e^{-\upsilon}$$

which follows from the following facts

$$\lim_{l \to \infty}\frac{l!}{(l - \upsilon)^r(l - r)!} = 1 \quad \text{and} \quad \lim_{l \to \infty}\left(1 - \frac{\upsilon}{l}\right)^l = e^{-\upsilon}$$

This completes the proof. $\qquad\square$

Optimization of parameter $\beta$ of the transition probabilities $P_\beta(b \mid a)$ corresponds to optimization of the mutation rate $\mu$. As mentioned earlier, this optimization can be performed by considering probability $P_\beta(m \mid n)$ of transition from sphere $S(\top, n)$ to sphere $S(\top, m)$.

**Proposition 4.2.** *The probability of transition between spheres of radii $n$ and $m$ around $\top$ in a Hamming space $\mathcal{H}_\alpha^l$ under simple mutation with rate $\mu \in [0, 1]$ is*

$$P_\mu(m \mid n) = \sum_{r=0}^{l} \frac{\nu(m \cap r \mid n)}{(\alpha - 1)^r} \mu^r (1 - \mu)^{l-r} \tag{21}$$

*where the measure*

$$\nu(m \cap r \mid n) = \sum (\alpha - 2)^{r_0} \binom{n - r_+}{r_0} (\alpha - 1)^{r_-} \binom{l - n}{r_-} \binom{n}{r_+} \qquad (22)$$

*is cardinality of the intersection $S(\top, m) \cap S(a, r)$ with $d_H(\top, a) = n$. The summation runs over $r_0$, $r_-$ and $r_+$ satisfying $r_- \in [0, (r + m - n)/2]$, $r_+ \in [0, (n - |r - m|)/2]$, $r_+ - r_- = n - \max\{r, m\}$ and $r_0 + r_- + r_+ = \min\{r, m\}$.*

**Proof.** Probability $P_\mu(m \mid n)$ is represented as follows

$$P_\mu(m \mid n) = \sum_{r=0}^{l} P(m \mid n, r) P_\mu(r \mid n)$$

where $P_\mu(r \mid n) = \binom{l}{r} \mu^r (1 - \mu)^{l-r}$ is binomial distribution with parameter $\mu$ (Proposition 4.1), and probability $P(m \mid n, r)$ is defined by equation (18) as the ratio of measures $\nu(m \cap r \mid n)$ and $\nu(r)$. Here, measure $\nu(r) = (\alpha - 1)^r \binom{l}{r}$ is the number of sequences in the sphere $S(a, r)$. Measure $\nu(m \cap r \mid n)$ is the number of sequences in the intersection of spheres $S(\top, m)$ and $S(a, r)$, which is obtained by counting the numbers $r_0$ of neutral, $r_-$ of deleterious, and $r_+$ of advantageous mutations in $r$ satisfying metric inequalities for $r$, $m$ and $n$ (e.g. $|n - m| \le r \le n + m$). $\qquad \square$

**Remark 4.1.** It has been shown recently how the alphabet $[1, \ldots, \alpha]$ can be extended so that the resulting algebra over the extended Galois field takes into account insertions and deletions of letters in sequences[17]. The resulting vector space is the infinite set of all sequences of variable length $l \in \mathbb{N}$. Finite sets $\mathcal{H}_\alpha^l$ and $\bigcup_{k=0}^{l} \mathcal{H}_\alpha^k$ are its finite subspaces. The space can be equipped with the Levenshtain metric counting the number of substitutions, insertions and deletions between two sequences. Expressions for probabilities (17), (18) and (19) for variable length sequences can be obtained using measures $\nu(r)$ and $\nu(m \cap r \mid n)$, which are cardinalities of a sphere and intersection of spheres in the Levenshtain space.

### 4.3. *Optimal control of mutation rates*

Probability $P_\mu(m \mid n)$ of transitions between spheres around optimal element $\top \in \mathcal{H}_\alpha^l$ can be used to optimize the mutation rate $\mu$ as a function of $n$. This optimization, however, can be done with respect to different criteria. For example, conditional expectation $\mathbb{E}\{m \mid n\} = \sum_{m=0}^{l} m P_\mu(m \mid n)$

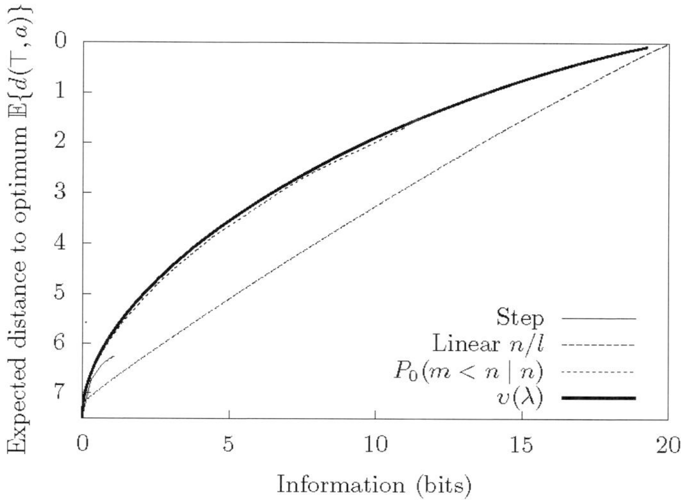



Figure 3.   Expected distance to optimum $\top \in \mathcal{H}_4^{10}$ as a function of information divergence $\lambda$ from initial distribution. Different curves correspond to different controls $\mu(n)$ of mutation rate; $v(\lambda)$ represents theoretical optimum.

is minimized if the mutation rates $\mu$ depend on $n$ as the following step function:

$$\mu(n) = \begin{cases} 0 & \text{if } n < l(1 - 1/\alpha) \\ \frac{1}{2} & \text{if } n = l(1 - 1/\alpha) \\ 1 & \text{otherwise} \end{cases} \tag{23}$$

The application $p(0) \mapsto Tp(0) = p(1)$ of the corresponding Markov operator $T_{\mu(n)}(\cdot) = \sum P_\mu(b \mid a)(\cdot)$ to the uniform distribution $p(0) = Q(a) = \alpha^{-l}$ of parent sequences produces the greatest decrease of the expected distance to the optimum $\top \in \mathcal{H}_\alpha^l$. Subsequent applications $p(0) \mapsto T_{\mu(n)}^t p(0) = p(t)$, $t > 1$, of this operator, however, do not produce the desirable evolution. Indeed, all parent sequences $a$ with $d_H(\top, a) < l(1 - 1/\alpha)$ do not 'move' closer to $\top$, and therefore $\lim_{t \to \infty} T_{\mu(n)}^t p(0) \neq \delta_\top(\omega)$. In fact, zero mutation rate corresponds to the identity function $\mathrm{id} : \Omega \to \Omega$ violating the constraint on mutual information in problem (15).

Another simple strategy is to maximize the probability of mutating directly into the optimal sequence:

$$P_\mu(m = 0 \mid n) = (\alpha - 1)^{-n} \mu^n (1 - \mu)^{l-n}$$

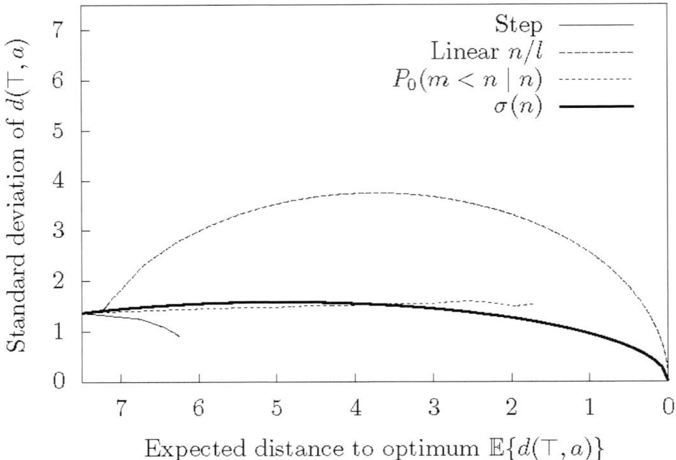

Figure 4. Standard deviation of distance to optimum $\top \in \mathcal{H}_4^{10}$ as a function of its expected value. Different curves correspond to different controls $\mu(n)$ of mutation rate; $\sigma(n)$ corresponds to mutation of $\top$ with $\mu = n/l$.

Taking its derivative over $\mu$ to zero gives condition $n - l\mu = 0$, which leads to the following linear function for the control of mutation rate:

$$\mu(n) = \frac{n}{l} \tag{24}$$

Because $\mu = 0$ if and only if $n = d_H(\top, a) = 0$ (i.e. if and only if $a = \top$), the only absorbing state of the Markov process $P_\mu(b \mid a)$ is $\top \in \mathcal{H}_\alpha^l$, and therefore $\lim_{t \to \infty} T_{\mu(n)}^t p(0) = \delta_\top(\omega)$. Analysis and experimental evaluation shows also that this Markov process has one of the shortest expected times of convergence. This evolution, however, is rather slow for small $t$, because mutation rates $\mu(n) = n/l$ are generally quite high even for sequences that are relatively close to $\top$. Experiments also show that the evolution $p(t) = T_{\mu(n)}^t p(0)$ is not optimal in the sense of problems (4) or (5).

Evolution that is optimal in these variational problems, that is evolution $p(t) \subset \mathcal{P}(\mathcal{H}_\alpha^l)$ maximizing expected value $f(\omega) = -d_H(\top, \omega)$ subject to constraint $I_{KL}(p, q) = \mathbb{E}_p\{\ln(p/q)\} \leq \lambda$ on KL-divergence of $p(t) = T_{\mu(n)}^t p(0)$ from $q := p(0)$, can be produced if the mutation rate is controlled by the following function:

$$\mu(n) = P_0(m < n \mid n) = \sum_{m=0}^{n-1} P_0(m) \tag{25}$$

where $P_0(m)$ is the probability of distance $m = d_H(\top, \omega)$ assuming a uniform distribution $P_0(\omega) = \alpha^{-l}$ of sequences in $\mathcal{H}_\alpha^l$. This probability can be obtained by counting sequences in the spheres $S(\top, m)$, and it is equivalent to $\top$ mutating with probability $\mu_0 = 1 - 1/\alpha$:

$$P_0(m) = \binom{l}{m} \left(1 - \frac{1}{\alpha}\right)^m \left(\frac{1}{\alpha}\right)^{l-m} = \binom{l}{m} \frac{(\alpha-1)^m}{\alpha^l}$$

We used a computer simulation to evaluate and compare different evolutions produced by Markov operators $T_{\mu(n)}$ parametrized by different mutation rate control functions $\mu(n)$. Figure 3 shows expected value of $f(\omega) = -d_H(\top, \omega)$ as function of information divergence $I_{KL}(p, q)$ for different evolutions of sequences in Hamming space $\mathcal{H}_4^{10}$. The chart also shows theoretical optimal value function $v(\lambda) = \sup\{\mathbb{E}_p\{f\} : I_{KL}(p, q) \leq \lambda\}$ (solid curve). One can see that the optimal values are achieved almost perfectly when mutation rate is controlled by function (25). Other controls do not achieve optimality in the sense of variational problem (5). Figure 4 shows standard deviations $\sigma$ from $\mathbb{E}_p\{d_H(\top, \omega)\} = -v$ as $v$ increases (i.e. populations getting closer to $\top$). The curve corresponding to optimal control function (25) matches closely theoretical curve $\sigma(v) = \sqrt{v(1 - v/l)}$, shown in bold, which is derived assuming simple mutation of $\top$ sequence with mutation rate $\mu = v/l$.

## 5. Discussion

We have considered the problem of optimization of replicator-mutator evolution by controlling the mutation rate parameter. Geometry of the space, in which quasispecies are represented, dictates the probability law of adaptation. This law allows one to derive optimal control functions parametrizing the semigroup of Markov operators acting on the simplex of probability measures of the populations. Specific choice of the control function depends on additional criteria. Thus, we have observed that short-term optimization in a Hamming space corresponds to a control of mutation rate by a step function; long-term optimization corresponds to a linear decrease of mutation rate. The minimum information distance criterion provides an interesting compromise between short- and long-term optimization, and the resulting Markov evolution has minimal variance of fitness values. The relation between geometry of a representation space and mutual information of optimal Markov transition kernels requires further investigation.

Although a control of mutation rate may significantly speed-up evolution, such an information processing function has to be represented in the DNA of an organism, which carries some cost due to increasing complexity. It is, therefore, not obvious if biological organisms have indeed evolved mutation rate controls. It is well-known, however, that modern organisms have evolved sophisticated DNA repair mechanisms allowing them to keep the mutation rates extremely low. A control of DNA repair may be equivalent to controlling the mutation rate, and therefore variations in DNA repair should be studied alongside variations in fitness.

## Acknowledgements

This work was supported by EPSRC grant EP/H031936/1. The author also expresses his gratitude to the organizers of the ICQBIC'11 conference, especially to Prof. Masanori Ohya, Prof. Noboru Watanabe and Prof. Satoru Miyazaki.

## References


1. 1. R. A. Fisher, The Genetical Theory of Natural Selection, Oxford University Press, Oxford, (1930).
2. R. V. Belavkin, A. Channon, E. Aston, J. Aston and C. G. Knight, Theory and practice of optimal mutation rate control in Hamming spaces of DNA sequences, in Advances in Artificial Life, ECAL 2011: Proceedings of the 11th European Conference on the Synthesis and Simulation of Living Systems, eds. T. Lenaerts, M. Giacobini, H. Bersini, P. Bourgine, M. Dorigo and R. Doursat, MIT Press, (2011).
3. R. V. Belavkin, Mutation and optimal search of sequences in nested Hamming spaces, in IEEE Information Theory Workshop, IEEE, (2011).
4. A. E. Eiben, R. Hinterding and Z. Michalewicz, IEEE Transactions on Evolutionary Computation 3, 124 (1999).
5. F. Vafaee, G. Turán and P. C. Nelson, Optimizing genetic operator rates using a Markov chain model of genetic algorithms, ACM, (2010).
6. A. N. Kolmogorov, Problems of Information Transmission (in Russian), 1 , 3 (1965).
7. C. T. Bergstrom and M. Lachmann, Shannon information and biological fitness, in Proceedings of the IEEE Workshop on Information Theoery, (2004).
8. C. E. Shannon, Bell System Technical Journal, 27, 379, (July and October (1948).
9. R. L. Stratonovich, Izvestiya of USSR Academy of Sciences, Technical Cybernetics (in Russian), 5, 3 (1965).

10. M. A. Nowak, Evolutionary Dynamics: Exploring the Equations of Life, Harvard University Press, Cambridge, London, (2006).
11. M. Eigen, J. McCaskill and P. Schuster, Journal of Physical Chemistry, 92, 6881 (1988).
12. S. Kullback, Information Theory and Statistics (John Wiley and Sons, 1959).
13. N. N. Chentsov, Statistical Decision Rules and Optimal Inference, Nauka, Moscow, U.S.S.R. (in Russian), (1972), English translation: Providence, RI: AMS, (1982).
14. S.I. Amari, Differential-Geometrical Methods of Statistics, Lecture Notes in Statistics, Vol. 25, Springer, Berlin, Germany, (1985).
15. R. L. Stratonovich, Information Theory, Sovetskoe Radio, Moscow, U.S.S.R. (in Russian), (1975).
16. 16. R. V. Belavkin, Journal of Global Optimization , 1 (2012).
17. R. Sánchez and R. Grau, International Journal of Biological and Life Sciences, 3, 89 (2007).

Quantum Bio-Informatics V
© 2013 World Scientific Publishing Co. Pte. Ltd.
pp. 117–125

# ON NON-MARKOVIAN QUANTUM EVOLUTION

DARIUSZ CHRUŚCIŃSKI* AND ANDRZEJ KOSSAKOWSKI

*Institute of Physics,*
*Nicolaus Copernicus University,*
*Grudziadzka 5/7, 87–100 Toruń, Poland*
**E-mail: darch@fizyka.umk.pl*

We analyze two measures of non-Markovianity: one based on the mathematical concept of divisibility of the dynamical map and the other one based on distinguishability of quantum states. We provide a simple example of qubit dynamic to show that these two measures need not agree. In addition, we discuss possible generalizations and intricate relations between these measures.

*Keywords* : Open systems; quantum dynamics; non-Markovian evolution.

## 1. Introduction

The non-Markovian dynamics of open quantum systems attracts nowadays increasing attention.[1,2,3] It is very much connected to the growing interest in controlling quantum systems and applications in modern quantum technologies such as quantum communication, cryptography and computation.[4,5] It turns out that the popular Markovian approximation which does not take into account memory effects is not sufficient for modern applications and todays technology calls for truly non-Markovian approach. Non-Markovian dynamics was recently studied in Refs. 6–21.

The usual approach to the dynamics of an open quantum system consists in applying the Markovian approximation, that leads to the following local master equation

$$\frac{d}{dt}\rho(t) = \mathcal{L}_{\mathrm{M}}\,\rho(t) \ , \quad \rho(t_0) = \rho_0 \ , \tag{1}$$

where $\rho(t)$ is the density matrix of the system investigated and $\mathcal{L}_{\mathrm{M}}$ the time-independent generator of the dynamical semigroup possessing the following well known representation [22]

$$\mathcal{L}_{\mathrm{M}}\rho = -i[H,\rho] + \sum_\alpha \left( V_\alpha \rho V_\alpha^\dagger - \frac{1}{2}\{V_\alpha^\dagger V_\alpha, \rho\} \right) . \tag{2}$$

The above structure of $\mathcal{L}_{\mathrm{M}}$ guarantees that dynamical map $\Lambda(t,t_0)$, defined by $\rho(t) = \Lambda(t,t_0)\rho_0$, is completely positive and trace preserving for $t \geq t_0$.

In this paper we analyze a general evolution governed by the following equation

$$\frac{d}{dt}\rho(t) = \mathcal{L}(t)\,\rho(t) \ , \quad \rho(0) = \rho_0 \ , \tag{3}$$

giving rise to the dynamical map $\Lambda(t,0)$, that is, $\rho(t) = \Lambda(t,0)\rho_0$.

Surprisingly, it turns out the concept of (non)Markovianity of the evolution $\Lambda(t,0)$ is not uniquely defined. One approach is based on the idea of the composition law which is essentially equivalent to the idea of divisibility. This approach was used recently by Rivas, Huelga and Plenio (RHP) in Ref. 24 to construct the corresponding measure of non-Markovianity, that is, RHP measure the deviation from divisibility. A different approach is advocated by Breuer, Laine and Piilo (BLP) in Ref. 23. BLP define non-Markovian dynamics as a time evolution for the open system characterized by a temporary flow of information from the environment back into the system and manifests itself as an increase in the distinguishability of pairs of evolving quantum states.

In this paper, following Ref. 25 we analyze these two approaches and show how these two concepts of Markovianity can be reconciled.

## 2. Divisible maps

The measure for non-Markovianity proposed by Rivas, Huelga and Plenio (RHP) [24] is based on a notion of divisibility:

**Definition 1.** A trace preserving completely positive map $\Lambda(t,0)$ is divisible if

$$\Lambda(t+\tau,0) = \Lambda(t+\tau,t)\Lambda(t,0) \ , \tag{4}$$

and $\Lambda(t+\tau,t)$ is completely positive for any $t,\tau > 0$.

According to RHP

Markovianity I 2. A map $\Lambda(t,0)$ is Markovian if and only if it is divisible.

Note that

$$\Lambda(t+\tau,t) = \Lambda(t+\tau,0)\Lambda^{-1}(t,0) \ , \tag{5}$$

satisfies composition law

$$\Lambda(s,t) = \Lambda(s,u)\Lambda(u,t) \ , \tag{6}$$

for any $s \geq u \geq t$, which is usually attributed for Markovian evolution. It is shown [24] that the quantity

$$g(t) = \lim_{\epsilon \to 0+} \frac{||\mathbb{1}_d \otimes \Lambda(t+\epsilon,t)P_d^+||_1 - 1}{\epsilon} \ , \tag{7}$$

enjoys $g(t) > 0$ if and only if the original map $\Lambda(t,0)$ is indivisible. $||\cdot||_1$ is a trace-norm defined by

$$||A||_1 := \mathrm{Tr}\sqrt{A^*A} \ .$$

As usual $\mathbb{1}_d$ denotes an identity map in $M_d(\mathbb{C})$, and $P_d^+$ denotes the maximally entangled state in $\mathbb{C}^d \otimes \mathbb{C}^d$ (we consider the evolution of a $d$-level system). Now, using the fact that any divisible (and differentiable) completely positive map satisfies a local in time master equation

$$\frac{d}{dt}\Lambda(t,0) = L(t)\Lambda(t,0) \ , \qquad \Lambda(0,0) = \mathbb{1}_d \ , \tag{8}$$

where the local generator $L(t)$ is a legitimate Markovian generator for any $t \geq 0$. The formula (7) may be equivalently rewritten in terms of $L(t)$:

$$g(t) = \lim_{\epsilon \to 0+} \frac{||[\mathbb{1}_d \otimes (\mathbb{1}_d + \epsilon L(t))]P_d^+||_1 - 1}{\epsilon} \ . \tag{9}$$

Finally one introduces [24] the following measure of non-Markovianity of dynamical map $\Lambda$:

$$\mathcal{N}_{\mathrm{RHP}}(\Lambda) = \int_0^\infty g(t)dt \ , \tag{10}$$

and the evolution is Markovian iff $\mathcal{N}_{\mathrm{RHP}}(\Lambda) = 0$.

## 3. Information flow vs. discrimination problem

A second criterion of non-Markovianity was proposed by Breuer, Laine and Piilo (BLP) in Ref. 23. The BLP criterion identifies non-Markovian dynamics with certain physical features of the system-reservoir interaction. They define non-Markovian dynamics as a time evolution for the open system characterized by a temporary flow of information from the environment back into the system. This backflow of information may manifest itself as an increase in the distinguishability of pairs of evolving quantum states. Suppose we are given one of two known states $\rho_1$ and $\rho_2$, each with equal probability. Our goal is to guess which one it is with minimal error probability. The solution of this problem is well known due to Helstrom [26]: the minimal error probability (MEP) reads as follows

$$\mathrm{MEP} = \frac{1 - D[\rho_1, \rho_2]}{2} \, , \tag{11}$$

where $D[\rho_1, \rho_2] = \frac{1}{2}||\rho_1 - \rho_2||_1$ is the distinguishability of $\rho_1$ and $\rho_2$. For any two states $\rho_1$ and $\rho_2$

$$0 \leq D[\rho_1, \rho_2] \leq 1 \, . \tag{12}$$

Recall, that $D[\rho_1, \rho_2] = 0$ if and only if $\rho_1 = \rho_2$, i.e. they are not distinguishable. Moreover, $D[\rho_1, \rho_2] = 1$ if and only if $\rho_1$ and $\rho_2$ are supported on mutually orthogonal subspaces, that is, they are distinguishable.

Now, if $\Lambda$ is a trace preserving positive map (in particular completely positive) then

$$D[\Lambda\rho_1, \Lambda\rho_2] \leq D[\rho_1, \rho_2] \, . \tag{13}$$

Let $\Lambda(t, 0)$ be a dynamical map and $\rho_k(t); +\Lambda(t, 0)\rho_k$. One defines [23] an information flow

$$\sigma(\rho_1, \rho_2; t) = \frac{d}{dt} D[\rho_1(t), \rho_2(t)] \, . \tag{14}$$

Note that if $\Lambda(t, 0)\rho = U(t)\rho U^*(t)$, then

$$\sigma(\rho_1, \rho_2; t) = 0 \, . \tag{15}$$

Moreover, if $\Lambda(t, 0) = e^{Lt}$, where $L$ is a generator of Markovian semigroup, then

$$\sigma(\rho_1, \rho_2; t) < 0 \, . \tag{16}$$

The authors of Ref. 23 propose the following

Markovianity II 3. The evolution $\Lambda(t, 0)$ is Markovian iff

$$\sigma(\rho_1, \rho_2; t) \leq 0 \, , \tag{17}$$

for any initial states $\rho_1$, $\rho_2$ and $t \geq 0$.

Finally, one introduces the corresponding measure

$$\mathcal{N}_{\mathrm{BLP}}(\Lambda) = \sup_{\rho_1,\rho_2} \int_{\sigma>0} \sigma(\rho_1,\rho_2;t)dt \ . \tag{18}$$

It is clear that $\mathcal{N}_{\mathrm{BLP}}(\Lambda)$ is rather difficult to compute in practice. It is easy to prove the following

**Proposition 1.** *If the map is divisible (i.e. Markovian I), then $\sigma(\rho_1,\rho_2;t) \le 0$ (i.e. Markovian II).*

Interestingly, the converse is not true: if $\sigma(\rho_1,\rho_2;t) \le 0$, then in general the dynamical map needs not be divisible.

## 4. Example: qubit dynamics

In this section we illustrate the relation between two concepts of Markovianity (cf. Definitions 2 and 3) on a simple example. Consider the following dynamics of a qubit

$$\begin{aligned}
\rho_{00}(t) &= \rho_{00}\, x_0(t) + \rho_{11}\left[1 - x_1(t)\right] , \\
\rho_{11}(t) &= \rho_{00}\left[1 - x_0(t)\right] + \rho_{11}\, x_1(t) , \\
\rho_{01}(t) &= \rho_{01}\, \gamma(t) ,
\end{aligned} \tag{19}$$

where $x_0(t), x_1(t) \in [0,1]$, and

$$|\gamma(t)|^2 \le x_0(t)x_1(t) \ . \tag{20}$$

The above conditions for $x_k(t)$ and $\gamma(t)$ guarantee that the dynamics is completely positive. One easily finds for the corresponding local generator

$$L\,\rho = -i\frac{\Omega}{2}[\sigma_z,\rho] + \sum_{k=0}^{1} a_k L_k \rho + \frac{\Gamma}{2} L_z \rho \ , \tag{21}$$

where

$$\begin{aligned}
L_0\rho &= \sigma_+\rho\sigma_- - \frac{1}{2}\{\sigma_-\sigma_+,\rho\} , \\
L_1\rho &= \sigma_-\rho\sigma_+ - \frac{1}{2}\{\sigma_+\sigma_-,\rho\} , \\
L_z\rho &= \sigma_z\rho\sigma_z - \rho \ .
\end{aligned} \tag{22}$$

The time-dependent coefficients $a_0$ and $a_1$ are defined by

$$a_0 = \frac{\dot{x}_0(1 - x_1) + \dot{x}_1 x_0}{1 - x_0 - x_1} \ , \tag{23}$$

$$a_1 = \frac{\dot{x}_0 x_1 + \dot{x}_1(1 - x_0)}{1 - x_0 - x_1} \ , \tag{24}$$

whereas $\Gamma(t)$ and $\Omega(t)$ read as follows

$$\Gamma(t) = -\frac{a_0(t) + a_1(t)}{2} - \operatorname{Re}\frac{\dot{\gamma}(t)}{\gamma(t)} , \tag{25}$$

$$\Omega(t) = \operatorname{Im}\frac{\dot{\gamma}(t)}{\gamma(t)} . \tag{26}$$

One has

**Proposition 2.** *The corresponding dynamical map $\Lambda(t,0)$ is divisible if and only if $a_k(t), \Gamma(t) \geq 0$ for all $t \geq 0$.*

On the other hand one finds the following formula

$$\sigma(\rho_1, \rho_2; t) = -\frac{2A(t)\Delta_{00}^2 + [A(t) + 4\Gamma(t)]|\Delta_{01}|^2}{\sqrt{\Delta_{00}^2 + |\Delta_{01}|^2}} , \tag{27}$$

where $A(t) = a_0(t) + a_1(t)$, and a $2 \times 2$ matrix $\Delta$ is defined by

$$\Delta = \rho_1 - \rho_2 . \tag{28}$$

In principle one may have $\sigma(\rho_1, \rho_2; t) \leq 0$ even if $a_0(t) < 0$ or $a_1(t) < 0$. Indeed, let us define

$$x_0(t) = \int_0^t f_0(\tau)d\tau , \quad x_1(t) = \int_0^t f_1(\tau)d\tau , \tag{29}$$

such that $0 \leq \int_0^t f_k(\tau)d\tau \leq 1$, for all $t \geq 0$. Now, following Ref. 25 let

$$f_0(t) = \kappa \sin t , \quad t \geq 0 , \tag{30}$$

and $f_1(t) = 0$ for $t \in [0, \pi]$ together with

$$f_1(t) = -\kappa \sin t , \quad t \geq \pi , \tag{31}$$

where $0 < \kappa < 1/2$. One finds

$$a_0(t) + a_1(t) = \frac{\kappa \sin t}{1 - \kappa + \kappa \cos t} , \quad t \in [0, \pi] , \tag{32}$$

and $a_0(t) + a_1(t) = 0$ for $t \geq \pi$. Note, that for $t \geq \pi$ one has

$$a_0(t) = -a_1(t) = \kappa \sin t , \tag{33}$$

which proves that $\Lambda(t,0)$ is not divisible.

## 5. Markovianity and the generalized discrimination problem

Consider now the following generalized discrimination problem: suppose we are given one of two known states $\rho_1$ and $\rho_2$, with probabilities $p$ and $1-p$, respectively. Our goal is to guess which one it is with minimal error probability. Again, the solution of this problem is well known due to Helstrom [26]: the $p$-dependent minimal error probability (MEP$(p)$) reads as follows

$$\text{MEP}(p) = \frac{1 - D_p[\rho_1, \rho_2]}{2} \, , \tag{34}$$

where

$$D_p[\rho_1, \rho_2] := ||(1-p)\rho_1 - p\rho_2||_1 \, . \tag{35}$$

Clearly, for the unbiased problem, that is $p = 1/2$, one reproduces (11). Now, following Ref. 25 let us introduce $p$-dependent information flow

$$\sigma_p(\rho_1, \rho_2; t) = \frac{d}{dt} D_p[\rho_1(t), \rho_2(t)] \, . \tag{36}$$

Finally, consider an extended dynamics $\widetilde{\Lambda}(t,0) = \Lambda(t,0) \otimes \mathbb{1}_d$ of the extended system living in $\mathcal{H} \otimes \mathcal{H}$ and let

$$\widetilde{\sigma}_p(\widetilde{\rho}_1, \widetilde{\rho}_2; t) = \frac{d}{dt} D_p[\widetilde{\rho}_1(t), \widetilde{\rho}_2(t)] \, , \tag{37}$$

where $\widetilde{\rho}_k$ are density operators in $\mathcal{H} \otimes \mathcal{H}$, and $\widetilde{\rho}_k(t) = \widetilde{\Lambda}(t,0)\widetilde{\rho}_k$. One proves (cf. Ref. 25) the following

**Theorem 1.** *A map $\Lambda(t,0)$ is divisible (Markovian I) if and only if $\widetilde{\sigma}_p(\widetilde{\rho}_1, \widetilde{\rho}_2; t) \leq 0$ for any density operators $\widetilde{\rho}_1$ and $\widetilde{\rho}_2$, $t \geq 0$, and any $p \in (0, 1)$.*

## 6. Conclusions

In conclusion, we have analyzed two concepts of Markovianity, one based on the divisibility property of the dynamical map (Markovianity I) and the other based upon the distinguishability of quantum states (Markovianity II). We have given very simple example for the dynamics of a single qubit where these two criteria do not coincide. Furthermore we proposed a way to make them equivalent, in the sense that Markovianity would be identified by divisibility, but keeping the interpretation in terms of flows of information.

## Acknowledgments

It is a pleasure to thank Professor Noboru Watanabe for his warm hospitality during QBIC 2011. Many thanks for his help after the earthquake on March 11.

## References


1. H.-P. Breuer and F. Petruccione, *The Theory of Open Quantum Systems* (Oxford Univ. Press, Oxford, 2007).
2. U. Weiss, *Quantum Dissipative Systems*, (World Scientific, Singapore, 2000).
3. R. Alicki and K. Lendi, *Quantum Dynamical Semigroups and Applications* (Springer, Berlin, 1987).
4. M. A. Nielsen and I. L. Chuang, *Quantum Computation and Quantum Information* (Cambridge Univ. Press, Cambridge, 2000).
5. R. Horodecki, P. Horodecki, M. Horodecki and K. Horodecki, Rev. Mod. Phys. **81**, 865 (2009).
6. J. Wilkie, Phys. Rev. E **62**, 8808 (2000); J. Wilkie and Yin Mei Wong, J. Phys. A: Math. Theor. **42**, 015006 (2009).
7. A. A. Budini, Phys. Rev. A **69**, 042107 (2004); *ibid.* **74**, 053815 (2006).
8. H.-P. Breuer, Phys. Rev. A **69** 022115 (2004); *ibid.* **70**, 012106 (2004).
9. S. Daffer, K. Wódkiewicz, J.D. Cresser, and J.K. McIver, Phys. Rev. A **70**, 010304 (2004).
10. A. Shabani and D.A. Lidar, Phys. Rev. A **71**, 020101(R) (2005).
11. S. Maniscalco, Phys. Rev. A **72**, 024103 (2005); S. Maniscalco and F. Petruccione, Phys. Rev. A **73**, 012111 (2006).
12. J. Piilo, K. Härkönen, S. Maniscalco, K.-A. Suominen, Phys. Rev. Lett. **100**, 180402 (2008); Phys. Rev. A **79**, 062112 (2009).
13. E. Andersson, J. D. Cresser and M. J. W. Hall, J. Mod. Opt. **54**, 1695 (2007).
14. A. Kossakowski and R. Rebolledo, Open Syst. Inf. Dyn. **14**, 265 (2007); *ibid.* **15**, 135 (2008); *ibid.* **16**, 259 (2009).
15. H.-P. Breuer and B. Vacchini, Phys. Rev. Lett. **101** (2008) 140402; Phys. Rev. E **79**, 041147 (2009).
16. M. Moodley and F. Petruccione, Phys. Rev. A **79**, 042103 (2009).
17. B. Vacchini and H.-P. Breuer, Phys. Rev. A **81**, 042103 (2010).
18. D. Chruściński, A. Kossakowski, and S. Pascazio, Phys. Rev. A **81**, 032101 (2010)
19. D. Chruściński and A. Kossakowski, Phys. Rev. Lett. **104**, 070406 (2010).
20. D. Chruściński, A. Kossakowski, P. Aniello, G. Marmo, F. Ventriglia, Open Systems and Inf. Dynamics **17**, 255-278 (2010).
21. D. Chruściński, A. Kossakowski and S. Pascazio, Phys. Rev. A **81**, 032101 (2010).
22. G. Lindblad, Comm. Math. Phys. **48**, 119 (1976); V. Gorini, A. Kossakowski, and E.C.G. Sudarshan, J. Math. Phys. **17**, 821 (1976).
23. H.-P. Breuer, E.-M. Laine, J. Piilo, Phys. Rev. Lett. **103**, 210401 (2009).
24. Á. Rivas, S.F. Huelga, and M.B. Plenio, Phys. Rev. Lett. **105**, 050403 (2010).

25. D. Chruściński, A. Kossakowski and Á. Rivas, Phys. Rev. A **83**, 052128 (2011).
26. C. W. Helstrom, *Quantum Detection and Estimation Theory* (Academic-Press, New York, 1976).

Quantum Bio-Informatics V
© 2013 World Scientific Publishing Co. Pte. Ltd.
pp. 127–142

# HIGH DENSITY LIMIT OF THE DISTRIBUTION OF THE OUTCOME OF EEG-MEASUREMENTS

K.-H. FICHTNER, L. FICHTNER,

*Unversity Jena, Institute of Applied Mathematics, 07743 Jena, Germany*
E-mail: fichtner@mathematik.uni-jena.de


W. FREUDENBERG

*Brandenburg Technical University Cottbus, Department of Mathematics,*
*03013 Cottbus, Germany.*  E-mail: freudenberg@math.tu-cottbus.de


M. OHYA

*Department of Information Science and Quantum Bio-Informatic Center,*
*Tokyo University of Science, Noda City, Chiba 278-8510, Japan*
E-Mail: ohya@rs.noda.tus.ac.jp

Using EEG measurements one gets information on the densities of excited neurons located in the regions of the brain. Up to now there exist different hypothesises concerning the distribution of the random outcomes of EEG measurements. Using classical models for describing brain activities it turned out to be difficult to explain the observed properties of these outcomes. We will describe the distribution of the random outcomes of EEG measurements and certain conditional distributions in terms of a high density limit. These considerations are based on a quantum statistical model of the process of recognition that was developed in the last years (cf. [1] – [10]).

## 1. Introduction

The procedure of recognition can be described as follows: There is a set of complex signals stored in the memory. Choosing one of these signals may be interpreted as generating a hypothesis concerning an "expected view of the world". Then the brain compares a signal arising from our senses with the signal chosen from the memory leading to a change of the state of both signals. Measurements of that procedure like EEG or MEG are based on the fact that recognition of signals causes a certain loss of excited neurons, i.e. the neurons change their state from *excited* to *non-excited*. For that reason a statistical model of the recognition process should reflect both the change of the signals and the loss of excited neurons. Physicists as R. Penrose or H. P. Stapp (cf. [12, 18, 19]) but also at an increasing rate specialists of modern brain research (cf. [15, 16, 14, 17]) are convinced that information processing in the brain cannot be described appropriately by models based on classical physics or classical stochastics. A first attempt to explain the process of recognition in terms of quantum statistics was given in [1]. In a (still incomplete) series of papers based on a quantum statistical model of the recognition process the procedures of creation of signals from the memory, amplification, accumulation and transformation of input signals, and measurements like EEG and MEG are treated in detail (cf. [1] – [10]). In the present paper it is not possible to present this approach in detail. In lieu we will sketch roughly a few of the basic ideas and structures of the proposed model of the recognition process. Our main purpose is to describe the distribution of the outcome of EEG-measurements. Using EEG-measurements one gets information on the densities of excited neurons located in the regions of the brain. Let us remark that these measurements are classical ones, and the results can be described by classical statistics though the underlying statistical model of the recognition process is a non-classical one. We will describe the distribution of the random outcomes of EEG-measurements and certain conditional distributions in terms of a high density limit. The proofs of these limit theorems will be given in a forthcoming paper.

## 2. The Space of Signals

In the present section we introduce briefly notions and notations needed in the sequel. For interpretation and motivation of the introduced notions we refer to the above mentioned papers. Starting point will be a set $G$ representing the space where the process of recognition and processing of the signals takes place. For the mathematical model the concrete structure of $G$ is irrelevant. To start with a general space $G$ has the advantage that it can be used as a model for very different aspects of the recognition processes in the brain. So let $G$ be an arbitrary complete separable metric space and $\mathfrak{G}$ its $\sigma$-algebra of Borel sets. Further, let $\mu$ be a fixed finite diffuse measure on $[G, \mathfrak{G}]$. Especially, we are concerned with the case where $G$ is a compact subset of $\mathbb{R}^d$ and $\mu$ is the $d-$dimensional Lebesgue measure restricted to $G$. The elements of the Hilbert space $\mathcal{H} = L_2(G, \mu)$ can be interpreted as *functions of the excited neurons*. From this space of square integrable complex-valued functions on $G$ we pass over to the Hilbert space

$L_2(M(G)) = L_2(M(G), F)$ where $M(G) := \{\{x_1, \ldots, x_n\} : x_j \in G, n \geq 0\}$ is the set of all finite symmetric point configurations (including the void configuration $\mathfrak{o}$). To avoid symmetrisation procedures it is convenient to identify the sets $\{x_1, \ldots, x_n\}$ with the corresponding counting measures $\varphi = \delta_{x_1} + \ldots + \delta_{x_n}$ where $\delta_x$ denotes the Dirac measure in $x$. So we can make the identification $M(G) = \{\varphi : \varphi \text{ is a counting measure on } G\}$. Within this notation $|\varphi| = \varphi(G)$ denotes the number of points in $\varphi$, and $x \in \varphi$ means that $\varphi - \delta_x \in M(G)$ (i. e. $x$ belongs to the the configuration $\varphi$). The measure $F$ defined on a canonical $\sigma$-algebra $\mathfrak{M}(G)$ over $M(G)$ restricted to $n$-point configurations is (up to symmetrisation) just the product measure $\frac{1}{n!}\mu^{\otimes n}$.

More precisely,

$$F(Y) := \mathbb{1}_Y(\mathfrak{o}) + \sum_{n \geq 1} \frac{1}{n!} \int_{G^n} \mathbb{1}_Y \Big( \sum_{j=1}^{n} \delta_{x_j} \Big) \mu^{\otimes n}(d[x_1, \ldots, x_n]) \quad (Y \in \mathfrak{M}(G)).$$

(1)

Hereby, $\mathbb{1}_Y$ denotes the indicator function of a set $Y$. Observe that $F(\{\mathfrak{o}\}) = 1$, $F(M(G)) = \exp\{\mu(G)\}$ and that $F$ is concentrated on the set of simple point configurations (without multiple points). One can show that the space $L_2(M(G))$ is isomorphic to the usual symmetric (bosonic) Fock space $\Gamma(L_2(G, \mu))$:

$$L_2(M(G)) \cong \Gamma(L_2(G, \mu))$$

For details we refer for instance to [8]. The space $L_2(M(G))$ consists of (complex-valued) *functions of configurations of excited neurons*, and will be interpreted as the *space of signals*. Later on we restrict ourselves to an essentially smaller subspace. A quantum state on a Hilbert space $\mathcal{H}$ is a positive linear normalized functional $\omega$ on the set $\mathcal{L}(\mathcal{H})$ of all bounded linear operators on $\mathcal{H}$. In our special case a general (normal) state $\omega$ on $L_2(M(G))$ can be written in the form

$$\omega(A) = \sum_{n=1}^{N} \alpha_n \langle \Psi_n, A\Psi_n \rangle \quad (A \in \mathcal{L}(L_2(M(G)))) \tag{2}$$

where $(\Psi_n)_{n=1}^{N}$ is a given orthonormal sequence from $L_2(M(G))$ and $(\alpha_n)_{n=1}^{N}$ is a sequence of non-negative numbers with $\sum_{n=1}^{N} \alpha_n = 1$, $N \leq \infty$. Normal states which are represented by one wave function ($N = 1$) are called *pure states*. So the *space of signals* can be identified with the *set of pure states* on $L_2(M(G))$. Analogously, sometimes we identify normal states $\omega$ given by (2) with their *density matrices* (positive normalised trace-class operators) $\varrho = \sum_{n=1}^{N} \alpha_n \langle \Psi_n, \cdot \Psi_n \rangle$.

We assume that $G$ decomposes into disjoint regions $G_1, \ldots, G_n$ being responsible for different tasks. So $L_2(G_k, \mu_{G_k})$ represents the space of the excited neurons in the region $G_k$. Hereby, $\mu_B$ is the restriction of $\mu$ to the set $B \subseteq G$. The main reason to choose the bosonic Fock space as basic space of signals is the possibility to identify the Fock space over $L_2(G)$ with the tensor product of the Fock spaces over $L_2(G_1), \ldots, L_2(G_n)$:

$$L_2(M(G_1 \cup \ldots \cup G_n)) \cong L_2(M(G_1)) \otimes \ldots \otimes L_2(M(G_n)).$$

Now, let us be given functions $f^r \in L_2(G_r)$, $r \in \{1, \ldots, n\}$. For $k \geq 1$ we define functions $f_k^r \in L_2(M(G_r))$ by

$$f_k^r(\varphi) := \begin{cases} \sqrt{k!} \cdot \prod_{x \in \varphi} f^r(x), & \varphi \in M(G_r), \ |\varphi| = k, \\ \\ 0 & \text{elsewhere.} \end{cases}$$

Further, we set $f_0^r(\varphi) := \mathbb{1}_0(\varphi)$. Observe that for each $r$ the sequence $(f_k^r)_{k \geq 0}$ is an orthogonal system in $L_2(M(G_r))$ (being orthonormal if $\|f^r\| = 1$). An especially important class of functions in the Fock space are the so-called *exponential vectors* $\exp\{f^r\}$ defined by

$$\exp\{f^r\} = \sum_{k=0}^{\infty} \frac{1}{\sqrt{k!}} f_k^r. \tag{3}$$

For $f \in L_2(G)$, we define the exponential vector $\exp\{f\}$ by

$$\exp\{f\}(\varphi) := \left( \exp\{f^1\} \otimes \ldots \otimes \exp\{f^n\} \right)(\varphi_1, \ldots, \varphi_n) \quad (\varphi \in M(G)) \tag{4}$$

where $f^r := \mathbb{1}_{G_r} \cdot f$ is the restriction of $f$ to $G_r$ and $\varphi_r = \varphi(\cdot \cap G_r)$ denotes the restriction of the configuration $\varphi$ to points from $G_r$.

The states

$$\omega_g := e^{-\|g\|^2} \cdot \langle \exp\{g\}, \cdot \exp\{g\} \rangle$$

are called *coherent states* on $L_2(M(G))$ if $g \in L_2(G)$ resp. on $L_2(M(G_r))$ if $g \in L_2(G_r)$. Hereby, $\langle \cdot, \cdot \rangle$ denotes the scalar product in the corresponding Hilbert space. Roughly speaking, coherent states describe states of systems of quantum particles where each particle is in the same one-particle state. Furthermore, $\omega_0 = \langle \exp\{0\}, \cdot \exp\{0\} \rangle$ is called the *vacuum state*.

Observe that $\langle \exp\{f\}, \exp\{g\}\rangle = e^{<f,g>}$ what implies $\|\exp\{af\}\|^2 = e^{|a|^2 \cdot \|f\|^2}$ for $a \in \mathbb{C}$, $f \in L_2(G)$. We denote by $\mathcal{N}$ the number operator:

$$\mathcal{N}\Psi(\varphi) := |\varphi| \cdot \Psi(\varphi) \qquad (\Psi \in L_2(M(G)),\ \varphi \in M(G)) \tag{5}$$

For a coherent state $\omega_g$ the value $\omega_g(\mathcal{N})$ is the expectation of the number of excited neurons supporting the signal $\exp\{g\}$. It is easy to prove that $\omega_g(\mathcal{N}) = \|g\|^2$. So we get for all $g \in L_2(G)$, $\|g\| = 1$

$$\omega_{a \cdot g}(\mathcal{N}) = |a|^2 \qquad (a \in \mathbb{C}). \tag{6}$$

Consequently, a large value $|a|$ indicates a high expectation of the number of excited particles. This will be important for our considerations of the high density limit in Section 6.

## 3. Time Evolution of Signals

We identify signals with pure states on $L_2(M(G))$, i. e. signals are represented by wave functions $\Psi \in L_2(M(G))$, $\|\Psi\| = 1$. We want to describe the time evolution $U_t\Psi$ of the signal $\Psi$ according to the unitary operators $U_t$ on $L_2(M(G))$. It is well-known (cf.[7]) that the brain acts parallel on the different regions $G_1, \ldots, , G_n$ and that regions can change their size in the long run. From this we may conclude that the unitary operators $U_t$, $t \geq 0$ on $L_2(M(G))$ should be second quantisations of unitary operators $\nu_t$ on $L_2(G)$ corresponding to a certain Hamilton operator $B$ on $L_2(G)$, i. e. $U_t = \Gamma(\nu_t) = e^{itd\Gamma(B)}$. In accordance with the decomposition of the Fock space the operator $B$ should be the orthogonal sum of Hamilton operators $B^{G_s}$ on $L_2(G_s)$, $s \in \{1, \ldots, n\}$. Consequently, we have the representation

$$U_t = \Gamma(\nu_t) = e^{itd\Gamma(B)} = \bigotimes_{s=1}^{n} e^{itd\Gamma(B^{G_s})}. \tag{7}$$

We assume that the decomposition of $G$ is maximal in the sense that for each reasonable $B^{\tilde{G}}$ the region $\tilde{G}$ will be the union of sets from the given decomposition $G_1, \ldots, G_n$.

Now, let us be given a wave function $\Psi \in L_2(M(G))$, $\|\Psi\| = 1$. Since exponential vectors are total in the Fock space we can restrict ourselves to the case

$$\Psi = \sum_k c_k \exp\{f_k\} \qquad (f_k \in L_2(G)). \tag{8}$$

We have $f_k = f_k^1 + \ldots + f_k^n$ where $f_k^s = f_k|_{G_s}$ denotes the restriction of $f_k$ from $G$ to $G_s$. These functions $f_k^s$ can be represented by the sequence

$(\hat{f}_j^s)_j$ of eigenfunctions of the Hamilton operator $B^{G_s}$, $s \in \{1, \ldots, n\}$:

$$f_k^s = \sum_j a_{kj}^s \hat{f}_j^s \qquad (s \in \{1, \ldots, n\}. \qquad (9)$$

So $\Psi$ given by (8) can be written as

$$\Psi = \sum_k c_k \left( \bigotimes_{s=1}^n \exp\{\sum_j a_{kj}^s \hat{f}_j^s\} \right). \qquad (10)$$

It is well-known that there are *"typical frequencies"* related to different regions [13]. To make it precise and also to simplify our expressions we make the additional assumption that in each region $G_s$ only one of the eigenfunctions $\hat{f}_j^s$ of $B^{G_s}$ occurs. We will denote this eigenfunction by $f^s$. In this case (10) simplifies to

$$\Psi = \sum_k c_k \left( \bigotimes_{s=1}^n \exp\{a_k^s f^s\} \right) = \sum_k c_k \exp\{a_k^1 f^1 + \ldots + a_k^n f^n\}. \qquad (11)$$

Denoting by $\alpha_s$ the eigenvalue corresponding to $f^s$, i. e. $B^{G_s} f^s = \alpha_s f^s$, $s \in \{1, \ldots, n\}$ for the representation of $U_t \Psi$ we finally get the expression

$$U_t \Psi = \sum_k c_k \left( \bigotimes_{s=1}^n \exp\{a_k^s e^{it\alpha_s} f^s\} \right) \qquad (12)$$
$$= \sum_k c_k \exp\{a_k^1 e^{it\alpha_1} f^1 + \ldots + a_k^n e^{it\alpha_n} f^n\} \qquad (t \geq 0).$$

Throughout the paper the regions $G_s$, the Hamilton operators $B^{G_s}$, the eigenfunctions $f^s$ of $B^{G_s}$ and the corresponding eigenvalues $\alpha_s$, $s \in \{1, \ldots, n\}$ will be fixed.

**Definition 3.1.** $\psi \in L_2(M(G)$, $\|\Psi\| = 1$ is called a *regular state of a signal* if there exists a finite sequences $(c_k)_{k=1}^m$, $(a_k^s)_{k=1}^m$, $s \in \{1, \ldots, n\}$ of complex numbers such that $\Psi$ can be represented in the form (11). The closure of the linear span of $\{\exp\{af^s : a \in \mathbb{C}\}\}$ will be called the *space of signals in the region $G_s$* and we will denote this space by $\mathcal{H}_s^{\mathrm{sig}}$ $s \in \{1, \ldots, n\}$. $\mathcal{H}^{\mathrm{sig}} := \mathcal{H}_1^{\mathrm{sig}} \otimes \ldots \otimes \mathcal{H}_n^{\mathrm{sig}}$ is called the *space of signals*.

Observe that regular states are *pure states* on $\mathcal{H}^{\mathrm{sig}}$. From (12) we conclude that the unitary transformations $U_t$ leave the signal space $\mathcal{H}^{\mathrm{sig}}$ invariant.

## 4. The Process of Recognition

The recognition process is based on a comparison of signals: one signal will be the input signal coming from our senses, the other one is taken from the memory. Both signals are modeled by the Hilbert space $\mathcal{H}^{\mathrm{sig}}$ introduced above. The memory space $\mathcal{H}_{\mathrm{mem}}$ is a further Hilbert space the structure of which we will not discuss in the present paper (cf. [9]. Also we will not describe here the mechanism how the signal is taken out from memory. A detailed discussion and interpretation one can find for instance in [7]. The whole processing procedure will take place step by step on the space

$$\mathcal{H}^{\mathrm{sig}}\otimes\mathcal{H}^{\mathrm{sig}}\otimes\mathcal{H}_{\mathrm{mem}} = \left(\mathcal{H}_1^{\mathrm{sig}}\otimes\ldots\otimes\mathcal{H}_n^{\mathrm{sig}}\right)\otimes\left(\mathcal{H}_1^{\mathrm{sig}}\otimes\ldots\otimes\mathcal{H}_n^{\mathrm{sig}}\right)\otimes\mathcal{H}_{\mathrm{mem}}.$$

The comparison procedure between the two mentioned signals is done with the aid of operators (projections) on $\mathcal{H} := \mathcal{H}^{\mathrm{sig}}\otimes\mathcal{H}^{\mathrm{sig}}$, and we concentrate our considerations to these first two spaces of the tensor product space.

Basic for our considerations will be the *symmetric beam splitter* being a well-known operator in quantum optics describing the splitting of coherent light into two beams.

**Definition 4.1.** Let $r \in \{1,\ldots,n\}$ be fixed. The linear operator

$$\mathcal{V}_r : L_2(M(G_r))\otimes L_2(M(G_r)) \longrightarrow L_2(M(G_r))\otimes L_2(M(G_r))$$

defined for $f,\ g \in L_2(G_r)$ by

$$\mathcal{V}_r(\exp\{f\}\otimes\exp\{g\}) := \exp\left\{\frac{1}{\sqrt{2}}\cdot(f+g)\right\}\otimes\exp\left\{\frac{1}{\sqrt{2}}\cdot(f-g)\right\} \quad (13)$$

we call *symmetric beam splitter* in the region $r$.

Since tensor products of exponential vectors are total in $L_2(M(G_r))\otimes L_2(M(G_r))$ the symmetric beam splitter $\mathcal{V}_r$ is fully characterized by formula (13). Properties of the symmetric beam splitter are described e.g. in [7] or [11]. In the sequel $Pr_\Psi := \langle\Psi,\cdot\rangle\Psi$ denotes the projection onto (the subspace generated by) $\Psi$. For $r \in \{1,\ldots,n\}$ define projections $T_1^r,\ \widehat{T_1^r},\ T_0^r,\ \widehat{T_0^r}$ on $\mathcal{H}_r^{\mathrm{sig}}\otimes\mathcal{H}_r^{\mathrm{sig}}$ by

$$T_1^r := \mathcal{V}_r\left(\mathbb{1}_{\mathcal{H}_r^{\mathrm{sig}}}\otimes Pr_{\exp\{0\}}\right)\mathcal{V}_r, \qquad T_0^r := \mathcal{V}_r^2 = \mathbb{1}_{\mathcal{H}_r^{\mathrm{sig}}}\otimes\mathbb{1}_{\mathcal{H}_r^{\mathrm{sig}}} =: I_r,$$

$$\widehat{T_0^r} := \mathcal{V}_r\left(I_r - \mathbb{1}_{\mathcal{H}_r^{\mathrm{sig}}}\otimes Pr_{\exp\{0\}}\right)\mathcal{V}_r = I_r - T_1^r = I_r - \widehat{T_1^r}$$

Observe that $T_1^r$ has the property

$$T_1^r\left(\exp\{af^r\}\otimes\exp\{bf^r\}\right) = \exp\left\{\frac{1}{2}(a+b)f^r\right\}\otimes\exp\left\{\frac{1}{2}(a+b)f^r\right\}.$$

So, $T_1^r$ corresponds to a projection onto the subspace of $\mathcal{H}_r^{\mathrm{sig}} \otimes \mathcal{H}_r^{\mathrm{sig}}$ with equal first and second factor of the tensor product. $T_1^r$ is interpreted as *recognition in the $r$-th region* whereas $\widehat{T}_0^r = I_r - T_1^r$ indicates that there was *no recognition* in the area $r$.

Now we join together these operators $T_1^r$ and $\widehat{T}_0^r$ of partial recognition resp. "no" recognition in the single areas to operators on the whole tensor product signal space $\mathcal{H} := \mathcal{H}^{\mathrm{sig}} \otimes \mathcal{H}^{\mathrm{sig}}$. We put

$$\Omega := \{0,1\}^n = \{\overline{\varepsilon} = (\varepsilon_1, \ldots, \varepsilon_n) : \ \varepsilon_k \in \{0,1\}, \ k \leq n\}$$

$$T_{(\varepsilon_1, \ldots, \varepsilon_n)} := \bigotimes_{r=1}^{n} T_{\varepsilon_r}^r, \qquad \widehat{T}_{(\varepsilon_1, \ldots, \varepsilon_n)} := \bigotimes_{r=1}^{n} \widehat{T}_{\varepsilon_r}^r, \qquad (\varepsilon_1, \ldots, \varepsilon_n) \in \Omega$$

Because of notational convenience we have put in the above definition $\widehat{T}_1^r := T_1^r$ (however $T_0^r \neq \widehat{T}_0^r$) We want to sketch briefly the process of recognition. A more detailed description one can find in [7]. Starting point is a state $\varrho_0$ on $\mathcal{H}$ representing a pair of signals the first arisen from the senses the second one created from the brain. Now, there is chosen a certain element $\overline{\varepsilon}^1 = (\varepsilon_1^1, \ldots, \varepsilon_n^1) \in \Omega$ indicating what happens in the first step of recognition: $\varepsilon_k^1 = 1$ indicates recognition in the $k$-th region, $\varepsilon_k^1 = 0$ means there is no recognition in the $k$-th region. The corresponding *event* will be identified with the projection $\widehat{T}_{(\varepsilon_1^1, \ldots, \varepsilon_n^1)}$. Like in the case of measurements the probability of the event $(\varepsilon_1^1, \ldots, \varepsilon_n^1)$ is given by $tr\left( \varrho_0 \cdot \widehat{T}_{(\varepsilon_1^1, \ldots, \varepsilon_n^1)} \right)$ where $tr(\cdot)$ denote the usual trace of an operator. Together with the probability $tr(\varrho_0 \cdot \widehat{T}_{\overline{\varepsilon}^1})$ representing the *subjective reduction of the state $\varrho$* caused by the measurement according to $\widehat{T}_{\overline{\varepsilon}^1}$ we can consider the following transformation of the state $\varrho_0$ representing *objective reduction* caused by the self-collapse indicated by $\overline{\varepsilon}^1$:

$$K_{\overline{\varepsilon}^1}(\varrho_0) := \frac{T_{\overline{\varepsilon}^1} \, \varrho_0 \, T_{\overline{\varepsilon}^1}}{tr(\varrho_0 \cdot T_{\overline{\varepsilon}^1})}$$

provided $tr(\varrho_0 \cdot T_{\overline{\varepsilon}^1}) > 0$.

This procedure can be repeated arbitrarily often - given an initial state $\varrho_0$ and a sequence $(\overline{\varepsilon}^k)_{k=1}^{\infty}$ one obtains this way a sequence $(\varrho_k)_{k=0}^{\infty}$ of states on $\mathcal{H}$ and a sequence $(tr(\varrho_{k-1} \cdot \widehat{T}_{\overline{\varepsilon}^k}))_{k=1}^{\infty}$ of probabilities of the events $(\overline{\varepsilon}^k)_{k=1}^{\infty}$ provided these expressions exist (i. e. if $tr(\varrho_{k-1} \cdot T_{\overline{\varepsilon}^k}) > 0$ for $k \geq 1$). A necessary condition for this to hold is that the sequence $(\overline{\varepsilon}^k)_{k=1}^{\infty}$ has to be in some sense increasing. More precisely, we require

$$\varepsilon_r^k \leq \varepsilon_r^{k+1} \qquad (r \in \{1, \ldots, n\}) \tag{14}$$

for all $k \in \mathbb{N}$. The relation (14) defines a semi-ordering of the elements in $\Omega$. We write $\bar{\varepsilon}^k \overset{s}{\leq} \bar{\varepsilon}^{k+1}$ to indicate relation (14). The sequence $(\varrho_k)_{k=0}^{\infty}$ is in accordance with this relation. Especially, we have

$$tr(\varrho_{k-1}\widehat{T}_{\bar{\varepsilon}^k}) = 0 \qquad \text{if} \quad \bar{\varepsilon}^{k-1} \overset{s}{\leq} \bar{\varepsilon}^k \quad \text{does not hold,} \qquad (15)$$

and

$$\varrho_2 = K_{\overline{\varepsilon^2}}(K_{\overline{\varepsilon^1}}(\varrho_0)) = K_{\overline{\varepsilon^2}}(\varrho_0) \qquad \text{if} \quad \bar{\varepsilon}^{k-1} \overset{s}{\leq} \bar{\varepsilon}^k. \qquad (16)$$

Summarizing we state that recognition process is characterised by an *increasing sequence* $(\bar{\varepsilon}^k)_{k=1}^{\infty} \subset \overline{\Omega}$ with corresponding sequence of states

$$\boxed{(m)_{m=1}^{\infty} = \Big(K_m(0)\Big)_{m=1}^{\infty}.}$$

Unfortunately, this sequence $(\bar{\varepsilon}^k)_{k=1}^{\infty}$ is unknown and even seems to be non-computable. Only in the case the sequences $(\bar{\varepsilon}^k)_{k=1}^{\infty}$ would be known one can identify the process of recognition with the sequence $(\varrho_k)_{k=0}^{\infty}$ of states. The way out of this dilemma is to consider the vectors $\bar{\varepsilon}^k$ as random and to consider a classical Markov chain describing this process of recognition. We replace $(\bar{\varepsilon}^k)_{k=1}^{\infty}$ by a sequence of random vectors $(X_k)_{k=1}^{\infty}$. Hereby, $X_k$ has values in $\Omega$ and is interpreted as the random outcome in the $k$-th step of the recognition process. The sequence $(\varrho_k)_{k=0}^{\infty})$ of states we replace by a new sequence $(\beta_k)_{k=0}^{\infty}$ of states on $\mathcal{H}$ given by

$$\beta_k(\cdot) := \sum_{\bar{\varepsilon} \in \Omega} P(X_k = \bar{\varepsilon}) \cdot tr(K_{\bar{\varepsilon}}(\varrho_0)(\cdot)) \qquad (k \geq 1). \qquad (17)$$

Hereby, $P(X_0 = (0,\ldots,0)) = 1$ and $\beta_0 = \varrho_0$ is the initial state on $\mathcal{H}$. Following the above arguments one can show (cf. [7]) that the sequence $(X_k)_{k=0}^{\infty}$ forms a classical homogenous Markov chain with initial distribution $P_{X_0} = \delta_{(0,\ldots,0)}$ and with transition probabilities

$$p(\bar{\varepsilon}^{k-1},\bar{\varepsilon}^k) = tr\left(K_{\bar{\varepsilon}^{k-1}}(\varrho_0)\widehat{T}_{\bar{\varepsilon}^k}\right) = \frac{tr(\varrho_0\widehat{T}_{\bar{\varepsilon}^k})}{tr(\varrho_0 T_{\bar{\varepsilon}^{k-1}})} \qquad (k \geq 0) \qquad (18)$$

where $\bar{\varepsilon}^0 = (0,\ldots,0)$. So the above probabilities are the conditional probabilities of passing in the $k$-th step to $\bar{\varepsilon}^k$ starting from $\bar{\varepsilon}^{k-1}$. More general, we have

$$p(\bar{\varepsilon}^1,\bar{\varepsilon}^2) = \frac{tr(\varrho_0 T_{\bar{\varepsilon}^1}\widehat{T}_{\bar{\varepsilon}^2})}{tr(\varrho_0 T_{\bar{\varepsilon}^1})} \qquad (\bar{\varepsilon}^1 \leq \bar{\varepsilon}^2, \ tr(\varrho_0 T_{\bar{\varepsilon}^1}) > 0). \qquad (19)$$

The $m$-step transition probabilities $p^m$ of the above Markov chain are given by

$$p^m(\overline{\varepsilon}^1, \overline{\varepsilon}^{m+1}) = p(\overline{\varepsilon}^1, \overline{\varepsilon}^{m+1}) \sum_{\overline{\varepsilon}^1 \overset{s}{\leq} \overline{\varepsilon}^2 \overset{s}{\leq} \ldots \overset{s}{\leq} \overline{\varepsilon}^{m+1}} \prod_{k=2}^{m} p(\overline{\varepsilon}^k, \overline{\varepsilon}^k), \tag{20}$$

$$p^m(\overline{\varepsilon}, \overline{\varepsilon}) = (p(\overline{\varepsilon}, \overline{\varepsilon}))^m \qquad (m \geq 1, \ \overline{\varepsilon} \in \Omega).$$

We can describe the states (17) explicitly in terms of the initial state $\varrho$ and the $m$-step transition probabilities $p^m$ of the above Markov chain:

$$\beta_k(A) := \sum_{\overline{\varepsilon} \in \Omega} tr\left(K_{\overline{\varepsilon}}(\varrho)A\right) \cdot p^k(\overline{0}, \overline{\varepsilon}) \qquad (k \geq 1, \ a \in \mathcal{L}(\mathcal{H})). \tag{21}$$

For more details and and a characterisation of this Markov chain we refer to [7].

## 5. EEG-Measurements

EEG measures the electric potential of the set of excited neurons. We consider the general case of $r$ recording electrodes and one reference electrode. First let us consider the case of one signal: Measurements according to the *recording electrodes* correspond to operators of multiplication $O_{g_{u_k}}$, $k \in \{1, \ldots, r\}$ on the space $\mathcal{H}^{\mathrm{sig}}$ of signals, and measurements according to the *reference electrode* correspond to the operator $O_{g_{u_0}}$. Hereby, the functions $u_k$ are real-valued functions defined on $G$, and the functions $g_{u_k} : M(G) \to \mathbb{R}$ are given by

$$g_{u_k}(\varphi) = \int u_k(x)\varphi(dx) = \sum_{x \in \varphi} u_k(x) \qquad (\varphi \in M(G), \ k \in \{0, \ldots, r\}).$$

**Remark 5.1.** Usually it is assumed that $G$ is a convex compact subset of $\mathbb{R}^d$ with $d = 2$ or $d = 3$ equipped with the Lebesgue measure, and the functions $u_k : G \to \mathbb{R}$ are given by

$$u_k(x) := u(x - y_k) \qquad (x \in G)$$

where $u$ is a function on $\mathbb{R}^d$, and $y_k$ represents the position of the $k$-th electrode. Observe that the electrodes are fixed on the scalp. Thus $y_k$ is not an element of G representing the physical body of the brain. Nevertheless we will assume that the functions $u_k$ are continuous and bounded on $G$.

Now, we put

$$\hat{u}_k := u_k - u_0, \qquad u_{\bar{t}} = \sum_{k=1}^{r} t_k \hat{u}_k \qquad (k \in \{1, \ldots, r\}, \ \bar{t} = (t_1, \ldots, t_r) \in \mathbb{R}^r).$$

$$(22)$$

One easily checks the following relations:

$$O_{g_{\hat{u}_k}} = O_{g_{u_k}} - O_{g_{u_0}}, \qquad (k \in \{1, \ldots, r\})$$

$$O_{g_{u_{\bar{t}}}} = \sum_{k=1}^{r} t_k O_{g_{\hat{u}_k}} \qquad (\bar{t} = (t_1, \ldots, t_r) \in \mathbb{R}^r).$$

An EEG-measurement corresponds to a measurement with respect to the *commuting* operators $O_{g_{\hat{u}_k}}$. The *characteristic function* $\gamma_\Psi(t_1, \ldots, t_r)$ of the random vector representing the output of the EEG-measurement in the case of a pure state $\Psi \in \mathcal{H}^{\mathrm{sig}}$ of *one* signal is the quantum mechanical expectation of $e^{iO_{g_{u_{\bar{t}}}}}$, i. e. for $\bar{t} = (t_1, \ldots, t_r) \in \mathbb{R}^r$

$$\gamma_\Psi(t_1, \ldots, t_r) := \left\langle \Psi, e^{i\sum_{k=1}^{r} t_k O_{g_{\hat{u}_k}}} \Psi \right\rangle = \left\langle \Psi, e^{iO_{g_{u_{\bar{t}}}}} \Psi \right\rangle. \qquad (23)$$

Now, we have to consider pairs of signals (one coming from the senses the other one taken from the memory). The pure state of such a pair is given by a normalised element $\Psi = \Psi_1 \otimes \Psi_2 \in \mathcal{H}^{\mathrm{sig}} \otimes \mathcal{H}^{\mathrm{sig}}$. We have to replace the operators $O_{g_{\hat{u}_k}}$ by the operators $A_{\hat{u}_k}$ on $\mathcal{H}^{\mathrm{sig}} \otimes \mathcal{H}^{\mathrm{sig}}$ where

$$A_u := O_{g_u} \otimes \mathbb{1}_{\mathcal{H}^{\mathrm{sig}}} + \mathbb{1}_{\mathcal{H}^{\mathrm{sig}}} \otimes O_{g_u}. \qquad (24)$$

The operators $A_u$ have important commutation properties:

**Proposition 5.1.** *For $\bar{\varepsilon} \in \Omega$ and bounded continuous functions $u : G \to \mathbb{R}$ the operators $T_{\bar{\varepsilon}}$ and $\hat{T}_{\bar{\varepsilon}}$ commute with $A_u$, $e^{A_u}$ and $e^{iA_u}$.*

The characteristic function $\gamma_\Psi(t_1, \ldots, t_r)$ of the random vector representing the output of the EEG-measurement in that case is given for $\bar{t} = (t_1, \ldots, t_r) \in \mathbb{R}^r$ by

$$\gamma_\Psi(t_1, \ldots, t_r) := \left\langle \Psi, e^{i\sum_{k=1}^{r} t_k A_{\hat{u}_k}} \Psi \right\rangle = \left\langle \Psi, e^{iA_{u_{\bar{t}}}} \Psi \right\rangle. \qquad (25)$$

Now, each pair of signals is supported by a certain set of excited neurons represented by a random point field $\Phi$ in $G$. If $\varrho$ denotes the state of such a pair of signals then the *characteristic functional* of the corresponding random point field $\Phi$ is given by

$$\mathbb{E}e^{ig_u(\Phi)} = tr(\varrho \cdot e^{iA_u}) \qquad (u \in L^\infty(G, \mathbb{R}))$$

i.e., the output of the EEG-measurement can be identified with the random vector

$$\overline{Y} = (Y_1, \ldots, Y_r) := (g_{\hat{u}_1}(\Phi), \ldots, g_{\hat{u}_r}(\Phi)). \tag{26}$$

Recognition in a special region $G_s$ of the brain causes a loss of excited neurons supporting the signal. More precisely, the excited neurons supporting the considered pair of signals located *inside* $G_s$ change their states from excited to nonexcited. Then only the remaining excited neurons *outside* $G_s$ will contribute to the measured electric potential. More general, under condition $X_k = \overline{\varepsilon}^k$ the characteristic functional of the system of the remaining excited neurons after the $k$-th step of recognition starting with the initial pair of signals being in the state $\varrho$ is given by

$$\omega_{\Phi|\overline{\varepsilon}^k}(u) = tr\left(K_{\overline{\varepsilon}^k}(\varrho \cdot e^{iA_u})\right).$$

Then in the special case $u = u_{\overline{t}}$ we obtain the characteristic function of the output of EEG-measurements under condition $X_k = \overline{\varepsilon}^k$ :

$$\gamma_{\varrho|\overline{\varepsilon}^k}(\overline{t}) = \omega_{\Phi|\overline{\varepsilon}^k}(u_{\overline{t}}) = tr\left(K_{\overline{\varepsilon}^k}(\varrho \cdot e^{iA_{u_{\overline{t}}}})\right).$$

## 6. High Density Limit in the Case of EEG-Measurements

Using units of time according to the process of recognition there exists a natural number $\tau$ ($\tau \approx 10^5$) such that the $l$-th EEG-measurement will be performed at time $\tau_l := \tau \cdot l \in \mathbb{N}$.

Now the operators $A_{\hat{u}_k}$, $T_{\overline{\varepsilon}}$, $\hat{T}_{\overline{\varepsilon}}$, $k \in \{1, \ldots, r\}, \overline{\varepsilon} \in \Omega$ commute (Proposition 5.1). For that reason we obtain finite sequences of random variables $(X_j)_{j=0}^{\infty}$ and $(\overline{Y}_l)_{l=0}^{\infty}$ related to an initial state $\varrho$ of the process of recognition. Hereby, $(X_j)_{j=0}^{\infty}$ is the Markov chain described by (18) indicating in which of the regions recognition occurs, and $\overline{Y}_l = (y_{l,1}, \ldots, y_{l,r})$, $l \geq 0$ represents the random output of the $l$-th EEG- measurement given by (26). EEG specialists are searching for information about $(X_j)_{j=0}^{\infty}$. But this sequence cannot be measured directly. For that reason several statistical methods are used to get information about the distribution of the $(X_j)_{j=0}^{\infty}$ based on the *observed values* $(\overline{Y}_l)_{l=0}^{\infty}$, i. e. one looks for information about the *conditional distribution* of the $(X_j)_{j=0}^{\infty}$ under condition $\overline{Y}_l = (y_{l,1}, \ldots, y_{l,r})$. Applying Bayes statistics one needs the conditional distribution of $\overline{Y}_l$ given the condition $(X_0, \ldots, X_{\tau l}) = (\overline{\varepsilon}^0, \ldots, \overline{\varepsilon}^{\tau l})$, $l \geq 1$. Now, we set for a function $u : G \to \mathbb{R}$ and $\overline{\varepsilon} \in \Omega$

$$u^{\overline{\varepsilon}} := \varepsilon_1 \mathbb{1}_{G_1} \cdot u + \ldots + \varepsilon_n \mathbb{1}_{G_n} \cdot u \tag{27}$$

where again $\mathbb{I}_B$ denotes the indicator function of a set $B$. Observe that

$$tr\left(K_{\overline{\varepsilon}}(\varrho)e^{iA_u}\right) = tr\left(K_{\overline{\varepsilon}}(\varrho)e^{iA_{u\overline{\varepsilon}}}\right)$$

Taking into account notation (22) we conclude that the characteristic function related to this conditional distribution $\varrho|(\overline{\varepsilon}^{\tau l-1}, \overline{\varepsilon}^{\tau l})$ is given by

$$\gamma_{\varrho|(\overline{\varepsilon}^{\tau l-1},\overline{\varepsilon}^{\tau l})}(\overline{t}) = \frac{tr\left(K_{\overline{\varepsilon}^{\tau l-1}}(\varrho)\hat{T}_{\overline{\varepsilon}^{\tau l}}e^{iA_{u\frac{\overline{\varepsilon}^{\tau l}}{t}}}\right)}{p(\overline{\varepsilon}^{\tau l-1},\overline{\varepsilon}^{\tau l})}. \tag{28}$$

Using (21) and (28) we get for the characteristic function of the distribution of $Y_l$

$$\gamma_{\beta_{\tau l}}(\overline{t}) = \sum_{\overline{\varepsilon}\in\Omega} p^{\tau l}(\overline{0},\overline{\varepsilon})\cdot tr\left(K_{\overline{\varepsilon}}(\varrho)e^{iA_{u\frac{\overline{\varepsilon}^{\tau l}}{t}}}\right) \tag{29}$$

$$= \sum_{\overline{\varepsilon}^1\leq\overline{\varepsilon}^2} p^{\tau l-1}(\overline{0},\overline{\varepsilon}^1)\cdot p(\overline{\varepsilon}^1,\overline{\varepsilon}^2)\cdot\gamma_{\varrho|(\overline{\varepsilon}^1,\overline{\varepsilon}^2)}(\overline{t})$$

At the beginning of the process of recognition the signal arisen from the senses and the signal created by the brain should be independent, i.e. the initial state is a tensor product of two states on $\mathcal{H}^{\mathrm{sig}}$. We will restrict ourselves to the case of regular signals (cf. Definition 3.1), i.e. we consider a pure state $\Psi = \Psi_1\otimes\Psi_2 \in \mathcal{H}^{\mathrm{sig}}\otimes\mathcal{H}^{\mathrm{sig}}$ with $\Psi_1$, $\Psi_2$ given in the form (11):

$$\Psi_1 = \frac{\sum_{k\in J_1} c_k\exp\left\{\sum_{l=1}^n a_k^l f^l\right\}}{\|\sum_{k\in J_1} c_k\exp\left\{\sum_{l=1}^n a_k^l f^l\right\}\|}, \tag{30}$$

$$\Psi_2 = \frac{\sum_{k\in J_2} d_k\exp\left\{\sum_{l=1}^n b_k^l f^l\right\}}{\|\sum_{k\in J_2} d_k\exp\left\{\sum_{l=1}^n b_k^l f^l\right\}\|}.$$

Now, for the expectation of the number of excited neurons supporting the elementary signal $\exp\left\{\sum_{l=1}^n a_k^l f^l\right\}$ we get

$$\left\langle \exp\{\sum_{l=1}^n a_k^l f^l\}, \mathcal{N}\exp\{\sum_{l=1}^n a_k^l f^l\}\right\rangle = \sum_{l=1}^n |a_k^l|^2\cdot\|f^l\|^2 \tag{31}$$

(cf. (6) and take into account that in the representation (11) the functions $f^1,\ldots,f^n$ are orthogonal). We see from (31) that the higher the $|a_k^l|$ the higher is this expectation. Usually that expectation is very high (the brain contains about $10^{11}$ neurons). In order to reflect that we choose large $m \in \mathbb{N}$ and replace $\Psi_1$, $\Psi_2$ by

$$\Psi_1^m = \frac{\hat{\Psi}_1^m}{\|\hat{\Psi}_1^m\|}, \qquad \Psi_2^m = \frac{\hat{\Psi}_2^m}{\|\hat{\Psi}_2^m\|} \tag{32}$$

140

with

$$\hat{\Psi}_1^m = \sum_{k \in J_1} c_k \exp\{\sum_{l=1}^n m \cdot a_k^l f^l\}, \tag{33}$$

$$\hat{\Psi}_2^m = \sum_{k \in J_2} d_k \exp\{\sum_{l=1}^n m \cdot b_k^l f^l\}.$$

On the other hand, the measured electric potential of a single neuron is very low. In order to reflect that we replace the function $u$ by $u/m^2$ , i.e., we consider the operators $A_{u^{\bar{\varepsilon}}/m^2}$ instead of $A_{u^{\bar{\varepsilon}}}$.

Depending on the value $m$ the transition probabilities (19) and characteristic functions (28) transition into

$$p_m(\bar{\varepsilon}^1, \bar{\varepsilon}^2) = \frac{\|\hat{T}_{\bar{\varepsilon}^2} T_{\bar{\varepsilon}^1} \hat{\Psi}_1^m \otimes \hat{\Psi}_2^m\|^2}{\|T_{\bar{\varepsilon}^1} \hat{\Psi}_1^m \otimes \hat{\Psi}_2^m\|^2} \qquad (\bar{\varepsilon}^1, \bar{\varepsilon}^2 \in \Omega, \ \bar{\varepsilon}^1 \leq \bar{\varepsilon}^2) \tag{34}$$

$$\gamma_{\Psi_1^m \otimes \Psi_2^m | (\bar{\varepsilon}^1, \bar{\varepsilon}^2)}(\bar{t}) = \frac{\left\langle \hat{\Psi}_1^m \otimes \hat{\Psi}_2^m, e^{iA_{(u^{\bar{\varepsilon}^2}/m^2)_{\bar{t}}}} \hat{T}_{\bar{\varepsilon}^2} T_{\bar{\varepsilon}^1} \hat{\Psi}_1^m \otimes \hat{\Psi}_2^m \right\rangle}{\|\hat{T}_{\bar{\varepsilon}^2} T_{\bar{\varepsilon}^1} \hat{\Psi}_1^m \otimes \hat{\Psi}_2^m\|^2} \tag{35}$$

$$(\bar{t} = (t_1, \ldots, t_r) \in \mathbb{R}^r, \ \bar{\varepsilon}^1, \bar{\varepsilon}^2 \in \Omega, \ \bar{\varepsilon}^1 \leq \bar{\varepsilon}^2)$$

We are interested in the *high density limit* of the transition probabilities and the states if $m \to \infty$.

**Theorem 6.1.** *Let* $u : G \to \mathbb{R}$ *be a bounded measurable function,* $\bar{\varepsilon} \in \Omega$, $J_1$, $J_2$ *finite index sets,* $(c_k)_{k \in J_1}$, $((a_k^1, \ldots, a_k^n))_{k \in J_1}$ *and* $(d_j)_{j \in J_2}$, $((b_j^1, \ldots, b_j^n))_{j \in J_2}$ *finite sequences of complex numbers resp. complex sequences such that there are no equal vectors in the sequence* $((a_k^1, \ldots, a_k^n))_{k \in J_1}$ *and in* $((b_j^1, \ldots, b_j^n))_{j \in J_2}$. *Then we get*

$$\lim_{m \to \infty} \|T_{\bar{\varepsilon}} \hat{\Psi}_1^m \otimes \hat{\Psi}_2^m\|^2 = \sum_{k \in J_1, j \in J_2} |c_k|^2 |d_j|^2 \prod_{s=1}^n (\delta_{a_k^s, b_j^s} \delta_{1, \varepsilon_s} + \delta_{0, \varepsilon_s})$$

$$\tag{36}$$

$$\lim_{m \to \infty} \|\hat{T}_{\bar{\varepsilon}} \hat{\Psi}_1^m \otimes \hat{\Psi}_2^m\|^2 = \sum_{k \in J_1, j \in J_2} |c_k|^2 |d_j|^2 \prod_{s=1}^n (\delta_{a_k^s, b_j^s} \delta_{1, \varepsilon_s} + (1 - \delta_{a_k^s, b_j^s}) \delta_{0, \varepsilon_s})$$

$$\lim_{m\to\infty} \left\langle \hat{\Psi}_1^m \otimes \hat{\Psi}_2^m, e^{iA_{(u^{\bar{\varepsilon}^2}/m^2)}} \hat{T}_{\bar{\varepsilon}^2} T_{\bar{\varepsilon}^1} \hat{\Psi}_1^m \otimes \hat{\Psi}_2^m \right\rangle = \sum_{k\in J_1, j\in J_2} |c_k|^2 |d_j|^2$$

$$\times \prod_{s=1}^{n} e^{i(|a_k^s|^2 + |b_j^2|^2)} < f^s, f^s u > \left( \delta_{a_k^s, b_j^s} \delta_{1,\varepsilon_s^2} + (1 - \delta_{a_k^s, b_j^s}) \delta_{0,\varepsilon_s^2} \right) \quad (37)$$

Hereby, $\delta_{a,b} = \begin{cases} 1 & \text{if } a = b \\ 0 & \text{if } a \neq b \end{cases}$ denotes the Kronecker delta.

**Remark 6.1.** Using (36) and (34) we get

$$\lim_{m\to\infty} p_m(\bar{\varepsilon}^1, \bar{\varepsilon}^2) = \frac{\sum_{k\in J_1, j\in J_2} |c_k|^2 |d_j|^2 \prod_{s=1}^{n} \left( \delta_{a_k^s, b_j^s} \delta_{1,\varepsilon_s^2} + (1 - \delta_{a_k^s, b_j^s}) \delta_{0,\varepsilon_s^2} \right)}{\sum_{k\in J_1, j\in J_2} |c_k|^2 |d_j|^2 \prod_{s=1}^{n} \left( \delta_{a_k^s, b_j^s} \delta_{1,\varepsilon_s^1} + \delta_{0,\varepsilon_s^1} \right)}$$

**Remark 6.2.** Analogously, using (35), (36) and (37) one obtains the high density limit $\lim_{m\to\infty} \gamma_{\Psi_1^m \otimes \Psi_2^m | (\bar{\varepsilon}^1, \bar{\varepsilon}^2)}(\bar{t})$ of the characteristic function of the output of the EEG-measurements.

## References


1. K.-H. Fichtner and L. Fichtner. Bosons and a quantum model of the brain. Jenaer Schriften zur Mathematik und Informatik Math/Inf/08/05, FSU Jena, Faculty of Mathematics and Informatics, Jena, 2005. 27 pages.
2. K.-H. Fichtner, L. Fichtner, W. Freudenberg, and M. Ohya. On a mathematical model of brain activities. In *Quantum Theory, Reconsideration of Foundations - 4*, volume 962 of *AIP Conference Proceedings*, pages 85 – 90, Melville, New York, 2007. American Institute of Physics.
3. K.-H. Fichtner, L. Fichtner, W. Freudenberg, and M. Ohya. On a quantum model of the recognition process. In L. Accardi, W. Freudenberg, and M.Ohya, editors, *Quantum Bio-Informatics*, volume XXI of *QP–PQ: Quantum Probability and White Noise Analysis*, pages 64 – 84, New Jersey London Singapore, 2008. World Scientifc.
4. K.-H. Fichtner, L. Fichtner, W. Freudenberg, and M. Ohya. On a quantum model of the brain activities. In L. Accardi, W. Freudenberg, and M.Ohya, editors, *Quantum Bio-Informatics III*, volume XXVI of *QP–PQ: Quantum Probability and White Noise Analysis*, pages 81–92, New Jersey London Singapore, 2010. World Scientifc.
5. K.-H. Fichtner, L. Fichtner, W. Freudenberg, and M. Ohya. Quantum models of the recognition process - mathematical prerequisites. Technical report, University Jena, 2010. 44 pages.
6. K.-H. Fichtner, L. Fichtner, W. Freudenberg, and M. Ohya. Quantum models of the recognition process - on a convergence theorem. *Open Systems & Information Dynamics*, 17 (2):161–187, 2010.

7. K.-H. Fichtner, L. Fichtner, W. Freudenberg, and M. Ohya. Self-collapses of quantum systems and brain activities. In *Quantum Bio-Informatics IV*, volume XXVIII of *QP–PQ: Quantum Probability and White Noise Analysis*, pages 101–115, New Jersey London Singapore, 2011. World Scientific.

8. K.-H. Fichtner and W. Freudenberg. Point processes and the position distribution of infinite boson systems. *J. Stat. Phys.*, 47:959—978, 1987.

9. K.-H. Fichtner and W. Freudenberg. The compound Fock space and its application to brain models. In L. Accardi, W. Freudenberg, and M. Ohya, editors, *Quantum Bio-Informatics II*, volume XXIV of *QP–PQ: Quantum Probability and White Noise Analysis*, pages 55 – 67, New Jersey London Singapore, 2009. World Scientifc.

10. K.-H. Fichtner and W. Freudenberg. On a quantum-like model of the recognition process. *AIP Conference Proceedings*, 1327(1):71–79, 2011.

11. M. Gäbler. *Fock Space, Factorisation and beam Splitting: Characterisation and Applications in the Natural Sciences*. PhD thesis, Brandenburg Technical University Cottbus, 2010.

12. Stuart Hameroff and Roger Penrose. Orchestrated objective reduction of quantum coherence in brain microtubules: the "orch or" model for consciousness. *Mathematics and Computer Simulation*, 40:453–480, 1996.

13. R. Hari and O. V. Lounasmaa. Neuromagnetism: tracking the dynamics of the brain. *Physical World*, pages 33–38, May 2000.

14. Wolf Singer. Consciousness and the structure of neuronal representations. *Phil. Trans. R. Soc. Lond.*, B 353:1829–1840, 1998.

15. Wolf Singer. Neuronal synchrony: a versatile code for the definition of relations? *Neuron*, 24:49–65, 1999.

16. Wolf Singer. Sriving for coherence. News and views. *Nature*, 397:391–393, 1999.

17. Wolf Singer. *Der Beobachter im Gehirn. Essays zur Hirnforschung*. Suhrkamp Verlag, Frankfurt a.M., 2002.

18. H. P. Stapp. *Mind, Matter and Quantum Mechanics*. Springer, Berlin Heidelberg, 2nd edition, 2003.

19. H. P. Stapp. A model of the quantum-classical and mind-brain connections, and the role of the quantum zeno effect in the physical implementation of conscious. *arXiv*, 0803.1633v1 (physics.gen-ph), 11 March 2008.

Quantum Bio-Informatics V
© 2013 World Scientific Publishing Co. Pte. Ltd.
pp. 143–157

# INTERNAL NOISE OF EEG-MEASUREMENTS AND CERTAIN BOSON SYSTEMS

KARL-HEINZ FICHTNER

*Friedrich-Schiller-Universitat Jena, School of Mathematics and Computer Science, Institute of Applied Mathematics, 07737 Jena, Germany. E-mail: fichtnerkh@web.de*

LARS FICHTNER

*Friedrich-Schiller-Universitat Jena, School of Social and Behavioral Sciences, Institute of Psychology, 07743 Jena, Germany. E-mail: lars.fichtner@uni-jena.de*

KEI INOUE

*Department of Electrical Engineering, Tokyo University of Science, Yamaguchi, Sanyo-Onoda, Yamaguchi 756-0884, Japan. E-mail: inoue@ed.yama.tus.ac.jp*

MASANORI OHYA

*Department of Information Sciences, Tokyo University of Science, Noda City, Chiba 278-8510, Japan. E-mail: ohya@rs.noda.tus.ac.jp*

Based on classical models of brain activities it seems to be difficult to explain the internal noise related to EEG-measurements. In this paper using a quantum model of the recognition process we consider the asymptotic behaviour of that internal noise.

## 1. Introduction

Specialists in modern brain research are convinced that signals in the brain should be coded by populations of excited neurons [1,17,21,22]. Considering models based on classical probability theory, states of signals should be identified with probability distributions of certain random point fields located inside the volume $G$ of the brain. As it was pointed out in [21,22] the space $G$ representing the whole volume of the brain can be divided into disjoint regions $G_1, \ldots, G_r$ responsible for different aspects of the signals. Then objects that the specialists observe are represented by a random vector $[\beta_1, \ldots, \beta_r]$ with $\beta_k = \xi_k^{\mathrm{ex}} + \xi_k^{\mathrm{sig}} + \xi_k^{\mathrm{mem}}$, $k = 1, \ldots, r$. What they want to have are just the second terms $\xi_k^{\mathrm{sig}}$ corresponding to the real signals. Concerning the remaining terms, the first terms $\xi_k^{\mathrm{ex}}$ are called "external noise" which are represented by the output of MEG-measurement, and the third terms $\xi_k^{\mathrm{mem}}$ are called "internal noise" which are represented by the output of EEG-measurement [19]. In general the external noises go to 0 as MEG-devices get improved. However the internal noises occur regardless of the performance of EEG-devices. Therefore one needs knowledge concerning the distribution of $\beta_k$, especially $\xi_k^{\mathrm{mem}}$. However, the specialists in EEG mapping have no idea how to explain the observed properties of the internal noise. Furthermore, researchers as for instance W. Singer [22] and H.P. Staap [23] expressed doubts whether classical models could give satisfying answers to fundamental problems of modern brain research, and plead for using quantum models.

As more general models than the classical models, the quantum models of the signals were studied in [2,3,7,8,9,10,11,6,18]. Now, [2] represents a first attempt to explain the process of recognition in terms of quantum statistics. Here (pure) states of signals are described by complex functions $\Psi$ of point configurations inside of $G$ where $|\Psi|^2$ is the probability density of a random point field. Thus we obtain again a probability distribution of a random point field called the position distribution corresponding to the quantum state. On the other hand, the probability of each random point field can be identified with the position distribution corresponding to a certain quantum state. In that sense the quantum models are more general than classical models, i.e. the use of quantum theory gives rise to a more detailed description of reality.

In this paper using the quantum models we consider the asymptotic behaviour of the internal noise caused by memory. The complete information (knowledge/expected view of the world) is represented by a certain signal which is characteristic for the individual at a fixed time. The memory consists of a large set of "copies" of that signal. Further, recognition is based on choosing one signal from the memory. That should not affect essentially the state of the memory. Taking into account the description of choosing a signal from the memory (cf. [2,3]) that motivates the choice of the state of the memory as a coherent state (cf. [11]).

Further details are discussed in [12].

## 2. Basic notions and notations

Concerning notions introduced in 2.1 and 2.2 we refer to [4,5], and also see [13,8,10].

### 2.1. *The Symmetric Fock Space*

Let $G$ be an arbitrary complete separable metric space and $\mathfrak{G}$ its $\sigma$-algebra of Borel sets. Further, let $\mu$ be a fixed finite diffuse measure on $[G, \mathfrak{G}]$, i.e. $\mu(G) < \infty$ and $\mu(\{x\}) = 0$ for all singletons $x \in G$.

Concerning brain models discussed in [2,3,7,8,9,10,11,6,18], $G$ represents the physical body of the brain or the surface of the brain, i.e. $G$ represents the space where recognition of signal and more general processing take place. Especially we are concerned with the case where $G$ is a compact subset of $\mathbf{R}^d$ ($d = 2, 3$) and $\mu$ is the $d$-dimensional Lebesgue measure restricted to $G$.

We denote by $\delta_x$ the Dirac measure in the point $x \in G$. Observe that a point $x$ may be equivalently described by the Dirac measure $\delta_x$. Analagously, a *symmetric* configuration $(x_1, \ldots, x_n)$ of $n$ points from $G$ is fully characterized by the *counting measure* $\varphi = \delta_{x_1} + \cdots + \delta_{x_n}$.

For each $n \in \mathbf{N}$ we denote by $M_n(G)$ the set of all symmetric $n$-particle configurations from $G$, i.e.

$$M_n(G) = \left\{\varphi = \delta_{x_1} + \cdots + \delta_{x_n} : x_1, \ldots, x_n \in G\right\}.$$

$\mathfrak{o}$ describes the empty configuration $\mathfrak{o}(G) = 0$, and we set

$$M(G) = \bigcup_{n=0}^{\infty} M_n(G) \tag{2.1}$$

with $M_0(G) = \{\mathfrak{o}\}$. The elements of $M(G)$ can be interpreted as finite (symmetric) configurations in $G$.

We equip $M(G)$ with its canonical $\sigma$-algebra $\mathfrak{M}(G)$, i.e. the smallest $\sigma$-algebra containing all sets of the form $\{\varphi \in M(G) : \varphi(K) = n\}$, $K \in \mathfrak{G}$, $n \in \mathbf{N}$. Observe that $\varphi(K) = n$ means that the configuration $\varphi$ has exactly $n$ points in the subset $K$ of the phase space $G$. On $[M(G), \mathfrak{M}(G)]$ we introduce a measure $F_\mu$ by setting

$$F_\mu(Y) := \mathbf{1}_Y(\mathfrak{o}) + \sum_{n \geq 1} \frac{1}{n!} \int_{G^n} \mathbf{1}_Y \left(\sum_{j=1}^{n} \delta_{x_j}\right) \mu^n(d[x_1, \ldots, x_n]), \quad Y \in \mathfrak{M}(G).$$

$$\tag{2.2}$$

Hereby, $\mathbf{1}_Y$ denotes the indicator function of a set $Y$, and $\mu^n = \otimes_{i=1}^{n} \mu$ denotes the $n$-fold product measure of $\mu$. Observe that $F_\mu$ restricted to $n$-

particle configurations is (up to symmetrization) just the product measure $\mu^n$. Further, $F_\mu$ is a finite measure. Indeed,

$$F_\mu(M(G)) = 1 + \sum_{n \geq 1} \frac{\mu^n(G^n)}{n!} = \sum_{n \geq 0} \frac{(\mu(G))^n}{n!} = e^{\mu(G)} < \infty.$$

Let us still mention that $F_\mu$ is concentrated on the set

$$M(G)_s := \{\varphi \in M(G) : \varphi(\{x\}) \leq 1 \text{ for all } x \in G\} \tag{2.3}$$

of so-called simple counting measures (i.e. without multiple points). This follows from the assumption that $\mu$ is diffuse. We denote by $L^2(M(G)) = L^2(M(G), \mathfrak{M}(G), F_\mu)$ the space of square integrable complex-valued functions on $G$, i.e.

$$L^2(M(G)) = \left\{ \Psi : M(G) \to \mathbf{C} : \int_{M(G)} |\Psi(\varphi)|^2 F_\mu(d\varphi) < \infty \right\}. \tag{2.4}$$

The scalar product in $L^2(M(G))$ is given by

$$\langle \Psi, \Phi \rangle := \int_{M(G)} \overline{\Psi(\varphi)} \cdot \Phi(\varphi) F_\mu(d\varphi),$$

where $\overline{z}$ denotes the complex conjugate of $z \in \mathbf{C}$.

**DEFINITION 1** *The space $L^2(M(G))$ is called the symmetric Fock space over $G$.*

Usually one defines the symmetric Fock space $\Gamma(\mathcal{H})$ over a Hilbert space $\mathcal{H}$ as the direct sum of the symmetrized tensor products $\mathcal{H}^{\otimes n}_{\text{symm}}$ of the underlying Hilbert space $\mathcal{H}$, i.e.

$$\Gamma(\mathcal{H}) = \bigoplus_{n=0}^{\infty} \frac{1}{n!} \mathcal{H}^{\otimes n}_{\text{symm}}.$$

In our case the basic Hilbert space $\mathcal{H}$ (the space of a single quantum particle) will be the space

$$L^2(G) = L^2(G, \mathfrak{G}, \mu) = \left\{ g : G \to \mathbf{C} : \int_G |g(x)|^2 \mu(dx) < \infty \right\}.$$

It was shown in [4] that $\Gamma(L^2(G))$ and $L^2(M(G))$ are isomorphic. So in the sequel we will identify both spaces

$$\Gamma(L^2(G)) = L^2(M(G), \mathfrak{M}(G), F_\mu) =: L^2(M(G)). \tag{2.5}$$

The above formula (2.5) shows that the Fock space $\Gamma(L^2(G))$ is again a $L^2$-space as a special Hilbert space. Later we will deal with Fock spaces over subspaces $B$ of $G$. To do this one just has to replace in the definitions above $[G, \mathfrak{G}, \mu]$ by $[B, \mathfrak{B}, \mu_{|B}]$ where $\mu_{|B}$ denoted the restriction of $\mu$ to $\mathfrak{B} = \mathfrak{G} \cap B$.

We want to sketch briefly a few basic notions from quantum theory which we need in the sequel. A single quantum particle in $G$ is described by the Hilbert space $L^2(G)$. The pure state of a particle is given by a function $g \in L^2(G)$ with $\|g\| = 1$ (called a *wave function*). A measurement $A$ is a (self-adjoint) linear operator on $L^2(G)$ and the quantum- mechanical expectation of the measurement $A$ in the state $g$ is given by the scalar product $\langle g, Ag \rangle$. Because of

$$1 = \|g\|^2 = \langle g, g \rangle = \int_G |g(x)|^2 \mu(dx),$$

$|g|^2$ is the density of a probability measure which is called the position distribution of the state $g$. This probability measure $P_g$ with

$$P_g(B) = \int_B |g(x)|^2 \mu(dx), \quad B \in \mathfrak{G}$$

describes the random position $\xi$ of the particle in the state $g$. For example, in the case $G \subset \mathbf{R}$, the quantum expectation of the position operator $P$, i.e.

$$Pf(x) = xf(x), \quad f \in L^2(G), \quad x \in G,$$

in the state $g$ is given by

$$\langle g, Pg \rangle = \int_G x|g(x)|^2 \mu(dx) = \mathrm{E}\xi.$$

So we obtain the classical expectation value of the random position $\xi$.

Analogously we can proceed in the symmetric Fock space $L^2(M(G))$ with respect to $F_\mu$ which describes random finite *systems of particles* localized in $G$. Pure states are given by wave functions $\Psi \in L^2(M(G))$ with $\|\Psi\| = 1$. The quantum-mechanical expectation of a measurement $A$ (being now a linear operator on $L^2(M(G))$ in the state $\Psi$ is given by the scalar product

$$\langle \Psi, A\Psi \rangle = \int_{M(G)} \overline{\Psi(\varphi)} \cdot A\Psi(\varphi) F_\mu(d\varphi).$$

Again $\|\Psi\|^2$ is the density of a probability measure $P_\Psi$ on $M(G)$, and we have

$$P_\Psi(Y) = \int_Y |\Psi(\varphi)|^2 F_\mu(d\varphi), \quad Y \in \mathfrak{M}(G). \tag{2.6}$$

**DEFINITION 2** *The probability measure $P_\Psi$ on $[M(G), \mathfrak{M}(G)]$ given by (2.6) is called position distribution of the quantum system in the state $\Psi$.*

We denote by $N$ the number operator on $L^2(M(G))$, i.e.

$$N\Psi(\varphi) = |\varphi| \cdot \Psi(\varphi), \quad \Psi \in L^2(M(G)), \quad \varphi \in M(G) \tag{2.7}$$

where $|\varphi| := \varphi(G)$ is the number of particles in the configuration $\varphi$. For each pure state $\Psi \in L^2(M(G))$ we obtain

$$\langle \Psi, N\Psi \rangle = \int_{M(G)} |\varphi| \cdot |\Psi(\varphi)|^2 F_\mu(d\varphi). \tag{2.8}$$

This is the quantum expectation of the number of particles in the state $\Psi$. We see that this value is the same as the classical expectation of the (random) number of particles distributed according to $P_\Psi$.

Further we define a finite measure $\Lambda_{P_\Psi}$ on $[G, \mathfrak{G}]$ by

$$\Lambda_{P_\Psi}(B) = \int_{M(G)} \varphi(B) P_\Psi(d\varphi), \quad B \subseteq G. \tag{2.9}$$

The measure $\Lambda_{P_\Psi}$ is called *intensity measure of the position distribution* $P_\Psi$.

We denote by $N_B$ the number operator corresponding to a subset $B$ of $G$ on $L^2(M(G))$, i.e.

$$N_B\Psi(\varphi) = \varphi(B) \cdot \Psi(\varphi), \quad \Psi \in L^2(M(G)), \quad \varphi \in M(G). \tag{2.10}$$

Measurements corresponding to the self-adjoint operator $N_B$ represent counting of the number of the particles inside the subset $B$ of $G$.

For each pure state $\Psi \in L^2(M(G))$ we obtain

$$\langle \Psi, N_B\Psi \rangle = \int_{M(G)} \varphi(B) \cdot |\Psi(\varphi)|^2 F_\mu(d\varphi) = \Lambda_{P_\Psi}(B). \tag{2.11}$$

Now, following the papers [2,3,7,8,9,10,11,6,18], we interpret the symmetric Fock space $\Gamma(L^2(G)) = L^2(M(G))$ as the *space of signals*. A pure state of such a signal is given by a wave function in this space, i.e. by a function $\Psi \in L^2(M(G))$ with $\|\Psi\| = 1$.

The corresponding position distribution represents the probability $\rho$ of the configuration of the excited neurons supporting the signals.

Following the papers [2,3,7,8,9,10,11,6,18], we assume the space $G$ where recognition and processing take place can be divided into disjoint regions responsible for different aspects of the signal, i.e. we consider a measurable decomposition $G_1, \ldots, G_n$ of $G$. For each $r \in \{1, \ldots, n\}$ the space

$$L^2(M(G_r)) = L^2(M(G_r), \mathfrak{M}(G_r), F_{\mu_{|G_r}}) = \Gamma(L^2(G_r))$$

can be identified with the subspace of functions from $L^2(M(G))$ being concentrated on $M(G_r)$, i.e.

$$L^2(M(G_r)) = \left\{ \Psi \in L^2(M(G)) : \Psi(\varphi) = 0 \text{ for } \varphi \notin M(G_r) \right\}.$$

Then $\varphi_{|B} \in M(G)$ represents the restriction of a configuration $\varphi \in M(G)$ to points inside the region $B \subseteq G$, i.e. we have

$$\varphi_{|B} = \varphi(B \cap (\cdot)) = \sum_{x \in B : \varphi(\{x\}) > 0} \delta_x.$$

Observe that

$$\varphi = \varphi_{|G_1} + \cdots + \varphi_{|G_n}, \quad \varphi \in M(G),$$

and

$$M(G) = \left\{ \varphi_1 + \cdots + \varphi_n : \varphi_k \in M(G_k), \ k \in \{1, \ldots, n\} \right\}.$$

This allows to represent each $\Psi \in L^2(M(G))$ as a function $\widetilde{\Psi} \in L^2(M(G_1) \times \cdots \times M(G_n))$ satisfying

$$\Psi(\varphi) = \widetilde{\Psi}(\varphi_{|G_1}, \cdots, \varphi_{|G_n}), \quad \varphi \in M(G). \tag{2.12}$$

This justifies the identification of the spaces:

$$L^2(M(G)) = L^2(M(G_1) \times \cdots \times M(G_n)) = L^2(M(G_1)) \otimes \cdots \otimes L^2(M(G_n)). \tag{2.13}$$

## 2.2. *Exponential Vectors*

An especially important class of functions in $L^2(M(G))$ are the so-called *exponential vectors*.

For measurable function $g : G \to \mathbf{C}$ we define $\exp\{g\} : M(G) \to \mathbf{C}$ by setting

$$\exp\{g\}(\varphi) = \begin{cases} \displaystyle\prod_{x \in \varphi} g(x) & \text{if } \varphi \neq \mathfrak{o}, \ \varphi \in M(G), \\ 1 & \text{if } \varphi = \mathfrak{o}. \end{cases} \tag{2.14}$$

The function $\exp\{g\}$ is called *exponential vector* corresponding to $g$. Hereby, $\mathfrak{o}$ again denotes the empty configuration ($\mathfrak{o}(G) = 0$), and $x \in \varphi$ stands for $\varphi(\{x\}) > 0$. So, the first line of (2.14) equivalently can be written as

$$\exp\{g\}(\delta_{x_1} + \cdots + \delta_{x_n}) = g(x_1) \cdots g(x_n), \quad x_1, \ldots, x_n \in G, \ n \in \mathbf{N}.$$

We make use of the following well-known properties of exponential vectors:

**LEMMA 1** *Let $f$ and $g$ be functions from $G$ to $\mathbf{C}$ and $\varphi, \varphi_1, \varphi_2$ be elements from $M(G)$. Then we have*

$$\exp\{f\}(\varphi_1 + \varphi_2) = \exp\{f\}(\varphi_1) \cdot \exp\{f\}(\varphi_2), \qquad (2.15)$$

$$\exp\{f + g\}(\varphi) = \sum_{\hat{\varphi} \subseteq \varphi} \exp\{f\}(\hat{\varphi}) \cdot \exp\{g\}(\varphi - \hat{\varphi}), \quad (2.16)$$

$$\exp\{f \cdot g\}(\varphi) = \exp\{f\}(\varphi) \cdot \exp\{g\}(\varphi), \qquad (2.17)$$

$$\| \exp\{g\} \|^2_{L^2(M(G))} = e^{\|g\|^2_{L^2(G)}}, \quad g \in L^2(G), \qquad (2.18)$$

$$\langle \exp\{f\}, \exp\{g\} \rangle_{L^2(M(G))} = e^{\langle f,g \rangle_{L^2(G)}}, \quad f, g \in L^2(G). \qquad (2.19)$$

The symbol $\hat{\varphi} \subseteq \varphi$ means that $\hat{\varphi}$ is a subconfiguration of $\varphi$, i.e. $\varphi - \hat{\varphi} \in M(G)$. Observe that $\exp\{g\} \in L^2(M(G))$ if and only if $g \in L^2(G)$. The importance of exponential vectors results mainly from the fact that these vectors are total in $L^2(M(G))$.

Exponential vectors can be decomposed into a tensor product of exponential vectors from $L^2(M(G_r))$. Indeed, from Lemma 1, formula (2.15) we obtain immediately

$$\exp\{g\} = \exp\{g_{|G_1}\} \otimes \cdots \otimes \exp\{g_{|G_n}\}. \qquad (2.20)$$

If $\psi \in L^2(M), \psi \neq \mathfrak{o}$ we can normalize $\psi$ putting

$$|\psi\rangle := \frac{1}{\|\psi\|} \psi.$$

One can check that it holds $\| \exp\{f\} \|^2 = e^{\|f\|^2}$ in [5]. For that reason we obtain the normalized element

$$|\exp\{f\}\rangle = \exp\{f\} e^{-\frac{1}{2}\|f\|^2}.$$

The corresponding state is called *the coherent state related to* $f \in L^2(G)$.

## 2.3. *The Memory Space*

Following the papers [2,3,7,8,9,10,11,6,18], we describe the memory using the Hilbert space $L^2(L^2(M^+(G)))$ where $M^+(G)$ is equal to $M(G) \setminus \{\mathfrak{o}\}$, i.e. the memory contains the information concerning sets of signals. An element $\Psi$ of $L^2(L^2(M^+(G)))$ is a function of point configurations $\Phi$ of point configurations $\varphi \in M(G)$, i.e. we have representations

$$\Phi = \sum_{\varphi \in \Phi} \delta_\varphi = \sum_{\varphi \in \Phi} \delta_{\sum_{x \in \varphi} \delta_x}. \qquad (2.21)$$

The space $L^2(L^2(M^+(G)))$ is called *compound Fock space over G*. In an analogous way to (2.14) we can define an exponential vectorEXP$\{\psi\}\in L^2(L^2(M^+(G)))$ $(\psi\in L^2(M^+(G))$.

The complete information (knowledge/expected view of the world) is represented by a certain signal which is characteristic for the individual (at fixed time-learning can change that). The memory consists of a large set of "copies" of that signal. Further, recognition is based on choosing one signal from the memory. That should not affect essentially the state of the memory. Taking into account the description of choosing a signal from the memory (cf. [2,3] ) that motivates the choice of the state of the memory as a coherent state (cf. [11])

### 2.4. *EEG-Measurements*

Using EEG-device one measures the electric potential of configurations of the excited neurons. Let $x$ and $y$ be the position of an excited neuron and the position of an electrode placed on the scalp, respectively. Then the measured potential of the neuron is a certain function $\tilde{u}(x-y)$ of the distance of these positions.

Now let $\varphi$ be the point system representing the positions of a set of excited neurons. For any function $u$ on $G$ we define a function $g_u$ on $M(G)$ putting

$$g_u(\varphi) = \sum_{x\in\varphi} u(x) = \int \varphi(dx)u(x).$$

Then the potential of $\varphi$ is given as $g_{\tilde{u}}(\varphi)$.

Now, the apparatus consists of $r+1$ electrodes. Let $(y_k)_{k=0}^r$ be the positions of $r+1$ electrodes where usually $r=2^m, m\in\mathbf{N}$. Putting

$$u_k(x) := \tilde{u}(x-y_k), \quad k=0,\ldots,r, \tag{2.22}$$

we get the measured potentials of a configuration of excited neurons related to the different electrodes:

$$g_{u_k}(\varphi) := \int \varphi(dx)u_k(x), \quad k=0,\ldots,r \tag{2.23}$$

**REMARK 1** *Usually specialists assume that G is a convex compact subset of $\mathbf{R}^d$ equipped with Lebesgue measure. Observe that the electrodes are fixed on the scalp. Thus $y_k$ is not an element of G representing the physical body of the brain. Nevertheless in the following we can assume that the functions $u_k$ are continuous (consequently bounded).*

Now we put

$$\hat{u}_k := u_k - u_0, \quad k = 1, \dots, r. \tag{2.24}$$

Using the apparatus the outcomes of the EEG-measurement are these differences of potentials

$$g_{\hat{u}_k}(\varphi) = g_{u_k}(\varphi) - g_{u_0}(\varphi), \quad k = 1, \dots, r. \tag{2.25}$$

Then the quantum measurement of the potential of the signal according to the $k$-th electrode is represented by the operator $O_{g_{u_k}}$ of multiplication corresponding to the function $g_{u_k}$ [2,9]. Consequently, the output of the quantum EEG-measurement corresponds to the differences of the operators

$$O_{\hat{g}_{u_k}} = O_{g_{u_k}} - O_{g_{u_0}}, \quad k = 1, \dots, r. \tag{2.26}$$

Now, for any function $u$ on $G$ we define a function $h_u$ on $M(M^+(G))$ putting

$$h_u(\Phi) := \int \Phi(d\varphi) g_u(\varphi) = \int \Phi(d\varphi) \int \varphi(dx) u(x), \quad \Phi \in M(M^+(G)). \tag{2.27}$$

Each signal is related to the set of excited neurons. The union of all these sets represents the set of excited neurons supporting the memory. For that reason, the outcomes of EEG measurements of the electric potentials of the excited neurons supporting the memory are given by $h_{\hat{u}_k}$ $(k = 1, \dots, r)$.

## 3. Main Results

We consider the asymptotic behaviour of the internal noise caused by the memory.

We consider the following pure state of the memory.

$$\Psi := |\mathrm{EXP}\{c\psi\}\rangle \tag{3.1}$$

where

$$\psi = \sum_{j \in J} c_j |\exp\{af_j\} - \exp\{0\}\rangle, \tag{3.2}$$

$0 \neq a, c \in \mathbf{C}, J \subseteq \mathbf{N}, (f_j)_{j \in J} \subseteq L^2(G)$ is an ONS, and $(c_j)_{j \in J}$ is a (finite) family of complex numbers with

$$\sum_{j \in J} |c_j|^2 = 1. \tag{3.3}$$

**REMARK 2** *From the assumption that $(f_j)_{j \in J} \subseteq L^2(G)$ is an ONS, it follows that*

$$(|\exp\{af_j\} - \exp\{0\}\rangle)_{j \in J}$$

*is an ONS (contained in $\Gamma(L^2(G))$) and*

$$\|\psi\| = 1. \tag{3.4}$$

Furthermore we assume

$$\langle f_j \,|\, u_k f_j \rangle = \langle f_j \,|\, u_0 f_j \rangle, \quad k = 1, \ldots, r, \quad j \in J. \tag{3.5}$$

**EXAMPLE 1** *Let $(G_r)_{r=1}^n$ be a measurable decomposition of $G \subseteq \mathbf{R}^d$. We fix $f^r \in L^2(G_r)$ such that $\sum_{r=1}^n |f^r|^2 \equiv 1$ on $G$ and define*

$$f_j := |G|^{-\frac{1}{2}} \sum_{r=1}^n a_j^r f^r \tag{3.6}$$

*where $a_j = [a_j^1, \ldots, a_j^n] \in \mathbf{C}$, $j \in J$ with*

$$|a_j^s| = 1, \quad s = 1, \ldots, n, \quad j \in J,$$

*and*

$$\sum_{s=1}^n \bar{a}_{j_1}^s a_{j_2}^s \|f^s\|^2 = 0, \quad j_1 \neq j_2.$$

**REMARK 3** *Let $(G_r)_{r=1}^n$ be a "maximal" measurable decomposition of $G \subseteq \mathbf{R}^d$ in the sense that the different regions are responsible for different tasks. It can be motivated by the postulates [3,9] that related to each of the regions $G_r$ we consider the motion of a noninteracting quantum mechanical particle systems("free motion") related to von Neumann boundary condition, i.e. the Hamiltonian operator is the second quantization of the corresponding one-particle operator $H_r$. The total Hamiltonian operator is given by the tensor product of these local Hamiltonian operators being again a second quantization.*

*It is often technical convenient when the regions are parallelepipeds to work with periodic boundary conditions. In order to avoid complicated formula in that case one can specialize $d = 1$, and $G_r := [z_r, z_r + L_r]$. Further*

154

*we put $\hbar = 1$. Then we have the following sequence of eigen functions of $H_r$ $(r = 1, \ldots, n)$*

$$f_{r,m}(x) := e^{i(x-z_r)2\pi m L_r^{-1}}, \quad m \geq 0, \quad x \in G_r \tag{3.7}$$

*related to the eigen values $(2\pi m L_r^{-1})^2$.*

*In general one should use a coherent function corresponding to a general superposition of these functions. But it is wellknown that the results of measurement of frequencies are non-random (up to external noise depending on the device). For that reason one should consider a coherent state related to a fixed eigen function $f_{r,m_r}$ in order to describe parts of the signal corresponding to the r-th region. Then inside of the regions the density of the excited neurons would be constant (we have $|f_{r,m}(x)| \equiv 1$). Further, it is wellknown that the density of excited neurons supporting the signal in the memory is constant on $G$, i.e. the size of the regions determines the expectation of the number of excited neurons. Thus the densities do not depend on the regions. Summarizing we obtain the representation considered in Example 1 putting*

$$f^r := f_{r,m_r}, \quad r = 1, \ldots, n \tag{3.8}$$

Our brain contains about 100 billions of neurons. Thus in reality the number of excited neurons supporting one signal and the number of signals stored in the memory are very large, i.e. the densities related to $|a|^2$ and $|c|^2$ are very high. On the other hand the values of the functions $u_s$ related to the electric potential of one excited neuron are very low. In order to reflect that we put

$$a := bm^{\frac{1}{4}}, \quad c := dm^{\frac{1}{4}}, \tag{3.9}$$

$$u_s := \frac{U_s}{\sqrt{m}}, \quad s = 0, \ldots, r. \tag{3.10}$$

(3.9) implies

$$|a|^2 = |b|^2 \sqrt{m}, \quad |c|^2 = |d|^2 \sqrt{m}. \tag{3.11}$$

Using (3.10) and (2.24) we get the representations

$$\hat{u}_k = \frac{\hat{U}_k}{\sqrt{m}}, \quad k = 1, \ldots, r \tag{3.12}$$

where

$$\hat{U}_k := U_k - U_0, \quad k = 1, \ldots, r. \tag{3.13}$$

Now, from (3.9) it follows that the density of the excited neurons supporting the memory is increasing of the order $m$. Taking into account (3.10) one can see that measuring the potential of all of the excited neurons supporting the signals stored in the memory by one electrode gives expected values increasing of the order $\sqrt{m}$ (very large and strongly depending on the individual). On the other hand concerning only one signal the expectation of the measured potential does not depend on $m$. For that reason it makes no sense to use the values related to the single electrode in order to obtain information concerning the process of recognition of signals. For that reason, as it was already done in 2.4, one uses the difference of the values related to the recording electrodes and the reference electrode (sometimes specialists consider pairs of electrodes). Practice shows that these differences give information on the process of recognition disturbed by an (internal) noise caused by the potential of the excited neurons supporting the signals stored in the memory.

One can use our main results (cf. Theorem 1) [12] in order to describe the statistical behaviour of that noise.

The operators $O_{h_{\hat{u}_k}}$ commute. For that reason the output of the EEG-measurement in the case of a pure state $\Psi$ of the memory gives the classical random vectors $\xi = [\xi_1, \ldots, \xi_r]$ with probability distribution $P_m$ depending on the level $m$.

**THEOREM 1** *Let $\eta = [\eta_1, \ldots, \eta_r]$ be a normal distributed with expectation*

$$E\eta_k = 0, \quad k = 1, \ldots, r$$

*and covariance matrix*

$$E\eta_k\eta_s = |d|^2|b|^2 \sum_{j \in J} |c_j|^2 \int |f_j|^2 \hat{U}_k \hat{U}_s d\mu, \quad k, s = 1, \ldots, r.$$

*Then the sequence $(P_m)$ of probability distribution converges weakly to $P_\eta$ of the random vector $\eta$.*

**REMARK 4** *The random vector $\eta$ can be represented by stochastic integration*

$$\eta_k := \int \hat{U}_k dB, \quad k = 1, \ldots, r$$

*with respect to a generalized Brownian motion $B$ with "noise intensity measure" being absolutely continuous with respect to $\mu$ with density*

$$|d|^2|b|^2 \sum_{j \in J} |c_j|^2 |f_j|^2.$$

*We can conclude that concerning EEG-measurement the memory causes a noise which can be asymptotically identified with the mentioned generalized Brownian motion. Considering the case of Remark 3 we have to specialize $|f_j|^2 \equiv |G|^{-1}$, and $\mu$ denotes the Lebesgue measure. In that case the parameters $|d|^2$ and $|b|^2$ depend on the individual.*

## References


1. A. K. Engel and W. Singer, Temporal binding and the neural correlates of sensory awareness, Trends in Cogn. Sci., **5(1)**, 16-25, 2001.
2. K.-H. Fichtner and L. Fichtner, Bosons and a quantum model of the brain, Jenaer Schriften zur Mathematik und Informatik, Math/Inf/08/05, FSU Jena, Faculty of Mathematics and Informatics, Jena ,2005, 27 pages.
3. K.-H. Fichtner and L. Fichtner, Quantum models of brain activities I - Recognition of signals, In J.C. Garcia, R. Quezada, and S.B. Sontz, editors, Quantum probability and Related topics, **XXIII** of QP-PQ: Quantum Probability and White Noise Analysis, 135-144, New Jersey London Singapore, 2008, World Scientific.
4. K.-H. Fichtner and W. Freudenberg, Point processes and the position distribution of infinite boson systems, J. Stat. Phys., **47**, 959-978, 1987.
5. K.-H. Fichtner and W. Freudenberg, Characterization of states of infinite boson systems, Comm. Math. Phys., **137**, 315-357, 1991.
6. K.-H. Fichtner and W. Freudenberg, The compound Fock space and its application to brain models, In L. Accardi, W. Freudenberg, and M. Ohya, editors, Quantum Bio-Informatics II, **XXIV** of QP-PQ: Quantum Probability and White Noise Analysis, 55-67, New Jersey London Singapore, 2009, World Scientific.
7. K.-H. Fichtner, L. Fichtner, W. Freudenberg, and M. Ohya, On a mathematical model of brain activities, In Quantum Theory, Reconsideration of Foundations-4, **962** of AIP Conference Proceedings, 85-90, Melville, New York, 2007. American Institute of Physics.
8. K.-H. Fichtner, L. Fichtner, W. Freudenberg, and M. Ohya, On a quantum model of the recognition process, In L. Accardi, W. Freudenberg, and M. Ohya, editors, Quantum Bio-Informatics, **XXI** of QP-PQ: Quantum Probability and White Noise Analysis, 64-84, New Jersey London Singapore, 2008, World Scientific.
9. K.-H. Fichtner, L. Fichtner, W. Freudenberg, and M. Ohya, On a quantum model of the brain activities, In L. Accardi, W. Freudenberg, and M. Ohya, editors, Quantum Bio-Informatics III, **XXVI** of QP-PQ: Quantum Probability and White Noise Analysis, 81-92, New Jersey London Singapore, 2010, World Scientific.
10. K.-H. Fichtner, L. Fichtner, W. Freudenberg, and M. Ohya, Quantum models of the recognition process – on a convergence theorem, Open Systems and Information Dynamics, **17(2)**, 161-187, 2010.
11. K.-H. Fichtner, L. Fichtner, W. Freudenberg, and M. Ohya, In L. Accardi, W.

Freudenberg, and M. Ohya, editors, to appear in Quantum Bio-Informatics IV, QP-PQ: Quantum Probability and White Noise Analysis, New Jersey London Singapore, 2011, World Scientific.

12. K.-H. Fichtner, L. Fichtner, K. Inoue, and M. Ohya, Internal noise caused by memory, Preprint.

13. K.-H. Fichtner, W. Freudenberg and V. Liebscher, Time evolution and invariance of boson systems given by beam splittings, Infinite Dimensional Analysis, Quantum Probability and Related topics, **1(4)**, 511-531, 1998.

14. K.-H. Fichtner, V. Liebscher and M. Ohya, A limit theorem for conditionally independent beam splittings, In M. Schürmann and U. Franz, editors, Quantum Probability and Infinite Dimensional Analysis, From Foundations to Applications, **XVIII** of QP-PQ: Quantum Probability and White Noise Analysis, 227-236, New Jersey London Singapore, 2005, World Scientific.

15. K.-H. Fichtner and M. Ohya, Quantum teleportation with entangled states given by beam splittings, Comm. Math. Phys., **222**, 229-247, 2001.

16. S. Hameroff and R. Penrose, Orchestrated objective reduction of quantum coherence in brain microtubules: the "Orch OR" model for consciousness, Mathematics and Computer Simulation, **40**, 453-480, 1996.

17. R. Hari and O.V. Lounasmaa, Neuromagnetism: tracking the dynamics of the brain, Physical World, 33-38, May 2000.

18. M. Ohya, Complexity in quantum system and its application to brain function, In T. Hida and K. Saito, editors, Quantum Information , **II**, 144-160, Singapore, 2000, World Scientific.

19. J.W. Philips, R.M. Leahy, and J.C. Mosher, Imaging neural activity using MEG and EEG, IEEE Engineering in Medicine and Biology, **16(3)**, 34-42, 1997.

20. J.M. Schwartz, H.P. Staap, and M. Beauregard, Quantum physics in neuroscience and psychology, Phil. Trans. Royal Soc. Lond., **B360(1458)**, 1309-1327, 2005.

21. W. Singer, Consciousness and the structure of neural representations, Phil. Trans. Royal Soc., **B353**, 1829-1840, 1998.

22. W. Singer, Der Beobachter in Gehiru. Essays zur Hirnforschung, Suhrkamp Verlag, Frankfurt a.M., 2002.

23. H.P. Staap, Mind, Matter and Quantum Mechanics, Springer, Berlin Heidelberg, 2nd edition, 2003.

24. H.P. Staap, A model of the quantum-classical and mind-brain connections, and the role of the quantum zeno effect in the physical implementation of conscious, arXiv, 0803.1663v1 (physics.gen-ph), 11 March 2008.

25. J. von Neumann, Mathematische Grundlagen der Quantenmechanik, Springer, Berlin, 1st edition, 1932.

Quantum Bio-Informatics V
© 2013 World Scientific Publishing Co. Pte. Ltd.
pp. 159–170

# SKEW INFORMATION AND UNCERTAINTY RELATION

SHIGERU FURUICHI

*World Scientific Publishing Co., Inc, 1060 Main Street, River Edge NJ 07661,
USA*
*E-mail: furuichi@chs.nihon-u.ac.jp*


KENJIRO YANAGI

*Division of Applied Mathematical Science, Graduate School of Science and
Engineering, Yamaguchi University, 2-16-1, Tokiwadai, Ube City, 755-0811,
Japan*
*E-mail: yanagi@yamaguchi-u.ac.jp*

This article is a short review on our recent results on uncertainty relations with
skew informations. In final section, we give the problem on uncertainty relations
with skew informations.

Keywords : Uncertainty relation, Wigner-Yanase skew information and
trace inequality

PACS numbers : 03.65.Ta and 03.67.-a

2010 Mathematics Subject Classification : 15A45, 47A63 and 94A17

## 1. Introduction

In quantum mechanical systems, the expectation value of an observable
(self-adjoint operator) $H$ in a quantum state (density operator) $\rho$ is expressed by $Tr[\rho H]$. Also, the variance for a quantum state $\rho$ and an observable $H$ is defined by $V_\rho(H) \equiv Tr[\rho (H - Tr[\rho H]I)^2] = Tr[\rho H^2] - Tr[\rho H]^2$.
We start from the famous Heisenberg uncertainty relations [10]

$$V_\rho(A)V_\rho(B) \geq \frac{1}{4}|Tr[\rho[A, B]]|^2 \tag{1}$$

for a quantum state $\rho$ and two observables $A$ and $B$. The further strong
result was given by Schrödinger [22]:

$$V_\rho(A)V_\rho(B) - |Re\{Cov_\rho(A, B)\}|^2 \geq \frac{1}{4}|Tr[\rho[A, B]]|^2, \tag{2}$$

where the covariance is defined by $Cov_\rho(A, B) \equiv Tr[\rho (A - Tr[\rho A]I) (B - Tr[\rho B]I)]$.

The variance has been considered that it contains both quantum correlation and classical mixture [17]. Therefore it was quite natural to establish a new uncertainty relation with quantity representing quantum correlation (without classical mixture) for a mixed state. The following inequality was one of the desired relations

$$I_\rho(A)I_\rho(B) \geq \frac{1}{4}|Tr[\rho[A, B]]|^2. \tag{3}$$

However, the above relation failed [14,11,27], because we have many counter-examples for the inequality (3). It is also notable that we have the following relation [14,11,27]. (Cf. Remark 2.4):

$$V_\rho(A)V_\rho(B) - |Re\{Cov_\rho(A, B)\}|^2 \geq I_\rho(A)I_\rho(B) - |Re\{Corr_\rho(A, B)\}|^2, \tag{4}$$

where $Corr_\rho(A, B)$ is the correlation measure defined in Eq.(13) and $I_\rho(H)$ is the Wigner-Yanase skew information [24] defined by

$$I_\rho(H) \equiv \frac{1}{2}Tr\left[\left(i[\rho^{1/2}, H_0]\right)^2\right], \tag{5}$$

which can be regarded as a degree for non-commutativity between a quantum state $\rho$ and an observable $H$, where $H_0 \equiv H - Tr[\rho H]I$.

It is well known that the convexity of the Wigner-Yanase-Dyson skew information

$$I_{\rho,\alpha}(H) \equiv \frac{1}{2}Tr\left[(i[\rho^\alpha, H_0])(i[\rho^{1-\alpha}, H_0])\right] = Tr[\rho H_0^2] - Tr[\rho^\alpha H_0 \rho^{1-\alpha} H_0], \tag{6}$$

which is a one-parameter extension of the Wigner-Yanase skew information $I_\rho(H)$, with respect to $\rho$ was successfully proven by E.H.Lieb in [13]. We have the relation between $I_\rho(H)$ and $V_\rho(H)$ such that $0 \leq I_\rho(H) \leq V_\rho(H)$ so it was quite natural to consider that we have the further sharpened uncertainty relation for the Wigner-Yanase skew information as (3).

In order to overcome this situation, S.Luo introduced a new quantity $U_\rho(H)$ representing a quantum uncertainty excluding the classical mixture:

$$U_\rho(H) \equiv \sqrt{V_\rho(H)^2 - (V_\rho(H) - I_\rho(H))^2}, \tag{7}$$

and then he successfully established a new Heisenberg-type uncertainty relation on $U_\rho(H)$ in [15]:

$$U_\rho(A)U_\rho(B) \geq \frac{1}{4}|Tr[\rho[A, B]]|^2. \tag{8}$$

As stated in the paper [15], the physical meaning of the quantity $U_\rho(H)$ can be interpreted as follows. For a mixed state $\rho$, the variance $V_\rho(H)$ has both classical mixture and quantum uncertainty. Also, the Wigner-Yanase skew information $I_\rho(H)$ represents a kind of quantum uncertainty [16,17]. Thus, the difference $V_\rho(H) - I_\rho(H)$ has a classical mixture so we can regard that the quantity $U_\rho(H)$ has a quantum uncertainty excluding a classical mixture. Therefore it is meaningful and suitable to study an uncertainty relation for a mixed state by use of the quantity $U_\rho(H)$.

Note that we have the following ordering among three quantities:

$$0 \leq I_\rho(H) \leq U_\rho(H) \leq V_\rho(H). \tag{9}$$

The inequality (8) is a refinement of the original Heisenberg's uncertainty relation (1) in the sense of the above ordering (9).

Recently, a one-parameter extension of the inequality (8) was given in [25]:

$$U_{\rho,\alpha}(A)U_{\rho,\alpha}(B) \geq \alpha(1-\alpha)|Tr[\rho[A,B]]|^2, \tag{10}$$

where

$$U_{\rho,\alpha}(H) \equiv \sqrt{V_\rho(H)^2 - (V_\rho(H) - I_{\rho,\alpha}(H))^2},$$

with the Wigner-Yanase-Dyson skew information $I_{\rho,\alpha}(H)$ is defined by Eq.(6).

In this short review, we show the further strong inequality (Schrödinger-type uncertainty relation) for the quantity $U_\rho(H)$ representing a quantum uncertainty. In addition, we give its generalizations using Wigner-Yanase-Dyson skew information and metric adjusted skew information. See our papers [2,3] for the details, since we omit all proofs for all results in this article.

## 2. Uncertainty relation with the quantity including Wigner-Yanase skew information

To show our main theorem, we prepare the definition for a few quantities.

**Definition 2.1.** For a quantum state $\rho$ and an observable $H$, we define the following quantities.

(i) The Wigner-Yanase skew information:

$$I_\rho(H) \equiv \frac{1}{2}Tr\left[(i[\rho^{1/2}, H_0])^2\right] = Tr[\rho H_0^2] - Tr[\rho^{1/2}H_0\rho^{1/2}H_0],$$

where $H_0 \equiv H - Tr[\rho H]I$ and $[X,Y] \equiv XY - YX$ is a commutator.

162

(ii) The quantity associated to the Wigner-Yanase skew information:

$$J_\rho(H) \equiv \frac{1}{2}Tr\left[\left(\left\{\rho^{1/2}, H_0\right\}\right)^2\right] = Tr[\rho H_0^2] + Tr[\rho^{1/2}H_0\rho^{1/2}H_0],$$

where $\{X, Y\} \equiv XY + YX$ is an anti-commutator.

(iii) The quantity representing a quantum uncertainty:

$$U_\rho(H) \equiv \sqrt{V_\rho(H)^2 - (V_\rho(H) - I_\rho(H))^2}.$$

For two quantities $I_\rho(H)$ and $J_\rho(H)$, by simple calculations, we have

$$I_\rho(H) = Tr[\rho H^2] - Tr[\rho^{1/2}H\rho^{1/2}H]$$

and

$$\begin{aligned}
J_\rho(H) &= Tr[\rho H^2] + Tr[\rho^{1/2}H\rho^{1/2}H] - 2(Tr[\rho H])^2 \\
&= 2V_\rho(H) - I_\rho(H),
\end{aligned} \tag{11}$$

which implies $I_\rho(H) \leq J_\rho(H)$.

**Theorem 2.2.** *(²) For a quantum state (density operator) $\rho$ and two observables (self-adjoint operators) $A$ and $B$, we have*

$$U_\rho(A)U_\rho(B) - |Re\{Corr_\rho(A, B)\}|^2 \geq \frac{1}{4}|Tr[\rho[A, B]]|^2, \tag{12}$$

*where the correlation measure is defined by*

$$Corr_\rho(X, Y) \equiv Tr[\rho X^*Y] - Tr[\rho^{1/2}X^*\rho^{1/2}Y] \tag{13}$$

*for any operators $X$ and $Y$.*

Theorem 2.2 improves the uncertainty relation (8) shown in [15], in the sense that the upper bound of the right-hand side of our inequality (12) is tighter than that of S.Luo's one (8).

**Remark 2.3.** For a pure state $\rho = |\varphi\rangle\langle\varphi|$, we have $I_\rho(H) = V_\rho(H)$ which implies $U_\rho(H) = V_\rho(H)$ for an observable $H$ and $Corr_\rho(A, B) = Cov_\rho(A, B)$ for two observables $A$ and $B$. Therefore our Theorem 2.2 coincides with the Schrödinger uncertainty relation (2) for a particular case that a given quantum state is a pure state, $\rho = |\varphi\rangle\langle\varphi|$.

**Remark 2.4.** The example given below

$$\rho = \frac{1}{4}\begin{pmatrix} 1 & 0 \\ 0 & 3 \end{pmatrix}, A = \begin{pmatrix} 2 & 1 \\ 1 & 2 \end{pmatrix}, B = \begin{pmatrix} 0 & 1 \\ 1 & 0 \end{pmatrix},$$

shows

$$V_\rho(A)V_\rho(B) - |Re\{Cov_\rho(A,B)\}|^2$$
$$- \left(U_\rho(A)U_\rho(B) - |Re\{Corr_\rho(A,B)\}|^2\right)$$
$$\simeq -0.232051.$$

The example given below

$$\rho = \frac{1}{10}\begin{pmatrix} 5 & 4 \\ 4 & 5 \end{pmatrix}, A = \begin{pmatrix} 4 & 4 \\ 4 & 1 \end{pmatrix}, B = \begin{pmatrix} 5 & -1 \\ -1 & 2 \end{pmatrix},$$

also shows

$$V_\rho(A)V_\rho(B) - |Re\{Cov_\rho(A,B)\}|^2$$
$$- \left(U_\rho(A)U_\rho(B) - |Re\{Corr_\rho(A,B)\}|^2\right)$$
$$\simeq 13.7862.$$

Therefore there is no ordering between $V_\rho(A)V_\rho(B) - |Re\{Cov_\rho(A,B)\}|^2$ and $U_\rho(A)U_\rho(B) - |Re\{Corr_\rho(A,B)\}|^2$ so we can conclude that neither the inequality (2) nor the inequality (12) is uniformly better than the other.

## 3. Generalizations of Theorem 2.2

### 3.1. *Uncertainty relation with Wigner-Yanase-Dyson skew information*

In this subsection, we give a direct generalization of the Schrödinger type uncertainty relation (12) by the use of the quantity $U_{\rho,\alpha}(H)$ defined by the Wigner-Yanase-Dyson skew information $I_{\rho,\alpha}(H)$.

**Theorem 3.1.** ([3]) *For $\alpha \in [1/2,1]$, a quantum state $\rho$ and two observables $A$ and $B$, we have*

$$U_{\rho,\alpha}(A)U_{\rho,\alpha}(B) \geq 4\alpha(1-\alpha)|Corr_{\rho,\alpha}(A,B)|^2. \tag{14}$$

*where the generalized correlation measure* [11,27] *is defined by*

$$Corr_{\rho,\alpha}(X,Y) \equiv Tr[\rho X^*Y] - Tr[\rho^\alpha X^* \rho^{1-\alpha} Y]$$

*for any operators $X$ and $Y$.*

Note that Theorem 3.1 recovers the inequality (12), if we take $\alpha = \frac{1}{2}$.

**Remark 3.2.** We take $\alpha = 0.1$ and

$$\rho = \frac{1}{3}\begin{pmatrix} 1 & 0 \\ 0 & 2 \end{pmatrix}, A = \begin{pmatrix} 2 & 2-i \\ 2+i & 1 \end{pmatrix}, B = \begin{pmatrix} 2 & i \\ -i & 1 \end{pmatrix},$$

then we have

$$U_{\rho,\alpha}(A)U_{\rho,\alpha}(B) - 4\alpha(1-\alpha)|Corr_{\rho,\alpha}(A,B)|^2 \simeq -0.28332.$$

Therefore the inequality (14) does not hold for $\alpha \in [0,1/2)$ in general.

**Corollary 3.3.** ($^3$) *Under the same assumptions with Theorem 3.1, we have the following inequality:*

$$\begin{aligned}
&U_{\rho,\alpha}(A)U_{\rho,\alpha}(B) \\
&-4\alpha(1-\alpha)\left(|Re\left\{Corr_{\rho,\alpha}(A,B)\right\}|^2 - |Im\left\{Tr[\rho^\alpha A\rho^{1-\alpha}B]\right\}|^2\right) \quad (15) \\
&\geq \alpha(1-\alpha)|Tr[\rho[A,B]]|^2. \quad\quad (16)
\end{aligned}$$

**Remark 3.4.** The following inequality does not hold in general for $\alpha \in [\frac{1}{2},1]$:

$$|Re\left\{Corr_{\rho,\alpha}(A,B)\right\}|^2 \geq |Im\left\{Tr[\rho^\alpha A\rho^{1-\alpha}B]\right\}|^2. \quad (17)$$

Because we have a counter-example as follows. We take $\alpha = \frac{2}{3}$ and

$$\rho = \frac{1}{7}\begin{pmatrix} 2 & 3 \\ 3 & 5 \end{pmatrix}, A = \begin{pmatrix} 2 & 2-i \\ 2+i & 1 \end{pmatrix}, B = \begin{pmatrix} 2 & i \\ -i & 1 \end{pmatrix},$$

then we have

$$|Re\left\{Corr_{\rho,\alpha}(A,B)\right\}|^2 - |Im\left\{Tr[\rho^\alpha A\rho^{1-\alpha}B]\right\}|^2 \simeq -0.0548142.$$

This shows Theorem 3.1 does not always refine the inequality (10) in general.

### 3.2. *A two-parameter generalization*

In this subsection, we introduce the parametric extended correlation measure $Corr_{\rho,\alpha,\gamma}(X,Y)$ by the convex combination between $Corr_{\rho,\alpha}(X,Y)$ and $Corr_{\rho,1-\alpha}(X,Y)$. Then we establish the parametric extended Schrödinger-type uncertainty relation applying the parametric extended correlation measure $Corr_{\rho,\alpha,\gamma}(X,Y)$.

**Definition 3.5.** ($^3$) We define the parametric extended correlation measure $Corr_{\rho,\alpha,\gamma}(X,Y)$ for two parameters $\alpha,\gamma \in [0,1]$ by

$$Corr_{\rho,\alpha,\gamma}(X,Y) \equiv (1-\gamma)Corr_{\rho,\alpha}(X,Y) + \gamma Corr_{\rho,1-\alpha}(X,Y) \quad (18)$$

for any operators $X$ and $Y$.

Note that we have $Corr_{\rho,\alpha,\gamma}(H,H) = I_{\rho,\alpha}(H)$ for any observable $H$. Then we can prove the following inequality.

**Theorem 3.6.** *([3]) If $0 \leq \alpha, \gamma \leq \frac{1}{2}$ or $\frac{1}{2} \leq \alpha, \gamma \leq 1$, then we have*

$$U_{\rho,\alpha}(A)U_{\rho,\alpha}(B) \geq 4\alpha(1-\alpha)|Corr_{\rho,\alpha,\gamma}(A,B)|^2$$

*for two observables $A$, $B$ and a quantum state $\rho$.*

**Corollary 3.7.** *([3]) For any $\alpha \in [0,1]$, two observables $A$, $B$ and a quantum state $\rho$, we have*

$$U_{\rho,\alpha}(A)U_{\rho,\alpha}(B) \geq 4\alpha(1-\alpha)|Corr_{\rho,\alpha,\frac{1}{2}}(A,B)|^2,$$

*where we call $Corr_{\rho,\alpha,\frac{1}{2}}(A,B)$ a symmetrized correlation measure.*

### 3.3. *Generalization with metric adjusted correlation measure*

In this subsection, we give a further generalization for Schrödinger-type uncertainty relation applying metric adjusted correlation measure introduced in [9]. We firstly give some notations according to those in [8]. Let $M_n(\mathbb{C})$ and $M_{n,sa}(\mathbb{C})$ be the set of all $n \times n$ complex matrices and all $n \times n$ self-adjoint matrices, equipped with the Hilbert-Schmidt scalar product $\langle A, B \rangle = Tr[A^*B]$, respectively. Let $M_{n,+}(\mathbb{C})$ be the set of all positive definite matrices of $M_{n,sa}(\mathbb{C})$ and $M_{n,+,1}(\mathbb{C})$ be the set of all invertible density matrices.

A function $f : (0, +\infty) \to \mathbb{R}$ is said operator monotone if the inequalities $0 \leq f(A) \leq f(B)$ hold for any $A, B \in M_{n,sa}(\mathbb{C})$ such that $0 \leq A \leq B$. An operator monotone function $f : (0, +\infty) \to (0, +\infty)$ is said symmetric if $f(x) = xf(x^{-1})$ and normalized if $f(1) = 1$. We represents the set of all symmetric normalized operator monotone functions by $\mathcal{F}_{op}$. We have the following examples as elements of $\mathcal{F}_{op}$:

**Example 3.8.** $(^{9,8,4,20})$

$$f_{RLD}(x) = \frac{2x}{x+1}, \quad f_{SLD}(x) = \frac{x+1}{2}, \quad f_{BKM}(x) = \frac{x-1}{\log x},$$

$$f_{WY}(x) = \left(\frac{\sqrt{x}+1}{2}\right)^2, \quad f_{WYD}(x) = \alpha(1-\alpha)\frac{(x-1)^2}{(x^\alpha-1)(x^{1-\alpha}-1)}, \quad \alpha \in (0,1).$$

The functions $f_{BKM}(x)$ and $f_{WYD}(x)$ are normalized in the sense that $\lim_{x\to 1} f_{BKM}(x) = 1$ and $\lim_{x\to 1} f_{WYD}(x) = 1$. Note that a simple proof of the operator monotonicity of $f_{WYD}(x)$ was given in [4]. See also [23] for the proof of the operator monotonicity of $f_{WYD}(x)$ by use of majorization.

**Remark 3.9.** ([8,12,19,20]) For any $f \in \mathcal{F}_{op}$, we have the following inequalities:

$$f_{RLD}(x) \le f(x) \le f_{SLD}(x), \quad x > 0.$$

For $f \in \mathcal{F}_{op}$ we define $f(0) = \lim_{x\to 0} f(x)$. We also denote the sets of regular and non-regular functions by

$$\mathcal{F}_{op}^{reg} = \{f \in \mathcal{F}_{op} | f(0) \neq 0\} \ \text{ and } \ \mathcal{F}_{op}^{non} = \{f \in \mathcal{F}_{op} | f(0) = 0\}.$$

**Definition 3.10.** ([6,8]) For $f \in \mathcal{F}_{op}^{reg}$, we define the function $\tilde{f}$ by

$$\tilde{f}(x) = \frac{1}{2}\left\{(x+1) - (x-1)^2 \frac{f(0)}{f(x)}\right\}, \quad (x > 0).$$

**Proposition 3.11.** ([6,4,21]) *The correspondence* $f \to \tilde{f}$ *is a bijection between* $\mathcal{F}_{op}^{reg}$ *and* $\mathcal{F}_{op}^{non}$.

We can use matrix mean theory introduced by Kubo-Ando in [12]. Then a mean $m_f$ corresponds to each operator monotone function $f \in \mathcal{F}_{op}$ by the following formula

$$m_f(A, B) = A^{1/2} f(A^{-1/2} B A^{-1/2}) A^{1/2},$$

for $A, B \in M_{n,+}(\mathbb{C})$. By the notion of matrix mean, we may define the set of the monotone metrics [18] by the following formula

$$\langle A, B \rangle_{\rho,f} = Tr[A m_f(L_\rho, R_\rho)^{-1}(B)],$$

where $L_\rho(A) = \rho A$ and $R_\rho(A) = A\rho$.

**Definition 3.12.** ([9,6]) For $A, B \in M_{n,sa}(\mathbb{C})$, $\rho \in M_{n,+,1}(\mathbb{C})$ and $f \in \mathcal{F}_{op}^{reg}$, we define the following quantities:

$$Corr_\rho^f(A, B) \equiv \frac{f(0)}{2} \langle i[\rho, A], i[\rho, B] \rangle_{\rho,f}, \quad I_\rho^f(A) \equiv Corr_\rho^f(A, A),$$

$$C_\rho^f(A, B) \equiv Tr[m_f(L_\rho, R_\rho)(A)B], \quad C_\rho^f(A) \equiv C_\rho^f(A, A),$$

$$U_\rho^f(A) \equiv \sqrt{V_\rho(A)^2 - (V_\rho(A) - I_\rho^f(A))^2}.$$

The quantity $I_\rho^f(A)$ is known as metric adjusted skew information [9] and the metric adjusted correlation measure $Corr_\rho^f(A,B)$ was also previously defined in [9]. Then we have the following proposition.

**Proposition 3.13.** ([6,8]) *For $A, B \in M_{n,sa}(\mathbb{C})$, $\rho \in M_{n,+,1}(\mathbb{C})$ and $f \in \mathcal{F}_{op}^{reg}$, we have the following relations, where we put $A_0 \equiv A - Tr[\rho A]I$ and $B_0 \equiv B - Tr[\rho B]I$.*

*(1)* $I_\rho^f(A) = Tr[\rho A_0^2] - Tr[m_{\tilde{f}}(L_\rho, R_\rho)(A_0)A_0] = V_\rho(A) - C_\rho^{\tilde{f}}(A_0)$.
*(2)* $J_\rho^f(A) = Tr[\rho A_0^2] + Tr[m_{\tilde{f}}(L_\rho, R_\rho)(A_0)A_0] = V_\rho(A) + C_\rho^{\tilde{f}}(A_0)$.
*(3)* $0 \le I_\rho^f(A) \le U_\rho^f(A) \le V_\rho(A)$.
*(4)* $U_\rho^f(A) = \sqrt{I_\rho^f(A)J_\rho^f(A)}$.
*(5)* $Corr_\rho^f(A,B) \quad = \quad \frac{1}{2}Tr[\rho A_0 B_0] \quad + \quad \frac{1}{2}Tr[\rho B_0 A_0] \quad - \quad Tr[m_{\tilde{f}}(L_\rho, R_\rho)(A_0)B_0] = \frac{1}{2}Tr[\rho A_0 B_0] + \frac{1}{2}Tr[\rho B_0 A_0] - C_\rho^{\tilde{f}}(A_0, B_0)$.

The following inequality is the further generalization of Corollary 3.7 by the use of the metric adjusted correlation measure.

**Theorem 3.14.** ([3]) *For $f \in \mathcal{F}_{op}^{reg}$, if we have*

$$\frac{x+1}{2} + \tilde{f}(x) \ge 2f(x), \tag{19}$$

*then we have*

$$U_\rho^f(A)U_\rho^f(B) \ge 4f(0)|Corr_\rho^f(A,B)|^2, \tag{20}$$

*for $A, B \in M_{n,sa}(\mathbb{C})$ and $\rho \in M_{n,+,1}(\mathbb{C})$.*

**Remark 3.15.** Under the same assumptions with Theorem 3.14, we have the following Heisenberg-type uncertainty relation [26]:

$$U_\rho^f(A)U_\rho^f(B) \ge f(0)|Tr[\rho[A,B]]|^2 \tag{21}$$

by the similar way to the proof of Theorem 3.14, since we have

$$|Tr[\rho[A,B]]| \le 2\sum_{j<k}|\lambda_j - \lambda_k||a_{jk}||b_{kj}|.$$

As stated in Remark 3.4, there is no ordering between the right hand side of the inequality (20) and that of the inequality (21), in general.

Since the function $f_{WYD}(x)$ satisfies the condition of Theorem 3.14, we have the following uncertainty relation.

**Corollary 3.16.** *(³) For $A, B \in M_{n,sa}(\mathbb{C})$ and $\rho \in M_{n,+,1}(\mathbb{C})$, we have*

$$U_\rho^{fWYD}(A)U_\rho^{fWYD}(B) \geq 4\alpha(1-\alpha)|Corr_{\rho,\alpha,\frac{1}{2}}(A,B)|^2.$$

Note that Corollary 3.7 coincides with Corollary 3.16, since we have $U_{\rho,\alpha}(A) = U_\rho^{fWYD}(A)$ which is obtained by the fact the function $f_{WYD}(x)$ corresponds to the Wigner-Yanase-Dyson skew information.

## 4. Conclusion

We have briefly reviewed our recent results on Schrödinger-type uncertainty relations. We here point out the following facts. The relation $x \not\geq y$ means that $x \geq y$ does not hold in general.

- We have the following relations. See the inequality (4) and [14,11,27].

$$V_\rho(A)V_\rho(B) - |Re\{Cov_\rho(A,B)\}|^2$$
$$\geq I_\rho(A)I_\rho(B) - |Re\{Corr_\rho(A,B)\}|^2$$
$$\not\geq \frac{1}{4}|Tr[\rho[A,B]]|^2.$$

- We also have the following relations in general. See Theorem 2.2 and Remark 2.4.

$$V_\rho(A)V_\rho(B) - |Re\{Cov_\rho(A,B)\}|^2$$
$$\not\geq U_\rho(A)U_\rho(B) - |Re\{Corr_\rho(A,B)\}|^2$$
$$\geq \frac{1}{4}|Tr[\rho[A,B]]|^2.$$

- We have $Corr_\rho(A,A) = I_\rho(A)$ and $Cov_\rho(A,A) = V_\rho(A)$. But $Corr_\rho(A,A) \neq U_\rho(A)$ in general.

Taking an account for the above discussions, we may have the following problem. That is, for two observables $A$, $B$ and a quantum state $\rho$, do we have a new quantity $\Lambda_\rho(A,B)$ satisfying the following relations?

$$V_\rho(A)V_\rho(B) - |Re\{Cov_\rho(A,B)\}|^2$$
$$\geq \Lambda_\rho(A,A)\Lambda_\rho(B,B) - |Re\{\Lambda_\rho(A,B)\}|^2$$
$$\geq \frac{1}{4}|Tr[\rho[A,B]]|^2.$$

## Acknowledgements

The first author was supported in part by the Japanese Ministry of Education, Science, Sports and Culture, Grant-in-Aid for Encouragement of Young Scientists (B), No.20740067. The second author was partially supported by the Japanese Ministry of Education, Science, Sports and Culture, Grant-in-Aid for Scientific Research (C), No.20540175 and No.23540208.

## References


1. R.Bhatia and C. Davis, A Cauchy-Schwarz inequality for operators with applications, Linear Alg. Appl., Vol.223/224(1995), pp.119-129.
2. S.Furuichi, Schrödinger uncertainty relation with Wigner-Yanase skew information, Phys. Rev. A, Vol.82(2010), 034101.
3. S.Furuichi and K.Yanagi, Schrödinger uncertainty relation, Wigner-Yanase-Dyson skew information and metric adjusted correlation measure, arXiv:1010.0392v3.
4. P.Gibilisco, F.Hansen and T.Isola, On a correspondence between regular and non-regular operator monotone functions, Linear Alg. Appl., Vol.430(2009), pp.2225-2232.
5. P.Gibilisco, F.Hiai and D.Petz, Quantum covariance, quantum Fisher information, and the uncertainty relations, IEEE Trans. Information Theory, Vol.55(2009), pp.439-443.
6. P.Gibilisoco, D.Imparatoand T.Isola, Uncertainty principle and quantum Fisher information, II, J. Math. Phys., Vol.48(2007), 072109.
7. P.Gibilisco, D.Imparato and T.Isola, A Robertson-type uncertainty principle and quantum Fisher information, Linear Alg. Appl., Vol.428(2008), pp.1706-1724.
8. P.Gibilisco and T.Isola, On a refinement of Heisenberg uncertainty relation by means of quantum Fisher information, J. Math. Anal. Appl., Vol.375(2011), pp.270-275.
9. F.Hansen, Metric adjusted skew information, Proc. Nat. Acad. Sci., U.S.A., Vol.105(2008), pp.9909-9916.
10. W.Heisenberg, Über den anschaulichen Inhalt der quantummechanischen Kinematik und Mechanik, Zeitschrift für Physik, Vol.43(1927), pp.172-198.
11. H.Kosaki, Matrix trace inequality related to uncertainty principle, Int. J. Math., Vol.16(2005), pp.629-646.
12. F. Kubo and T. Ando, Means of positive linear operators, Math. Ann., Vol.246(1980), pp.205-224.
13. E.H.Lieb, Convex trace functions and the Wigner-Yanase-Dyson conjecture, Adv. Math., Vol.1(1973), pp.267-288.
14. S.Luo and Q.Zhang, On skew information, IEEE Trans. Information Theory, Vol.50(2004), pp.1778-1782, and Correction to "On skew information", IEEE Trans. Information Theory, Vol.51(2005), p.4432.
15. S.Luo, Heisenberg uncertainty relation for mixed states, Phys. Rev.A, Vol.72(2005), 042110.

16. S.Luo and Q.Zhang, Informational distance on quantum-state space, Phys. Rev.A, Vol.69(2004), 032106.

17. S.Luo, Quantum versus classical uncertainty, Theoret. Math. Phys., Vol.143(2005), pp.681-688.

18. D.Petz, Monotone metrics on matrix spaces, Linear Alg. Appl., Vol.244(1996), pp.81-96.

19. D. Petz, Quantum information theory and quantum statistics, Springer, Berlin, Heidelberg, 2008.

20. D.Petz and C.Ghinea, Introduction to quantum Fisher information, arXiv:1008.2417v1.

21. D. Petz and V.E.S. Szabó, From quasi-entropy to skew information, Int. J. Math. Vol.20(2009),pp.1421-1430.

22. E.Schrödinger, Proceedings of The Prussian Academy of Sciences, Physics-Mathematical Section 19(1930), pp.296-303.

23. M.Uchiyama, Majorization and some operator monotone functions, Linear Alg. Appl., Vol.432 (2010), pp.1867-1872.

24. E.P.Wigner and M.M.Yanase, Information content of distribution, Proc. Nat. Acad. Sci U.S.A., Vol.49(1963), pp.910-918.

25. K.Yanagi, Uncertainty relation on Wigner-Yanase-Dyson skew information, J. Math. Anal. Appl., Vol.365(2010), pp.12-18.

26. K.Yanagi, Metric adjusted skew information and uncertainty relation, J. Math. Anal. Appl., Vol.380(2011), pp.888-892.

27. K.Yanagi, S.Furuichi and K.Kuriyama, A generalized skew information and uncertainty relation, IEEE Trans. Information Theory, Vol.51(2005), pp.4401-4404.

Quantum Bio-Informatics V
© 2013 World Scientific Publishing Co. Pte. Ltd.
pp. 171–180

# MULTIPLE-PHOTON ABSORPTION ATTACK ON ENTANGLEMENT-BASED QUANTUM KEY DISTRIBUTION PROTOCOLS

G. ADENIER

*Tokyo University of Science, 2641 Yamazaki, Noda, Chiba 278-8510, Japan*
*University of Oslo, P.O. Box 1048 Blindern, 0316 Oslo, Norway*
*E-mail: adenierg@fys.uio.no*


N. WATANABE

*Tokyo University of Science, 2641 Yamazaki, Noda, Chiba 278-8510, Japan*


I. BASIEVA AND A. YU. KHRENNIKOV

*Linnaeus University, Vejdes plats 7, SE-351 95 Växjö, Sweden*

We propose a multiple-photon absorption attack on Quantum Key Distribution protocols. In this attack, the eavesdropper (Eve) is in control of the source and sends pulses correlated in polarization (but not entangled) containing several photons at frequencies for which only multiple-photon absorptions are possible in Alice's and Bob's detectors. Whenever the number of photons from one pulse are dispatched in insufficient number to trigger a multiple-photon absorption in either channel, the pulse remains undetected. We show that this simple feature is enough to reproduce the type of statistics on the detected pulses that are considered as indicating a secure quantum key distribution in entangled-based protocols, even though the source is controlled by Eve, and we discuss possible countermeasures.

*Keywords* : Quantum Key Distribution; Bell inequalities; Fair sampling.

## 1. Introduction

In actual implementations of Quantum Key Distribution (QKD) protocols, the imperfections of the optical components used can be exploited by Eve to extract some information about the key [1]. Here, we propose a multiple-photon absorption attack on entangled based quantum cryptography protocols, in particular the Ekert [2] and BBM92 protocols [3].

The attack [4] consists in replacing the source of entangled photons altogether by a controlled source of separable states. Eve sends pulses containing several photons, all photons in a pulse having the same polarization $|\lambda\rangle$ known and controlled by Eve. The essential feature that Eve must implement to perform the attack is that she must be able to prevent single photons from triggering Alice's and Bob's detectors. Eve can for instance set the frequency of the photons such that the only way to get a click in a detector is when all the photons from a pulse are absorbed in one same detector, through a multiple-photon absorption [5]. The principle of the attack is general and would also work if Eve can tamper Alice's and Bob's detectors to prevent them from detecting single photons.

To simplify the discussion and highlight the principle of the attack, we assume that Eve can control precisely the number of photons in each pulse, their polarization and their frequency. Eve sends $n$-photon pulses to Alice, the photons having a frequency $\nu_n$ such that only an $n$-photon absorption is possible in Alice's detectors, and she sends $m$-photon pulses to Bob, the photons having a frequency $\nu_m$ such that only an $m$-photon absorption is possible in Bob's detectors.

## 2. Single-photon absorption attack $n = m = 1$

In this first attempt, Eve sends single photon pulses at a frequency $\nu_1$ where the dominant absorption is a standard single-photon absorption. Each pulse contains exactly one photon that is linearly polarized in a direction $\lambda_i$ chosen by Eve. It can thus be described in its own Hilbert subspace $\mathcal{H}_i$ by a two-level state of the form

$$|\lambda_i\rangle = \cos \lambda_i |0\rangle + \sin \lambda_i |1\rangle, \tag{1}$$

where $|0\rangle$ and $|1\rangle$ are the eigenvectors of $\hat{\sigma}_z$ with eigenvalues $-1$ and $+1$ respectively. The pair of pulses can then be described in the tensor product Hilbert space $\mathcal{H}_{12} = \mathcal{H}_1 \otimes \mathcal{H}_2$ as $|\Lambda_{12}\rangle = |\lambda_1\rangle \otimes |\lambda_2\rangle$.

Alice and Bob, who are performing local measurements respectively in $\mathcal{H}_1$ and $\mathcal{H}_2$, want to measure the statistical correlation of the pairs they receive from Eve when they perform some rotations $\hat{R}(\theta_A)$ and $\hat{R}(\theta_B)$ on their respective pulses, followed by a measurement of the observable $\hat{\sigma}_z$.

Since Eve wants them to obtain an as good (anti)correlation as possible with separable states, she sends orthogonal states to Alice and Bob, that is, such that $\lambda_1 = \lambda$ and $\lambda_2 = \lambda + \frac{\pi}{2}$, and the initial state sent to Alice and Bob can be written:

$$|\Lambda_{12}\rangle = |\lambda\rangle \otimes |\lambda + \frac{\pi}{2}\rangle \tag{2}$$

For a pair initially represented by this state (2), the state after local rotations on each sides becomes

$$|\Lambda_{AB}\rangle = |\lambda + \theta_A\rangle \otimes |\lambda + \frac{\pi}{2} + \theta_B\rangle. \tag{3}$$

Keeping the condition of orthogonality, Eve is randomizing the parameter $\lambda$ associated to each pair. From the point of view of Alice and Bob, the polarization state of each pulse prepared by Eve is unknown and can therefore be described quantum mechanically by a mixture. We can characterize the probability of obtaining a state between $|\lambda\rangle$ and $|\lambda + d\lambda\rangle$ by a probability density distribution $\rho(\lambda)$ on a single probability space $(\Omega, \mathcal{F}, P)$. The mixture $\hat{\rho}$ describing the set of pairs after rotation is therefore

$$\hat{\rho}_{AB} = \int_\Omega \rho(\lambda) d\lambda \, |\Lambda_{AB}\rangle\langle\Lambda_{AB}|, \tag{4}$$

with $\int_\Omega \rho(\lambda) d\lambda = 1$.

After performing their respective rotation on the received pulses, Alice and Bob perform a joint measurement $\sigma_z \otimes \sigma_z$ on each pair. The expectation value of this measurement is:

$$E_{AB} = \langle \hat{\sigma}_z \otimes \hat{\sigma}_z \rangle_{\hat{\rho}_{AB}} = \mathrm{Tr}(\hat{\rho}_{AB}\hat{\sigma}_z \otimes \hat{\sigma}_z), \tag{5}$$

with $\hat{\sigma}_z = |0\rangle\langle 0| - |1\rangle\langle 1|$, so that

$$\hat{\sigma}_z \otimes \hat{\sigma}_z = |00\rangle\langle 00| - |10\rangle\langle 10| - |01\rangle\langle 01| + |11\rangle\langle 11|,$$

and where Tr is the trace, defined as the sum of the diagonal matrix elements in any orthonormal basis, which in our case is $\mathrm{Tr}\hat{O} = \sum_{i,j=0}^{1}\langle ij|\hat{O}|ij\rangle$, where $\hat{O}$ is an operator in $\mathcal{H}_{12}$.

By linearity of the trace, we can write

$$E_{AB} = P_{00} - P_{10} - P_{01} + P_{11},$$

with

$$P_{kl} = \mathrm{Tr}(\hat{\rho}_{AB}|kl\rangle\langle kl|),$$

where $k$ and $l$ are 0 or 1.

The $P_{kl}$ are the joint probabilities, and we can now express them explicitly, using Eq. (4), as

$$P_{kl} = \sum_{i,j=0}^{1} \int_\Omega \rho(\lambda) d\lambda \, \langle ij|\Lambda_{AB}\rangle\langle\Lambda_{AB}|kl\rangle\langle kl|ij\rangle,$$

which, since $\{|00\rangle, |10\rangle, |01\rangle, |11\rangle\}$ forms an orthonormal basis in $\mathcal{H}_{12}$, simplifies as

$$P_{kl} = \int_\Omega \rho(\lambda)d\lambda \, \left|\langle kl|\Lambda_{\mathrm{AB}}\rangle\right|^2. \tag{6}$$

Note that, owing to the separability of $|\Lambda_{\mathrm{AB}}\rangle$ as expressed by Eq. (3), we can factorize the integrand in the above joint probability into a product, that is,

$$P_{kl} = \int_\Omega \rho(\lambda)d\lambda \, \left|\langle k|\lambda + \theta_{\mathrm{A}}\rangle\langle l|\lambda + \frac{\pi}{2} + \theta_{\mathrm{B}}\rangle\right|^2. \tag{7}$$

Denoting the probability to get a photon in a channel $i$ for a local state $|\lambda\rangle$ and a measurement direction $\theta$ as

$$P_i(\lambda, \theta) = \left|\langle i|\lambda + \theta\rangle\right|^2, \tag{8}$$

which is explicitly

$$\begin{aligned}
P_0(\lambda, \theta) &= \cos^2(\lambda + \theta), \\
P_1(\lambda, \theta) &= \sin^2(\lambda + \theta),
\end{aligned} \tag{9}$$

we can rewrite Eq. (7) as

$$P_{kl} = \int_\Omega \rho(\lambda)d\lambda \, P_k(\lambda, \theta_{\mathrm{A}}) \, P_l(\lambda + \frac{\pi}{2}, \theta_{\mathrm{B}}), \tag{10}$$

which can be simply interpreted as the integral on all possible $\lambda$ of the product of the probability to detect a photon in channel $k$ on Alice's side for a local state $|\lambda\rangle$ by the probability to get a photon in channel $l$ on Bob's side for a local state $|\lambda + \frac{\pi}{2}\rangle$.

Assuming $\rho(\lambda)$ is a uniform distribution on the interval $[0, 2\pi[$, we obtain

$$\begin{aligned}
P_{00} = P_{11} &= \frac{1}{8}(2 - \cos 2(\theta_{\mathrm{A}} - \theta_{\mathrm{B}})), \\
P_{10} = P_{01} &= \frac{1}{8}(2 + \cos 2(\theta_{\mathrm{A}} - \theta_{\mathrm{B}})),
\end{aligned} \tag{11}$$

and finally, we get

$$E_{\mathrm{AB}} = -\frac{1}{2}\cos 2(\theta_{\mathrm{A}} - \theta_{\mathrm{B}}). \tag{12}$$

This result differs essentially from the prediction for the singlet state by its visibility of $1/2$ instead of $1$. With this correlation function, the maximum of the CHSH function [6], defined as

$$S = |E(\theta_{\mathrm{A}}, \theta_{\mathrm{B}}) + E(\theta'_{\mathrm{A}}, \theta_{\mathrm{B}}) + E(\theta_{\mathrm{A}}, \theta'_{\mathrm{B}}) - E(\theta'_{\mathrm{A}}, \theta'_{\mathrm{B}})| \tag{13}$$

is $S = \sqrt{2}$, which is clearly below 2 and therefore insufficient as an attack on Ekert protocol.

Similarly, this attack fails against the BBM92 protocol. The QBER can be estimated [7,8] from the visibility V as QBER $= \frac{1-V}{2}$. With V $= 1/2$, the QBER measured by Alice and Bob is as high as 25% in this single-photon attack, which Alice and Bob would not fail to reject.

## 3. Two-photon absorption attack

We now consider a two-photon absorption attack. As in the previous single photon attack, Eve sends pairs of pulses described by a mixture of separable states (4) to Alice and Bob, but each pulse contains now two photons (with identical polarization) instead of one, and they must not be able to trigger a click individually. For this purpose, Eve can prepare them at a lower frequency $\nu_2$ such that the only way to get a click is through a two-photon absorption. We assume for simplicity that whenever two photons hit the same detector simultaneously, the probability that they trigger a click in the detector is 1. When they are dispatched to different channels instead, no detection can occur. This possibility brings us in the realm of the detection loophole [9,10,11,12,13]. It is the essential reason for the appearance of the violation of Bell inequalities that Alice and Bob are going to obtain.

Assuming independence between the photons, for a local state $|\lambda\rangle$ and a measurement angle $\theta$ on either side, the probability that both photons from one pulse end up in the same channel $i$ is simply the square of the probability $P_i(\lambda, \theta)$ to see one such photon exit the PBS through this channel $i$, that is: $P_i^2(\lambda, \theta)$

So, similarly to what we had in the single photon case in Eq.(10), the probability $P_{kl}^{(2)}$ to get a click in channel $k$ for Alice and in channel $l$ for Bob in a two-photon absorption process is therefore the integral over all possible state of the product of the probability $P_k^2(\lambda, \theta_A)$ for Alice to get a click in channel $k$ by the probability $P_l^2(\lambda + \frac{\pi}{2}, \theta_B)$ for Bob to get a click in channel $l$:

$$P_{kl}^{(2)} = \int_\Omega \rho(\lambda)d\lambda \, P_k^2(\lambda, \theta_A) \, P_l^2(\lambda + \frac{\pi}{2}, \theta_B). \tag{14}$$

For a rotationally invariant source, $\lambda$ is uniformly distributed on the interval $[0, 2\pi[$, we obtain:

$$\begin{aligned}
P_{00}^{(2)} &= P_{11}^{(2)} = \frac{1}{128}(18 - 16\cos 2(\theta_A - \theta_B) + \cos 4(\theta_A - \theta_B), \\
P_{01}^{(2)} &= P_{10}^{(2)} = \frac{1}{128}(18 + 16\cos 2(\theta_A - \theta_B) + \cos 4(\theta_A - \theta_B).
\end{aligned} \tag{15}$$

Note that these four probabilities no longer add up to 1, because of the cases involving a non detection on either side or both, which are discarded by Alice and Bob. So, as is standard in optical EPR experiment and in entanglement-based QKD with photons, the correlation function has to be normalized by the sum $\sum_{k,l} P_{kl}^{(2)}$:

$$E_{AB}^{(2)} = \frac{P_{00}^{(2)} - P_{10}^{(2)} - P_{01}^{(2)} + P_{11}^{(2)}}{P_{00}^{(2)} + P_{10}^{(2)} + P_{01}^{(2)} + P_{11}^{(2)}}. \tag{16}$$

We obtain explicitly

$$E_{AB}^{(2)} = -\frac{16 \cos 2(\theta_A - \theta_B)}{18 + \cos 4(\theta_A - \theta_B)},$$

which lead to a violation of Bell inequalities for $\theta_A = \{0, \frac{\pi}{4}\}$ and $\theta_B = \{-\frac{\pi}{8}, \frac{\pi}{8}\}$ of

$$S^{(2)} = \frac{16}{18} \, 2\sqrt{2} \approx 2.51,$$

which is clearly above 2.

Here the visibility is $V = 0.842$, so that a BBM92 protocol would give a QBER of 7.9%, which is already well below the security bound of 11% against coherent attacks [8,14,15].

## 4. Multiple-photon absorption attack

Quite generally, a multi-photon absorption process can lead to a violation of Bell inequalities as large as desired within the algebraic limit of 4, the only limit being the order of the multiple-photon absorptions that Eve can drive into Alice's and Bob's detectors.

Of particular interest can be the case when Eve drives a two-photon absorption attack on one side and a three-photon absorption on the other side, since it leads to a violation of Bell inequalities for $\theta_A = \{0, \frac{\pi}{4}\}$ and $\theta_B = \{-\frac{\pi}{8}, \frac{\pi}{8}\}$ of exactly

$$S^{(2,3)} = 2\sqrt{2},$$

which is the maximum violation predicted by Quantum Mechanics for entangled states.

The performances of the various multiple-photon absorption attacks are displayed in Table 1.

## 5. Eve's knowledge of the key

In a multiple-photon absorption attack, Eve has a full knowledge of the local states that she sends to Alice and Bob. Each pair is characterized by state of the form Eq.(2), where the polarization $\lambda$ characterizing each pair is chosen by Eve. This perfect knowledge of the local state sent to Alice and Bob does not however mean that Eve has a perfect knowledge of the key that they will generate with these pairs, because of the inherent probabilistic nature of Quantum predictions whenever a measurement is performed in another basis than the one in which a state was prepared.

The key is generated in those cases in which Alice and Bob are performing identical measurement, that is $\theta_A = \theta_B = \theta$. Knowing $\alpha = |\lambda + \theta|$ for each pulse, Eve knows the probabilities for 0 and 1 to be realized on each side, and she simply bets for the one that has the higher probability.

Consider the cases for which $0 < |\lambda + \theta| < \pi/4$. We then have $\cos^2(\lambda + \theta) > \sin^2(\lambda + \theta)$, so that $P_0(\lambda, \theta) > P_1(\lambda, \theta)$ for Alice and $P_1(\lambda + \pi/2, \theta) > P_0(\lambda + \pi/2, \theta)$ for Bob. Eve therefore bets that the bit measured by Alice is 0 and that the bit measured by Bob is 1.

Assuming independence between the photons, the probability that all the photons from an $n-$photon pulse on Alice's side end up in channel 1, and thus generate a click opposite to Eve's guess, is $P_1^n(\lambda, \theta)$. Just the same, the probability that all the photons from an $m-$photon pulse on Bob's side end up in channel 0, and thus generate a click opposite to Eve's guess, is $P_0^m(\lambda + \pi/2, \theta)$. The probability that Eve guesses incorrectly a bit shared by Alice and Bob when $0 < |\lambda + \theta| < \pi/4$ is therefore

$$P_1^n(\lambda, \theta) P_0^m(\lambda + \pi/2, \theta) = \sin^{2(n+m)}(\lambda + \theta).$$

Similarly, in the cases for which $\pi/4 < |\lambda + \theta| < \pi/2$, the probability that Eve guesses incorrectly is

$$P_0^n(\lambda, \theta) P_1^m(\lambda + \pi/2, \theta) = \cos^{2(n+m)}(\lambda + \theta).$$

The same reasoning can be extended to higher values of $|\lambda + \theta|$, leading either to a dependence on $\sin^{2(n+m)}(\lambda + \theta)$ or $\cos^{2(n+m)}(\lambda + \theta)$, so that on average, for a uniform distribution of $\lambda$ on the circle, the probability that Eve guesses incorrectly a bit shared by Alice and Bob is

$$P_{\text{error}}^{(n,m)} = \frac{2}{\pi}\left( \int_0^{\pi/4} d\alpha\, \sin^{2(n+m)} \alpha + \int_{\pi/4}^{\pi/2} d\alpha\, \cos^{2(n+m)} \alpha \right).$$

Note that the pulses for which Eve has the least information are those where $\cos^{2n}(\lambda + \theta) \approx \sin^{2n}(\lambda + \theta)$, that is $|\lambda + \theta| \approx \pi/4 + k\pi/2$. It is

| (n,m) | $\eta$ | S | V | QBER | $P_{\text{error}}^{(n,m)}$ |
|---|---|---|---|---|---|
| (1,1) | 100% | $\sqrt{2}$ | 0.50 | 25% | 5.67% |
| (2,2) | 75% | 2.51 | 0.84 | 7.9% | 0.82% |
| (2,3) | 68.75% | $2\sqrt{2}$ | 0.91 | 4.5% | 0.17% |
| (3,3) | 62.5% | 3.17 | 0.96 | 2.1% | 0.14% |

also for these pulses that Alice and Bob are the most likely to not end up with the same bit, thus increasing the measured QBER. Unfortunately for Alice and Bob, as soon as $n > 1$ or $m > 1$, it is also for these pulses that the probability of detection is the smallest for a fixed $n$ and $m$. Since this probability decreases for higher-order multiple-photon absorption, it also means that, rather counter-intuitively, the probability $P_{\text{error}}^{(n,m)}$ that Eve makes a mistake guessing a bit shared by Alice and Bob decreases with higher violation of Bell inequality and with lower QBER (see Table 1).

## 6. Discussion and countermeasures

An unwanted feature that could betray Eve's attack is that the sum of coincidences depends on the measurement settings $\theta_A$ and $\theta_B$. However, Eve can remove this unwanted effect entirely by driving different detection patterns for Alice and Bob, in a similar way to what was done by Larsson [10] and Gisin [11] in their hidden-variable models. The simplest method would be to alternatively drive a single photon absorption on one side, and a multiple-photon absorption on the other side. The sampling is then always fair on the side driven to a single photon absorption, and the total number of coincidences becomes independent of the measurement angles.

Monitoring the single counts as such would not betray Eve's attack, because all the channels are treated on equal footing by the attack. Similarly, monitoring the rates of coincidences is also unlikely to betray Eve's attack because they are all rotationally invariant: they depend only on the angle difference $|\theta_B - \theta_B|$, as is the case for a genuine singlet state. The correlation function does differ slightly from the $-V\cos 2(\theta_A - \theta_B)$ expected for a singlet state in a lossy channel, but this difference is small and would require more measurement settings than in the standard Ekert and BBM92 protocols.

However, a multiple-photon absorption is a second order phenomenon, and is as such very inefficient. Eve would thus need to send more pairs per second to compensate for this additional inefficiency. The trouble for Eve would be that, assuming independent errors, the singles scale linearly with the efficiency whereas the coincidences scale with the square of the efficiency. Alice and Bob would thus be able to spot the attack by monitoring the absolute efficiency, calculated as the ratio of the coincidences over the singles [16].

Using frequency filters so that only a known range of restricted frequencies can reach the detectors is another obvious way to limit the possibility of multiple-photon absorption in the ideal case, but Alice and Bob still need to guarantee that the dominant way to get a click in their necessarily imperfect detector is through a single-photon absorption, even at those frequencies allowed by the filters.

It should be emphasized that the principle of the attack is not bound to multiple-photon absorption. The essential feature is really that Eve can prevent single photons to be detected by Alice and Bob, regardless of the technique used for this purpose. This issue is quite critical considering the existing faked-state attacks that have already been successfully implemented against QKD protocols [17,18], by forcing the detectors to exit the single-photon sensitive Geiger mode [18]. It is conceivable that this type of blinding attack could be tailored to allow only a $n$-photon absorption in the detectors used by Alice and Bob. Such a combination of blinding detector attack together with our source designed for a multiple-photon absorption attack could then constitute a robust attack even against Ekert protocol, something that the blinding attack alone could not do [17].

An active method for Alice and Bob to detect the attack would be to implement a fair sampling test [19] adapted to quantum key distribution with entangled states [20]. This fair sampling test does not introduce any loss and can be performed locally and unilaterally on either side during the production of the key. The setup used for the Fair Sampling test would also make the other attacks against QKD protocols more complicated (if not impossible) to implement [21].

## 7. Conclusion

To our knowledge the multiple-photon absorption attack is in fact the first explicit attack against Ekert protocol with light sources.

The multiple-photon absorption attack does not require Eve to be physically located between Alice and Bob and to intercept what was intended for either of them, as in a typical intercept-and-resend attack [17,18]. The only requirement is that Eve has managed at some point in the past to replace the source of entangled state by her own mixture of separable states. She just needs to know the state $\lambda$ of each pulse and that is something that Eve could have set deterministically in the source. Once this is done, the security of the QKD protocol is compromised by anyone who happens to know the polarization of the pulses as a function of time $\lambda(t)$.

By mimicking the statistics of an entangled state well enough to pass the security checks normally undertaken by Alice and Bob, the attack can in principle work against any entanglement-based quantum key distribution protocol, independently of which protocol is chosen by Alice and Bob.

## References


1. V. Scarani, et al *Rev. Mod. Phys.* **81** 1301 (2009).
2. A.K. Ekert, *Phys. Rev. Lett.* **67** 661 (1991).
3. C.H. Bennett, G. Brassard and N.D. Mermin, *Phys. Rev. Lett.* **68** 557 (1992).
4. G. Adenier, I. Basieva, A. Khrennikov and N. Watanabe, arXiv:1011.4740
5. R. Braunstein, *Phys. Rev.* **125** 475 (1961).
6. J. F. Clauser, M. A. Horne, A. Shimony and R. A. Holt *Phys. Rev. Lett.* **23** 880 (1969).
7. T. Scheidl *et al, New J. Phys.* **11** 085002 (2009).
8. C. Branciard, N. Gisin, B. Kraus and V. Scarani, *Phys. Rev. A.* **72** 032301 (2005).
9. P.M. Pearle, *Phys. Rev. D* **2** 1418 (1970).
10. J.-Å. Larsson, *Phys. Rev. A* **57** 3304 (1998).'
11. N. Gisin and B. Gisin, *Phys. Lett. A* **260** 323 (1999).
12. G. Adenier, *Am. J. Phys.* **76** 147 (2008).
13. G. Adenier, *AIP Conf. Proc.* **1101** 8 (2009).
14. X. Ma, Chi-HangFred Fung, H.K. Lo, *Phys. Rev. A* **76** 012307 (2007).
15. C. Erven, C. Couteau, R. Laflamme, and G. Weihs, *Opt. Express* **16** 16840 (2008).
16. D. N. Klyshko, *Sov. J. Quantum Electron* **10**, 1112-1116 (1981).
17. V. Makarov and J. Skaar, *Quant. Inf. Comp.* **8** 0622 (2008).
18. L. Lydersen, C. Wiechers, C. Wittmann, D. Elser, J. Skaar and V. Makarov, *Optics Express* **18** 27938 (2010).
19. G. Adenier, *J. Russ. Laser Res.* **29** 409 (2008).
20. G. Adenier, N. Watanabe and A. Yu. Khrennikov, arXiv:1004.1242
21. G. Adenier, I. Basieva, A. Yu. Khrennikov, M. Ohya, N. Watanabe, arXiv:1102.3366

Quantum Bio-Informatics V
© 2013 World Scientific Publishing Co. Pte. Ltd.
pp. 181–185

# PROTEIN SEQUENCE ALIGNMENT TAKING THE STRUCTURE OF PEPTIDE BOND

TOSHIHIDE HARA*, KEIKO SATO and MASANORI OHYA

*Department of Information Sciences, Tokyo University of Science,
2641 Yamazaki, Noda City, Chiba, Japan
*E-mail: hara@is.noda.tus.ac.jp*

In a previous paper[1] we proposed a new method for performing pairwise alignment of protein sequences. The method, called MTRAP, achieves the highest performance compared with other alignment methods such as ClustalW2[2,3] on two benchmarks for alignment accuracy.

In this paper, we introduce a new measure between two amino acids based on the formation of peptide bonds. The measure is implemented into MTRAP software to further improve alignment accuracy.

Our alignment software is available at
"http://www.rs.noda.tus.ac.jp/%7Eohya-m/".

*Keywords*: Sequence analysis; Sequence alignment; Dihedral angle; Ramachandran plot; Peptide bond.

## 1. Introduction

A protein is a polypeptide chain made up of amino acid residues. Each amino acid residue in a protein has the phi and psi dihedral angles.[4] Dihedral angles of amino acid residues are of importance in understanding protein structure[5,6] because they specify the backbone conformation of a protein. In this paper, we introduce a new measure of difference between a pair of amino acid sequences based on dihedral angles of amino acid residues. Recently, we developed a high accurate alignment method called MTRAP.[7] The measure is implemented into MTRAP software to further improve alignment accuracy.

## 2. The measure of difference between a pair of sequences used in MTRAP

We use the following notations. Let $\Omega$ be the set of all 20 amino acids, $\Omega^*$ be the union of $\Omega$ and $\{*\}$ (where $*$ is indel (gap)), and $[\Omega^*]^n$ be the set of sequences of length $n$ consisting of elements of $\Omega^*$. We call an element of $\Omega$ a residue and an element of $\Omega^*$ a symbol. In addition, let $\Gamma \equiv \Omega \times \Omega$ be the direct product of two $\Omega$s and $\Gamma^* \equiv \Omega^* \times \Omega^*$.

MTRAP uses two types of functions as quantitative measures of difference between two sequences $A = (a_1, a_2, \cdots, a_n)$ and $B = (b_1, b_2, \cdots, b_n) \in [\Omega^*]^n$ : $d_{\mathrm{sub}}(A, B)$ and $d_{\mathrm{trans}}(A, B)$. The function $d_{\mathrm{sub}}(A, B)$ is defined as follows:

$$d_{\mathrm{sub}}(A, B) = \sum_{i=1}^{n} \tilde{s}(a_i, b_i),\tag{1}$$

where $\tilde{s}(a, b)$ is a normalized difference of symbols $a$ and $b$ expressed as a value between 0 and 1. This normalize difference is derived by converting a substitution matrix score such as PAM[8] and BLOSUM.[9] Gaps require special treatment (see ref. 7 for details).

Furthermore, we consider the measure of difference between a pair of sequences based on two consecutive pairs of sites. For the two sequences $A$ and $B$, the function $d_{\mathrm{trans}}(A, B)$ is defined as follows:

$$d_{\mathrm{trans}}(A, B) = \sum_{i=1}^{n-1} \tilde{t}(u_i, u_{i+1}),\tag{2}$$

where $u_i = (a_i, b_i) \in \Gamma^*$ and $\tilde{t}(u, v)$ is a normalized difference of sites $u$ and $v$. This difference takes value in the range of $[0, 1]$ with 0 indicating that the correlation between $u$ and $v$ is strong (see ref. 8 for details).

The difference measure used in the MTRAP is defined by a weighted mixture of $d_{\mathrm{sub}}(A, B)$ and $d_{\mathrm{trans}}(A, B)$:

$$d_{\mathrm{MTRAP}}(A, B) = (1 - \varepsilon)\, d_{\mathrm{sub}}(A, B) + \varepsilon d_{\mathrm{trans}}(A, B).\tag{3}$$

Here $\varepsilon$ is a degree of mixture set to 0.775 by default.


## 3. A new measure based on dihedral angles of amino acids

Let us introduce a new measure of difference mentioned in Sec. 1.

We use Ramachandran plot stored in DASSD.[10] Each dot on the Ramachandran plot $((\varphi, \psi) = (-180°, 180°) \times (-180°, 180°))$ shows the phi (x-axis) and psi (y-axis) value for an amino acid. We first divide the each axe into $N$ parts equally: $\varphi = \cup_{i=1}^{N} \triangle_i^{\varphi}$, $\psi = \cup_{i=1}^{N} \triangle_i^{\psi}$, $\triangle_i^{\varphi} \cap \triangle_j^{\varphi} = \phi$, $\triangle_i^{\psi} \cap \triangle_j^{\psi} = \phi$, then we count the number of dots in each region. Let $n_a(i, j)$ and $n_b(i, j)$ be the number of dots in each $\triangle_i^{\varphi} \times \triangle_j^{\psi}$ region for amino acid $a$ and $b$ (where

$i, j = 1, 2, \cdots, N$). Then we define the following difference of dihedral angle between amino acid $a$ and $b$:

$$\tilde{r}(a, b) = \frac{\sum_{i=1}^{N} \sum_{j=1}^{N} \left| \frac{n_a(i,j)}{N_a} - \frac{n_b(i,j)}{N_b} \right|}{2}, \tag{4}$$

where $N_a = \sum_{i=1}^{N} \sum_{j=1}^{N} n_a(i, j)$, $N_b = \sum_{i=1}^{N} \sum_{j=1}^{N} n_b(i, j)$, and $0 \leq \tilde{r}(a, b) \leq 1$. Using $\tilde{r}(a, b)$, a difference between the two sequence $A$ and $B$ is defined as follows:

$$d_{\text{ang}}(A, B) = \sum_{i=1}^{n} \tilde{r}(a_i, b_i). \tag{5}$$

Finally, we incorporate the function $d_{\text{ang}}(A, B)$ into equation (3), the difference measure defined in this paper is a weighted mixture of $d_{\text{sub}}, d_{\text{trans}}$ and $d_{\text{trans}}$:

$$d_{\text{NEW}}(A, B) = (1 - \varepsilon) \left\{ (1 - \gamma) d_{\text{sub}}(A, B) + \gamma d_{\text{ang}}(A, B) \right\} + \varepsilon d_{\text{trans}}(A, B), \tag{6}$$

where $\gamma$ is a degree of mixture.

We calculate the above $\tilde{r}$ with $N = 72$ using the data-set obtained from DASSD[10] which contains the values of dihedral angles $\varphi, \psi$ and the "secondary structure details" for short fragments of amino acids having 1, 3 and 5 lengths. In this paper, we use the all data of length 1 and identify $\gamma = 0.1$ as the value minimizing cross-validation error.

## 4. Results and discussion

We compare the accuracy of new method with ClustalW2,[3] MAFFT,[11] MUSCLE,[11] TCoffee[12] and MTRAP[7] by using the PREFAB benchmark test.[13] We use the all 1682 pairwise alignments on PREFAB as reference alignments. The accuracy of generated alignment is measured by Q score.[13] The Q score is defined as the ratio of the number of residue pairs correctly aligned in the test alignment (i.e., the alignment generated by a specified method such as MTRAP) to the total number of residue pairs aligned in the reference alignment. When all pairs are correctly aligned, the score have a maximum value 1, and when no-pairs are correctly aligned, the score have a minimum value 0. The performance is evaluated on the three different range of sequence identity of the references: 0-15%, 15-30% and 0-100%. The number of alignments in each range is 423, 911 and 1682, respectively.

In Table 1 we report the accuracy of the new method, the original MTRAP,[7] ClustalW2[3] ver. 2.0.12, MAFFT[11] ver. 6.864b, MUSCLE[13] ver. 3.7 and TCoffee[12] ver. 8.47.

The new method achieves the best results in all ranges. However, the alignment accuracy is similar to the original MTRAP. The new method shows a slight improvement in alignment accuracy.

Table 1.   Median and average of Q scores on PREFAB database

| Method | Median / Average of Q scores for each %ID range (PREFAB) | | |
| --- | --- | --- | --- |
| | 0-15%(423) | 15-30%(911) | 0-100%(1682) |
| New Method | *0.190 / 0.250* | *0.750 / 0.678* | *0.729 / 0.616* |
| MTRAP | 0.188 / *0.250* | *0.750* / 0.676 | 0.728 / 0.615 |
| ClustalW2 | 0.123 / 0.199 | 0.727 / 0.647 | 0.701 / 0.585 |
| MAFFT | 0.102 / 0.170 | 0.686 / 0.627 | 0.650 / 0.568 |
| MUSCLE | 0.138 / 0.205 | 0.701 / 0.636 | 0.675 / 0.581 |
| TCoffee | 0.133 / 0.198 | 0.706 / 0.645 | 0.680 / 0.585 |

*Note*: The highest values in each %ID range are shown in bold style. The number in parentheses denotes the number of alignments in each sequence identity range. ClustalW2, MUSCLE and TCoffee run with their default parameters. MAFFT runs with "linsi" option.

## 5.  Conclusion

Although the new method shows the best performance in comparison with the other method, it gives slightly better accuracy than the original MTRAP contrary to our expectation. From this result, we come to the conclusion that to consider correlations between two consecutive pairs of sites might covers to consider dihedral angles of amino acid residues.

## References


1. T. Hara, K. Sato and M. Ohya, *QP-PQ: Quantum Probability and White Noise Analysis (Quantum Bio-Informatics IV)* **28**, 129 (2011).
2. J. D. Thompson, D. G. Higgins and T. J. Gibson, *Nucleic Acids Res.* **22**, 4673(Nov 1994).
3. M. A. Larkin, G. Blackshields, N. P. Brown, R. Chenna, P. A. McGettigan, H. McWilliam, F. Valentin, I. M. Wallace, A. Wilm, R. Lopez, J. D. Thompson, T. J. Gibson and D. G. Higgins, *Bioinformatics* **23**, 2947(Nov 2007).
4. G. Ramachandran, C. Ramakrishnan and V. Sasisekharan, *Journal of molecular biology* **7**, p. 95 (1963).
5. R. Laskowski, M. MacArthur, D. Moss and J. Thornton, *Journal of applied crystallography* **26**, 283 (1993).
6. R. Hooft, C. Sander and G. Vriend, *Computer applications in the biosciences: CABIOS* **13**, 425 (1997).
7. T. Hara, K. Sato and M. Ohya, *BMC Bioinformatics* **11**, p. 235 (2010).

8. M. O. Dayhoff, R. M. Schwartz and B. C. Orcutt, *Atlas of Protein Sequence and Structure* **5(3)**, 345 (1978).
9. S. Henikoff and J. G. Henikoff, *Proc. Natl. Acad. Sci. U.S.A.* **89**, 10915(Nov 1992).
10. S. Dayalan, N. Gooneratne, S. Bevinakoppa and H. Schroder, *Bioinformation* **1**, p. 78 (2006).
11. K. Katoh, K. Misawa, K. Kuma and T. Miyata, *Nucleic Acids Res.* **30**, 3059(Jul 2002).
12. C. Notredame, D. G. Higgins and J. Heringa, *J. Mol. Biol.* **302**, 205(Sep 2000).
13. R. C. Edgar, *Nucleic Acids Res.* **32**, 1792 (2004).

Quantum Bio-Informatics V
© 2013 World Scientific Publishing Co. Pte. Ltd.
pp. 187–191

# SPACE - TIME - NOISE (RAUM - ZEIT - RAUSCHEN)

TAKEYUKI HIDA

*Professor Emeritus, Nagoya University, Japan*
*E-mail: takeyuki@math.nagoya-u.ac.jp*

## 1. Introduction

In the mathematical study of a random complex phenomenon, there is an ideal road map, by which we start with Reduction. This means that it would be fine if we can find a system of independent random variables that contains the same information as the given random phenomena, and it is expressed as a (non-random) function of those random variables. The random variables may be ideal random functions which are assumed to be i.i.d. (independent identically distributed) and atomic. Such a system of idealized elemental random variables will be called a noise.

The well-known good examples are the (Gaussian) white noise $\dot{B}(t)$ and the Poisson noise $\dot{P}(t)$. They are depending on the time parameter $t$.

The time derivative of a compound Poisson process is not quite a noise, since it is decomposed into (formally speaking) atomic Poisson noises,

In this note we shall show that another kind of noise, in fact, depending not on the time, but on a space parameter $u$, does arise naturally. We shall also give a plausible probabilistic interpretation on this fact, and discuss its properties of new noise by comparing with those depending on time parameter; some properties are in parallel and some others are in a dual manner.

Some of other related properties will be discussed as much as space-time permits.

If the family of random variables are parametrized by a discrete parameter, either time or space, then we consider a sequence of i.i.d. random variables. However, if we come to a continuous parameter case, a continuous analogue of this is not acceptable. For one thing, in this case the probability distribution associated with such family can not be an abstract Lebesgue space. So, calculus does not follow smoothly.

We now change our eyes slightly. Instead of i.i.d. we take sum of random variables with stationary independent increments. As a counter part of this we can consider the continuous parameter case, for which we take an additive process with stationary increments.

With this understanding we proceed to find possible noises.


## 2. Two kinds of noise

We shall show that two different kind of noises can arise naturally from a limit theorem in probability theory. In fact, our interpretation comes from very basic and elementary background.

Case I. Take the unit interval $I = [0, 1]$, Let $\Delta^n = \{\Delta_j^n, 1 \leq j \leq 2^n\}$ be the partition of $I$. To fix the idea, we assume that $|\Delta_j^n| = 2^{-n}$.

To each subinterval $\Delta_j^n$ we associate a random variable $X_j^n$. We assume that $\{X_j^n\}$ are i.i.d. with mean 0 and finite variance $v$.

Let $n$ be larger, Then, we can appeal to the central limit theorem to have a standard Gaussian distribution by observing

$$S_n = (\sum_1^{2^n} X_j^n)/\sqrt{2^n v}.$$

If we take subinterval $[a, b]$ of $I$, then the same trick gives us a Gaussian distribution $N(0, b - a)$.

We may therefore consider $S_n$ as an approximation of a Brownian motion and each $X_j^n$ approximates elemental random variable. Because of the central limit theorem, Gaussian noise is involved.

Case II. The interval $I$ and its partition $\Delta^n$ are the same as in $I$. The independent random variables $X_j^n$ are i.i.d., but all are subject to a probability distribution such that

$$P(X_j^n = 1) = p_n, \ \ P(X_j^n = 0) = 1 - p_n.$$

Let $p_n$ be getting small as $n$ is getting large keeping the relation

$$2^n p_n = \lambda.$$

for some constant $\lambda > 0$. Then, by the law of small probability, the sum $S_n$ is an approximation of a Poisson random variable $P(\lambda)$ with intensity $\lambda$.

We then come to a partition of $I$ as in Case I, however the idea to form the partial sums of random variables is different. Keep the random variables $X_j^n$ just as above and divide the sum $S_n$ into partial sums in a manner that

$$S_n = \sum_k S_n^k,$$

where $S_n^k = \sum_{j(k)+1}^{j(k+1)} X_j^n$, with $1 < j(1) < j(2) < \cdots j(m) = 2^n$.

Then, assuming tacitly that each $j(k)$ is large, we see that each $S_n^k$ is an approximation of Poisson random variable $P(\lambda_k)$. From the method of forming the partial sum we can easily prove the following result.

**Theorem 2.1.** *The Poisson random variables $P(\lambda_k)$ thus obtained are mutually independent and*

$$P(\lambda) = \sum_k P(\lambda_k).$$

Before we come to further observation on the above results, we should like to mention some short remarks.

Remark. On the type of probability distributions.

We claim that all the Gaussian distributions are of the same type (a constant, the exceptional Gaussian variable is excluded. It is also excluded in Case I.)

On the other hand, Poisson type distributions with different intensities are not of the same type. This can be proved by the formula of characteristic function. By the way, Poisson type distribution means a distribution of $uP(\lambda) + c$, $c$ may be ignored. Namely, we can compare two characteristic functions

$$\varphi_1(z) = e^{\lambda_1(e^{iz}-1)}$$
$$\varphi_2(z) = e^{\lambda_2(e^{iz}-1)}$$

These two functions of $z$ can not be exchanged by any affine transformation of $z$ if $\lambda_1 \neq \lambda_2$. - end of Remark -.

We can say that 1) We have a freedom to choose $\lambda$ arbitrary. Hence, we can form, by the sum of i.i.d. random variables, random variables with different type.

2) The intensity is taken to be a space parameter. This also comes from our construction. The above construction also shows it is additive in $\lambda$

3) Multiplication by a constant to Poisson type variable is possible but the constant can be a label. So, take a constant $u = u(\lambda)$ as a label of the intensity. The function $u(\lambda)$ is therefore univalent. In view of this fact, we can form an inverse function $\lambda = \lambda(u)$ which is to be monotone.

## 3. A noise with space parameter

It is meaningful to consider a system of Poisson type random variables with different intensities. From the viewpoint of information theory.

There is an important property that is deserved to be mentioned. This is a simple rephrasement of the swell-known Raikov theorem

**Proposition 3.1.** *A Poisson type random variable is atomic, i.e. it is elemental.*

Proof. Suppose a Poisson variable $X$ were decomposed into two independent random variables

$$X = Y + Z.$$

Then, both $Y$ and $Z$ are Poisson variable. Except a trivial case, the sigma-field generated by $X$ is strictly smaller than that generated by two variables.

We shall now think of the sum of independent Poisson random variables with different intensities like

$$X = \sum u(\lambda_j) P(\lambda_j),$$

where $\lambda_j$'s are different and so are $u(\lambda_j)$'s.

Then, the characteristic function $\varphi_X(z)$ of $X$ is given by

$$\varphi(z) = \prod e^{\lambda_j (e^{izu(\lambda_j)} - 1)}.$$

By using the one-to-one correspondence between $u$ and $\lambda$ we have

$$\varphi(z) = \prod e^{\lambda(u_j)\left(e^{izu_j} - 1\right)}.$$

The passage to infinity gives us a characteristic functional $C(\xi)$, It is of the form

$$C(\xi) = \exp\left[\int \left(e^{iu\xi(u)} - 1\right) \lambda(u)\right] du.$$

which has been discussed by Si Si in this conference [10].

## 4. Concluding remarks

[I] The idea of "Reduction" can be seen in the lecture notes by J. -L. Lions [7]. Also, we note that a concrete expression of realizing the idea of reduction can be seen through Lévy's stochastic infinitesimal equation. See [6].

[II] The term "noise" is more familiar in the communication theory in electrical engineering. There a noise is a disturbance against smooth transmission of message. On the other hand, it, e.g. white noise, contains maximum information ( in terms of entropy) under the limited power.

## References


1. T. Hida, Analysis of Brownian functionals. Carleton Math. Lecture Notes. no. 13, July, 1975.
2. T. Hida and Si Si, Lectures on white noise functionals. World Scientific Pub. Co. 2008.
3. T. Hida, The concept of infinite dimension in the analysis of white noise generalized functionals. RIMS Ojima Conf. Oct. 2007. (in Japanes.)
4. P. Lévy, Processus stochastiques et mouvement brownien. Gauthier-Villars, 1948, 2éme ed. 1965.
5. P. Lévy, Problèmes concrets d'analyse fonctionnelle. Gauthier-Villars, 1951.
6. P. Lévy, Univ. of California Publication, 1953,
7. J. -L. Lions, The earth, planet the role of mathematics and of super computers. (in Spanish), Inst. Espana. Espas a Calpe. 1990.
8. J. Mikusinski, On the square of the Dirac Delta-distribution. Bull. de l'Academie Polonaise des Sciences. Ser. Sci. math. astr. et Phys. XIV, no. 9 (1966) 511-513.
9. Si Si, An aspect of quadratic Hida distributions in the realization of a duality between Gaussian and Poisson noises. Infinite Dim. Analysis, Quantum Prob. and Related Topics. vol. 11.(2008), 109-118.
10. Si Si, A new noise depending on a space parameter and its application. QBIC2011 Conf. paper. 2011.
11. L. Streit, Feynman integrals as generalized functions on path space: Things done and open problems.
12. H. Weyl, Raum Zeit Materie; English translation.by H.L.Brose. Dover Pub.1922.

Quantum Bio-Informatics V
© 2013 World Scientific Publishing Co. Pte. Ltd.
pp. 193–202

# QUANTUM ALGORITHM FOR PROTEIN FOLDING AND ITS COMPUTATIONAL COMPLEXITY

S.IRIYAMA, M.OHYA AND I.YAMATO

*Tokyo university of science*

We have studied applications of quantum algorithm for the problems in bio-informatics. The protein folding problem is known as one of the difficult problems which we do not solve in polynomial time of input data. We constructed a quantum algorithm for the protein folding problem using our new quantum binary search algorithm, and discussed its computtational complexity. We prove that the computational complexity is polynomial of the number of molecular in the simulation.

## 1. Introduction

As one of main themes of QBIC, we have discussed applications of quantum algorithm for the problems of bio-informatics. There are many problems which we do not know how to solve them in polynomial time, for example, alignment problem of amino acid sequence and protein folding problem. Protein folding problem is to know why proteins can fold into a specific conformation within a finite time. In fact proteins fold spontaneously into a well-ordered three dimensional structure depending on its amino acid sequence encoded by the gene. Yamato et.al. have studied the way to solve it using a classical parallel computer for many years. They constructed the method which uses molecular dynamics with Brownian motion, and it improved a time scale to simulate in nano second[6]. However it is still a short time because the protein folding appears around milli second.

In this paper, we consider a quantum algorithm to solve this problem using a technique of quantum binary search[5] and discuss the computational complexity of it.

## 2. Quantum Algorithm

First, we explain foundations of quantum algorithm. A quantum algorithm is constructed by the following steps:

(1) Prepare a Hilbert space $\mathcal{H} = \mathbb{C}^{\otimes n}$

(2) Construct an initial state $|\psi_{in}\rangle \in \mathcal{H}$.

(3) Construct unitary operators $U$ to solve the problem.

(4) Apply them for the initial state and obtain a result state $|\psi_{out}\rangle = U |\psi_{in}\rangle$.

(5) If necessary, amplify the probability of correct result.

(6) Measure an observable with the result state.

In the first step, we define the Hilbert space depending on the problem. Let $\mathbb{C}^2$ be a Hilbert space spanned by $|0\rangle = \binom{1}{0}$ and $|1\rangle = \binom{0}{1}$, a normalized vector $|\psi\rangle = \alpha |0\rangle + \beta |1\rangle$ on this space is called a qubit. Since we can use a superposition of $|0\rangle$ and $|1\rangle$ as an initial state vector, the quantum algorithm is more effective than classical one.

One can apply Hadamard transformation

$$U_H = \frac{1}{\sqrt{2}} \begin{pmatrix} 1 & 1 \\ 1 & -1 \end{pmatrix}$$

to create a superposition. Hadamard transformation has a very important role in a quantum algorithm.

Applying $(U_H)^{\otimes n}$ to the vector $|\psi\rangle = |0\rangle \otimes \cdots \otimes |0\rangle \in \left(\mathbb{C}^2\right)^{\otimes n}$, we have

$$(U_H)^{\otimes n} |\psi\rangle = \frac{1}{\sqrt{2^n}} \sum_{i=0}^{2^n - 1} |e_i\rangle ,$$

where $\{|e_i\rangle\}$ is a complete orthonormal system of $\left(\mathbb{C}^2\right)^{\otimes n}$ defined as

$$|e_0\rangle = |0\rangle \otimes \cdots \otimes |0\rangle$$
$$|e_1\rangle = |1\rangle \otimes |0\rangle \otimes \cdots \otimes |0\rangle$$
$$\cdots$$
$$|e_{2^n - 1}\rangle = |1\rangle \otimes \cdots \otimes |1\rangle$$

One can represent any positive integers less than $2^n$ by $|e_i\rangle$.

Here we introduce unitary gates, which are NOT gate, Controlled gate. We call these gates fundamental gates. We can also construct AND and OR gate by considering the product of them and some implementations[3]. The NOT gate $U_{NOT}$ is defined on a Hilbert space $\mathbb{C}^2$ as

$$U_{NOT} = |1\rangle \langle 0| + |0\rangle \langle 1| .$$

Controlled gate $U_C (U)$ for a unitary operator $U$ is given by

$$U_C (U) = |0\rangle \langle 0| \otimes I + |1\rangle \langle 1| \otimes U$$

Controlled-Controlled NOT gate $U_{CC}(U_{NOT})$ are constructed by the above gates, and it can calculate logical And with dust qubits

$$U_{CC}(U_{NOT})|x\rangle \otimes |y\rangle \otimes |0\rangle = |x\rangle \otimes |y\rangle \otimes |x \wedge y\rangle$$

We define the computational complexity of quantum algorithm as the number of fundamental gates in it.

**Definition 2.1.** We say the computational complexity is $n$ if the unitary operator $U$ is constructed by a product of $n$ fundamental gates.

Moreover, we say the algorithm is efficient if the computational complexity is a polynomial of $\log N$ where $N$ is a size of input. Note that the decomposition of a given unitary gate into a product of fundamental gates is not unique.

### 2.1. *OV algorithm*

Let $X$ be a set of literals, and a truth assignment $t$ on $X$ is given by a function from $X$ to $\{0, 1\}$ holding

$$t(x) = 1 - t(\bar{x})$$

for all $x \in X$. Let $f$ be a conjunction of clauses which are disjunctions of literals. For a set of clauses $C = \{C_i\}$ and a truth assignment $t$, $f(t)$ is defined by

$$f(t) = \bigwedge_{C_i \in C} \bigvee_{x_j \in C_i} t(x_j)$$

$f$ is said satisfiable if there exists a truth assignment such that $f(t) = 1$. SAT problem is given by the following.

**Definition 2.2.** (SAT problem) Decide whether $f$ is satisfiable or not.

This problem is called a decision problem. For all input $x$, we have to check whether $f(x) = 1$ or not.

Ohya and Volovich proposed an efficient algorithm for SAT problem, an NP complete problem[1,2,3]. Let $P_0$ and $P_1$ be projections to the state vectors $|0\rangle$ and $|1\rangle$, respectively, and they generate an Abelian algebra. The final state of the OV algorithm is given by

$$\rho = \left(1 - q^2\right) P_0 + q^2 P_1$$

where

$$q^2 = card\{t | f(t) = 1\}$$

In this algorithm, they considered the following quantum channel for Abelian state $\rho$,

$$\Lambda_{CA}^* (\rho) \equiv \frac{I + g_a (\rho) \sigma_3}{2}$$

where $\sigma_3$ is the z-component of Pauli matrices, and $g_a$ is a logistic map with a parameter $a \in [0, 4]$ defined by

$$g_a (x) = ax (1 - x)$$

$\Lambda_{CA}^*$ is so-called a Chaotic Amplifier. To find a proper value $k$, we measure the value of $\sigma_3$ in the state $\rho_k = (\Lambda_{CA}^*)^k (\rho)$ such that

$$M_k \equiv \mathrm{tr}\rho_k \sigma_3 \geq \frac{1}{2}$$

The following theorem is proven in [1,2,3].

**Theorem 2.1.** *For a number of literals $n$, $k$ satisfies the following inequality if $q^2 \neq 0$,*

$$k \leq \left[ \frac{5}{4} (n - 1) \right].$$

### 2.2. *Language classes*

Here we review language classes defined by Turing machine.

**Definition 2.3.** NP: A problem $X$ is in a class NP if it is verifiable in polynomial time by a deterministic Turing machine.

**Definition 2.4.** NP-complete: The problems which all problems in the class NP are reduced to in polynomial time, and belong to NP (e.g., SAT problem).

**Definition 2.5.** NP-intermediate: The problems which are in the class NP but are neither in the class P nor NP-complete (e.g., prime factorization).

**Definition 2.6.** NP-hard: A problem $X$ is called NP-hard if and only if there exists the NP-complete problem which is reduced to $X$. (e.g., search problem).

## 3. Quantum Algorithm for Binary Search

In this section, we briefly explain a quantum algorithm for binary search. The details are given in the paper[5].

Let $X = Y = \{0, 1, \cdots, 2^n - 1\}$, and $f$ a function from $X$ to $Y$. Here we assume that $f$ is given as a unitary operator $U_f$ on the Hilbert space $\mathcal{H} = \left(\mathbb{C}^2\right)^{\otimes n}$.

Let $m$ be a positive number, and we assume that $m$ is written by a polynomial of $n$. Let $\mathcal{H} = \left(\mathbb{C}^2\right)^{\otimes n+m+1}$ be a Hilbert space. $m$ qubits are used for the computation of $f$, so that this quantum algorithm has dust qubits. However, this algorithm will be done in polynomial time if a size of dust qubits $m$ and the computational complexity of $f$ are polynomial of $n$.

We construct the following algorithm $M_1$, and let $|\psi_{1,in}\rangle = |0^n\rangle \otimes |0^m\rangle \otimes |0\rangle \in \mathcal{H}$ be an initial vector.

[Phase 1]

Step1: Apply Hadamard gates from 2nd qubit to $n$-th qubit.

$$I \otimes U_H^{\otimes n-1} \otimes I^{m+1} |\psi_{1,in}\rangle = \frac{1}{\sqrt{2^{n-1}}} |0\rangle \otimes \left(\sum_{i=0}^{2^{n-1}-1} |e_i\rangle\right)$$

$$\otimes |0^m\rangle \otimes |0\rangle$$

$$= |\psi_{1,1}\rangle$$

Step2: Apply the unitary gate $U_f$ to compute $f$ for the superposition made in Step1, and store the result in $n + m + 1$-th qubit.

$$U_f |\psi_{1,1}\rangle = \frac{1}{\sqrt{2^{n-1}}} |0\rangle \otimes \left(\sum_{k=0}^{2^{n-1}-1} |e_k\rangle \otimes |z_k\rangle \otimes |f(0, e_k)\rangle\right)$$

$$= |\psi_{1,2}\rangle$$

where $z_k$ is dust qubits depending on $i$.

Step3: Take a partial trace except for the last qubit.

$$\mathrm{tr}_{1,\cdots,n+m}\rho = (1 - p) |0\rangle \langle 0| + p |1\rangle \langle 1|$$

$$\rho = |\psi_{1,2}\rangle \langle \psi_{1,2}|$$

where $p = card\left(f(x) = 1\right)/2^{n-1}$.

Step4: Apply the Chaos Amplifier to check whether the last qubit is 0 or 1.

After this algorithm, we know that the 1st bit denoted by $x_1$ is 0 (resp. 1) if the last qubit is 1 (resp. 0). We call $x_1$ a result of Phase 1.

Next we repeat this algorithm with the initial vector $|\psi_{2,in}\rangle = |x_1, 0^{n-1}\rangle \otimes |0^m\rangle \otimes |0\rangle$ with a modified Step1 as

[Phase 2]

Step1: Apply Hadamard gates from 3rd qubit to $n$-th qubit.

Then we obtain the 2nd bit $x_2$, and call $x_2$ the result of Phase 2.

In general, we write the algorithm $M_i\left(x_1, x_2, \cdots, x_{i-1}, 0^{n-i+1}\right)$ for an initial vector $|\psi_{i,in}\rangle = |x_1, x_2, \cdots, x_{i-1}, 0^{n-i+1}\rangle \otimes |0^m\rangle \otimes |0\rangle$ as the following:

[Phase $i$]

Step1: Apply Hadamard gates from $i+1$-th to $n$-th qubits.

$$I^{\otimes i} \otimes U_H^{\otimes n-i} \otimes I^{m+1} |\psi_{i,in}\rangle = \frac{1}{\sqrt{2^{n-i}}} |x_1, \cdots, x_{i-1}\rangle$$

$$\otimes \left( \sum_{k=0}^{2^{n-i}-1} |e_k\rangle \right) \otimes |0^m\rangle \otimes |0\rangle$$

$$= |\psi_{i,1}\rangle$$

Step2: Apply the unitary gate to compute $f$ for the superposition made in Step1, and store the result in $n+m+1$-th qubit.

$$U_f |\psi_{i,1}\rangle = \frac{1}{\sqrt{2^{n-i}}} |x_1, \cdots, x_{i-1}\rangle$$

$$\otimes \left( \sum_{k=0}^{2^{n-1}-1} |e_k\rangle \otimes |z_k\rangle \otimes |f\left(x_1, \cdots, x_{i-1}, e_k\right)\rangle \right)$$

$$= |\psi_{i,2}\rangle$$

Step3: Take a partial trace except for the last qubit.

$$\mathrm{tr}_{1,\cdots,n+m}\rho = (1-p)|0\rangle\langle 0| + p|1\rangle\langle 1|$$

$$\rho = |\psi_{i,2}\rangle\langle\psi_{i,2}|$$

Step4: Apply the Chaos Amplifier to find that last qubit is 1.

After this algorithm $M_i\left(x_1, x_2, \cdots, x_{i-1}, 0^{n-i+1}\right)$, we know the $i$-th bit such that $f(x) = 1$. Each $M_i$, $i \geq 2$ use the result of all $M_k$, $k < i$ to prepare the initial vector.

Note that this quantum algorithm allows to use dust qubits to calculate $f$. In the paper[8], we applied this algorithm to prime factorization, and discussed the computational complexity rigorously by constructing unitary gates.

## 4. Molecular Dynamics

### 4.1. *Governing equation*

We simulate motions of atoms according to the Newton equations. Let $N$ be a number of atoms in a system, $r_i \in \mathbb{R}^3$ and $m_i$ the position and mass of atom $i$. The Newton equations are given as

$$m_i \frac{d^2 r_i(t)}{dt^2} = F_i(t)$$

where $i = 1, 2, \cdots, N$, and $F_i(t)$ is the force on atom $i$. For the potential energy $V_{total}(r_1(t), r_2(t), \cdots, r_N(t))$ of the system depending on the position of $N$ atoms, $F_i(t)$ is written as

$$F_i(t) = -\nabla_i V_{total}(r_1(t), r_2(t), \cdots, r_N(t))$$

The potential energy $V_{total}$ is sum of the following five interaction terms[6].

$$V_{total} = V_{bond} + V_{angle} + V_{dihedral} + V_{van\ der\ Waals} + V_{electrostatic}$$

$$V_{bond} = \sum_{bonds} K_b(r - r_0), \quad V_{angle} = \sum_{angles} K_\theta(\theta - \theta_0)^2,$$

$$V_{dihedral} = \sum_{dihedrals} K_\phi(1 + \cos(n\varphi - \gamma))$$

$$V_{van\ der\ Waals} = \sum_{\substack{van\ der\ Waals \\ i,j\ pairs}} \left( \frac{A_{ij}}{r_{ij}^{12}} - \frac{B_{ij}}{r_{ij}^{6}} \right),$$

$$V_{electrostatic} = \sum_{\substack{electrostatic \\ i,j\ pairs}} \frac{q_i q_j}{\varepsilon r_{ij}}$$

where $K_b, K_\theta, K_\phi, A_{ij}, B_{ij}$ and $\varepsilon$ are the force constants.

### 4.2. *Verlet algorithm*

According to the Verlet algorithm, for a short time $h$, the position of $i$-th atom becomes

$$r_i(t + h) = 2r_i(t) - r_i(t - h) + h^2 F_i(t)$$

## 5. Quantum Algorithm for Protein Folding

Let $m$ be the positive integer representing the position of atoms, the Hilbert space for computation is

$$\mathcal{H} = \left(\mathbb{C}^2\right)^{\otimes 2mN + \mu}$$

where $\mu$ is the size of workspace. An initial vector is given by

$$|\psi_0\rangle = |r_1(0)\rangle \otimes |r_2(0)\rangle \otimes \cdots \otimes |r_N(0)\rangle \otimes \left|0^{mN+\mu}\right\rangle$$

where $r_i(0)$ $(i = 1, \cdots, N)$ is the initial position of $i$-th atom. Here we construct the unitary operator which calculate the position of time $t + h$ for all atoms. Using the information of the positions at time $t$, we know the algorithm to calculate $F_i(t)$. By the Verlet algorithm we can construct the unitary operator such that

$$
\begin{aligned}
U_{t,h} |\psi_t\rangle = {}& |r_1(t)\rangle \otimes |r_2(t)\rangle \otimes \cdots \otimes |r_N(t)\rangle \\
& \otimes |r_1(t+h)\rangle \otimes |r_2(t+h)\rangle \otimes \cdots \otimes |r_N(t+h)\rangle \otimes |x\rangle
\end{aligned}
$$

where $|x\rangle$ is the workspace for computation.

Therefore, we measure the position at time $t + h$, and repeat the above algorithm for the initial vector

$$|\psi_{t+h}\rangle = |r_1(t+h)\rangle \otimes |r_2(t+h)\rangle \otimes \cdots \otimes |r_N(t+h)\rangle \otimes \left|0^{mN+\mu}\right\rangle$$

We call this algorithm as a quantum MD(QMD).

In the previous studies[6], the most difficult part of protein folding is to calculate $F_i(t)$ because of taking summation running over the interactions of all pairs of atoms. Then we consider the following classical algorithm:

Let $M$ be the positive integer such as $M = O(\log N)$, and $\varepsilon$ the positive small number.

Step1: Split $N$ atoms into the $M$ atom groups for approximation

Step2: Calculate $\tilde{r}(t) = (\tilde{r}_1(t), \cdots, \tilde{r}_M(t))$ which is the positions of the center of atoms in each groups

Step3: Calculate the potential energy $\tilde{V}_{total}$ for $M$ atom groups according to the way of splitting.

Step4: Calculate the position $\tilde{r}(t+h)$ at time $t + h$, and recover the position $r'(t+h)$ of atoms

Step5: If $|r(t+h) - r'(t+h)| < \varepsilon$, we use this $M$ groups for simulation for a while.

Step6: After time $10h$, repeat this algorithm to update the splitting of molecular.

In this algorithm, each atoms belongs to one group. We keep the relative positions of each atoms and the center of the group.

The total number of way of splitting is at most $N!$, so that we propose the following quantum algorithm:

Step1: Construct a superposition of all the way of splitting according to [7]

Step2: Translate the way of splitting into the vector of positions $\tilde{r}(t)$

Step3: Calculate the potential energy $\tilde{V}_{total}$

Step4: Run the QMD for $M$ atoms, the potential $\tilde{V}_{total}$ and the vector of positions $\tilde{r}(t)$

Step5: If there exists the way of splitting such that
$|r(t+h) - r'(t+h)| < \varepsilon$ , change the flag qubit $|1\rangle$

Step6: Using the quantum binary search, we obtain the group of molecular

By this algorithm we can obtain $M$ groups of molecular for which we can make a good approximation to run the MD.


## 6. Computational Complexity

We define the computational complexity of quantum algorithm as the number of fundamental gates, $U_{NOT}, U_{CN}$ and $U_{CCN}$[1,2,3,4].


**Theorem 6.1.** *The computational complexity of QMD is polynomial of N*


**Proof.** In Step1 it requires $\lceil \log N \rceil N$ Hadamard gates, $N^2$ Toffoli gates and at most $N^2 \log N$ $U_{NOT}$ gates to construct the superposition.

In Step2 we construct the unitary operator $U_r$ to translate and its computational complexity is $O(N \log N)$

The computational complexity of Step3 and Step4 are same as the classical ones.

In Step5 it requires $N \log N$ gates for the calculation of $|r(t+h) - r'(t+h)|$ and at most $2N \log N$ for comparison with $\varepsilon$. Therefore Step5 needs $O(N \log N)$ gates.

In Step6 we use the same way of quantum binary search and the computational complexity is $O(N^3 \log 2N)$.

Therefore the total number of quantum gates is at most $O(N^3 \log 2N) \sim poly(N)$. $\square$

## 7. Conclusion

In this study we construct three algorithms, QMD, a classical approximation and a quantum search for the approximation. For QMD, we construct the unitary operator $U_{t,h}$, apply it to the position vector $|\psi_t\rangle$, and obtain $|\psi_{t+h}\rangle$. The computational complexity of classical algorithm for the approximation is $N!$, while the quantum algorithm is polynomial of $N$.

Once we obtain the good approximation, the computational complexity of each steps of QMD becomes $O\left(\log N\right)$.

As the further research, we consider the more effective quantum algorithm where we divide atoms into proteins and water part, and uses different algorithms to calculate each potentials.

## References


1. M.Ohya and I.V.Volovich, *Quantum computing and chaotic amplification*, J. opt. B, **5**,No.6 639-642, 2003.
2. M.Ohya and I.V.Volovich, *New quantum algorithm for studying NP-complete problems*, Rep.Math.Phys., **52**, No.1,25-33 2003.
3. M.Ohya and I.V.Volovich, Mathematical Foundation of Quantum Computers, Teleportations and Cryptography, Springer, 2011.
4. S.Iriyama and M.Ohya (2008) Language Classes Defined by Generalized Quantum Turing Machine, Open System and Information Dynamics 15:4, 383-396.
5. S.Iriyama,M.Ohya, I.V.Volovich, Quantrum Algorithm for Binary Search and Its Computation Complexity, TUS preprint, 2012
6. T.Ando,I.Yamato, Basics of Molecular Simulation and Its Application to Biomolecules, QP-PQ 24, Quantum Bio-Informatics II, World Scientific, 2009
7. K.Goto,S.Iriyama,M.Ohya, Quantum Algorithm for Hamilton Path Problem, TUS preprint, 2010
8. K.Goto, S.Iriyama and M.Ohya, On Quantum Algorithm for Binary Search and Its Computational Complexity, TUS preprint, 2012

Quantum Bio-Informatics V
© 2013 World Scientific Publishing Co. Pte. Ltd.
pp. 203–216

# ON EFFECTIVE PROCEDURES IN ANALYZING OF QUANTUM OPERATIONS AND PROCESSES

ANDRZEJ JAMIOŁKOWSKI

*Institute of Physics,*
*Nicolaus Copernicus University,*
*87–100 Toruń, Poland,*
*E-mail: jam@fizyka.umk.pl*

In this paper we discuss some constructive procedures which can be used in certain characterizations of quantum operations and quantum channels. In particular, an effective method of checking if a given quantum operation is irreducible or not is presented. Also the question of the existance of decoherence–free subspaces for a given quantum operation is discussed. Our methods make use of an explicit form of a fixed quantum operation in its Kraus representation. Some examples based on the theory of open quantum systems are discussed.

## 1. Introduction

In general, decoherence is understood as a non-unitary evolution of an open quantum system that is a consequence of system–environment coupling. As we know, non-classical correlations among parts of a composite quantum system, known as entanglement, can be precisely defined mathematically but, at least in the case of mixed states, in a rather ineffective way. Here, by an effective way (effective procedure) we mean a method which uses only a finite number of arithmetic operations on elements of a given density matrix (representing a quantum state) and allows us to answer the question: is a given composite quantum system in a separable or entangled state? A similar question: whether a given quantum channel possess a decoherence free subspace should also be discussed from the point of view of effective methods.

It appears that the questions of the above type are naturally connected with some problems of changes in the space of quantum states, that is with some aspects of quantum dynamics and properties of dynamical maps. Here, by dynamical maps we understand linear transformations that take one density operator to another. Moreover, this kind of connection between linear transformations and states of composed quantum systems is of a one-to-one type. At first sight this seems strange that two entirely different issues, namely, the inner structure of states and their properties and, on the other hand, their dynamics are so strongly connected. But in fact there exists an intricate and strong link between them — there exists an isomorphism between positive maps on quantum states (such maps are very often called superoperators on the space of quantum states) and some of density matrices representing composite quantum systems [1,2,3]. At present, various methods of study of entangled systems based on properties of positive maps are discussed in hundreds of papers and many monographs and textbooks (e.g. [4,5,6,7]).

In this paper we will discuss some effective methods of study of certain properties of quantum operations and we will use some methods and results which are typical for the research area known as *noncommutative Perron-Frobenius theory*. In particular, in our study of dynamical maps we will use the natural identifications

$$M_k(\mathbb{C}) \otimes B(\mathcal{H}) \cong M_k(B(\mathcal{H})), \tag{1}$$

where $\mathcal{H}$ denotes a finite dimensional Hilbert space, $B(\mathcal{H})$ represents the set of all linear operators on $\mathcal{H}$ and $M_k(\mathcal{A})$ for $k = 2, 3, \dots$ denotes $k \times k$ matrices with elements in an algebra $\mathcal{A}$. On the other hand we have also the identification

$$M_k(\mathbb{C}) \otimes B(\mathcal{H}) \cong B\left(\bigoplus_{i=1}^{k} \mathcal{H}\right). \tag{2}$$

In our approach we will concentrate on a quantum analogy of the classical theory of positive maps also known as Perron-Frobenius theory. Perron theorem on positive matrices [8] and its generalization by Frobenius on non-negative matrices [9] have attracted attention of mathematicians since their announcement at the beginning of last century. Later these theorems have been generalized to operators on partially ordered real Banach spaces [10]. This has motivated several authors to consider linear maps on a finite dimensional space which leave a fixed cone invariant (cf., e.g. [11,12]).

Almost at the same time it appeared that some natural questions connected with fundamentals of quantum mechanics (more precisely, with the theory of open quantum systems) lead to investigations of linear maps in a real Banach space of self-adjoint operators on a fixed Hilbert space [13,14]. This concept of a Banach space with the partial order defined by a specific cone, namely, the cone of positive semidefinite operators, constitutes a basic idea in the description of open quantum systems. Summing up, as the fundamental objects in modern quantum theory one considers the set of states

$$S\left(\mathcal{H}\right) := \left\{\rho : \mathcal{H} \to \mathcal{H}; : \rho \geq 0, : \mathrm{Tr}\rho = 1\right\}, \tag{3}$$

and the set of bounded hermitian (self-adjoint) operators

$$\mathcal{B}_{\star}\left(\mathcal{H}\right) := \left\{Q : \mathcal{H} \to \mathcal{H}; : Q = Q^{\star}\right\}. \tag{4}$$

Time evolutions of systems are governed by linear master equations of the form (in the so-called Schrödinger picture)

$$\frac{d\rho\left(t\right)}{dt} = \mathbb{K}\rho\left(t\right), \tag{5}$$

or in the dual form (in the so–called Heisenberg picture)

$$\frac{dQ\left(t\right)}{dt} = \mathbb{L}Q\left(t\right), \tag{6}$$

where superoperators $\mathbb{K}$ and $\mathbb{L}$ act on operators from the sets $S\left(\mathcal{H}\right)$ and $\mathcal{B}_{\star}\left(\mathcal{H}\right)$, respectively. They represent dual forms of the same physical idea. Both sets $S\left(\mathcal{H}\right)$ and $\mathcal{B}_{\star}\left(\mathcal{H}\right)$ can be considered as subsets of the vector space $\mathcal{B}\left(\mathcal{H}\right)$ of all bounded linear operators on $\mathcal{H}$ and they can be treated as scenes on which problems of quantum mechanical systems should be discussed. As in this paper we will consider finite-dimensional Hilbert spaces, so in fact $\mathcal{B}\left(\mathcal{H}\right)$ denotes the set of *all* linear operators on $\mathcal{H}$, $\dim : \mathcal{H} = d$. If we introduce the scalar product in $\mathcal{B}\left(\mathcal{H}\right)$ by the equality

$$\langle A, B \rangle := \mathrm{Tr}\left(A^{\star}B\right), \tag{7}$$

then $\mathcal{B}\left(\mathcal{H}\right)$ can be regarded as yet another inner product space, namely the *Hilbert–Schmidt* space. It is not difficult to see that $\mathcal{B}_{\star}\left(\mathcal{H}\right)$ with scalar product defined by (7) is a real vector space and $\dim \mathcal{B}\left(\mathcal{H}\right) = d^{2}$. It should be stressed that in this case we consider linear maps on $B(\mathcal{H})$ not only as maps on a vector space but also as maps on $B(\mathcal{H})$ equipped with a natural structure of an algebra. Such linear maps on the set of operators are often called *superoperators* and their general form is well known. Namely, for a given superoperator $\Phi$,

$$\Phi : B(\mathcal{H}) \to B(\mathcal{H}), \tag{8}$$

there always exists an operator-sum representation given by

$$\Phi(X) = \sum_{i=1}^{\kappa} A_i X B_i, \tag{9}$$

where $A_i, B_i$ are elements of $B(\mathcal{H})$. A particular class of such maps, the so-called completely positive maps (or in physical terminology — *quantum operations*), plays a prominent role in formulation of *evolution* of open quantum systems and in the theory of *quantum measurement*. In the case of completely positive maps we have in the above formula $B_i = A_i^{\star}$ for all $i = 1, \ldots, \kappa$. A comprehensive description of these problems from the physical point of view can be found in books [2,15].

It is important to observe that most of the papers devoted to the classical Perron-Frobenius theory of nonnegative linear maps are concerned mainly with the existence problems. The purpose of this paper is, on the one hand, to connect the Frobenius theory of irreducible linear operators with the so-called Kraus representation of the completely positive linear maps and, on the other hand, to show that there exist some effective procedures which allow us to verify if a given quantum operation (i.e. a given completely positive map) is irreducible or not. Here by irreducibility we mean the natural generalization of this concept, introduced by Frobenius in his famous paper of 1912 year ([9]) and based on the specific block representation of nonnegative matrices, to a geometric approach formulated in terms of the invariance of faces of a fixed cone. In this context, by an effective procedure we understand a method which uses only a finite number of arithmetic operations on matrices $A_i$ (Kraus coefficients) which define a fixed quantum operation (completely positive map)

$$\Phi(X) = \sum_{i=1}^{\kappa} A_i^{*} X A_i. \tag{10}$$

Another important question may be formulated in the following way: for a fixed quantum operation (a superoperator) $\Phi$ defined by a set of Kraus operators $A_1, \ldots, A_{\kappa}$ there exists a decoherence-free subspace or not. By definition, a decoherence-free subspace (DFS) it is a subspace of the system's Hilbert space $\mathcal{H}$ which that is invariant under non-unitary evolution. Alternatively formulated, DFS is this part of the system Hilbert space where the system is decoupled from the environment and thus its evolution is completely unitary. For a given quantum operation $\Phi$ defined by (10) DFS can exist or not. The questions is: can we check this in an effective way?

The paper is organized as follows. In Section 2 we will review the concepts needed for our discussion based on geometric and operator theoretic formulation of the Perron-Frobenius theory and, on the other hand, we describe the structure of the cone of positive definite operators defined on a given Hilbert space. Moreover, some properties of faces of this cone are analyzed. Section 3 describes some properties of amplifications of general positive maps (superoperators). The main results of the paper are discussed in Section 4, where we formulate an effective method of checking whether a given superoperator, i.e., a fixed quantum operation is irreducible or not and whether for a given superoperator $\Phi$ there exists DFS or not.

## 2. Properties of maps which have invariant cones

First, let us recall some results of the classical theory of nonnegative matrices. In this theory a prominent role is played by so-called irreducible matrices. By definition, a matrix $A$ is irreducible if no permutation matrix $P$ exists such that

$$P^T A P = \begin{pmatrix} X & Z \\ 0 & Y \end{pmatrix}, \tag{11}$$

where $X$ and $Y$ are square matrices and $0$ denotes a block of zeros. There are three main categories of results in Perron and, respectively, Frobenius approach to linear operators which preserve the nonnegative orthant $\mathbb{R}^n_+$:

**C I.** If $A$ is a strictly positive matrix, $A > 0$, i.e., all entries of $A$ satisfy the inequality $a_{ij} > 0$, then
  **a)** the spectral radius of the matrix $A$, $r(A)$, is a simple eigenvalue of $A$, greater than the magnitude of any other eigenvalue;
  **b)** there exists the corresponding eigenvector which is positive (componentwise);
  **c)** if $A \leq B$ and $A \neq B$, then $r(A) < r(B)$.
**C II.** If $A$ is a nonnegative matrix, $A \geq 0$, that is some entries $a_{ij}$ can be equal to zero, then
  **a)** the spectral radius $r(A)$ of the matrix $A$ is an eigenvalue of $A$;
  **b)** there exists the corresponding eigenvector which is nonnegative;
  **c)** if $A \leq B$ , then $r(A) \leq r(B)$.
**C III.** If $A$ is irreducible and nonnegative, $A \geq 0$, then we have
  **a)** $r(A)$ is a simple eigenvalue;
  **b)** there exists the corresponding eigenvector which is positive;
  **c)** if $A \leq B$ and $A \neq B$, then $r(A) < r(B)$.

Simple proofs of the Perron-Frobenius results for matrices can be found, e.g., in [18]. Let us observe that the assumption of irreducibility introduced by Frobenius for nonnegative matrices gives us a reproduction of results obtained by Perron for positive matrices. The generalizations mentioned in the Introduction involve two important extensions of the classical theory. First, the considered spaces are very often assumed to be infinite-dimensional and second, positivity (nonnegativity) is replaced by the assumption that the considered operator leaves a fixed cone invariant. In this paper we concentrate on this second case and, in order to obtain stronger results, important for quantum information theory, we consider only finite-dimensional cases.

Now, let us recall some notions and results important in analyzing partially ordered vector spaces. Let $V$ be a normed and partially ordered real linear space with a fixed cone $K$, i.e., $K$ is a nonempty closed subset of $V$ which satisfies : $K + K \subseteq K$, $\lambda K \subseteq K$ for all $\lambda \geq 0$, $K \cap (-K) = \{\mathbf{0}\}$, where $\mathbf{0}$ denotes the zero element of $V$. If we have the equality $V = K + (-K)$, then the cone $K$ is called *generating* or *reproducing* (sometimes one also uses the term *full cone*).

If $\Phi$ is a linear transformation on $V$, $\Phi : V \to V$, then we denote by $r(\Phi)$ the *spectral radius* of $\Phi$, i.e.,

$$r(\Phi) := \max\{|\lambda| \, ; \, \lambda \in \sigma(\Phi)\}, \tag{12}$$

where $\sigma(\Phi)$ denotes the spectrum of $\Phi$. For any cone $K$ we let $K^\circ$ denote the interior of $K$ and by $\partial K$ we denote its boundary.

As is well known, any fixed cone $K$ in $V$ determines a partial order in $V$. For this order we use the following terminology:

**1.** $x$ is *nonnegative*, $x \geq 0$, iff $x \in K$;
**2.** $x$ is *positive*, $x > 0$, iff $x \geq 0$ and $x \neq 0$;
**3.** $x$ is *strictly positive*, $x \gg 0$, iff $x \in K^\circ$.

Now, let us define the concept of face which plays a basic role in the theory of irreducible operators. Let $K$ be a cone in $V$. By a *face* $F$ of $K$ one understands a subset of $K$ which is a cone and satisfies an extra condition: if $0 \leq y \leq x$ and $x \in F$, then $y$ also belongs to $F$.

If $E \subset K$, then we will denote by $\Omega(E)$ the intersection of all faces containing $E$. It is easily seen that $\Omega(E)$ is a face. It is called the face *generated by $E$*.

The set of all operators $\Phi : V \to V$ such that $\Phi(K) \subseteq K$ we will denote by $\Pi(K)$. Let $B(V)$ be the set of all linear operators on $V$. Then we have

$$\Pi(K) := \{\Phi \,;\, \Phi(K) \subseteq K\} \subseteq B(V) \tag{13}$$

and $\Pi(K)$ is a cone in $B(V)$. The elements of the set $\Pi(K)$ are said to be $K$- *nonnegative operators*. In particular, the operator $\Phi$ is called $K$-*positive* in case $\Phi(K \setminus \{0\}) \subseteq K^o$. The set of all $K$-positive maps will be denoted by $\Pi^+(K)$.

Now we introduce one of the main ideas of the Perron-Frobenius theory both in classical and quantum cases. For a fixed $K$ in $V$ a natural generalization of the concept of an irreducible matrix is the following: $\Phi$ is $K$-*irreducible* if and only if $\Phi$ leaves invariant no face of $K$ except $\{0\}$ and $K$ itself. In other words, an operator in $\Pi(K)$ is $K$-reducible iff it leaves invariant a nontrivial face of $K$.

Another, strictly equivalent, definition of $K$-irreducibility can be given by the following theorem: An operator $\Phi \in \Pi(K)$ is $K$-irreducible if and only if no eigenvector of $\Phi$ lies on the boundary of $K$. In fact, one can say even more: An operator $\Phi \in \Pi(K)$ is $K$-irreducible if and only if $\Phi$ has exactly one (up to scalar multiples) eigenvector in $K$ and this vector belongs to $K^o$. Moreover, for any proper cone $K$ we have $\Pi^+(K) \subseteq \widetilde{\Pi}(K) \subseteq \Pi(K)$, where $\widetilde{\Pi}(K)$ denotes the set of all $K$-irreducible operators. If $K = \mathbb{R}^n_+$, then both definitions, Frobenius one and the above, coincide. For details, see e.g. [12,16,17,18].

Some important spectral properties of K-nonnegative operators are summarized in the following theorems, which one can consider as natural generalizations of CI, CII and CIII.

**Theorem 1.** *Let* $\Phi \in \Pi^+(K)$. *Then we have*

    *a)* *the spectral radius of the operator* $\Phi$ *is a simple eigenvalue of* $\Phi$, *greater than the magnitude of any other eigenvalue;*

    *b)* *an eigenvector of* $\Phi$ *corresponding to* $r(\Phi)$ *belongs to* $K^o$;

    *c)* *no other eigenvector of* $\Phi$ *(up to scalar multiples) belongs to* $K$.

**Theorem 2.** *Let* $\Phi \in \Pi(K)$. *Then the following hold*

    *a)* $r(\Phi)$ *is an eigenvalue of* $\Phi$ ;

    *b)* $K$ *contains an eigenvector of* $\Phi$ *corresponding to* $r(\Phi)$ ;

    *c)* *if* $\Phi \leq \Psi$, *then* $r(\Phi) \leq r(\Psi)$.

**Theorem 3.** *Let* $\Phi \in \widetilde{\Pi}(K)$. *Then the following hold* *a)* $r(\Phi)$ *is a simple eigenvalue of* $\Phi$*;*

*b)* *no eigenvector of* $\Phi$ *lies on the boundary of* $K$ *;*

*c)* $\Phi$ *has exactly one (up to scalar multiples) eigenvector in* $K$ *and this vector belongs to* $K^o$*;*

*d)* $(I + \Phi)^{n-1} \in \Pi^+(K)$*, where* $n = \dim V$*.*

For proofs of the above theorems consult ([11,12,18]). Let us observe that for the case $K = \mathbb{R}^n_+$, the whole story reduces to the classical Perron-Frobenius theory and in physics we use this theory for the so-called mesoscopic description of classical systems.

The case $K = M_d^+(\mathbb{C})$, where $M_d^+(\mathbb{C})$ denotes the set of all semipositive matrices on the space $\mathcal{H} = \mathbb{C}^d$ plays a fundamental role in the description of representations of states for open quantum systems. In this case, $B(\mathcal{H})$ is a $d^2$- dimensional vector space. According to (??) we denote by $B_*(\mathcal{H})$ the set of all self-adjoint operators on $\mathcal{H}$ which can be naturally considered as a $d^2$-dimensional real Banach space. At the same time, $B(\mathcal{H})$ can be regarded as a Hilbert space with the scalar product defined by (7). The vector space $B_*(\mathcal{H})$ of all Hermitian (self-adjoint) operators on $\mathcal{H}$ constitutes an $d^2$- dimensional, real subspace of the Hilbert-Schmidt space. One can use the "internal structure" of vectors in $B_*(\mathcal{H})$ to define a positive cone. By definition, a *semipositive* element of $B_*(\mathcal{H})$ is an operator A on $\mathcal{H}$ such that $\langle \psi | A | \psi \rangle$ is real and nonnegative for all vectors $|\psi\rangle$ from $\mathcal{H}$. Of course, one can equivalently define a positive element of $B(\mathcal{H})$ as a self-adjoint operator with nonnegative eigenvalues. The set of all semipositive operators on $\mathcal{H}$ we will denote by $B_*^+(\mathcal{H})$ or PSD. In particular, if we have the inequality $\langle \psi | A | \psi \rangle > 0$ for all $|\psi\rangle$ from $\mathcal{H}$, then we say that $A$ is *positive*.

## 3. Positive maps on PSD

It is well known that if a linear map $\Phi : B(\mathcal{H}) \to B(\mathcal{H})$ sends the set $B_*(\mathcal{H})$ of all hermitian elements of $B(\mathcal{H})$ into itself, then $\Phi$ can be represented in the form

$$\Phi(X) = \sum_{i=1}^{\kappa} a_i A_i^\star X A_i, \tag{14}$$

where $A_i \in B(\mathcal{H})$, and $a_i$ for $i = 1, 2, ..., \kappa$ are real numbers [2,22]. In general, all maps of the above form are hermitian-preserving. However, the representation (14) is not unique. In general, for a given $\Phi$ , there exist many possible representations of the form (14). For a given $\Phi$ the smallest $\kappa$ in (14) is called the *minimal length* of $\Phi$ and this minimal length is always smaller or equal to $(\dim \mathcal{H})^2$. If we assume that the operators $A_i$ for $i = 1, 2, \ldots, \kappa$ are linearly independent, then $\kappa$ in (14) must be minimal.

According to the general definition introduced in Section 2 a *positive map* $\Phi$ (*PSD-positive*) is a linear map from $B(\mathcal{H})$ into itself, which leaves PSD invariant. Now, $\Phi$ is called *k-positive* if its *k-amplification* $\Phi_{(k)} := \mathbb{I}_k \otimes \Phi$ that is the map

$$\mathbb{I}_k \otimes \Phi : M_k(\mathbb{C}) \otimes B(\mathcal{H}) \to M_k(\mathbb{C}) \otimes B(\mathcal{H}) \qquad (15)$$

is positive. Here $M_k(\mathbb{C})$ denotes as usual the set of all $k \times k$ complex matrices. It is not difficult to observe that we can identify the set $M_k(\mathbb{C}) \otimes \mathcal{A}$, where for simplicity we denote the algebra $B(\mathcal{H})$ by $\mathcal{A}$, with the set of all $k \times k$ matrices $M_k(\mathcal{A})$ with entries from $\mathcal{A}$. In such notation one can represent $\Phi_{(k)} : M_k(\mathcal{A}) \to M_k(\mathcal{A})$ by

$$\Phi_{(k)} \begin{pmatrix} & \vdots & \\ \cdots & X_{ij} & \cdots \\ & \vdots & \end{pmatrix} := \begin{pmatrix} & \vdots & \\ \cdots & \Phi\left(X_{ij}\right) & \cdots \\ & \vdots & \end{pmatrix}. \qquad (16)$$

The map $\Phi$ is called completely positive if it is $k$-positive for all $k = 1, 2, \ldots$. This terminology goes back to Stinespring [21], cf. also [6]. It is well known that for $d$-dimensional Hilbert space $\mathcal{H}$, $d$-positive maps on $B(\mathcal{H})$ are already completely positive [22].

Let us observe that all hermiticity-preserving maps which are not only positive but completely positive can be written in the form (14) with positive $a_i$, $i = 1, \ldots, \kappa$, i.e. by

$$\Phi(X) = \sum_{i=1}^{\kappa} K_i^\star X K_i, \qquad (17)$$

where $K_i := \sqrt{a_i} A_i$, and $\kappa \leq d^2$. The above expression is called the Kraus representation of a completely positive map $\Phi$ and, in case of the finite-dimensional Hilbert space $\mathcal{H}$, can be regarded also as a definition of the completely positive map. This representation is very useful in quantum information theory. In particular, completely positive maps are used to describe quantum operations, quantum channels and to model quantum devices.

It was shown by R. Timoney [23], that a positive map $\Phi$ (positive superoperator) which is $m$-positive, where $m = \lfloor \sqrt{\kappa} \rfloor$ must be completely positive. Here $\lfloor \sqrt{\kappa} \rfloor$ denotes the integer part of the number $\sqrt{\kappa}$. In other words, if a positive map $\Phi : B(\mathcal{H}) \to B(\mathcal{H})$, $\dim \mathcal{H} = d$, has a minimal length $\kappa$ and $\Phi$ is $m$-positive, for some $m < d$ such that $(m+1)^2 > \kappa$, then $\Phi$ is already completely positive (cf. also [19]).

Now, let us observe that one can apply the results stated in Theorems 1 – 3 of Section 2, to the particular cone PSD in $B(\mathcal{H})$, $\dim \mathcal{H} = d$. In particular, Theorem 3 describes properties of irreducible superoperators and according to this theorem we have: for irreducible $\Phi$, the spectral radius of $\Phi$ is a simple eigenvalue of the superoperator and $\Phi$ has exactly one (up to scalar coefficient) eigenvector in PSD and this vector belongs to $PSD \setminus \partial PSD$. One can say even more. Let $\mathcal{P}_d$ denote the set of all orthogonal projections, i.e. $A \in \mathcal{P}_d$ iff $A^2 = A$ and $A = A^\star$. Then we have [24]

**Theorem 4.** *[Farenick] The following statements are equivalent for a positive map on PSD.*

*1.)    There is a nontrivial (that is different from $\{0\}$ and PSD) face of PSD that is invariant under $\Phi$;*

*2.)    There is a nontrivial projection $P \in \mathcal{P}_d$ and a positive real number $\lambda > 0$ such that $\Phi(P) \leq \lambda P$;*

*3.)    There is a nontrivial projection $P \in \mathcal{P}_d$ such that the subalgebra $P(PSD)P$ is invariant under $\Phi$.*

In order to produce nontrivial examples of irreducible positive maps and in certain cases to characterize all irreducible maps within the class of completely positive maps we will use the following consequences of the Kraus representations of completely positive maps.

A family of closed subspaces of a given Hilbert space is called trivial if this family contains only $\{0\}$ and $\mathcal{H}$. For a fixed operator $X \in B(\mathcal{H})$ we will denote by $\mathrm{Inv}\,(X)$ the set of all invariant subspaces of $X$. Now, we can state the following theorem which is a reformulation of some results from [24].

**Theorem 5.** *Let $\Phi$ denote a superoperator on $B(\mathcal{H})$ which is PSD-positive. If $\Phi$ is completely positive, then there exist some operators $A_1, \ldots, A_\kappa$ such that $\Phi(X) = \sum_j A_j^\star X A_j$. Completely positive $\Phi$ is irreducible if and only if the Kraus operators $A_j$ do not have a nontrivial common invariant subspace in $\mathcal{H}$.*

To better understand the above theorem, let us observe that if Kraus operators do not have a common invariant subspace, i.e., are such that $\cap_j \mathrm{Inv}(A_j)$ is trivial and $\Phi(P) \leq \lambda P$ for some $\lambda \geq 0$ and $P \in \mathcal{P}_n$, then we have

$$\langle\Phi(P)\psi|\psi\rangle = \sum_{j=1}^{\kappa}\langle PA_j\psi|A_j\psi\rangle \le \lambda\langle P\psi|\psi\rangle. \tag{18}$$

The left-hand side of the above equality is nonnegative for all $\psi \in \mathcal{H}$. On the other hand for $\psi \in \ker P$, we have $\langle P\psi|\psi\rangle = 0$ on the right-hand side. In this way the equality (18) implies that $\langle PA_j\psi|A_j\psi\rangle = 0$ for each $j = 1,\ldots,\kappa$. This means that $\langle PA_j\psi|PA_j\psi\rangle = 0$, if we remember that $P^2 = P$, $P^* = P$. In consequence, $\ker P \in \cap_j \mathrm{Inv}(A_j)$, that is $\ker P$ is either $\{0\}$ or $\mathcal{H}$.

## 4. Criteria for existence of DFS and irreducibility of completely positive maps

According to Theorem 5 any completely positive map $\Phi$ (superoperator $\Phi$) on $B(\mathcal{H})$ can be represented by a set of operators $A_1,\ldots,A_\kappa$ and the superoperator $\Phi$ is irreducible if the operators $A_j$ $(j = 1,\ldots,\kappa)$ do not have a nontrivial common invariant subspace in the Hilbert space $\mathcal{H}$. On the other hand, the questions connected with existence of DFS with a fixed dimension can be formulated in the following way: Is it possible to verify whether operators $A_1,\ldots,A_\kappa$ have - or do not have - a common invariant subspace of dimension $m < d$, by an effective procedure? For $m = 1$ an answer to this question was given by Dan Shemesh in 1984 [25].

**Theorem 6.** *Let $K_1$ and $K_2$ denote two matrices acting on $H = \mathbb{C}^d$. A common eigenvector of $K_1$ and $K_2$ exists if and only if the subspace defined by*

$$\mathcal{M}_1 := \bigcap_{\alpha,\beta}^{d-1} ker\left[K_1^\alpha,\, K_2^\beta\right] \tag{19}$$

*is nontrivial, that is $\mathcal{M}_1 \ne \{\mathbf{O}\}$ or, in other word, $\dim\mathcal{M}_1 > 0$. Here the symbol $[\cdot,\cdot]$ denotes the commutator of the matrices.*

As it was stressed in [26], the genuine meaning of the subspace $\mathcal{M}_1$, can be stated as follows.

**Theorem 7.** *A subspace $\mathcal{M}_1$ is invariant with respect to both matrices $K_1$ and $K_2$, and moreover, $K_1$ and $K_2$ commute on $\mathcal{M}_1$. Every subspace of $\mathcal{H}$, which is invariant under $K_1$ and $K_2$ and on which $K_1$ and $K_2$ commute is contained in $\mathcal{M}_1$.*

The condition of Theorem 7, that is $\dim \mathcal{M}_1 > 0$, can be formulated in a constructive form. To this end let us define the matrix

$$\Omega := \sum_{\alpha,\beta}^{d-1} \left[ K_1^{\alpha}, K_2^{\beta} \right]^* \left[ K_1^{\alpha}, K_2^{\beta} \right]. \tag{20}$$

The matrices $K_1$ and $K_2$ have a common eigenvectors if and only if the matrix $\Omega$ is singular, i.e $\det \Omega = 0$. Now, using the concept of the so-called standard polynomials and the Amitsur–Levitzki theorem [27,28], we can generalize the Theorem 7. Recall that the standard polynomial of degree $n$ is the polynomial in non-commuting variables $X_1, \ldots, X_n$ of the form

$$S_r \left( X_1, \ldots, X_n \right) := \sum \operatorname{sign}\left(\sigma\right) X_{\sigma(1)} \cdots X_{\sigma(n)}. \tag{21}$$

The summation here is assumed over all permutations of $1, \ldots, n$. The importance of the standard polynomials is underlined by the following Amitsur-Levitzki theorem

**Theorem 8.** *The full matrix algebra $\mathbb{M}_d\left(\mathbb{C}\right)$ satisfies the standard identity $S_{2d}\left(X_1, \ldots, X_{2d}\right) \equiv 0$. Moreover, the algebra $\mathbb{M}_d\left(\mathbb{C}\right)$ does not satisfy any polynomial of degree less than $2d$.*

Let us observe that according to the above theorem, the algebra $\mathbb{M}_{d+1}\left(\mathbb{C}\right)$ cannot satisfy the identity for $n = 2d$. In other words, the algebra $\mathbb{M}_k\left(\mathbb{C}\right)$ satisfies the identity $S_{2d} = 0$ for $k \leq d$ and does not satisfy this identity for $k \geq d+1$.

Now, we are ready to discuss a generalization of the Shemesh's theorem. Namely, we introduce the family of the subspaces

$$\mathcal{M}_k := \bigcap \ker\left[S_{2k}\left(N_1, \ldots, N_{2k}\right) N_{2k+1}\right], \tag{22}$$

where $S_{2k}$ denotes the standard polynomial of degree $2k$ and the intersection is taken over all $(2k+1)$ - tuples of matrices from the algebra $\mathcal{A}$ generated by two elements (Kraus operators $K_1$ and $K_2$). One can prove:

**Theorem 9.** *Assume that $\mathcal{M}_k$ satisfies $\dim \mathcal{M}_k > 0$. Then $\mathcal{M}_k$ is an invariant subspace of $\mathcal{A}$ and elements of this algebra restricted to $\mathcal{M}_k$ satisfy the identity $S_{2k} \equiv 0$; i.e. for all $N_1, \ldots, N_{2k}$ form $\mathcal{A}$ and $X \in \mathcal{M}_k$ we have*

$$S_{2k}\left(N_1, \ldots, N_{2k}\right) X = 0. \tag{23}$$

It is not obvious from (22) that $\mathcal{M}_k$ can be constructed by an effective procedure; however, it is possible to show that these subspaces can be constructed by a finite number of arithmetic operations [29].

As a conclusion we can say that if in the Kraus representation of any fixed completely positive map $\Phi$ at least two Kraus operators do not have a common eigenvector, then the map $\Phi$ is irreducible. Moreover, using the generalization of Shemes theorem we can check in an effective way whether the algebra $\mathcal{A}$ generated by Kraus operators has a decoherence-free subspace of dimension $m \geq 2$, or not.

## References


1. A. Jamiołkowski, Rep. Math. Phys. **3**, 275 (1972); ibidem **5**, 415 (1975).
2. I. Bengtsson, K. Życzkowski, *Geometry of Quantum States*, Cambridge Univ. Press, 2008.
3. M. Schlosshaur, *Decoherence: and the Quantum–to–Classical Transition*, Springer, 2008.
4. M. D. Choi, Lin. Alg. Appl. **10**, 85 (1975).
5. D. Bruss, G. Leuchs (Eds.), *Lectures on Quantum Information*, Wiley-VCH, 2007.
6. V. Vedral, *Introduction to Quantum Information Science*, Oxford Univ. Press, 2006.
7. F. Benatti, *Dynamics, Information and Complexity in Quantum Systems*, Springer, 2009.
8. O. Perron, Math. Ann. **64**, 248 (1907).
9. G. Frobenius, S. B. Preuss. Akad. Wiss. (Berlin), 256 (1912).
10. M. G. Krein, M. A. Rutman, *Linear Operators Leaving Invariant a Cone in a Banach Space*, American Mathematical Society Translations **26**, (1950).
11. G. Birkhoff, Amer. Math. Monthly **74**, 274 (1967).
12. J. S. Vandergraft, SIAM J. App. Math. **16**, 1208 (1968).
13. K. Kraus, Ann. Phys. **64**, 119, (1971).
14. A. Kossakowski, Rep. Math. Phys. **3**, 243 (1972).
15. M. A. Nielsen, I. Chuang, *Quantum Computation and Quantum Information*, Cambridge Univ. Press., 2000.
16. G.P. Barker, Linear Alg. Appl. **39**, 263, (1981).
17. R.S. Varga, *Matrix Iterative Analysis*, Springer, 2000.
18. A. Berman, R.J. Plemmons, *Nonnegative Matrices in the Mathematical Sciences*, Academic Press, 1979 .
19. A. Jamiołkowski, Open Sys. Infor. Dyn. **11** 385 (2004).
20. A. Jamiołkowski, pp. 185-197, in: *Quantum Bio-Informatics IV*, World Scientific (2011).
21. W. F Stinespring, Proc. AMS **6**, 211, (1955).
22. M. D. Choi, Linear Alg. Appl. **10**, 285, (1975); ibidem **12**, 95 (1975).
23. R. M. Timoney, Bull. London Math. Soc. **32**, 229, (2000).

24. D. R. Farenick, Proc. AMS **124**, 3381, (1996).
25. D. Shemesh, Linear Alg. Appl. **62**, 11, (1984).
26. Yu. A. Alpin, Kh. D. Ikramov, J. of Math. Sciences **114**, 1757, (2003).
27. S. A. Amitsur, J. Levitzki, Proc. AMS **1**, 449 (1950).
28. V. Drensky, E. Formanek, *Polynomial Identity Rings*, Birkhauser, 2004.
29. Y. Alpin et al. , Linear Alg. App. **312**, 115, (2000).

Quantum Bio-Informatics V
© 2013 World Scientific Publishing Co. Pte. Ltd.
pp. 217–228

# ON NUMERICAL RANGES OF OPERATORS

JACEK JURKOWSKI

*Institute of Physics, Nicolaus Copernicus University*
*Grudziadzka 5, 87–100 Toruń, Poland*
*E-mail: jacekj@fizyka.umk.p*

We recall the notions of numerical ranges of some operator acting on a final-dimensional Hilbert space and we show how they can be used in quantum theory of composite systems. We emphasize their role in entanglement detection based on entanglement witnesses. As an instructive example we analyze a local numerical range for a family of circulant observables and states of composite 2?d systems.

## 1. Restricted Numerical Ranges

The well-established notion used in describing some properties of an operator $A$ acting on some Hilbert space is its *numerical range* which describes the set of *all* possible expectation values of $A$ [1,2]. In what follows we generalize this notion and show some connections with the theory of positive maps and entanglement witnesses [3]. To this end let us introduce the final dimensional Hilbert space $\mathcal{H}$ and define the set of quantum states

$$\mathcal{S}(\mathcal{H}) = \{\, \rho \, : \, \rho \geq 0, \ \mathrm{Tr}\rho = 1 \,\}$$

acting on $\mathcal{H}$. Now, let $A$ be an arbitrary linear operator acting on $\mathcal{H}$. One defines a *numerical range* (NR) of $A$ as

$$\Lambda(A) = \{\, \mathrm{Tr}(\rho A) \, : \, \rho \in \mathcal{S}(\mathcal{H}) \,\} = \left\{\, \frac{\langle\psi|A|\psi\rangle}{\langle\psi|\psi\rangle} \, : \, |\psi\rangle \in \mathcal{H} \,\right\}.$$

Some obvious properties of $A$ are the following: $\Lambda(A)$ is a compact, connected and convex subset of the complex plane $\mathbb{C}$.

Generalizing this notion, recently [4,5], in a context of control theory one introduces a *C-numerical range* with respect to a subgroup $G$ of a unitary group, where $C$ is some matrix, as follows

$$\Lambda_G^C(A) = \{\, \mathrm{Tr}(C^* U A U^*) \, : \, U \in G \subset \mathrm{U}(N) \,\}.$$

Actually, if $C = |\psi\rangle\langle\psi|$ and $G = \mathrm{U}(N)$ is the full unitary group, then $\Lambda_G^C(A) = \Lambda(A)$.

218

Even more generally, one can consider a *restricted numerical range* (RNR) [6,7] defined as

$$\Lambda^{\mathcal{R}}(A) = \{\,\mathrm{Tr}(\rho A)\,:\,\rho \in \mathcal{R} \subset \mathcal{S}(\mathcal{H})\,\},$$

where $\mathcal{R}$ denotes some subset of quantum states. According to possible applications, one can choose for $\mathcal{R}$

- the set of real projections: $\mathcal{R} = \{\,|\psi\rangle\langle\psi|\,:\,|\psi\rangle \in \mathbb{R}^N\,\}$,
- the set of projections onto $\mathrm{SU}(N)$-invariant coherent states

$$\mathcal{R} = \{\,|\psi\rangle\langle\psi|\,:\,|\psi\rangle - \mathrm{SU}(N)\text{-invariant coherent states}\,\},$$

- various subsets of the set of composite quantum states acting on $\mathcal{H} = \mathcal{H}_A \otimes \mathcal{H}_B$.

Obviously, $\Lambda^{\mathcal{R}}$ is a compact, connected, but *not necessary convex* subset of $\mathbb{C}$. Let us concentrate on the case when $\mathcal{R}$ is some subset of composite quantum states. One defines *product numerical range* (PNR) [6,7] as

$$\Lambda^{\otimes}(A) = \{\,\mathrm{Tr}(\rho A)\,:\,\rho \in \mathcal{R}_{\otimes}\,\}$$

where

$$\mathcal{R}_{\otimes} = \{\,|\psi_{AB}\rangle\langle\psi_{AB}|\,:\,|\psi_{AB}\rangle = |\psi_A\rangle \otimes |\psi_B\rangle,\ |\psi_A| = 1,\ |\psi_B| = 1\,\}.$$

is the set of projections onto product states, and *separable numerical range* (SNR)

$$\Lambda^{\mathrm{sep}}(A) = \{\,\mathrm{Tr}(\rho A)\,:\,\rho \in \mathcal{R}_{\mathrm{sep}}\,\}$$

where

$$\mathcal{R}_{\mathrm{sep}} = \left\{\,\rho\,:\,\rho = \sum_i p_i |\psi_A^{(i)}\rangle\langle\psi_A^{(i)}| \otimes |\psi_B^{(i)}\rangle\langle\psi_B^{(i)}|\,\right\},$$

is the set of separable quantum states (see for instance [3,8,9] for a definition of separability). In principle, some other interesting choices specific for quantum information theory and theory of positive maps are possible.

As far as PNR is concerned, some of its properties are the following:

(1) $\Lambda^{\otimes}(A)$ is a connected, but, in general, not a convex subset of $\mathbb{C}$,
(2) $\Lambda^{\otimes}(A + B) \subset \Lambda^{\otimes}(A) + \Lambda^{\otimes}(B)$,
(3) $\Lambda^{\otimes}(A + \alpha\mathbb{1}) = \alpha + \Lambda^{\otimes}(A), \quad \Lambda^{\otimes}(\alpha A) = \alpha\Lambda^{\otimes}(A), \quad \alpha \in \mathbb{C}$,
(4) $\Lambda^{\otimes}((U_1 \otimes U_2)A(U_1 \otimes U_2)^*) = \Lambda^{\otimes}(A)$,
(5) $\Lambda^{\mathrm{sep}}(A) = \mathrm{co}(\Lambda^{\otimes}(A))$, where "co" means the convex hall of a set,

(6) $\Lambda^{\otimes}(A \otimes B) = \Lambda(A) \boxtimes \Lambda(B)$, where $Z_1 \boxtimes Z_2$ denotes the Minkowski product of $Z_1$ and $Z_2$ defined as

$$Z_1 \boxtimes Z_2 = \{\, z = z_1 \cdot z_2 \,:\, z_1 \in Z_1,\ z_2 \in Z_2 \,\},$$

(7) $\Lambda(A \otimes B) = \mathrm{co}(\Lambda^{\otimes}(A \otimes B))$ for normal $A$, i.e. $[A, A^*] = 0$.

The situation is much more simpler for Hermitian operators, i.e. when $A = A^*$. Obviously, in this case

$$\Lambda(A) = [\lambda_{\min}(A), \lambda_{\max}(A)] \subset \mathbb{R}$$

is an interval and

$$\Lambda^{\otimes}(A) = [\lambda^{\otimes}_{\min}(A), \lambda^{\otimes}_{\max}(A)] \subset \Lambda(A)\,.$$

One can also prove that $\Lambda^{\mathrm{sep}}(A) = \Lambda^{\otimes}(A)$ and, if $A$ and $C$ are Hermitian, then $\Lambda^{C}(A) = [\gamma_{\min}, \gamma_{\max}]$, where

$$\gamma_{\min} = \sum_j a_j c_{n-j}\,, \qquad \gamma_{\max} = \sum_j a_j c_j,$$

with $a_j$ and $c_j$ being eigenvalues of $A$ and $C$, respectively, in non-increasing order.

The notion of a restricted numerical range can be related to various topics in quantum optics, in particular, to the theory of positive maps, completely positive maps, entanglement witnesses and others [6,7]. Let us mention some of the obvious connections. If a map $\Phi : B(\mathcal{H}_B) \to B(\mathcal{H}_A)$ is *positive* then the PNR of its Choi matrix $D_\Phi = (\mathbb{1} \otimes \Phi)|\psi_+\rangle\langle\psi_+|$, where $|\psi_+\rangle\langle\psi_+|$ is the projection onto maximally entangled state, is contained in the positive semi-line, i.e.,

$$\Lambda^{\otimes}(D_\Phi) \subset [0, +\infty).$$

Similarly, if a map is *completely positive*, then the full NR of its Choi matrix is contained in the positive semi-line, i.e.,

$$\Lambda(D_\Phi) \subset [0, +\infty).$$

Finally, we mention that, for every positive operator $A$ and $\gamma \geq \lambda^{\otimes}_{\max}(A)$, $W = \gamma I - A$ is an entanglement witness.

In what follows we will consider PNR for so-called circulant operators [10,11,12,13].

## 2. Circulant Operators in $\mathbb{C}^2 \otimes \mathbb{C}^d$

Let $\mathcal{H} = \mathbb{C}^2 \otimes \mathbb{C}^d$ and let $\{|g_i\rangle \otimes |f_k\rangle\}$ ($i = 1, 2$, $k = 1, \ldots, d$) be an orthonormal product basis in $\mathcal{H}$. One defines the family of 2-dimensional subspaces $\Sigma_k$ in $\mathcal{H}$:

$$\Sigma_1 = \mathrm{span}\Big\{|g_1\rangle \otimes |f_1\rangle, |g_2\rangle \otimes |f_2\rangle\Big\},$$

$$\Sigma_2 = \mathrm{span}\Big\{|g_1\rangle \otimes |f_2\rangle, |g_2\rangle \otimes |f_3\rangle\Big\}$$

$$\vdots$$

$$\Sigma_d = \mathrm{span}\Big\{|g_1\rangle \otimes |f_d\rangle, |g_2\rangle \otimes |f_1\rangle\Big\}.$$

It is clear that $\Sigma_k$ give rise to the direct sum decomposition [10,11]

$$\mathbb{C}^2 \otimes \mathbb{C}^d = \bigoplus_{k=1}^{d} \Sigma_k. \tag{1}$$

We shall call (1) a *circulant decomposition*. Now, we call a linear operator $\mathcal{O} \in \mathcal{B}(\mathcal{H})$ to be *circulant operator* with respect to a circulant decomposition (1) iff

$$\mathcal{O} = \mathcal{O}_1 \oplus \ldots \oplus \mathcal{O}_d, \tag{2}$$

where $\mathcal{O}_k$ is supported on $\Sigma_k$, that is,

$$\mathcal{O}_k = \sum_{i,j=1}^{2} a_{ij}^{(k)} |g_i\rangle \langle g_j| \otimes |f_{i+k}\rangle \langle f_{j+k}| \quad \mathrm{mod}\, d, \tag{3}$$

and $||a_{ij}^{(k)}||$ is a $2 \times 2$ complex matrix. In particular, for $d = 2$ and $d = 3$ we obtain the following matrix representations of the circulant operators (in the basis $|g_i\rangle \otimes |f_k\rangle$)

$$
\begin{pmatrix}
a_{11}^{(2)} & \cdot & \cdot & a_{12}^{(2)} \\
\cdot & a_{11}^{(1)} & a_{12}^{(1)} & \cdot \\
\cdot & a_{21}^{(1)} & a_{22}^{(1)} & \cdot \\
a_{21}^{(2)} & \cdot & \cdot & a_{22}^{(2)}
\end{pmatrix},
\quad
\begin{pmatrix}
a_{11}^{(3)} & \cdot & \cdot & \cdot & a_{12}^{(3)} & \cdot \\
\cdot & a_{11}^{(1)} & \cdot & \cdot & \cdot & a_{12}^{(1)} \\
\cdot & \cdot & a_{11}^{(2)} & a_{12}^{(2)} & \cdot & \cdot \\
\cdot & \cdot & a_{21}^{(2)} & a_{22}^{(2)} & \cdot & \cdot \\
a_{21}^{(3)} & \cdot & \cdot & \cdot & a_{22}^{(3)} & \cdot \\
\cdot & a_{21}^{(1)} & \cdot & \cdot & \cdot & a_{22}^{(1)}
\end{pmatrix},
\tag{4}
$$

where to make the picture more transparent we replaced all zeros by dots. Interestingly for $d = 2$ the circulant matrix displays characteristic X-shape. Such 2-qubit states have been recently investigated [14,15,16,17,18].

In the following we limit ourselves to circulant states and observables, i.e. hermitian circulant matrices, only. Let us introduce a more convenient notation and denote by

$$ w_{ik} = a_{ii}^{(k-i)} \mod d \ , $$

for $i = 1, 2$, $k = 1, \ldots, d$, and

$$ a_{12}^{(k+2)} = u_k e^{i\alpha_k} \ , $$

where $u_k = |a_{12}^{(k+2)}| \geq 0$, and $\alpha_k \in (-\pi, \pi]$. As a consequence, the general circulant observable reads

$$ \mathcal{O} = \sum_{i=1}^{2} \sum_{k=1}^{d} w_{ik} |g_i\rangle\langle g_i| \otimes |f_k\rangle\langle f_k| + \left( \sum_{k=1}^{d} u_k e^{i\alpha_k} |g_1\rangle\langle g_2| \otimes |f_k\rangle\langle f_{k+1}| + \text{h.c.} \right) , \tag{5} $$

where as usual h.c. stands for hermitian conjugation.

Now, let $\mathcal{O}$ be an hermitian circulant operator living in $\mathbb{C}^2 \otimes \mathbb{C}^d$. We proved [19] the following proposition:

**Proposition 2.1.** *There exists an orthonormal product basis $\{|g_i'\rangle \otimes |f_k'\rangle\}$ such that*

*(1) $\mathcal{O}$ is circulant with respect to the circulant decomposition constructed out of $\{|g_i'\rangle \otimes |f_k'\rangle\}$,*
*(2) matrix elements of $\mathcal{O}$ with respect to $\{|g_i'\rangle \otimes |f_k'\rangle\}$ satisfy:*

$$ w_{ik}' = w_{ik} \ , \qquad a_{12}^{(k+2)'} = |a_{12}^{(k+2)}| = u_k \ . \tag{6} $$

We will call the corresponding matrix representation of $\mathcal{O}$ with respect to $\{|g_i'\rangle \otimes |f_k'\rangle\}$ *real representation*.

## 3. Product Numerical Range for a Circulant Operator

Let $\mathcal{O}$ be an hermitian circulant operator with respect to a fixed basis $|g_i\rangle \otimes |f_k\rangle$ in $\mathbb{C}^2 \otimes \mathbb{C}^d$, and let us define

$$ F(x, y) = \frac{\langle x \otimes y | \mathcal{O} | x \otimes y \rangle}{\langle x \otimes y | x \otimes y \rangle} \ . \tag{7} $$

Now to provide $\text{LNR}(\mathcal{O})$ one has to find $\gamma_{\min} = \inf F(x, y)$ and $\gamma_{\max} = \sup F(x, y)$. Let

$$ \gamma_{\min} = F(x^-, y^-) \ , \qquad \gamma_{\max} = F(x^+, y^+) \ . \tag{8} $$

One has the following proposition [19]:

**Proposition 3.1.** *The corresponding vectors $|x^\pm\rangle \in \mathbb{C}^2$ and $|y^\pm\rangle \in \mathbb{C}^d$ have the following components with respect to basis $|g_i'\rangle$ and $|f_k'\rangle$ provided in Proposition 1*

$$|x^\pm\rangle = (x_1^\pm, x_2^\pm) \ , \quad |y^\pm\rangle = (y_1^\pm, \ldots, y_d^\pm) \ , \tag{9}$$

*where*

$$x_i^\pm \geq 0 \ , \quad y_k^\pm \geq 0 \ . \tag{10}$$

Hence, essentially $\mathrm{LNR}(\mathcal{O})$ calculations can be done in $\mathbb{R}^2 \otimes \mathbb{R}^d$ instead of $\mathbb{C}^2 \otimes \mathbb{C}^d$ . Unfortunately, solving the set of $d + 2$ polynomial equations is in general very hard (see [19] for some details). Here, we limit ourselves to a simple example to emphasize some characteristic features arising in LNR calculations.

**Example 3.1.** Let us consider circulant hermitian operator $\mathcal{O}$ in $\mathbb{C}^2 \otimes \mathbb{C}^2$ represented in the standard computational basis by the following real matrix

$$M_\mathcal{O} = \begin{pmatrix} 2 & 0 & 0 & 1 \\ 0 & 1 & -1 & 0 \\ 0 & -1 & 1 & 0 \\ 1 & 0 & 0 & 2 \end{pmatrix} . \tag{11}$$

The spectrum of $M_\mathcal{O}$ is $\{0, 1, 2, 3\}$. As a consequence, $\mathrm{NR}(\mathcal{O}) = [0, 3]$, whereas, as we shall see, $\mathrm{LNR}(\mathcal{O}) = [0.5, 2.5]$. Moreover, the upper bound $\gamma_{max}$ is achieved at complex vectors $|x\rangle = \frac{1}{\sqrt{2}}(1, i)$ and $|y\rangle = \frac{1}{\sqrt{2}}(1, -i)$ and when calculating expectation values on normalized vectors from $\mathbb{R}^2 \otimes \mathbb{R}^2$ we do not go beyond 2.

In order to proof that the upper bound of $\mathrm{LNR}(\mathcal{O})$ is indeed 2.5, let us bring the observable $\mathcal{O}$ into the real form by a local unitary transformation (which does not change the ranges but does change the extremal vectors),

$$U_1 = D[1, -i] \ , \quad U_2 = D[1, i] \ . \tag{12}$$

giving rise to

$$M_\mathcal{O}' = U_1 \otimes U_2 M_\mathcal{O} U_1^\dagger \otimes U_2^\dagger = \begin{pmatrix} 2 & 0 & 0 & 1 \\ 0 & 1 & 1 & 0 \\ 0 & 1 & 1 & 0 \\ 1 & 0 & 0 & 2 \end{pmatrix} . \tag{13}$$

Now, it is easy to show that for $|x\rangle = (x_1, x_2) \in \mathbb{C}^2$ and $|y\rangle = (y_1, y_2) \in \mathbb{C}^2$ we get

$$\langle x \otimes y | M'_{\mathcal{O}} | x \otimes y \rangle = 2(|x_1|^2 |y_1|^2 + |x_2|^2 |y_2|^2) + |x_1|^2 |y_2|^2 + |x_2|^2 |y_1|^2$$

$$+ 4 Re(x_1 x_2^*) Re(y_1 y_2^*) \tag{14}$$

$$\leq 2(|x_1|^2 |y_1|^2 + |x_2|^2 |y_2|^2) + |x_1|^2 |y_2|^2 + |x_2|^2 |y_1|^2 + 1$$

due to $Re(x_1 x_2^*) \leq 1/2$ which follows from the normalization condition $|x_1|^2 + |x_2|^2 = 1$. Equality in (14) is achieved for $|x_1| = |x_2| = \frac{1}{\sqrt{2}}$ and $|y_1| = |y_2| = \frac{1}{\sqrt{2}}$ and therefore $\langle x \otimes y | \mathcal{O} | x \otimes y \rangle = 2.5$. Similar proof can be carried out for the lower bound $\gamma_{min}$.

### 3.1. *Product Numerical Range for $d = 2$*

Consider now a 2-qubit case corresponding to $d = 2$ and let us introduce a normalized vector $|q\rangle = (q_1, q_2, q_3, q_4) \in \mathbb{R}^4$. It is separable if and only if $q_1 q_4 = q_2 q_3$. Next, we define

$$\widetilde{G}(q) = \langle q | M'_{\mathcal{O}} | q \rangle - \lambda_1 \left( \sum_{j=1}^{4} q_j^2 - 1 \right) - 2\lambda_2 (q_1 q_4 - q_2 q_3),$$

where the matrix $M'_{\mathcal{O}}$ represents $\mathcal{O}$ in the basis $|g'_i\rangle \otimes |f'_k\rangle$, that is,

$$M'_{\mathcal{O}} = \begin{pmatrix} w_{11} & 0 & 0 & u_1 \\ 0 & w_{12} & u_2 & 0 \\ 0 & u_2 & w_{21} & 0 \\ u_1 & 0 & 0 & w_{22} \end{pmatrix}, \qquad u_1, u_2 \geq 0, \quad w_{ij} \in \mathbb{R}. \tag{15}$$

Now, $d\widetilde{G} = 0$ leads to a linear matrix equation

$$\mathcal{M} |q\rangle = |0\rangle, \tag{16}$$

where

$$\mathcal{M} = \begin{pmatrix} -\lambda_1 + w_{11} & 0 & 0 & u_1 - \lambda_2 \\ 0 & w_{12} - \lambda_1 & u_2 + \lambda_2 & 0 \\ 0 & u_2 + \lambda_2 & -\lambda_1 + w_{12} & 0 \\ u_1 - \lambda_2 & 0 & 0 & w_{22} - \lambda_1 \end{pmatrix}.$$

Obviously, this way we arrive at two separate two-dimensional linear problems. In order to obtain nonzero solutions the following condition should be fulfilled:

$$\det \mathcal{M} = d_1(\lambda_1, \lambda_2) \cdot d_2(\lambda_1, \lambda_2) = 0,$$

224

where

$$d_1(\lambda_1, \lambda_2) = u_2^2 - w_{12}w_{21} + (w_{12} + w_{21})\lambda_1 - \lambda_1^2 + 2u_2\lambda_2 + \lambda_2^2, \quad (17)$$

$$d_2(\lambda_1, \lambda_2) = u_1^2 - w_{11}w_{22} + (w_{11} + w_{22})\lambda_1 - \lambda_1^2 - 2u_1\lambda_2 + \lambda_2^2. \quad (18)$$

Now, assuming

$$\begin{cases} d_1(\lambda_1, \lambda_2) = 0 \\ d_2(\lambda_1, \lambda_2) \neq 0 \end{cases} \quad \text{or} \quad \begin{cases} d_1(\lambda_1, \lambda_2) \neq 0 \\ d_2(\lambda_1, \lambda_2) = 0 \end{cases}$$

and using the separability condition, we get four possible product vectors $|g_i\rangle \otimes |f_j\rangle$,

$$\left\{ \begin{pmatrix} 0 \\ 1 \end{pmatrix} \otimes \begin{pmatrix} 1 \\ 0 \end{pmatrix}, \begin{pmatrix} 1 \\ 0 \end{pmatrix} \otimes \begin{pmatrix} 0 \\ 1 \end{pmatrix}, \begin{pmatrix} 0 \\ 1 \end{pmatrix} \otimes \begin{pmatrix} 0 \\ 1 \end{pmatrix}, \begin{pmatrix} 1 \\ 0 \end{pmatrix} \otimes \begin{pmatrix} 1 \\ 0 \end{pmatrix} \right\}, \quad (19)$$

whereas solving

$$\begin{cases} d_1(\lambda_1, \lambda_2) = 0 \\ d_2(\lambda_1, \lambda_2) = 0 \end{cases} \quad (20)$$

we obtain two solutions $(\lambda_1^+, \lambda_2^+)$ and $(\lambda_1^-, \lambda_2^-)$ which inserted into (16) imply the following conditions:

$$\begin{cases} q_1 = a_{\pm}q_4 \\ q_2 = b_{\pm}q_3 \\ q_1^2 + q_2^2 + q_3^2 + q_4^2 = 1 \\ q_1 q_4 = q_2 q_3 . \end{cases} \quad (21)$$

Solving (21) and factorizing $|q\rangle = |x\rangle \otimes |y\rangle$ we arrive at

$$|q\rangle = \begin{pmatrix} \sqrt{\frac{\xi_{\pm}}{1+\xi_{\pm}}} \\ \frac{1}{\sqrt{1+\xi_{\pm}}} \end{pmatrix} \otimes \begin{pmatrix} \sqrt{\frac{\kappa_{\pm}}{1+\kappa_{\pm}}} \\ \frac{1}{\sqrt{1+\kappa_{\pm}}} \end{pmatrix}, \quad (22)$$

where

$$a_{\pm} = \frac{l_{\pm}}{m_{\pm}}, \qquad b_{\pm} = \frac{g_{\pm}}{h_{\pm}}, \qquad \kappa_{\pm} = \frac{a_{\pm}}{b_{\pm}}, \qquad \xi_{\pm} = a_{\pm} \cdot b_{\pm},$$

and

$$\begin{aligned}
l_\pm &= 2u_1^4 + 8u_1^3 u_2 + 2u_2^4 \pm (u_1 + u_2)(-w_{11} + w_{12} + w_{21} - w_{22})\sqrt{\Delta} \\
&\quad + u_2^2\left(-w_{11}^2 - 2w_{12}w_{21} + w_{11}(w_{12} + w_{21}) + (w_{12} + w_{21})w_{22} - w_{22}^2\right) \\
&\quad + 2u_1 u_2\left(4u_2^2 - w_{11}^2 - 2w_{12}w_{21} + w_{11}(w_{12} + w_{21}) + (w_{12} + w_{21})w_{22} - w_{22}^2\right) \\
&\quad + u_1^2\left(12u_2^2 - w_{11}^2 - 2w_{12}w_{21} + w_{11}(w_{12} + w_{21}) + (w_{12} + w_{21})w_{22} - w_{22}^2\right), \\[4pt]
m_\pm &= (u_1 + u_2)\Big(\pm 2(u_1 + u_2)\sqrt{\Delta} + (w_{11} - w_{12})(w_{11} - w_{21})(w_{11} - w_{12} - w_{21} + w_{22}) \\
&\quad + u_1^2(-3w_{11} + w_{12} + w_{21} + w_{22}) + 2u_1 u_2(-3w_{11} + w_{12} + w_{21} + w_{22}) \\
&\quad + u_2^2(-3w_{11} + w_{12} + w_{21} + w_{22})\Big), \\[4pt]
g_\pm &= 2u_1^4 + 8u_1^3 u_2 + 2u_2^4 \pm (u_1 + u_2)(w_{11} - w_{12} - w_{21} + w_{22})\sqrt{\Delta} \\
&\quad + u_2^2\left(-w_{12}^2 - w_{21}^2 + w_{11}(w_{12} + w_{21} - 2w_{22}) + (w_{12} + w_{21})w_{22}\right) \\
&\quad + 2u_1 u_2\left(4u_2^2 - w_{12}^2 - w_{21}^2 + w_{11}(w_{12} + w_{21} - 2w_{22}) + (w_{12} + w_{21})w_{22}\right) \\
&\quad + u_1^2\left(12u_2^2 - w_{12}^2 - w_{21}^2 + w_{11}(w_{12} + w_{21} - 2w_{22}) + (w_{12} + w_{21})w_{22}\right), \\[4pt]
h_\pm &= (u_1 + u_2)\Big(\pm 2(u_1 + u_2)\sqrt{\Delta} + (w_{11} - w_{12})(w_{12} - w_{22})(w_{11} - w_{12} - w_{21} + w_{22}) \\
&\quad + u_1^2(w_{11} - 3w_{12} + w_{21} + w_{22}) + 2u_1 u_2(w_{11} - 3w_{12} + w_{21} + w_{22}) \\
&\quad + u_2^2(w_{11} - 3w_{12} + w_{21} + w_{22})\Big),
\end{aligned}$$

with

$$\Delta = \left((u_1 + u_2)^2 + (w_{11} - w_{21})(w_{12} - w_{22})\right)\left((u_1 + u_2)^2 + (w_{11} - w_{12})(w_{21} - w_{22})\right).$$

Note that, in order to have real components of $|q\rangle$,

$$\xi_\pm \geq 0, \qquad \kappa_\pm \geq 0, \tag{23}$$

should be fulfilled. As a consequence, either both $a_\pm$, $b_\pm$ are nonnegative or both are non-positive. Finally, for $|q\rangle$ given by (22) we obtain

$$\begin{aligned}
\langle q|M'_{\mathcal{O}}|q\rangle \;\equiv\; F_\pm &= \frac{2(u_1 + u_2)\sqrt{\xi_\pm \kappa_\pm} + \xi_\pm \kappa_\pm w_{11} + \xi_\pm w_{12} + \kappa_\pm w_{21} + w_{22}}{(1 + \kappa_\pm)(1 + \xi_\pm)} \\
&= \frac{2(u_1 + u_2)|a_\pm| + a_\pm^2 w_{11} + \xi_\pm w_{12} + \kappa_\pm w_{21} + w_{22}}{1 + \xi_\pm + \kappa_\pm + a_\pm^2}. \tag{24}
\end{aligned}$$

Taking into account vectors (19) one obtains

$$\langle g_i \otimes f_j|M'_{\mathcal{O}}|g_i \otimes f_j\rangle = w_{ij}. \tag{25}$$

Hence, LNR of the circulant observable $\mathcal{O}$ is given by $[\gamma_{\min}, \gamma_{\max}]$, where

$$\gamma_{\min} = \min\left\{w_{ij}, F_\pm\right\} \tag{26}$$

$$\gamma_{\max} = \max\left\{w_{ij}, F_\pm\right\} \tag{27}$$

To summarize, in order to calculate LNR for a given $\mathbb{C}^2 \otimes \mathbb{C}^2$ circulant operator, we propose the following procedure

(1) if in a given basis a matrix representation of an operator $M_\mathcal{O}$ has complex or negative off-diagonal entries then change the basis due to Proposition 1 and bring the matrix to the real form,

(2) determine real vectors $|x\rangle$ and $|y\rangle$ (see (22)) together with $F_\pm$ and compare these values with diagonal elements of $M'_\mathcal{O}$. Then LNR$\mathcal{O}) = [\gamma_{\min}, \gamma_{\max}]$, where $\gamma_{\min}$ and $\gamma_{\max}$ are defined in (26) and (27), respectively.

**Example 3.2.** As an illustration let us consider a two-parameter family of matrices $Q_{t,s}$, $t, s \geq 0$, analyzed in the paper [6],

$$Q_{t,s} = \begin{pmatrix} 2 & 0 & 0 & t \\ 0 & 1 & s & 0 \\ 0 & s & -1 & 0 \\ t & 0 & 0 & -2 \end{pmatrix}.$$

Denoting by $p = t + s \geq 0$ one obtains

$$\Delta = (1 + p^2)(9 + p^2)$$

$$a_\pm = \frac{4p \pm \sqrt{\Delta}}{p^2 - 3}$$

$$b_\pm = \frac{2p \pm \sqrt{\Delta}}{p^2 + 3}$$

$$\kappa_\pm = \frac{p^4 + 2p^2 + 9 \pm 2\sqrt{\Delta}}{(p^2 - 3)(p^2 + 3)}$$

$$\xi_\pm = \frac{p^4 + 18p^2 + 9 \pm 6\sqrt{\Delta}}{(p^2 - 3)(p^2 + 3)}.$$

Note that $b_+ \geq 0$ and $b_- \leq 0$, hence $a_+ \geq 0$ and $a_- \leq 0$. Finally, $|x\rangle$ and $|y\rangle$ are real under the condition $p \geq \sqrt{3}$ (see (23)) and using (24) we arrive at

$$F_\pm = \pm\frac{\sqrt{\Delta}}{2p}.$$

Because the maximal and minimal values of $w_{ij}$ are equal to 2 and $-2$, respectively, due to (26) and (27) we get

$$\gamma_{\max} = \begin{cases} 2 & \text{for } 0 \leq p < \sqrt{3} \\ \dfrac{1}{2p}\sqrt{\Delta} & \text{for } p \geq \sqrt{3} \end{cases}$$

and $\gamma_{\min} = -\gamma_{\max}$ in complete agreement with the result of calculations presented in the paper [6].

## Acknowledgments

The author would like to thank all the Organizers of QBIC 2011 for a warm hospitality during his stay in Japan.

## References


1. R. A. Horn, C. R. Johnson, *Topics in Matrix Analysis*, Cambridge Univ. Press, 1992.
2. P. D. Lax, *Linear Algebra and its Applications*, Wiley, 2007.
3. M. A. Nielsen and I. L. Chuang, *Quantum computation and quantum information*, Cambridge University Press, Cambridge, 2000.
4. T. Schulte-Herbrüggen, G. Dirr, U. Helmke, S. J. Glaser, *The significance of the C-numerical range and the local C-numerical range in quantum control and quantum information*, Linear and Multilinear Algebra **56**, 3 (2008).
5. G. Dirr, U. Helmke, M. Kleinsteuber, T. Schulte-Herbrüggen, *Relative C-numerical ranges for applications in quantum control and quantum information*, Linear and Multilinear Algebra **56**, 27 (2008).
6. P. Gawron, Z. Puchała, J. A. Miszczak, Ł. Skowronek, and K. Życzkowski, *Restricted numerical range: a versatile tool in the theory of quantum information*, J. Math. Phys. 51, 102204 (2010).
7. Z. Puchała, P. Gawron, J. A. Miszczak, L. Skowronek, Man-Duen Choi, and K. Życzkowski, *Product numerical range in space with tensor product structure*, Lin. Alg. Appl. 434, 327 (2011).
8. O. Gühne, G. Toth, *Entanglement detection*, Phys. Reports **474**, 1–75 (2009).
9. R. Horodecki, P. Horodecki, M. Horodecki, K. Horodecki, *Quantum entanglement*, Rev. Mod. Phys. **81**, 865–942 (2009).
10. D. Chruściński and A. Kossakowski, *Circulant states with positive partial transpose*, Phys. Rev. A 76, 032308 (2007).
11. D. Chruściński, A. Kossakowski, K. Młodawski, and T. Matsuoka, *A class of Bell diagonal states and entanglement witnesses*, Open Sys. Information Dyn. **17**, 235 (2010).
12. D. Chruściński and A. Pittenger, *Generalized Circulant Densities and a Sufficient Condition for Separability*, J. Phys. A: Math. Theor. **41** (2008) 385301.
13. D. Chruściński and A. Kossakowski, *Multipartite Circulant States with Positive Partial Transpose*, Open Sys. Information Dyn. **15** (2008) 189-212.
14. A. R. P. Rau, *Algebraic characterization of X-states in quantum information*, J. Phys. A: Math. Gen. **42**, 412002 (2009).
15. F. F. Fanchini, T. Werlang, C. A. Brasil, L. G. E. Arruda, A. O. Caldeira, *Non-Markovian Dynamics of Quantum Discord*, Phys. Rev. A. **81**, 052107 (2010).
16. M. Ali, A. R. P. Rau, G. Alber, *Quantum discord for two-qubit X-states*, Phys. Rev. A **81**, 042105 (2010).

17. B. Bylicka, D. Chruściński, *Witnessing quantum discord in* $2 \times N$ *systems*, Phys. Rev. A **81**, 062102 (2010).

18. Y. S. Weinstein, *Entanglement Sudden Death in Three Qubit X-States*, Phys. Rev. A **82**, 032326 (2010).

19. J. Jurkowski, A. Rutkowski, D. Chruściński, *Local Numerical Range for a Class of* $2 \times d$ *Hermitian Operators*, Open Sys. Information Dyn. 17, 347 (2010).

Quantum Bio-Informatics V
© 2013 World Scientific Publishing Co. Pte. Ltd.
pp. 229–236

# PARTIAL ROC REVEALS SUPERIORITY OF MUTUAL RANK OF PEARSON'S CORRELATION COEFFICIENT AS A COEXPRESSION MEASURE TO ELUCIDATE FUNCTIONAL ASSOCIATION OF GENES*

TAKESHI OBAYASHI*, KENGO KINOSHITA†

*Graduate School of Information Sciences, Tohoku University,
6-3-09, Aramaki-Aza-Aoba, Aoba-ku, Sendai, 980-8579, Japan
*e-mail: obayashi@ecei.tohoku.ac.jp
†e-mail: kengo@ecei.tohoku.ac.jp*

Gene coexpression analysis is a powerful approach to elucidate gene function. We have established and developed this approach using vast amount of publicly available gene expression data measured by microarray techniques. The coexpressed genes are used to estimate gene function of the guide gene or to construct gene coexpression networks. In the case to construct gene networks, researchers should introduce an arbitrary threshold of gene coexpression, because gene coexpression value is continuous value. In the viewpoint to introduce common threshold of gene coexpression, we previously reported rank of Pearson's correlation coefficient (PCC) is more useful than the original PCC value. In this manuscript, we re-assessed the measure of gene coexpression to construct gene coexpression network, and found that mutual rank (MR) of PCC showed better performance than rank of PCC and the original PCC in low false positive rate.

Keywords: Gene function prediction, microarray, gene coexpression, database.

## 1. Introduction

Since near the end of the 20th century, genome sequences of model species have been completed. Although the number of genes in each species is diverse, they commonly include vast amount of functional unknown genes. Some of the functional unknown genes

---

* This work was supported by MEXT, Japan [Grants-in-Aid for Publication of Scientific Research Results (No. 238052 to T.O. and No. 237003 to K.K.), and a Grant-in-Aid for Innovative Areas 'HD physiology' (No. 22136005) to K.K.].

do not have any homologous genes, while some other genes have weak homology to a large gene family, such as CYP, glucosidase etc. In both cases, computational predictions of gene functions are required for systematic identification of gene function.

Gene expression analysis is the easiest way to investigate the cellular context in which gene is activated. Especially, microarray technology can simultaneously measure gene expression of almost all genes in a genome [1]. In each experiment, researchers focus on differentially expressed genes among the samples used in the experiment. Such genes are key to understanding the situation of samples. In addition to the individual experiments, multiple microarray experiments make it possible to cluster genes for the selected experiments. Clustering analysis can identify functional group of genes under the experiments of interest.

Now, huge amount of microarray data are produced, and they are accumulated in the public repositories such as Gene Expression Omnibus [2] and ArrayExpress [3]. Such microarray data are freely downloadable and thus reanalyzed by many bioinformatics expertise. However, when we use too many samples for clustering analyses, interpretation of each cluster becomes difficult with ambiguous borders of clusters. Therefore, instead of clustering analysis, we directly used similarity of gene expression profiles. Gene pairs with similar expression profiles will have the same cellular context and thus the gene pairs can have the similar function. Such gene pairs are called as *coexpressed gene pairs*.

Recently many databases have been constructed to represent coexpressed gene lists with the coexpression strength from one guide gene to the other genes in a genome. We have also constructed and developed coexpression database, ATTED-II [4-6] for plants and COXPRESdb [7,8] for animals.

Table 1 shows an example of gene coexpression in ATTED-II. The guide gene is one of histone H4 genes in Arabidopsis (At3g53730), and most of coexpressed genes are other histone subunit genes. Apparently these coexpressed genes will work together with the guide gene. Although the function of the 23rd gene in the list is still unknown, we can expect that this gene will also function with the histone genes. Such gene coexpression data are available for all the genes in a genome with expression data, and they are so powerful to identify gene function (see reviews of gene coexpression analyses [9-11]). To calculate gene coexpression, namely similarity of gene expression profiles, Pearson's correlation coefficient (PCC) is often used as the coexpression measure. Since the selection of the coexpression measure strongly affects coexpressed gene lists, we have investigated appropriate coexpression measures for functional predictions and for construction of gene networks.

Table 1. Top 30 coexpressed genes for a histone H4 gene (At3g53730) in Arabidopsis based on ATTED-II (version Ath.c4-1).

|  | Gene ID | Function | Coexpression strength (MR) |
|---|---|---|---|
| 1 | At4g27230 | Histone H2A | 1.4 |
| 2 | At1g07790 | Histone H2B | 3.5 |
| 3 | At5g10980 | Histone H3 | 5.0 |
| 4 | At1g06760 | Histone H1 | 6.0 |
| 5 | At5g65360 | Histone H3 | 7.5 |
| 6 | At3g45980 | Histone H2B | 7.8 |
| 7 | At3g46030 | Histone H2A | 8.1 |
| 8 | At5g27670 | Histone H2B | 8.6 |
| 9 | At5g22880 | Histone H2B | 10.7 |
| 10 | At3g27360 | Histone H3 | 11.5 |
| 11 | At2g30620 | Histone H1.2 | 13.2 |
| 12 | At2g28740 | Histone H4 | 15.1 |
| 13 | At5g59910 | Histone H2B | 15.4 |
| 14 | At5g59870 | Histone H2A | 17.0 |
| 15 | At3g53650 | Histone H2B | 19.0 |
| 16 | At1g51060 | Histone H2A | 19.6 |
| 17 | At1g09200 | Histone H3 | 20.2 |
| 18 | At3g20670 | Histone H2A | 22.4 |
| 19 | At1g08880 | Histone H2A | 25.7 |
| 20 | At5g10390 | Histone H3 | 26.6 |
| 21 | At5g10400 | Histone H3 | 33.2 |
| 22 | At5g23420 | HMGB6 | 43.6 |
| 23 | At5g38690 | unknown | 54.5 |
| 24 | At2g37470 | Histone H2B | 67.9 |
| 25 | At5g43080 | CYCA3;1 | 68.9 |
| 26 | At5g61000 | RPA70D | 70.2 |
| 27 | At5g65350 | Histone H3 | 78.0 |
| 28 | At1g15660 | CENP-C | 80.2 |
| 29 | At1g04020 | BARD1 | 80.8 |
| 30 | At5g08020 | RPA70B | 83.8 |

## 2. A common threshold of gene coexpression to construct gene network

Constructing gene networks is one of the fundamental topics in systems biology. Large-scale gene function annotations are used to construct gene networks. One of the most direct annotations for gene functional network is protein-protein interaction (PPI) data, which are accumulated in some public databases such as HPRD for human [12] and AtPIN for Arabidopsis [13]. Since PPI data is physical relationship data to represent whether binding or not, PPI network can be directly constructed as a summation of each PPI data. On the other hand, gene coexpression is not physical relationship but just a degree of functional association, so that we have to set a threshold to decide coexpressed or not (Figure 1).

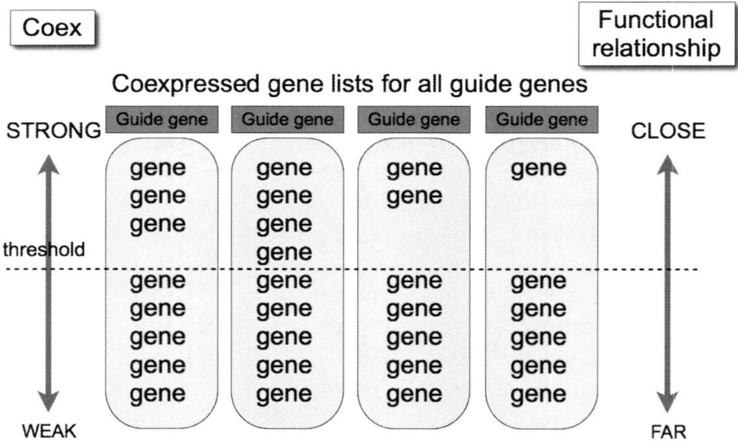


Figure 1. A schematic view to construct coexpressed gene network from coexpressed gene list.

Lower coexpression threshold make denser gene networks with larger size, but deciding the optimal threshold is difficult. Aoki et al. focused on the network density to optimize gene coexpression threshold. For various PCC thresholds, the threshold to produce gene network with minimum network density was selected as the best coexpression threshold [9].

We studied this problem by focusing on the performance of coexpressed genes in the prediction of the function of guide genes [14]. We first introduced a threshold to judge if gene pairs are coexpressed or not (Figure 1). Then the functional annotations of coexpressed genes were applied as the function of each guide gene. By changing the coexpression thresholds, we counted the numbers of true positives, false positives, false negatives and true negatives of the functional prediction of guide genes, and we wrote ROC and calculated AUCs (area under the ROC curves).

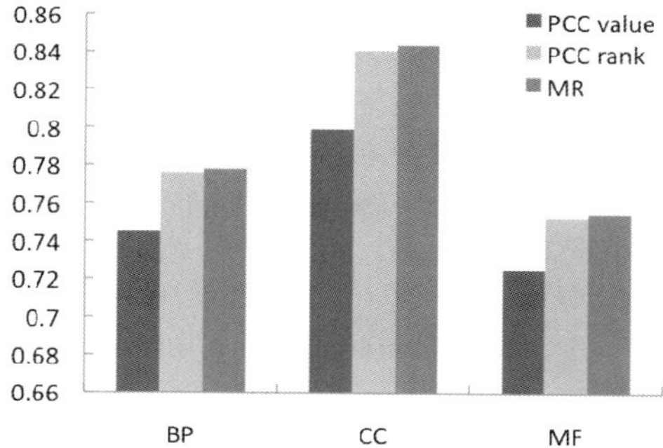


Figure 2. Effect of rank-based coexpression measures to predict GO annotation (monochrome display of our previous report, Figure 4 in Obayashi and Kinoshita, 2009b). The y-axis indicates AUCs to predict three types of GO annotations (BP, CC and MF) from gene coexpression data represented by PCC value, PCC rank and MR. BP, biological process; CC, cellular component; MF, molecular function.

As shown in Figure 2, we found that rank or mutual rank (MR) of PCC were better measure than the original PCC value to introduce common threshold for all guide genes (Figure 2) [14]. However, there are two points to be considered for this assessment. (1) Predictive performance under high false positive rate is almost meaningless to design actual experiments. Since stronger relationships can be verified more easily by experiments, we should pay attention to the performance under low positive rate. (2) This procedure did not consider redundant annotations in the coexpressed genes. Since we used GO annotations associated with similar number of genes, the number of redundancy of functional annotations in coexpressed genes may indicate prediction reliability. To include these two points, we performed another assessment to complement our previous assessment shown in Figure 2.

## 3. New assessment of coexpression measures using partial ROC

Instead of coexpressed gene list from a guide gene (Table 1), we focused on each gene pair to assess the relationship of coexpression and gene function separately. There are two types of gene pairs. One is gene pair with one or more common GO annotation(s), and the other is gene pair without any common GO annotations. Please note that multiple common GO annotations between a gene pair are not considered in this study, namely gene pairs with one common GO and those with two common GO annotations are not discriminated, although this point was considered in the previous assessment (Figure 2). To assess the selection power of coexpressed gene pairs with and without common GO annotations, we used $ROC_{0.01}$; the ROC where the false positive rate (FPR) is in the rage of 0 to 0.01, which we think realistic range for the actual experimental validation studies. Here we call the $ROC_{0.01}$ as partial ROC. Figure 3 shows the partial ROC curves of gene pairs with and those without common GO annotations for three GO categories. As in the usual ROC, curves in the left upper region indicates powerful measure, and thus the MR shows the highest selection power and PCC rank is second good measure. This result is basically consistent with our previous report. However, in this time the difference between MR and PCC rank is observed, while MR and PCC rank showed the same performance in our previous report with assessment using gene functional prediction (Figure 2). The superiority of MR to PCC rank is prominent only in the low FPR area, and thus focusing on low FPR area highlight the strength of MR, which was not observed in Figure 2, because we considered usual AUC values for the full FPR range.

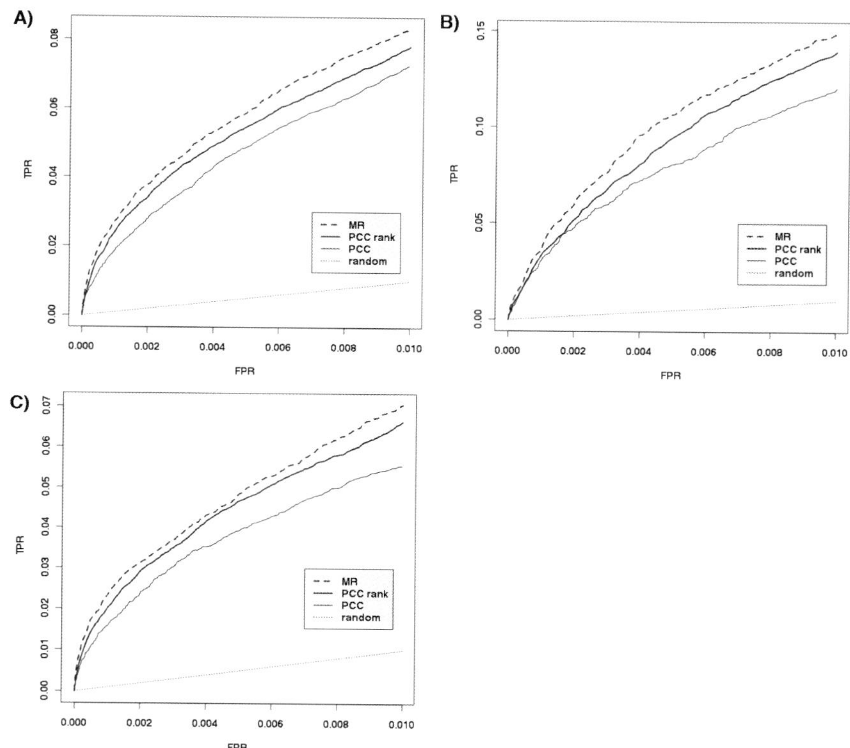

Figure 3. Partial ROCs to evaluate the power of discrimination of gene pairs with at least one common GO annotation and those without any common annotation for GO BP (A). (B) Same for GO CC, and (C) for GO MF.

## 4. Conclusion

Coexpression analysis is widely applied to identify gene function or gene functional partners with many experimental validation reports of the predicted gene function. However, goodness of the measure of coexpression is still an open question, even though it strongly affects coexpression performance. In this study, we re-assessed the superiority of MR over PCC as a coexpression measure. In this new assessment, we focused on gene pair rather than coexpressed gene list differently in the previous assessment, and found that MR is better than PCC rank. In addition, MR shows better performance than PCC rank especially when the low false positive rate condition are considered.

236

## Acknowledgments

Super-computing resources were provided by the Human Genome Center, Institute of Medical Science, The University of Tokyo.

## References

1. M.B. Eisen et al., *Proc. Natl. Acad. Sci. USA.* **95**, 14863-14868 (1998).
2. T. Barrett et al., *Nucleic Acids Res.*, **39**, D1005-D1010 (2011)
3. H. Parkinson et al., *Nucleic Acids Res.*, **39**, D1002-D1004
4. T. Obayashi et al., *Nucleic Acids Res.*, **35**, D863-D869 (2007)
5. T. Obayashi et al., *Nucleic Acids Res.*, **37**, D987-D991 (2009a)
6. T. Obayashi et al., *Plant Cell Physiol*, **53**, 213-219 (2011)
7. T. Obayashi et al., *Nucleic Acids Res.*, **36**, D77-D82 (2008)
8. T. Obayashi and K. Kinoshita, *Nucleic Acids Res.*, **35**, D1016-D1022 (2011)
9. K. Aoki et al., *Plant Cell Physiol.* **48**, 381-390 (2007).
10. B. Usadel et al, *Plant Cell Environ*, **32**, 1633-1651 (2009)
11. T. Obayashi and K. Kinoshita, *J. Plant Res.*, **123**, 311-319 (2010)
12. T.S. Keshava Prasad et al., *Nucleic Acids Res.*, **37**, D767-772 (2011)
13. M.M. Brandão et al., *BMC Bioinformatics*, **10**, 454 (2009)
14. T. Obayashi and K. Kinoshita, *DNA Research*, **16**, 249-260 (2009b)

Quantum Bio-Informatics V
© 2013 World Scientific Publishing Co. Pte. Ltd.
pp. 237–252

# QFT AND HADRONIC WORLD
# AS DYNAMICAL BASES OF NATURAL HISTORY

IZUMI OJIMA

*RIMS, Kyoto University*

Aiming at a consistent understanding of **repeatable laws** and their **historical developments without repetitions** as seen typically in the cosmological and biological evolutions, we try to examine the basic features of hadronic world consisting of the highly dynamical behaviours of strongly interacting hadrons from the viewpoint of Micro-Macro and Micro-Micro dualities. As a byproduct, we notice here that the essence of Haag-GLZ formula for S-matrix functionals can be found in the Ward-Takahashi identities associated with a symmetry breaking caused by the interactions.

## 1. Introduction: Hadrons and Bacteria as "Unsung Heros" behind Nature

In the "history of nature", **hadronic world** is characterized by its **extreme activity** and its **longest history of existence** (of the level as a whole but not at the level of individual units); we cannot imagine and verify the possibility of historical period without its activities and existence; for instance, the history of universe with evolution of stars starts from **protons** as the most typical and simplest kind of hadrons. In the biological context, similar features can be found in the **bacterial** levels, as was emphasized by Stephen Jay Gould in his book, "Full House – The Spread of Excellence from Plato to Darwin" (Harmony Books, 1996)[a].

I have chosen the expression, "unsung heros"[b], as the one corresponding to "en no shita no chikara-mochi" in Japanese to express the invisible

---

[a]This interesting book was brought to my attention by Prof. I. Yamato in August 2010, with whom Prof. T. Matsuoka and myself have enjoyed extremely interesting discussions on the basic points about the repeatable laws in nature and the historical developments in physical, biological, cognitive, social and even humanity sciences. Let me express my cordial thanks to both Professors.

[b]unsung hero = a person who makes a substantive yet unrecognized contribution; a person whose bravery is unknown or unacknowledged (IEEE: Dictionary.com, in Dictionary.com's 21st Century Lexicon. Source)

238

existence whose supports are indispensable for the subsistence of all the visible levels above it. This corresponds just to the actual roles played by the hadronic and bacterial levels, respectively, in the physical and biological worlds. I believe that this kind of aspects are crucial for our purposes in Quantum-Bio-Informatics, where satisfactory understanding is indispensable of the **consistency** between **repeatable laws** and their **historical developments without repetitions**, as seen in the cosmological and biological evolutions. At the end (unfortunately!!) of this challenging project of QBIC, let me try to examine the above issue in the realm of quantum fields and hadrons, from the viewpoint of **quadrality scheme** based on **Micro-Macro duality** [1], which is expected hopefully to be useful for considering the various bridges among Quantum, Biological and Informatic aspects in nature. After explaining Micro-Macro duality and qudrality scheme, we recollect the formulation of scattering amplitudes (= S-matrix functional) in terms of quantum fields, on the basis of which basic features of hadrons are examined.

## 2. Micro-Macro Duality and Quadrality Scheme

The essence of Micro-Macro duality can be seen in the following scheme where sectors are defined by quasi-equivalence classes of factor states [1]:

| $\longleftarrow$ Visible *Macro* | of | *independent objects* | | $\cdots \longrightarrow$ | | *Inter-sectorial* |
|---|---|---|---|---|---|---|
| $\cdots \quad \gamma_N$ | | *sectors* $\gamma$ | $\gamma_2$ | $\gamma_1$ | | $Sp(3)$ |
| $\vdots$ | | $\vdots$ | $\vdots$ | $\vdots$ | | $\uparrow$ *Intra-sectorial* |
| $\cdots \quad \pi_{\gamma_N}$ | | $\pi_\gamma$ | $\pi_{\gamma_2}$ | $\pi_{\gamma_1}$ | | $\parallel$ |
| $\vdots$ | | $\vdots$ | $\vdots$ | $\vdots$ | | $\downarrow$ invisible *Micro* |

On the basis of this duality, "Quadrality Scheme" can be formulated as duality of duality pairs involving **emergence from Rep to Spec** [2] and **co-emergence from Dyn to Alg**. Here, emergence and co-emergence explain, respectively, the physical origin of space-time structure and the formation of objects from perpetual motions:

| [Macro] | $\boldsymbol{Spec} = \dfrac{\text{particles in}}{\text{spacetime}}$ | |
|---|---|---|
| emergence $\nearrow$ | | |
| $\boldsymbol{Rep} = \dfrac{\text{asymptotic}}{\text{fields } \phi^{in/out}}$ | $\leftrightarrows$ | $\boldsymbol{Alg} =$ Heisenberg fields $\varphi_H$ with stable hadrons $\pi - N$ in $\boldsymbol{Regge\ traj.'s}$ |
| | | $\nearrow$ co-emergence |
| | $\boldsymbol{Dyn} =$ | $\boldsymbol{Hadronic\ world}$ [Micro] |

## 3. QFT and S-Matrix

To choose and fix the appropriate vocabulary for describing phenomena in a given domain in nature, it is vital for us to select the most relevant notions and constituents suitable for providing the "0-th approximation" of the phenomena and processes to be described. In the present context of our discussion concerning quantum fields and their behaviours, we need to focus our attention to the physical origin & meaning of independence [3] (of tensor type) in relativistic QFT as follows [4]:

1) Particles are *visible* units of independence, characterized by the *on-shell condition* $p^2 = m^2$ due to a "Central Limit Theorem" as *emergence of independence* via *asymptotic condition*[c], $\varphi_H(x) \overset{x^0 = t \to \mp\infty}{\longrightarrow} \phi^{in/out}(x)$, from *non-independent* interacting Heisenberg fields $\varphi_H$ to *independent* free *asymptotic fields* $\phi^{as} = \phi^{in/out}$ *& asymptotic states*:

Precise meaning of this "Central Limit Theorem" can be found

---

[c]While the asymptotic condition itself is very old, due to [5], its reinterpretation in relation with "independence" in quantum probability is new [4].

in *"Micro-Macro Duality" in QFT* [4]:

| [[Macro]] | |
|---|---|
| ***Asymptotic fields*** $\phi^{as}$ as ***independent*** objects s.t. $(\Box+m^2)\phi^{as}= 0 \Longleftrightarrow$ $p^2 = m^2$: on-shell cond. | $\diagdown$ ***asymptotic condition*** $= CLT$ |
| GLZ-Fock expansion of $\varphi_H$ $\searrow$ | ***Interacting fields*** $\varphi_H$ Yang-Feldman eqn: $\varphi_H= \Delta_{ret}*J_H+\phi^{in}$ |
| ***Coupling*** $J_H := (\Box+m^2)\varphi_H$ | [[Micro]] |

or,

| Macro | S-matrix & PCT | Micro |
|---|---|---|
| $\phi^{as}$: universal | ***asymp.cond.*** $\leftrightharpoons$ | $\varphi_H$ : generic |
| $p^2 = m^2$ $\Longleftrightarrow$ $(\Box+m^2)\phi^{as}= 0$ | GLZ-Fock expansion of $\varphi_H$ in $\phi^{as}$ | $(\Box+m^2)\varphi_H= J_H$ $\Longleftrightarrow$ $\varphi_H= \Delta_{ret}*J_H+\phi^{in}$ |

2) ***Independent*** "units" serve as ***vocabulary*** for describing ***state changes due to interactions*** (=***violations of independence***) in scattering processes = [***asymptotic in-states*** $\stackrel{\text{S-matrix}}{\Longrightarrow}$ ***out-states***].

In spite of the common belief that quantum free fields $\phi(x)$ with $a^*(\vec{p}), a(\vec{p})$ are sufficient for describing wave-particle dualism inherent in elementary particles, the perpetual creation and annihilation processes of particles, however, require ***interactions*** among elementary particles, which is not consistent with the linearity of free field equation. Therefore, Einstein's famous equivalence $E = mc^2$ of energy $E$ and mass $m$ should be seen to give only *partial* information for dynamical descriptions of relativistic quantum fields, with ***off-shell*** aspects being totally neglected in spite of their vital importance for non-trivial scattering processes, particle decays and productions, etc., etc. [4]! Since the ***on-shell*** asymptotic fields $\phi^{as}$ can***not*** by themselves ignite scattering processes for lack of interactions, the ***dependent off-shell interacting Heisenberg fields*** $\varphi_H$ is indispensable for describing non-trivial scattering processes due to the interactions among quantum fields.

Moreover, the famous ***Haag theorem*** [6,7,8] tells us that Poincaré (or even, Galilei)-covariant quantum fields related to free fields by a unitary transformation are only free fields. Thus, it is meaningless to treat interacting Heisenberg fields in terms of a unitary transformation of free fields (as is common in perturbative approaches), which is in sharp contrast to quantum systems with finite degrees of freedom.

To describe relativistic scattering processes of elementary particles, we need at least the following three items: Poincaré-covariant quantum fields/ their interactions/ free fields. Free fields are necessary because it provide us with indispensable vocabulary for the description of scattering processes, where an initial state with incoming free particles is changed into a final one with outgoing particles. Instead of the direct relation between interacting Heisenberg and free fields forbidden by the above Haag theorem, the **unitary S-matrix appears between two free asymptotic fields,** $\phi^{in}(x)$ and $\phi^{out}(x)$ in the form of a basis change $S_{\beta,\alpha} := \langle \beta, out | \alpha, in \rangle$ between in-state basis $|\alpha, in\rangle$ and out-state basis $|\beta, out\rangle$ [5]:

| out | S-matrix & PCT | in | |
|---|---|---|---|
| $Ad\Theta^{out} \curvearrowright \phi^{out}(x)$ | $\begin{array}{c} AdS \ \& \ Ad\Theta \\ \rightleftarrows \\ AdS^{-1} \ \& \ Ad\Theta \end{array}$ | $\phi^{in}(x) \curvearrowleft Ad\Theta^{in}$ | Asymp. fields : Macro |
| $t \to +\infty \nwarrow$ | asymp.cond. | $\nearrow t \to -\infty$ | |
| | $\varphi_H(x) \curvearrowleft Ad\Theta$ | | Interacting fields : Micro |

## 4. Asymptotic Condition & Yang-Feldman Equation

Applying the ergodic theory to the dynamical system $\mathcal{O} \longmapsto \mathcal{P}(\mathcal{O})$ consisting of the algebra $\mathcal{P}(\mathcal{O})$ of polynomials in Heisenberg fields $\varphi_H$ supported in spacetime regions $\mathcal{O} \subset \mathbb{R}^4$ under the Poincaré group action $\mathcal{P}(\mathcal{O}) \overset{\alpha_{(a,\Lambda)}}{\hookrightarrow} \mathcal{P}(\Lambda\mathcal{O} + a)$ with $(a, \Lambda) \in \mathcal{P}_+^\uparrow$, we see the equivalence [7,8] among [cyclicity $\overline{\mathcal{P}(\mathbb{R}^4)\Omega} = \mathfrak{H}$ of vacuum vector $\Omega$] $\Longleftrightarrow$ [irreducibility of $\mathcal{P}(\mathbb{R}^4)$ in the vacuum representation characterized by spectral condition] $\Longleftrightarrow$ [uniqueness of vacuum vector (: $U(x)\Psi = \Psi \Longrightarrow \Psi \propto \Omega$)] $\Longleftrightarrow$ [validity of cluster property]:

$$|\omega_0(A(x)B(y)) - \omega_0(A)\omega_0(B)| \to 0 \quad \text{as } (\vec{x} - \vec{y})^2 \to \infty,$$

where $\omega_0 = \langle \Omega | \cdots \Omega \rangle$ is the vacuum state, and $A(x) := \alpha_x(A) = U(x)AU(x)^*$, $B(y) := \alpha_y(B)$ are spacetime translates by $x, y \in \mathbb{R}^4$ of local observables $A, B \in \mathcal{P}(\mathcal{O})$, respectively. The last relation follows from the partition of unity due to spectral resolution of spacetime translations

242

$$U(x) = \int_{p\in\overline{V_+}} \exp(ipx)\,dE(p):$$

$$1 = |\Omega\rangle\langle\Omega| \;+\; \sum_i (\textit{1-particle singularities on mass-shell } p^2 = m_i^2)$$

$$+\ (\text{absolutely continuous } p\text{-spectra}).$$

From this cluster property combined with local commutativity:

$$\langle\Omega|A\alpha_{\vec{x}}(B)\Omega\rangle \overset{\vec{x}\to\infty}{\longrightarrow} \langle\Omega|A\Omega\rangle\langle\Omega|B\Omega\rangle,$$

the asymptotic condition $\varphi_H(x) \overset{x^0=t\to\mp\infty}{\longrightarrow} \phi^{in/out}(x)$ (as weak convergence) follows. In sharp contrast to this *asymptotic factorization* for interacting Heisenberg fields valid in the asymptotic limit, the asymptotic fields $\phi^{as}$ materialize the *kinematical* factorization (= independence) of correlations *without* taking asymptotic limit, which is just equivalent to the validity of "Wick theorem":

$$\omega_0(\phi^{as}\phi^{as}\cdots\phi^{as}) = \sum \omega_0(\phi^{as}\phi^{as})\cdots\omega_0(\phi^{as}\phi^{as}),$$

as the expansion of $n$-point functions into the sum of products of 2-point functions. This is nothing but the "quasi-freeness" of $\omega_0$ w.r.t. $\phi^{as}$ constituting the contents of independence of Gaussian type. As a quantized solution of Klein-Gordon equation $(\Box + m^2)\phi = 0$ free quantum field $\phi^{as}(x) =: \phi(x)$ describes "particle pictures" in terms of creation and annihilation operators: $\phi(x) \rightleftarrows$ creation and annihilation operators $a(\vec{p}), a^*(\vec{q})$:

$$\phi(x) = \int \frac{d^3p}{\sqrt{(2\pi)^3 2\omega_{\vec{p}}}}(a(\vec{p})\exp(-ip_\mu x^\mu) + h.c.),$$

$$a^*(f) := i\int \phi(x)\overleftrightarrow{\partial_0} f(x)\,d^3x = \int a^*(\vec{p})\tilde{f}(\vec{p})\,d^3p = [a(f)]^*,$$

$$[a(f), a^*(g)] = \int \overline{\tilde{f}(\vec{p})}\tilde{g}(\vec{p})\,d^3p = \langle\tilde{f}, \tilde{g}\rangle,$$

$$[\phi(x), \phi(y)] = \int \frac{d^4p}{(2\pi)^3}\varepsilon(p^0)\delta(p^2 - m^2)\exp(-ip(x-y))$$

$$=: i\Delta(x - y; m^2),$$

with $\omega_{\vec{p}} := \sqrt{\vec{p}^2 + m^2}$ in the "natural unit system" with $\hbar = c = 1$.

Thus, the independence embodied by asymptotic fields $\phi^{as}$ is seen to emerge from interacting Heisenberg fields $\varphi_H$ via asymptotic condition as a kind of central limit theorem. In this context, what corresponds to

"Langevin equation" is the **Yang-Feldman equation** to connect Heisenberg field $\varphi_H(x)$ and asymptotic field $\phi^{as}(x)$:

$$\varphi_H(x) = \int \Delta_{ret}(x-y; m^2) J_H(y) d^4 y + \phi^{in}(x)$$

$$= \int \Delta_{adv}(x-y; m^2) J_H(y) d^4 y + \phi^{out}(x)$$

$$= [\Delta_{ret} * J_H + \phi^{in}](x) = [\Delta_{adv} * J_H + \phi^{out}](x),$$

where $J_H = (\Box + m^2)\varphi_H$: **Heisenberg source current** (of interactions causing deviations from $E = mc^2/\sqrt{1 - v^2/c^2}$), $\Delta_{ret/adv}(x-y; m^2)$: retarded/advanced Green's functions (i.e., principal solutions) of Klein-Gordon equation defined by

$$(\Box_x + m^2)\Delta_{ret/adv}(x-y; m^2) = \delta(x-y),$$

$$\Delta_{ret/adv}(x-y; m^2) = 0 \quad \text{for } x_0 \lessgtr y_0.$$

In this Yang-Feldman equation, the asymptotic fields $\phi^{as}(x)$ and the Heisenberg source current $J_H$ appear, respectively, as *residue* and *quotient* in the division of $\varphi_H$ by $\Delta_{ret/adv}$: $\varphi_H = \Delta_{ret} * J_H + \phi^{in} = \Delta_{adv} * J_H + \phi^{out}$.

More important is: creation and annihilation operators $a(\vec{p}), a^*(\vec{q})$ are **infinite number of conserved quantities** corresponding to a **symmetry** of $\phi^{as}$ **unbroken in the absence of interactions** [$\Longleftrightarrow S = 1$], validating Wick theorem (= **quasi-freeness = Gaussianity**) for the vacuum state, i.e., independence as a symmetry: for any solution $f(x)$ of the Klein-Gordon equation $(\Box + m^2)f = 0$, we have

$$J_\mu(f) := i\phi(x)\overleftrightarrow{\partial_\mu} f(x);$$

$$\partial^\mu J_\mu(f) = i\partial^\mu[\phi(x)\overleftrightarrow{\partial_\mu} f(x)] = i\phi(x)\overleftrightarrow{\Box} f(x) = 0,$$

among which we find $a(\vec{p}) = \int dS^\mu J_\mu(f)$, $a^*(\vec{p}) = \int dS^\mu J_\mu(\bar{f})$ for $f(x) := \exp(-ip_\mu x^\mu)$. The **breakdown** of this symmetry **due to the interaction** [$\Longleftrightarrow S \neq 1$] describes the scattering processes caused by $J_H$, which gives **residues at the on-shell pole** $\dfrac{1}{p^2 - m^2}$ to determine scattering amplitudes as "Ward-Takahashi identities associated to broken symmetry" as will be seen below.

## 5. Micro-Macro Duality in "Central Limit Theorem" in QFT

Now we discuss the **universality** aspect due to the Haag-GLZ expansion [6,9], similar to "Fock expansion" in WNA [10]. In terms of **S-matrix** $S$

defined by

$$S :=: \exp(\phi^{in}(\Box + m^2)\frac{\delta}{\delta J}) : \omega_0(T(\exp(iJ\varphi_H)) \restriction_{J=0}$$

$$=: (\omega_0 \otimes id)(T \exp(iJ_H \otimes \phi^{in})) :=: (\omega_0 \otimes id)(T \exp(iJ_H \otimes \phi^{out})) :$$

$$:= (\omega_0 \otimes id)(W),$$

the Haag-GLZ expansion can be derived (from LSZ reduction formulae):

$$SA =: (\omega_0 \otimes id)(T[(A \otimes 1) \exp(iJ_H \otimes \phi^{in})]) : \quad = S(AS)S^{-1},$$
$$AS =: (\omega_0 \otimes id)(T[(A \otimes 1) \exp(iJ_H \otimes \phi^{out})]) : \quad = S^{-1}(SA)S,$$
$$A = S^{-1} : (\omega_0 \otimes id)(T[A \otimes 1] \exp(iJ_H \otimes \phi^{in})) :$$
$$=: (\omega_0 \otimes id)(T[A \otimes 1] \exp(iJ_H \otimes \phi^{out}) : S^{-1}.$$

S-matrix $S$ is seen to be an ***intertwiner*** between two free fields, in-coming $\phi^{in}$ and out-going $\phi^{out}$:

$$\phi^{in}(x)S = S\phi^{out}(x).$$

Noting the short exact sequence corresponding to Yang-Feldman equation $\varphi_H = \Delta_{ret} * J_H + \phi^{in}$,

$$\{\phi^{in}\} \hookrightarrow \{\varphi_H\} \overset{\Box+m^2}{\twoheadrightarrow} \{J_H\},$$

we see now that scattering processes can be interpreted as ***low-energy theorem via non-linear realization*** and also as a ***long exact sequence*** in homological algebra mediated by ***connecting morphisms supplied by Heisenberg source current*** $J_H$. (In this context, Yang-Feldman equation $\varphi_H = \Delta_{ret} * J_H + \phi^{in}$ can be compared with $1/(x - a \mp i\varepsilon) = P(1/(x-a))\pm\pi i\delta(x-a)$, according to which $\varphi_H$ may be viewed as composite system of divisor $J_H$ and zero $\phi^{as}$ w.r.t. the spectrum of $\Box + m^2$.) This constitutes the essence of Micro-Macro duality in "Central Limit Theorem" in QFT and the mutual relation between $\phi^{as}$ and $\varphi_H$:

$$\phi^{as} \overset{\text{asymp.cond.}}{\underset{\text{Haag-GLZ-Fock}}{\rightleftarrows}} \varphi_H$$

is just "***Micro-Macro duality***":

|  | ***Micro-Macro duality*** |  |
|---|---|---|
| Macro: $\phi^{as}$ | $\rightleftarrows$ | $\varphi_H$: Micro |
|  | K-T operator $W$ |  |

controlled by **Kac-Takesaki cocycle operator** $W$ given by

$$W \overset{\text{def}}{=} : T(\exp(iJ_H \otimes \phi^{in}) :$$

$$=: T(\exp(i \int d^3x [\phi^{in} - \phi^{out}] \otimes \overleftrightarrow{\partial_\mu} \phi^{as}) :$$

and characterized by the pentagonal relation

$$W_{12}(W^{as})_{23} = (W^{as})_{23} W_{13} W_{12}$$

($W^{as}$: K-T operator of CCR $\phi^{as}$ in regular representation)[d].

Its essence is seen in that free field/particle $\phi^{as}$ : emerges from interacting $\varphi_H$ via the asymptotic condition, and that $\varphi_H$ is reconstructed from $\phi^{as}$ by the Haag-GLZ-Fock expansion. Thus, the meaning of "central" (limit thm) as **universality** of $\phi^{as}$ is guaranteed by the Haag-GLZ-Fock expansion, similarly to the Fock expansion in WNA.

In view of free field equation $(\Box + m^2)\phi^{as} = 0$, the operator $J_H \otimes \phi^{as} := \int d^4x J_H(x) \otimes \phi^{as}(x)$ in $W$ can be rewritten as:

$$\begin{aligned}
J_H \otimes \phi^{as} &= (\Box + m^2)\varphi_H \otimes \phi^{as} \\
&= (\Box + m^2)\varphi_H \otimes \phi^{as} - \varphi_H \otimes (\Box + m^2)\phi^{as} \\
&= -\partial^\mu[\varphi_H \otimes \overleftrightarrow{\partial_\mu}\phi^{as}] = \partial^\mu j_\mu^{as} \\
&= \int d^4x \partial^\mu j_\mu^{as}(x) = \int d^3x [\phi^{in} - \phi^{out}] \otimes \overleftrightarrow{\partial_\mu}\phi^{as}(x),
\end{aligned}$$

where

$$j_\mu^{as} := -\varphi_H \otimes \overleftrightarrow{\partial_\mu}\phi^{as}.$$

Thus, Haag-GLZ formula for S-matrix functionals:

$$SA =: (\omega_0 \otimes id)(T[(A \otimes 1)\exp(iJ_H \otimes \phi^{in})])$$

$$=: (\omega_0 \otimes id)(T[(A \otimes 1)\exp(i\partial^\mu j_\mu^{as})])$$

can be seen as **broken WT identities** (unbroken for $j_\mu^{as} := -\phi^{in/out} \otimes \overleftrightarrow{\partial_\mu}\phi^{as}$ and broken for $j_\mu^{as} := -\varphi_H \otimes \overleftrightarrow{\partial_\mu}\phi^{as}$).

---

[d]Note again that, while the Haag-GLZ formula itself is very old, its reformulation in relation with K-T operator is new [4], without which the physical essence of S-matrix as a measurement process described by the coupling via a K-T operator $W$ between Heisenberg fields $\varphi_H$ and asympototic fields $\phi^{in/out}$ cannot be seen. This point has been emphasized by Prof.A.Hosoya and Mr.Y.Shikano, to whom I am very grateful.

In this way, the coupling term $\partial^\mu j_\mu^{as} = J_H \otimes \phi^{as} \neq 0$ between Heisenberg $\varphi_H$ and asymptotic $\phi^{as}$ responsible for non-trivial scattering processes can be interpreted as the effect of symmetry breakdown due to $(\Box + m^2)\varphi_H = J_H \neq 0$ (which corresponds to a shift from $Hom$ to $Ext^1$ in homological algebra [cf. M. Sato's formulation of solutions of linear PDE]). Moreover, the relevance of harmonic-analytic duality is evident from the Lie-algebraic structure of **Heisenberg source currents** (due to Bogoliubov, et al. [8]):

$$J_H^i(x) = (\Box + m^2)\varphi_H^i = S^{-1} \frac{\delta}{i\delta\phi_i^{in}(x)} S;$$

$$\frac{\delta J_H^i(x)}{\delta\phi^j(y)} - \frac{\delta J_H^j(y)}{\delta\phi^i(x)} = i[J_H^i(x), J_H^j(y)].$$

## 6. PCT and Modular Structure

Modular-theoretical aspects described by PCT symmetry due to the (weak) local commutativity [7,8] can be seen in the scattering theoretical context as follows. First, the existence of positive/ negative energy solutions $E = \pm\sqrt{(\vec{p}c)^2 + (m_0 c^2)^2}$ of $(\frac{E}{c})^2 - (\vec{p})^2 = m_0^2 c^2$ corresponds to the creation & annihilation operators, particle-antiparticle pairs, and time reversal $\boldsymbol{T}$ and **PCT invariance**, satisfying the equivalence relation: (weak) local commutativity $\Longleftrightarrow$ PCT invariance $\Longleftrightarrow$ S-matrix $\Longleftrightarrow$ Borchers classes of mutually local fields. From the definition of the PCT transformation $\theta$:

$$\theta(\varphi_H(x)) = \gamma\varphi_H(-x)^* \quad \text{(with } \gamma \in \mathbb{T}),$$

the vacuum $\omega_0$ is invariant under $\theta$: $\omega_0 \circ \theta = \omega_0$ owing to the local commutativity of $\varphi_H$. Thus, there exists an anti-unitary PCT operator $\Theta$ to implement $\theta$:

$$\theta(\varphi_H(x)) = \Theta\varphi_H(x)\Theta,$$

$$\Theta\Omega = \Omega.$$

While $\varphi_H, \phi^{in}, \phi^{out}$ are not mutually local, the PCT invariance of their common vacuum state $\Omega$ implies the existence of such PCT antiunitary operators $\Theta, \Theta^{in}, \Theta^{out}$, corresponding, respectively, to $\varphi_H, \phi^{in}, \phi^{out}$ as satisfying the relations:

$$\Theta\phi^{out}(x)\Theta = \gamma\phi^{in}(-x)^* = \Theta^{in}\phi^{in}(x)\Theta^{in},$$

$$\Theta\phi^{in}(x)\Theta = \gamma\phi^{out}(-x)^* = \Theta^{out}\phi^{out}(x)\Theta^{out}.$$

From $S\phi^{out}(x)S^{-1} = \phi^{in}(x) = \Theta\gamma^{-1}\phi^{out}(-x)^*\Theta = \Theta\Theta^{out}\phi^{out}(x)\Theta^{out}\Theta$, we have

$$S = \Theta^{in}\Theta = \Theta\Theta^{out},$$

when $\phi^{as}$: irreducible.

| out | S-matrix & PCT | in | |
|---|---|---|---|
| $Ad\Theta^{out}\curvearrowright\phi^{out}(x)$ | $AdS$ & $Ad\Theta$ $\rightleftarrows$ $AdS^{-1}$ & $Ad\Theta$ | $\phi^{in}(x)\curvearrowleft Ad\Theta^{in}$ | Asymp. fields:  Macro |
| $t \to +\infty \searrow$ | asymp.cond. | $\nearrow t \to -\infty$ | |
| | $\varphi_H(x)\curvearrowleft Ad\Theta$ | | Interacting fields:  Micro |

The conclusion that quantum fields with the **same PCT operator** $\Theta$ have the same S-matrix $S = \Theta^{in}\Theta = \Theta\Theta^{out}$ is usually interpreted as the "ambiguities" in **Heisenberg fields interpolating** the same S-matrix, which is in close relation with the notion of **Borchers classess** of mutually local fields sharing the same PCT operator $\Theta$ [8]. This kind of consideration is crucial for the "inverse problem" to reconstruct interacting Heisenberg fields $\varphi_H$ from the knowledge of asymptotic fields $\phi^{as}$ and the S-matrix $S$ intertwining them. In this way, the gaps between **idealized (= approximate) world of independence** and **realistic interacting world of dependence** are to be filled up by the **coupling = interactions** [4].

## 7. Duality in Hadronic World of Strong Coupling

In view of the strong logical coherence among scattering cross section, S-matrix and asymptotic fields, the above scheme is almost a unique solution as a systematic description of scattering processes of quantum fields. But, its pertinence highly depends on the validity of the asymptotic condition = central limit theorem and on the small deviations of the Heisenberg fields from the corresponding asymptotic fields. These crucial conditions do not seem to fit to strongly interacting Hadronic World, whose basic feature lies in strong couplings and "dependence", characterized by "Micro-Micro self-duality" between resonance poles[: objects] and Regge poles[: interactions]:

1) almost all hadrons are highly unstable **resonance states**, showing up transiently in the intermediate states of scattering processes of low-lying stable (or, almost stable) hadrons. This is just the

essential features of ***dependence*** whose sector structure can be understood as follows:

| El.mg.(diag)-Weak(off-diag) <br> interactions+ ***flavour sym.*** $\curvearrowright$ $\longleftarrow$ $\cdots$ $\longrightarrow$ | | | ***Independence*** <br> =Inter-hadronic |
|---|---|---|---|
| $\gamma_N$    Light-mass hadrons <br> $\gamma \in Sp(3_\pi)$ | $\gamma_2$ $\gamma_1$ | | ***Dependence*** <br> =Intra-hadronic |
| $\pi_{\gamma_N}$    $\pi_\gamma$ ***Resonances*** | $\pi_{\gamma_2}$ $\pi_{\gamma_1}$ | $\uparrow$   $\downarrow$ $\overset{dual}{\simeq}$ | $=$ Inside of <br> ***hadronic sectors*** <br> ***Regge*** ***trajectories*** <br> $\simeq \circlearrowleft$ ***Strong interactions*** |

2) It is remarkable that the basic statistical features of the resonance states can be found in the ***Cauchy distributions*** $\propto$
$$\frac{1}{(E - E_i)^2 + (\Gamma_i/2)^2} = \left| \frac{1}{E - E_i - i\Gamma_i/2} \right|^2 \quad \text{w.r.t. the energy vari-}$$
able $E$ which correspond to the ***exponential decays*** of unstable particle states $\propto |\exp(-it(E - E_i - i\Gamma_i/2))|^2 \propto \exp(-t\Gamma_i)$ in time. Because of this instability= dependence caused by the strong interactions, the controllable stable states can be found only in ***hadrons with the lowest masses*** among those with the same (internal) quantum numbers such as $P$(: proton), $N$(: neutron), and $\pi$(: pions) and so on.

3) The word ***"dual"*** in the above diagram means the ***duality*** inside a hadronic sector($\simeq$ ***Regge trajectory*** of hadron poles with energy-dependent angular momenta $\alpha(s) = \alpha_0 + \alpha's$) between its ***resonance poles*** $\leftrightarrows$ ***Regge poles***, the former appearing in ***time-like*** momentum region of S-matrix (called ***s-channel***) and the latter in ***spacelike*** ones called ***t-*** or ***u-channels***:

The former describes particle-like ***unstable*** modes (=***pseudo-independence***) whose ***lowest level members*** only such as proton, neutron, pions, etc., can exist in (meta-)stable ways, and the latter yields the *interaction terms between hadrons* reflecting the aspect of *dependence.*

4) The meaning of *duality* here should properly be understood in the following two ways, in contrasts to other contexts: first, the basic common features found in the mixed moments in all kinds of independences is the coexistence of all the above three terms of types, s-, t-, u-channels. In sharp contrast, these three-type diagrams are here *mutually transformed from one to another without co-existence.* Perhaps, this can be interpreted as one of the most essential features of *dependence* inherent in the hadronic processes.

Next, in contrast to the usual kinds of dualities valid between objects living at different levels, the duality appearing here holds at one and the same level of strongly interacting hadrons, connecting the aspect of *independent* objects and that of dynamical processes as *coupling and dependence.* Thus, this duality can be seen as *Micro-Micro duality.*

5) In relation with Wigner's construction representations of Poincaré group $\mathcal{P}_+^\uparrow = \mathcal{H}_2(\mathbb{C}) \lhd SL(2,\mathbb{C})$ (or, $\mathbb{R}^4 \lhd L_+^\uparrow$), we can understand the above duality (which motivated the investigation of dual resonance and string models) as the interchange between the little group $SU(2)$ (or $SO(3)$) at timelike momentum $p, p^2 > 0$ and that $SU(1,1)$ (or $SL(2,\mathbb{R})$) at spacelike momentum $p, p^2 < 0$.

Reformulating this Wigner-Mackey machinery of induction based on *little groups* $H$ of a semi-direct product group $G = N \lhd L$, we can see the close relation of the present hadronic duality with the (spontaneous) symmetry breakdown, degenerate vacua unified into the notion of *augmented algebras* [1] and with the spacetime *emergence* [2] as follows: the regular C*-group algebra $C_r^*(G)$ of $G$ is isomorphic to the crossed product $C_0(\hat{N}) \rtimes L$ of $C_0(\hat{N})$ (= C*-group algebra of abelian group $N$) by the action of $L$ on $N$, which is also related with the covariant representations of the dynamical system $C_0(\hat{N}) \curvearrowleft L$. For each character $\chi \in \hat{N}$ of $N$, we can consider the *little group* $H_\chi$ of a pure state $\delta_\chi$ of $C_0(\hat{N})$ defined by the isotropy subgroup of the latter, which can be viewed as the group of the *remaining unbroken symmetry* in the pure state $\delta_\chi$.

In the case of the Poincaré group $\mathcal{P}_+^\uparrow = G$, the existence of the well-known four types of the orbits $p^2 > 0, = 0, < 0, p_\mu \equiv 0$ having little groups

$SU(2), E(2), SU(1,1)$ and $SL(2,\mathbb{C})$, respectively, can be interpreted as a kind of **phase transitions** taking place in the process of space-time emergence, whose different **phases** can be mutually connected through the **analytic continuation** of the dynamical system $C_0(\hat{N}) \curvearrowright L$:

$$
\begin{array}{cc}
SU(2) & E(2) \\
\diagdown \quad \text{at } p^2 > 0 \quad \diagup & \text{at } p^2 = 0 \\
: \text{elliptic} & : \text{parabolic}
\end{array}
$$

$$
p^2 < 0 \quad
\begin{array}{cc}
SL(2,\mathbb{C}) & SU(1,1) \\
\text{at } p \equiv 0 & \text{at } p^2 < 0 \\
: \text{vacuum} \quad \diagdown & : \text{hyperbolic}
\end{array}
$$

$$
\diagup \qquad p^2 > 0 \qquad \diagdown
$$

It would be useful and interesting to re-view the **hadronic dual-resonance aspects** encoded in the string model, from the viewpoint of **quantum probability & its complex analysis**, in close relationship with the **independence, coupling and dependence.**

6) In view of the **dominant contributions** to the S-matrix coming from the low-lying hadrons with lightest masses, it would be interesting to examine the relevance of **monotone independence to this context of factorization of dominant components** w.r.t. the (inverse of) energy variable $s$ (or $1/s$) as "time parameter" in the quantum probability theory of monotone independence [11]. Along this line and also taking account of a kind of duality relation between monotone and free independences (as was informed to me by Dr.Saigo and Mr.Hasebe), we may be attracted by the possible relations of **monotone and/or free independences** with the energy-**level statistics of nuclei** (among dominant figures of low-lying hadrons) formulated by **random matrices** which are closely related with the **quantum chaos** and also with the **free probability with Wigner's semi-circle law** as its CLT. In this special situation, the mutual **relation between the shell model and the liquid one** of nuclei could possibly be understood as a kind of **duality**, similarly to that between resonance and Regge poles of hadrons.

## Acknowledgments

On this occasion, I would like to express my special thanks to Prof. M. Ohya, all his collaborators who have supported him, and all the participants to this project, for his and their indefatigable initiatives and efforts in promoting the fundamental researches to connect and unify very different areas, mathematics, information, physics and biology, which have turn out evidently to constitute visible core disciplines of Q(uantum)B(io)I(nformatics)C(enter). It has been really exciting experiences for me to be invited to all the series of these illuminating international conferences for these five years, which have been exceptional occasions to be inspired by unexpected deep connections with different fields and without which I would have not been able to develop my own ideas. I am also very grateful to Dr.Saigo and Mr.Hasebe for instructive discussions on the relevance of quantum probabilistic notions of independence to the hadronic and/or nuclear physics from a renewed viewpoint.

## References


1. Ojima, I., A unified scheme for generalized sectors based on selection criteria –Order parameters of symmetries and of thermal situations and physical meanings of classifying categorical adjunctions–, Open Sys. Info. Dyn. **10**, 235-279 (2003); Micro-macro duality in quantum physics, 143-161, Proc. Intern. Conf. "Stochastic Analysis: Classical and Quantum", World Sci., 2005; Ojima, I. and Takeori, M., How to observe and recover quantum fields from observational data? –Takesaki duality as a Micro-macro duality–, Open Sys. Info. Dyn. **14**, 307–318 (2007); Ojima, I. and Harada, R., A unified scheme of measurement and amplification processes based on Micro-Macro Duality – Stern-Gerlach experiment as a typical example –, Open Sys. Info. Dyn. **16**, 55–74 (2009).
2. Ojima, I., Space(-Time) Emergence as Symmetry Breaking Effect, Quantum Bio-Informatics IV, 279 - 289 (2011) (arXiv:math-ph/1102.0838 (2011)); Micro-Macro Duality and Space-Time Emergence, Proc. Intern. Conf. "Advances in Quantum Theory", 197 – 206 (2011); New Interpretation of Equivalence Principle in General Relativity from the viewpoint of Micro-Macro duality, Invited talks at International Conference, "Foundations of Probability and Physics 6", Linnaeus University, Sweden and at the 43th Symposium on Mathematical Physics, Nicolaus Copernicus University, Poland, June 2011.
3. Muraki, N., Five independences as quasi-universal products, Inf. Dim. Anal. Quantum Probab. Rel. Topics **5**, 113-134 (2002); Barndorff-Nielsen, O.E., Franz, U., Gohm, R. Kümmerer, B. and Thorbjørnsen, S., Quantum Independent Increment Processes II, Lecture Notes in Math., Vol. 1866, Springer-Verlag, 2006.

4. Ojima, I., Roles of asymptotic conditions and S-matrix as Micro-Macro Duality in QFT, Quantum Probability and WNA **26**, 277 - 290 (2010).

5. Lehmann, H., Symanzik, K. and Zimmermann, W., Zur Formulierung quantisierter Feldtheorien, Nuovo Cim., **1**, 205 (1955); The formulation of quantized field theories. II., Nuovo Cim., **6**, 319 (1957).

6. Haag, R., On quantum field theories, Kgl. Danske Videnskab. Selskab. Mat.-fys. Medd., **29**, no.12 (1955).

7. Streater, R.F. and Wightman, A.S., PCT, Spin and Statistics and All That, Benjamin, 1964 (1st ed.), 1978 (2nd ed.).

8. Bogoliubov, N.N., Logunov, A.A. and Todorov, I.T., Introduction to Axiomatic Quantum Field Theory, Benjamin, 1975.

9. Glaser, V., Lehmann, H. and Zimmermann, W., Field operators and retarded functions, Nuovo Cim., **6**, 1122 (1957); Kugo, T. and Ojima, I., Suppl. Prog. Theor. Phys. no.66 (1979), Appendix.

10. Obata, N., "White Noise Calculus and Fock Space", Lect. Notes in Math. Vol. 1577, Springer–Verlag, 1994.

11. Muraki, N., Monotonic independence, monotonic central limit theorem and monotonic law of small numbers, Inf. Dim. Anal. Quantum Probab. Rel. Topics **4** (2001) 39-58; Monotonic convolution and monotonic Lévy-Hinčin formula, preprint, 2000; Hasebe, T., On monotone convolution and monotone infinite divisibility, Master thesis (2009) and Inf. Dim. Anal. Quantum Probab. Rel. Topics **13**, 111-131 (2010).

Quantum Bio-Informatics V

pp. 253–269

# LONG-RANGE PROPERTY IN TIME-DEPENDENT INTERACTION WITH THREE-BODY STRUCTURE AND NEW ASPECT

SHINSHO ORYU

*Department of Physics, Tokyo University of Science,*
*2641 Yamazaki, Noda-city, Chiba 278-8510, Japan*
*E-mail: oryu@ph.noda.tus.ac.jp*

This is where the abstract should be placed. It should consist of one paragraph and give a concise summary of the material in the article below. Replace the title, authors, and addresses within the curly brackets with your own title, authors, and addresses. You may have as many authors and addresses as you wish. It's preferable not to use footnotes in the abstract or the title; the acknowledgments for funding bodies etc. are placed in a separate section at the end of the text.

keywords: Long range force; Efimov effect; Three-body potential; Quasi-two-body potential.

## 1. Introduction

At the three-body threshold, Efimov pointed out, in early 1970's, when the two-body potential has a large scattering length, three-particle bound states emerging one after the other [1]. Attempts to observe such a phenomenon in nature have not been fully successful. Recently, the strength of the potential was varied artificially. Kraemer *et al.* claim evidence for Efimov quantum states in an ultracold gas of cesium atoms [2]. Efimov physics has been studied theoretically [3−8]. Also, some report experimental evidence for the existence of Efimov states [2,9−11].
The non relativistic three-body Faddeev equations[12] can be solved when formulated as the Alt-Grassberger-Sandhas (AGS) equations [13], which consist of the particle exchange Born term or the quasi-two-body potential of the multi-channel two-body Lippmann-Schwinger (LS) equation [14], and the kernel. The kernel is given by the quasi-two-body potential times the quasi-Green's function. The three-body AGS Born term is given by the two-body form factor which is energy independent, and the three-body free Green's

function with its energy dependence. Therefore, the Green's function leads to the logarithmic singularity which is similar to the Coulomb singularity at the three-body threshold. It may give rise to a special property like Efimov effects. We found that such a property could appear in the quasi-two-body potential which contains a particle exchange term.

Our theory will be applied to the $\pi$NN three-body system, and we predict that the one-pion-exchange Yukawa potential is automatically accompanied by an additional longer range interaction. Such a potential generates a level condensation near at zero energy in which the rms radius becomes very large. This fact could make a large contribution to the nuclear physics including nuclear fusion reaction. Moreover, such a level could be available to control the nuclear reaction with very low energies.

## 2. Fourier Transform of the Three-Body Born Term

The Born term of the AGS equation in the non-relativistic system has the form (see Fifure1)

$$
\begin{aligned}
& Z_{\alpha n, \beta m}(\mathbf{q}_\alpha, \mathbf{q}_\beta; E) \\
&= g_{\alpha n}(\mathbf{p}_\alpha) G_0(\mathbf{q}_\alpha, \mathbf{q}_\beta : E) g_{\beta m}(\mathbf{p}_\beta) \overline{\delta}_{\alpha,\beta} \\
&= \frac{g_{\alpha n}(\mathbf{p}_\alpha) g_{\beta m}(\mathbf{p}_\beta) \overline{\delta}_{\alpha,\beta}}{E - q_\alpha^2/2m_\alpha - q_\beta^2/2m_\beta - (\mathbf{q}_\alpha + \mathbf{q}_\beta)^2/2m_\gamma},
\end{aligned}
\tag{1}
$$

where $\alpha$, $\beta$, and $\gamma$ are the three-body channels: channel-1, channel-2 denote $a_1$-$(b_2 c_3)$ and $b_2$-$(c_3 a_1)$ and channel-3 is $c_3$-$(a_1 b_2)$ with the notation $\overline{\delta}_{\alpha\beta} \equiv 1-\delta_{\alpha\beta}$. In eq.(1) the separable potentials for the two-body interactions with form factors $g_1(\mathbf{p}_1), g_2(\mathbf{p}_2), g_3(\mathbf{p}_3)$ were utilized for the two-body relative momenta: $\mathbf{p}_1$, $\mathbf{p}_2$ and $\mathbf{p}_3$. Here the subscripts $n, m$ and $s$ are the physical states for the corresponding channels. $m_1$, $m_2$ and $m_3$ are the masses.

In order to reconstruct a *quasi-two-body* potential from eq.(1), one has to replace the three-body momenta by using the relative momentum between a particle and a pair which is a bound state with binding energy $\epsilon_B (> 0)$,

$$
\mathbf{q}_\alpha \to -\mathbf{q}, \quad \mathbf{q}_\beta \to \mathbf{q}', \quad \mathbf{q}_\gamma = -\mathbf{q}_\alpha - \mathbf{q}_\beta \to \mathbf{q} - \mathbf{q}',
\tag{2}
$$

where $\mathbf{q}_\gamma$ is the momentum transfer with respect to the three-body center-of-mass system, *i.e.*, $\mathbf{q}_\alpha + \mathbf{q}_\beta + \mathbf{q}_\gamma = 0$.

Furthermore, the three-body free energy $E$ in eq.(1) should be replaced with $E + \epsilon_B$ by using the energy conservation law $E_i = E_{int} = E_f$ where $E_i$, $E_f$ and $E_{int}$ are the initial, the final and the intermediate state energies,

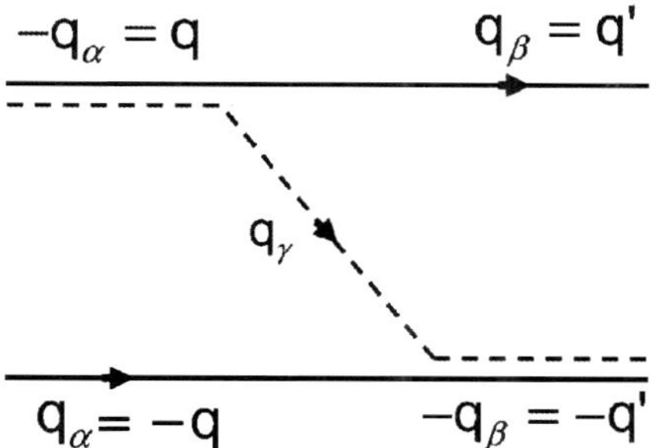



Figure 1. The AGS Born term diagram. Solid lines denote nucleons, and the dotted line is the pion.

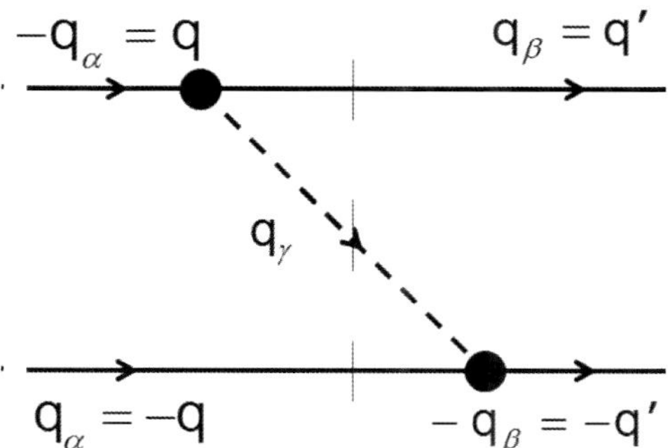



Figure 2. The quasi-two-body potential diagram. Solid lines denote nucleons, and the dotted line is the pion.

respectively. Because all the particle lines in Figure 2 are outer-lines which are free from the integral. Therefore, $\mathcal{E} \equiv E + \epsilon_B = q^2/2\mu_N (= E_i) = q'^2/2\mu_N (= E_f)$ should be equivalent to the intermediate three-body energy $E_{int}$, and $\mu_N = m_\alpha(m_\beta + m_\gamma)/(m_\alpha + m_\beta + m_\gamma)$ is the three-body reduced mass.

Here, it should be noted that the pair (or quasi-particle) is tightly bound, then the two-body total energy $\mathcal{E} \equiv E + \epsilon_B$ dominates the system, where the momenta $\mathbf{q}$ and $\mathbf{q}'$ are the integral variable which moves from $q'' = \sqrt{2\mu_N(E + \epsilon_B)} = 0$ to $\infty$, although $\mathbf{q}_\alpha$, $\mathbf{q}_\beta$ and $\mathbf{q}_\gamma$ are available only in the three-body free system, and the integral variable moves from $q''_\gamma =$

$\sqrt{2\mu_N E} = 0$ to $\infty$.

In this way, eq.(1) is replaced by the *quasi-two-body* (QTB) potential,

$$Z_{\alpha n,\beta m}(-\mathbf{q},\mathbf{q}';E) = \frac{-2g_{\alpha n}(\mathbf{p})m_\gamma g_{\beta m}(\mathbf{p}')\overline{\delta}_{\alpha,\beta}}{-2m_\gamma(E+\epsilon_B) + \Lambda_\alpha q^2 + \Lambda_\beta q'^2 - 2\mathbf{q}\,\mathbf{q}'}$$

$$= \frac{C^\gamma_{\alpha n,\beta m}(\mathbf{q},\mathbf{q}')}{2qq'(\chi'-x)}, \tag{3}$$

with the form factor function $C^c_{a,b}$, and the energy momentum core $\chi'$,

$$C^\gamma_{\alpha n,\beta m}(\mathbf{q},\mathbf{q}') \equiv -2g_{\alpha n}(\mathbf{p})m_\gamma g_{\beta m}(\mathbf{p}')\overline{\delta}_{\alpha,\beta}, \tag{4}$$

$$\chi' = \frac{-2m_\gamma(E+\epsilon_B) + \Lambda_\alpha q^2 + \Lambda_\beta q'^2}{2qq'}; \tag{5}$$

with

$$\begin{aligned}
\mathbf{q}\mathbf{q}' &= qq'x, \\
\Lambda_\alpha &= (1+m_\gamma/m_\alpha) = 1 + \Delta_\alpha, \\
\Lambda_\beta &= (1+m_\gamma/m_\beta) = 1 + \Delta_\beta.
\end{aligned} \tag{6}$$

The AGS equation with eq.(1) has the threshold at $q_{min} = \sqrt{2\mu_N E} = 0$ where the pinching singularity occurs in the kernel which gives rise to the Efimov effect. On the other hand, the multi-channel LS equation with eq.(3) has the threshold at $q_{min} = \sqrt{2\mu_N(E+\epsilon_B)} = 0$ where the integral kernel could cause similar pinching singularity at the threshold.

Here, let us adopt $m_\gamma < m_\alpha = m_\beta$ system without lacking generality. Then, we have

$$\begin{aligned}
\Delta_\alpha &= \Delta_\beta \equiv \Delta < 1 \\
\Lambda_\alpha &= \Lambda_\beta \equiv \Lambda
\end{aligned} \tag{7}$$

and also, $\Lambda_\alpha = \Lambda_\beta = \Lambda \equiv 1 + \Delta$.

Therefore, eq.(5) becomes

$$\chi' = \frac{-2m_\gamma(E+\epsilon_B)/\Lambda + q^2 + q'^2}{2qq'}\Lambda$$

$$= \frac{\sigma^2 + q^2 + q'^2}{2qq'}\Lambda \equiv \chi\Lambda \tag{8}$$

$$= \chi + \Delta\chi \tag{9}$$

with

$$\chi \equiv \frac{\sigma^2 + q^2 + q'^2}{2qq'}, \tag{10}$$

$$\sigma^2 \equiv -\frac{2m_\gamma(E+\epsilon_B)}{\Lambda} \geq 0. \tag{11}$$

In eq.(3), the term $(\chi' - x)^{-1}$ is given by the partial wave expansion,

$$\frac{1}{\chi' - x} = \sum_{l=0}^{\infty}(2l + 1)Q_l(\chi')P_l(x)$$

$$= \sum_{l=0}^{\infty}(2l + 1)Q_l(\chi + \Delta\chi)P_l(x), \tag{12}$$

where $P_l$ and $Q_l$ are the first and the second Legendre functions. By adopting a small value of parameter $\lambda$ which satisfies $0 < \lambda(\Delta\chi) < 1$, one can obtain

$$Q_l(\chi + \lambda\Delta\chi) = \sum_{k=0}^{\infty}\frac{1}{k!}Q_l(\chi)^{(k)}(\lambda\Delta\chi)^k, \tag{13}$$

where $Q_l(\chi)^{(k)}$ is the $k$-th derivative of $Q_l(\chi)$ with respect to $\chi$. Therefore, eq.(13) could be converged when $Q_l(\chi)$ takes a finite value.
Putting $\chi'(\lambda) \equiv \chi + \lambda\Delta\chi$ in eq.(12), we obtain by using eq.(A.5), (see Appendix A).

$$\frac{1}{\chi'(\lambda) - x} = \sum_{l=0}^{\infty}(2l + 1)\left\{\sum_{j=0}^{\infty}\frac{1}{j!}Q_l(\chi)^{(j)}(\lambda\Delta\chi)^j\right\}P_l(x)$$

$$= \sum_{j=0}^{\infty}\frac{1}{j!}(\lambda\Delta)^j\left\{\sum_{l=0}^{\infty}(2l + 1)\chi^j Q_l(\chi)^{(j)}P_l(x)\right\}$$

$$= \sum_{j=0}^{\infty}\frac{1}{j!}(\lambda\Delta)^j\left\{\frac{(-\chi)^j j!}{(\chi - x)^{j+1}}\right\} \tag{14}$$

$$= 2qq'\sum_{j=0}^{\infty}\frac{(-\lambda\Delta)^j(\sigma^2 + q^2 + q'^2)^j}{[\sigma^2 + (\mathbf{q} - \mathbf{q}')^2]^{j+1}}. \tag{15}$$

Eq.(14) indicates that the both sides of the function are a finite or bounded by $|x| \leq \chi$ and never diverges by the reason of the mass expansion with $\Delta$. By using the momentum relations: $\mathbf{q}' - \mathbf{q} = 2\mathbf{K}$ and $\mathbf{q}' + \mathbf{q} = \mathbf{Q}$, and also $\sigma^2 + 2K^2 + Q^2/2 = [(\sigma^2 + 4K^2) + (\sigma^2 + Q^2)]/2$, the Born term: eq.(3) is rewritten with eq.(15), (see also Appendix B)

$$
\begin{aligned}
Z_{\alpha n,\beta m}(-\mathbf{q},\mathbf{q}';\lambda,E) &= C^{\gamma}_{\alpha n,\beta m}(\mathbf{q},\mathbf{q}') \sum_{j=0}^{\infty} \frac{(-\lambda\Delta)^j(\sigma^2+q^2+q'^2)^j}{[\sigma^2+(\mathbf{q}-\mathbf{q}')^2]^{j+1}} \\
&= C^{\gamma}_{\alpha n,\beta m}(\mathbf{q},\mathbf{q}') \sum_{j=0}^{\infty}(-\lambda\Delta)^j \frac{(\sigma^2+2K^2+Q^2/2)^j}{(\sigma^2+4K^2)^{j+1}} \\
&= \frac{C^{\gamma}_{\alpha n,\beta m}(\mathbf{q},\mathbf{q}')}{(\sigma^2+4K^2)} \sum_{j=0}^{\infty}\left(\frac{-\lambda\Delta}{2}\right)^j\left[1+\frac{\sigma^2+Q^2}{\sigma^2+4K^2}\right]^j \\
&= C^{\gamma}_{\alpha n,\beta m}(\mathbf{q},\mathbf{q}') \\
&\quad \times \sum_{k=0}^{\infty}\sum_{j=0}^{\infty} {}_{j+k}C_k \left(\frac{-\lambda\Delta}{2}\right)^{j+k} \frac{(\sigma^2+Q^2)^k}{(\sigma^2+4K^2)^{k+1}}. \quad (16)
\end{aligned}
$$

Because of the binomial formula eq.(B.3) which is given in Appendix, we have

$$
\sum_{j=0}^{\infty} {}_{j+k}C_k\, x^{j+k} = \frac{x^k}{(1-x)^{k+1}} \tag{17}
$$

and putting $x = (-\lambda\Delta/2)$, the Born term becomes by using eq.(17),

$$
\begin{aligned}
Z_{\alpha n,\beta m}(-\mathbf{q},\mathbf{q}';E) &= \lim_{\lambda\to 1} \frac{2C^{\gamma}_{\alpha n,\beta m}(\mathbf{q},\mathbf{q}')}{2+\lambda\Delta} \sum_{k=0}^{\infty}\left(\frac{-\lambda\Delta}{2+\lambda\Delta}\right)^k \frac{(\sigma^2+Q^2)^k}{(\sigma^2+4K^2)^{k+1}}. \\
&= \frac{2C^{\gamma}_{\alpha n,\beta m}(\mathbf{q},\mathbf{q}')}{2+\Delta} \sum_{k=0}^{\infty}\left(\frac{-\Delta}{2+\Delta}\right)^k \frac{(\sigma^2+Q^2)^k}{(\sigma^2+4K^2)^{k+1}}. \quad (18)
\end{aligned}
$$

Therefore, the result indicates that the mass expansion for the AGS-Born term can safely converge for any $\Delta$ values.

The Fourier transform of the above result in eq.(18) is given by,

$$
\begin{aligned}
\mathcal{F}&\{Z_{\alpha n,\beta m}(-\mathbf{q},\mathbf{q}';E)\}\\
&= <\mathbf{x}_\alpha|Z_{\alpha n,\beta m}(-\mathbf{q},\mathbf{q}';E)|\mathbf{x}_\beta>\\
&= \sum_{k=0}^{\infty}\left(\frac{-\Delta}{2+\Delta}\right)^{k}\int\int\frac{d\mathbf{q}_\alpha}{(2\pi)^3}\frac{d\mathbf{q}_\beta}{(2\pi)^3}\frac{2C_{\alpha n,\beta m}^{\gamma}(-\mathbf{q}_\alpha,\mathbf{q}_\beta)}{2+\Delta}\\
&\quad\times\frac{(\sigma^2+Q^2)^k}{(\sigma^2+4K^2)^{k+1}}e^{i\mathbf{q}_\alpha\mathbf{x}_\alpha}e^{-i\mathbf{q}_\beta\mathbf{x}_\beta}\\
&= \sum_{k=0}^{\infty}\left(\frac{-\Delta}{2+\Delta}\right)^{k}\int\int\frac{d\mathbf{K}}{(2\pi)^3}\frac{d\mathbf{Q}}{(2\pi)^3}\\
&\quad\times\frac{2C_{\alpha n,\beta m}^{\gamma}(\mathbf{K}-\mathbf{Q}/2,\mathbf{K}+\mathbf{Q}/2)}{2+\Delta}\frac{(\sigma^2+Q^2)^k}{(\sigma^2+4K^2)^{k+1}}e^{-i\mathbf{Q}\mathbf{R}}e^{-i\mathbf{K}\mathbf{r}},\quad(19)
\end{aligned}
$$

where we adopted some wellknown coordinate relations by using eqs.(2),

$$2\mathbf{K}\equiv\mathbf{q}_\alpha+\mathbf{q}_\beta=\mathbf{q}'-\mathbf{q},\tag{20}$$

$$\mathbf{Q}=\mathbf{q}'+\mathbf{q},\tag{21}$$

$$\mathbf{q}'=\frac{\mathbf{Q}}{2}+\mathbf{K}=\mathbf{q}_\beta,\tag{22}$$

$$\mathbf{q}=\frac{\mathbf{Q}}{2}-\mathbf{K}=-\mathbf{q}_\alpha\tag{23}$$

$$
\begin{aligned}
q^2+q'^2&=\left(\frac{\mathbf{Q}}{2}-\mathbf{K}\right)^2+\left(\frac{\mathbf{Q}}{2}+\mathbf{K}\right)^2\\
&=2\left(K^2+\frac{Q^2}{4}\right)
\end{aligned}\tag{24}
$$

$$
\begin{aligned}
\mathbf{q}_\alpha\mathbf{x}_\alpha-\mathbf{q}_\beta\mathbf{x}_\beta&=-\mathbf{q}\mathbf{x}_\alpha-\mathbf{q}'\mathbf{x}_\beta\\
&=-\left(\frac{\mathbf{Q}}{2}-\mathbf{K}\right)\mathbf{x}_\alpha-\left(\frac{\mathbf{Q}}{2}+\mathbf{K}\right)\mathbf{x}_\beta\\
&=-\mathbf{Q}\left(\frac{\mathbf{x}_\alpha+\mathbf{x}_\beta}{2}\right)-\mathbf{K}(\mathbf{x}_\beta-\mathbf{x}_\alpha)\\
&=-\mathbf{Q}\mathbf{R}-\mathbf{K}\mathbf{r}
\end{aligned}\tag{25}
$$

with the center-of-mass coordinate $\mathbf{R}$, and the relative momentum $\mathbf{r}$,

$$\mathbf{R}\equiv\frac{\mathbf{x}_\beta+\mathbf{x}_\alpha}{2},\tag{26}$$

$$\mathbf{r}\equiv\mathbf{x}_\beta-\mathbf{x}_\alpha,\tag{27}$$

where one finds that $\mathbf{R}$ is the center-of-mass coordinate and the corresponding momentum is $\mathbf{Q}$, and $\mathbf{r}$ is the relative coordinate with the corresponding momentum $\mathbf{K}$. In terms of the new coordinates, the Jacobian is unity, *i.e.*,

$$d\mathbf{q}_\alpha d\mathbf{q}_\beta = \frac{\partial(\mathbf{q}_\alpha, \mathbf{q}_\beta)}{\partial(\mathbf{K}, \mathbf{Q})} d\mathbf{K} d\mathbf{Q} = d\mathbf{K} d\mathbf{Q}.$$

Here the form factor function $C^\gamma_{\alpha n, \beta m}(\mathbf{q}, \mathbf{q}')$ is a monotonic function with respect to the integral variables in small $\sigma$ value. Since we can take $C^\gamma_{\alpha n, \beta m}(\mathbf{q}, \mathbf{q}')$ as a constant $C_{\alpha n, \beta m}$, then eq.(19) becomes,

$$\mathcal{F}\{Z_{\alpha n, \beta m}(-\mathbf{q}, \mathbf{q}'; E)\} = \frac{2C_{\alpha n, \beta m}}{2 + \Delta} \sum_{k=0}^{\infty} \left(\frac{-\Delta}{2 + \Delta}\right)^k \int \int \frac{d\mathbf{K}}{(2\pi)^3} \frac{d\mathbf{Q}}{(2\pi)^3}$$

$$\times \frac{(\sigma^2 + Q^2)^k}{(\sigma^2 + 4K^2)^{k+1}} e^{-i\mathbf{Q}\mathbf{R}} e^{-i\mathbf{K}\mathbf{r}}, \tag{28}$$

Therefore, the integral part of the eq.(19) can be separated into the center of mass and the relative momentum parts,

$$F^k(\sigma) = \int (\sigma^2 + Q^2)^k \frac{e^{-i\mathbf{Q}\mathbf{R}} d\mathbf{Q}}{(2\pi)^3} \int \frac{e^{-i\mathbf{K}\mathbf{r}}}{(\sigma^2 + 4K^2)^{k+1}} \frac{d\mathbf{K}}{(2\pi)^3}$$

$$\equiv I(\sigma; k) J(\sigma; k) \tag{29}$$

with

$$I(\sigma; k) = \int (\sigma^2 + Q^2)^k e^{-i\mathbf{Q}\mathbf{R}} \delta(\mathbf{Q}) d\mathbf{Q} = \sigma^{2k} \tag{30}$$

$$J(\sigma; k) = \int \frac{e^{-i\mathbf{K}\mathbf{r}}}{(\sigma^2 + 4K^2)^{k+1}} \frac{d\mathbf{K}}{(2\pi)^3} \tag{31}$$

Finally, the Fourier transform of the Born term is given by,

$$\mathcal{F}\{Z_{\alpha n, \beta m}(-\mathbf{q}, \mathbf{q}'; E)\} = \frac{2C_{\alpha n, \beta m}}{2 + \Delta} \sum_{k=0}^{\infty} (-1)^k \left(\frac{\Delta}{2 + \Delta}\right)^k$$

$$\times \mathcal{F}\{(\sigma^2 + Q^2)^k\} \mathcal{F}\{1/(\sigma^2 + 4K^2)^{k+1}\}$$

$$= \frac{C_{\alpha n, \beta m}}{8\pi(2 + \Delta)} \sum_{k=0}^{\infty} \left(\frac{\Delta}{2 + \Delta}\right)^k \frac{\sigma^{2k}}{k!} \frac{\partial^k}{\partial(\sigma^2)^k} \left(\frac{e^{-r\sigma/2}}{r}\right)$$

$$= \frac{C_{\alpha n, \beta m}}{8\pi(2 + \Delta)} U(\Delta, \sigma; r) \tag{32}$$

with

$$U(\Delta, \sigma; r) = \sum_{k=0}^{\infty} \left(\frac{\Delta}{2+\Delta}\right)^k \frac{\sigma^{2k}}{k!} \frac{\partial^k}{\partial(\sigma^2)^k} \left(\frac{e^{-r\sigma/2}}{r}\right)$$

$$\equiv \sum_{k=0}^{\infty} U^{(k)}(\Delta, \sigma; r)$$

$$= \left\{ \frac{1}{r} e^{-r\sigma/2} + \left(\frac{\Delta}{2+\Delta}\right) \frac{\sigma^2}{1!} \left(\frac{-1}{4\sigma}\right) e^{-r\sigma/2} \right.$$

$$+ \left(\frac{\Delta}{2+\Delta}\right)^2 \frac{\sigma^4}{2!} \left(\frac{(-1)^2(r\sigma/2 + 1)}{(2\sigma)^3}\right) e^{-r\sigma/2}$$

$$\left. + \left(\frac{\Delta}{2+\Delta}\right)^3 \frac{\sigma^6}{3!} \left(\frac{(-1)^3(\sigma^2 r^2/2 + 3r\sigma + 6)}{(2\sigma)^5}\right) e^{-r\sigma/2} + \cdots \right\}$$

$$= \left\{ \frac{1}{r} e^{-r\sigma/2} + \left(\frac{\Delta}{2+\Delta}\right) \frac{\sigma}{1!} \frac{(-1)}{4} e^{-r\sigma/2} \right.$$

$$+ \left(\frac{\Delta}{2+\Delta}\right)^2 \frac{\sigma}{2!} \left(\frac{(-1)^2(r\sigma/2 + 1)}{2^3}\right) e^{-r\sigma/2}$$

$$\left. + \left(\frac{\Delta}{2+\Delta}\right)^3 \frac{\sigma}{3!} \left(\frac{(-1)^3(\sigma^2 r^2/2 + 3r\sigma + 6)}{2^5}\right) e^{-r\sigma/2} + \cdots \right\}.$$

$$\tag{33}$$

This means that the first term is the Yukawa-type potential with the energy dependence, and the higher terms are proportional to an exponential function with respect to $r$ and $\sigma$ (with $\sigma^2 = -2m_\gamma(E + \epsilon_B)/\Lambda$). It should be mentioned that the zero-energy limit, with which we are concerned, becomes a Coulomb like potential.

## 3. An energy average by a Laplace transform of the three-body Born term

It is perhaps difficult to understand that the Coulomb-type potential can be generated in the three-hadron system, because the historical three-hadron Faddeev calculations didn't include such a phenomenon. The origin of these curious phenomena are caused by the energy dependence of the particle transfer interaction in the AGS's Born term and the kernel in the Faddeev formalism.

It is not easy to obtain the energy independent potential from these three-body formalisms. We propose to take the "statistical average" by using the *probability density function* with respect to the possible energy range, which also represents effects of the structure or the form factors of the composite particles,

$$P_\sigma = \sigma^{2\gamma+1} e^{-a\sigma} \bigg/ \int_0^\infty \sigma^{2\gamma+1} e^{-a\sigma} d\sigma \equiv \frac{\sigma^{2\gamma+1} e^{-a\sigma}}{\rho}, \tag{34}$$

with

$$\rho = \int_0^\infty \sigma^{2\gamma+1} e^{-a\sigma} d\sigma = \frac{\Gamma(2\gamma+2)}{a^{2\gamma+2}},$$

where the weight function (or the probability density function) $\sigma^{2\gamma+1}$ is adopted, and $e^{-a\sigma}$ is the damping factor when $\sigma = \sqrt{-2m_\gamma(E + \epsilon_B)}/\Lambda$ or the three-body energy $|E|$ increases. The weight function comes from the dispersion theory, however, it is not main purpose of this paper, then the details will be presented elsewhere.

By using the probability density function, the expectation value of the energy dependent potential becomes an energy independent potential. This is the Laplace transform or the Euler integral of the second kind in the first term of eq.(33). Therefore, by using eqs.(34) the first order potential with $k = 0$ in (33) is

$$\mathcal{L}\{U^{(0)}(\Delta, \sigma; r)\} = \frac{1}{\rho} \int_0^\infty \sigma^{2\gamma+1} e^{-a\sigma} \frac{e^{-\sigma r/2}}{r} d\sigma$$

$$= \frac{a^{2\gamma+2}}{r(r/2 + a)^{2\gamma+2}}, \tag{35}$$

where $\gamma = 3/2$ means the Van der Waals potential of the London-type for $r \gg a$, which is similar to the potential between two atoms, and the Yukawa-type for $a \gg r$,

$$\mathcal{L}\{U^{(0)}(\Delta, \sigma; r)\} = \frac{a^5}{r(r/2 + a)^5} = \frac{a_0^5}{r(r + a_0)^5}$$

$$\approx \frac{a_0^5}{r^6} \quad \text{for} \quad r \gg a, \tag{36}$$

$$\approx \frac{e^{-5r/a_0}}{r} \quad \text{for} \quad a \gg r, \tag{37}$$

with $2a = a_0$, and also the Casimir-type for $\gamma = 2$, with $r \gg a$, and the Yukawa-type for $a \gg r$,

$$\mathcal{L}\{U^{(0)}(\Delta, \sigma; r)\} = \frac{a^6}{r(r/2 + a)^6} = \frac{a_0^6}{r(r + a_0)^6}$$

$$\approx \frac{a_0^6}{r^7} \qquad \text{for} \quad r \gg a, \tag{38}$$

$$\approx \frac{e^{-6r/a_0}}{r} \qquad \text{for} \quad a \gg r. \tag{39}$$

Such a short range potential like the Yukawa-type comes from a particle transfer in the three elementary particle system, although the short range potential has the repulsive core.

If and only if the weight function is replaced by

$$\sigma^{2\gamma+1} e^{-a\sigma} \rightarrow \delta(\sigma - 2\mu_0) \tag{40}$$

with the meson range $1/\mu_0$, then the potential is a Yukawa potential

$$\mathcal{L}\{U^{(0)}(\Delta, \sigma; r)\} = \frac{1}{\rho} \int_0^\infty \delta(\sigma - 2\mu_0) \frac{e^{-\sigma r/2}}{r} d\sigma$$

$$= \frac{e^{-\mu_0 r}}{r}, \tag{41}$$

$$\because \rho = \int_0^\infty \delta(\sigma - 2\mu_0) d\sigma = 1. \tag{42}$$

However, the (40) condition seems to be *intentional* in this situation.

In eqs.(18) and (19), if the form factors are energy and momentum independent or monotonic, the weight function $\sigma^{2\gamma+1}$ could be a constant which means $\gamma = -1/2$ in eq.(35).

We prefer this type which is more natural in this system. Therefore, the energy (or $\sigma$) independent Born term of eq.(32) is given by using the coupling constant $C_{\alpha n,\beta m}/[8\pi(2+\Delta)]$ for the three-body center of mass system,

$$V_{\alpha n,\beta m}(\Delta;r) \equiv \mathcal{LF}\{Z_{\alpha n,\beta m}(-\mathbf{q},\mathbf{q}';\mathbf{E})\}$$

$$= \frac{C_{\alpha n,\beta m}}{8\pi(2+\Delta)}\mathcal{L}\{U(\Delta,\sigma;r)\} \tag{43}$$

$$= \frac{C_{\alpha n,\beta m}}{8\pi(2+\Delta)}\frac{1}{\rho}\int_0^\infty U(\Delta,\sigma;r)e^{-a\sigma}\,d\sigma \tag{44}$$

$$= \frac{C_{\alpha n,\beta m}}{8\pi(2+\Delta)}\left\{\frac{a_0}{r(r+a_0)} + \frac{(-1)}{1!2^2}\left(\frac{\Delta}{2+\Delta}\right)\frac{2a_0}{(r+a_0)^2}\right.$$

$$+ \frac{(-1)^2}{2!2^3}\left(\frac{\Delta}{2+\Delta}\right)^2\left[\frac{2a_0}{(r+a_0)^2} + \frac{2^2 r a_0}{(r+a_0)^3}\right]$$

$$+ \frac{(-1)^3}{3!2^5}\left(\frac{\Delta}{2+\Delta}\right)^3\left[\frac{6\cdot 2a_0}{(r+a_0)^2} + \frac{6\cdot 2^2 r a_0}{(r+a_0)^3}\right.$$

$$\left.\left. + \frac{3\cdot 2^3 r^2 a_0}{(r+a_0)^4}\right] + \cdots\right\}. \tag{45}$$

One can easily find that the potential converges to $1/r^2$ for $r \to \infty$, while for $r < a_0$ it reaches the Yukawa-type potential plus an exponential one. The latter case is not quantitatively correct for the very low energy case; however, it is correct for the long range case, $i.e.$,

$$\lim_{r\to\infty} V_{\alpha n,\beta m}(\Delta;r) = \frac{C_{\alpha n,\beta m}}{8\pi(2+\Delta)}\left\{1 - \frac{1}{2}\left(\frac{\Delta}{2+\Delta}\right)\right.$$

$$\left. + \frac{3}{8}\left(\frac{\Delta}{2+\Delta}\right)^2 - \frac{5}{16}\left(\frac{\Delta}{2+\Delta}\right)^3 + \cdots\right\}\frac{a_0}{r^2}$$

$$= \frac{C_{\alpha n,\beta m}}{8\pi(2+\Delta)}\frac{S a_0}{r^2} \equiv \frac{V_0 a_0}{r^2}, \tag{46}$$

where $S = 0.8634$ is given by $\Delta = 1/3$ for the three identical particles, and also $S = 0.96741$ corresponds to $\Delta = m_\gamma/m_\alpha = M_\pi/M_N = 0.14703$ for the $\pi$NN-system.

This means that the first order potential approximation in eq.(46) is 86.34% correct for three equal mass case, while 96.74% correct for NN$\pi$ system in one pion transfer.

It is well known that $1/r^2$-potential has the energy and the r.m.s. relations,

$$\frac{E_n}{E_{n+1}} = \left(\frac{r_{n+1}}{r_n}\right)^2 = e^{2\pi/\mu} \tag{47}$$

$$\therefore\ r_{n+1} = r_n e^{\pi/\mu}, \tag{48}$$

with

$$\mu^2 = \frac{2m}{\hbar^2}|V_0|a_0 - (l + \frac{1}{2})^2 > 0. \tag{49}$$

The most prospect state is the triplet NN system and the deuteron bound state mainly depends on the D-state. However, if we could fit the S-wave parameters $V_0$ and $a_0$ to the deuteron binding energy, we obtain $\mu = 1.2221$. It gives $e^{2\pi/\mu} = 170.98$ and $e^{\pi/\mu} = 13.076$, and the first excited state ($n = 1$) of the deuteron could be

$$E_1 \approx E_0/e^{2\pi/\mu} = -2.2246/170.98 \approx -13 \text{ keV}, \tag{50}$$

$$r_1 \approx r_0 \times e^{\pi/\mu} = 1.97 \times 13.076 \approx 26 \text{ fm}. \tag{51}$$

Furthermore the lower quantum number states (or the lower excited states $n = 1, 2, 3$) have some small errors, because one needs the full wave function to fit the deuteron binding energy. Nevertheless, it should be pointed out that such an energy region ($|E| \leq 50$ keV) for the bound states, and the phase shift below the energy 100 keV have not been measured yet [15]. One of the possible experiments will be cold neutron absorption with proton where the $\gamma$-ray measurement of the deuteron peak will display a certain width which indicates the accumulation of the condensed energy spectra near the threshold.

## 4. Summary and discussion

We pointed out the discrepancy between the traditional three-body AGS equation and the LS equation with the QTB potential which causes singularity at the threshold. As the results, the latter equation causes new phenomena which are similar to the Efimov effect but more interesting in the nuclear and particle physics. In our discussion, we found that the local potential expansion of the QTB potential with the constant two-body form factors results in the $1/r^2$ type potential in the long range limit. After the Fourier transform of the potential, an energy average with respect to $\sigma$ by a distribution function $\sigma^{2\gamma+1}$ leads to a $1/r^2$ type potential for $a_0 \ll r$ in $\gamma = -1/2$, where the form factor is monotonic, however, $\gamma = 3/2$ generates the Van der Waals type potential. By this reason, one could say that the distribution function contains the characteristics of the two-body form factors. We have to admit the possibility of the generation of long range components in the three-body problem where the particle transfer exists. In this paper, we apply the prediction to the three-body $NN\pi$ system, where $(\pi N) + N \rightarrow N + (\pi N)$ channel generates the $1/r^2$ potential at the longer range which is more important than the heavy particle transfer: $(\pi N) + N \rightarrow \pi + (NN)$, because the latter channel contributes to rather shorter range, and being neglected in this paper. We can conclude that the Yukawa potential is accompanied by the long range potentials which are the $1/r^2$-type or the Van der Waals-type or whatever is in the energy dependent particle transfer interaction.

The deuteron binding energy is used for the scaling and estimating the series of the higher states by the rule of $E_{n+1} = e^{-2\pi/\mu} E_n$, and also the rms radius: $r_{n+1} = e^{\pi/\mu} r_n$.

If the $1/r^2$ potential does not support the existence of bound states by reason of $\mu^2 \le 0$ for the solution of the Schrödinger equation *i.e.* the modified Bessel function, such a potential could interfere with the centrifugal potential and the Coulomb potential. Furthermore, the three-body force at very low energy could originate from this long range potential, because the explanation in terms of the $\Delta$-isobar-origin of the Fujita-Miyazawa three-body force seems to be hard to justify in the very low energy region.

Finally, this kind of long range $1/r^2$ type potential structure could universally occur, not only in three-body systems but also in many-body systems [16−18] by means of the particle transfer type potentials.

In addition, the $1/r^2$ potential could affect not only for the final-state interaction in the break-up experiments with respect to nn, pp and np, but also very wide fields such as the neutron- or proton-rich nuclei [19], and the nuclear fusion problems, and nuclear reactions in the universe, and moreover the atomic- and molecular physics.

If one could find the predicted energy levels near at threshold, the nuclear reaction could possibly be controled by the chemical reaction size small energy.

## Acknowledgement

The author would like to express his thanks to Dr Takashi Watanabe, Dr. Tetsuo Sawada, Dr. Shoji Nakamura, and Mr. Yasuhisa Hiratsuka for their valuable discussions. This work is supported by Quantum Bio-Informatics Center of Tokyo University of Science ("Hi-Tech Research Center" National Project for Private Universities supported by MEXT in 2006).

## Appendix A.   Partial wave expansion

Eq.(14) is proved by the aid of partial differentiation,

$$\frac{1}{\chi - x} = \sum_{l=0}^{\infty} (2l+1) Q_l(\chi) P_l(x) \tag{A.1}$$

$$\frac{\partial}{\partial \chi} \frac{1}{\chi - x} = \frac{-1}{(\chi - x)^2} = \sum_{l=0}^{\infty} (2l+1) Q_l'(\chi) P_l(x)$$

$$\frac{-\chi}{(\chi - x)^2} = \sum_{l=0}^{\infty} (2l+1) \chi Q_l'(\chi) P_l(x) \tag{A.2}$$

$$\frac{\partial^2}{\partial \chi^2} \frac{1}{\chi - x} = \frac{2}{(\chi - x)^3} = \sum_{l=0}^{\infty} (2l+1) Q_l''(\chi) P_l(x)$$

$$\frac{2\chi^2}{(\chi - x)^3} = \sum_{l=0}^{\infty} (2l+1) \chi^2 Q_l''(\chi) P_l(x) \tag{A.3}$$

$$\frac{\partial^3}{\partial \chi^3} \frac{1}{\chi - x} = \frac{\partial}{\partial \chi} \frac{2}{(\chi - x)^3} = \frac{-6}{(\chi - x)^4}$$

$$= \sum_{l=0}^{\infty} (2l+1) Q_l'''(\chi) P_l(x)$$

$$\frac{-6\chi^3}{(\chi - x)^4} = \sum_{l=0}^{\infty} (2l+1) \chi^3 Q_l'''(\chi) P_l(x). \tag{A.4}$$

The general form is given by

$$\frac{j!(-\chi)^j}{(\chi - x)^{j+1}} = \sum_{l=0}^{\infty} (2l+1) \chi^j Q_l^{(j)}(\chi) P_l(x) \tag{A.5}$$

## Appendix B. Binomial Series

In order to prove eq.(17), let us consider some relations

$$\sum_{k=0}^{\infty} x^k (1+y)^k = 1 + x(1+y) + x^2(1+y)^2$$
$$+ x^3(1+y)^3 + \cdots + x^n(1+y)^n + \cdots$$
$$= 1 + x + xy + x^2 \left[{}_2C_0 y^0 + {}_2C_1 y^1 + {}_2C_2 y^2\right]$$
$$+ x^3 \left[{}_3C_0 y^0 + {}_3C_1 y^1 + {}_3C_2 y^2 + {}_3C_3 y^3\right]$$
$$+ \cdots + x^n \sum_{l=0}^{n} {}_nC_l y^l + \cdots \tag{B.1}$$
$$= y^0 \left\{1 + x + x^2 + x^3 + \cdots\right\}$$
$$+ y \left\{ {}_1C_1 x + {}_2C_1 x^2 + {}_3C_1 x^3 + \cdots \right\}$$
$$+ y^2 \left\{ {}_2C_2 x^2 + {}_3C_2 x^3 + {}_4C_2 x^4 \cdots \right\}$$
$$+ y^3 \left\{ {}_3C_3 x^3 + {}_4C_3 x^4 + {}_5C_3 x^5 \cdots \right\}$$
$$+ y^4 \left\{ {}_4C_4 x^4 + {}_5C_4 x^5 + {}_6C_4 x^6 + \cdots \right\}$$
$$+ \cdots + y^l \sum_{k=0}^{\infty} {}_{k+l}C_l x^{k+l} + \cdots$$

Therefore, we have

$$\sum_{k=0}^{\infty} x^k (1+y)^k = \sum_{l=0}^{\infty} \left[ \sum_{k=0}^{\infty} {}_{k+l}C_l x^{k+l} \right] y^l = \sum_{l=0}^{\infty} \frac{x^l}{(1-x)^{l+1}} y^l \tag{B.2}$$

Eq.(B.2) is obtained from the formula of the negative binomial series with $s - 1 = n$,

$$\frac{1}{(1-x)^s} = \sum_{k=0}^{\infty} {}_{k+s-1}C_k x^k \equiv \sum_{k=0}^{\infty} {}_{k+s-1}C_{s-1} x^k$$
$$= \frac{1}{x^n} \sum_{k=0}^{\infty} {}_{k+n}C_n x^{k+n} \tag{B.3}$$

Putting $x = -\Delta/2$,

$$\sum_{k=0}^{\infty} {}_{k+n}C_n x^{k+n} = \frac{x^n}{(1-x)^{n+1}} = \left(\frac{2}{2+\Delta}\right)\left(\frac{-\Delta}{2+\Delta}\right)^n. \tag{B.4}$$

# References


1. V. Efimov, Phys. Lett. B**33** (1970) 563; ibid. Nucl. Phys. **A210**, 157-188 (1973).
2. T. Kraemer, M. Mark, P. Waldburger, J. G. Danzl, C. Chin, B. Engeser, A. D. Lange, K. Pilch, A. Jaakkola, H.-C. Nägerl, and R. Grimm, Nature **440**, 315-318 (2006).
3. R. D. Amado and J. V. Noble, Phys. Lett. **35B** 25-27 (1971).
4. S. K. Adhikari and L. Tomio, Phys. Rev. C**26**, 83-86, (1982); S. K. Adhikari, A. C. Fonseca, and L. Tomio, Phys Rev. C**26**, 77-82 (1982); *ibid*, Phys. Rev. C **27**, 1826-1829 (1983).
5. P. F. Bedaque, H.-W. Hammer, and U. van Kolck, Phys. Rev. Lett. **82**, 463 (1999); Nucl. Phys. A**646**, 444 (1999); P. F. Bedaque, E. Braaten, and H.-W. Hammer, Phys. Rev. Lett. **85**, 908 (2000); E. Braaten and H.-W. Hammer, ibid. **87** 160407 (2001).
6. J. P. D'Incao, H. Suno, and B. D. Esry, Phys. Rev. Lett. bf 93, 123201-1-4, (2004).
7. S. Floerchinger, R. R. Schmidt, S. Moroz, and C. Wetterich, Phys. Rev. A **79**, 013603 (2009).
8. E. Braaten, and H.-W. Hammer, Phys. Rep. **428**, 259 (2006).
9. S. Knoop, F. Ferlaino, M. Mark, M. Berninger, H. Schoebel, H. Naegerl, and R. Grimm, Nat. Phys. **5**, 227-230 (2009).
10. M. Zaccanti, B. Deissler, C. D'errico, M. Fattori, M. Jona-Lasino, S. Müller, G. Roati, M. Inguscio and G. Modugno, Nat. Phys. **5**, 586-591 (2009).
11. G. Barontini, C. Weber, F. Rabatti, J. Catani, G. Thalhammer, M. Inguscio, and F. Minardi, Phys. Rev. Lett., **103**, 043201-1-4 (2009).
12. L. D. Faddeev, Zh. Eksp. i Teor. Fiz. **39** (1960) 1459, English transl.; Soviet Phys.-JETP **12** (1961) 1014.
13. E. O. Alt, P.Grassberger, and W. Sandhas, Nucl. Phys., **B2**, 167-180 (1967).
14. B. A. Lippmann and J. Schwinger, Phys. Rev. **79**, 469 (1950).
15. J. R. Bergervoet, P. C. van Campen, W. A. van der Sanden, and J. J. de Swart, Phys. Rev. C **38**, 15 (1988).
16. M. T. Yamashita, L. Tomio, A. Delfino and T. Frederico, Europhys. Lett. **75** 555-561 (2006).
17. F. Ferlaino, S. Knoop, M. Berninger, W. Harm, J. P. D'Incao, H.-C. Nägerl, and R. Grimm, Phys.Rev. Lett, **102** 140401-4 (2009).
18. J. von Stecher, J. P. D'Incao and Chris H. Greene, Nat. Phys. **5**, 417-421 (2009).
19. A. E. A. Amorim, T. Frederico, and L.Tomio, Phys. Rev. C**56** R2378-R2381, (1997); M. T. Yamashita, T. Frederico, and L.Tomio, Phys. Rev. Lett., **99** 269201-1 (2007).

Quantum Bio-Informatics V

pp. 271–280

# KINETIC ISOTOPE EFFECT ON TRANSPORT MEDIATED BY CLC-TYPE $H^+$/$CL^-$ EXCHANGERS

ALESSANDRA PICOLLO

*Department of Anesthesiology, Weill Cornell Medical College, New York, NY 10065, USA*

MATTIA MALVEZZI

*Department of Physiology and Biophysics, Weill Cornell Medical College, New York, NY 10065, USA*

ALESSIO ACCARDI

*Department of Anesthesiology, Department of Physiology and Biophysics, Department of Biochemistry, Weill Cornell Medical College, New York, NY 10065, USA*

CLC transporters mediate the stoichiometric exchange of 2 $Cl^-$ ions for 1 $H^+$ across the membranes of cellular compartments, mostly endosomes and lysosomes. Despite intense biophysical, structural and electrophysiological scrutiny the $H^+$ transfer mechanism of these exchangers remains largely unknown. Previous work showed that two conserved Glutamates define the extremities of the $H^+$ pathway in CLC exchangers. However, we don't know whether $H^+$ transfer between these residues takes place along a series of protonatable moieties, via a Grotthuss mechanism and by diffusion of an $H_3^+O$ cation and if at any step $H^+$ tunneling plays a role. To differentiate between these possible mechanisms we measured the deuterium kinetic isotope effect on the transport rate of CLC-ec1 and CLC-5, respectively a prokaryotic and a eukaryotic CLC exchanger. We found that transport mediated by both proteins is slowed by ~20-40% when $H_2O$ is replaced by $D_2O$. This result suggests that the rate limiting step for $H^+$ transport takes place along a hydrogen-bonded pathway, possibly formed by water molecules. However, we found that the voltage dependence of CLC-5 inhibition by extracellular $H^+$ is eliminated by this substitution. This suggests that the voltage dependence of this process arises from a mechanism that is exquisitely sensitive to particle mass such as proton tunneling.

## 1. Introduction

The idea that quantum mechanical effects can underlie essential steps in biological processes has fascinated physicists since the inception of quantum theory. In the last decades these ideas have evolved from the observation of quantum steps during catalytic reactions mediated by enzymes [1-3] to speculations on the nature of consciousness and the origin of life [4]. These latter ideas are fascinating and have sparked a lively debate, however the present manuscript focuses on a much more circumscribed question: is it possible to experimentally detect quantum mechanical effects in ion transport across biological membranes?

The controlled movement of ions and other solutes is essential to sustain all forms of life; nearly 30% of the energy consumption of cells is devoted to the creation and sustenance of ionic gradients, most notably of $Na^+$, $K^+$, $H^+$, $Ca^{2+}$ and $Cl^-$ [5]. These gradients are expended to accumulate nutrients, export xenobiotics, mediate cellular signaling and maintain homeostasis. Disruption of these processes can be the result of naturally occurring mutations or the effect of foreign compounds and with potential outcomes ranging from pathologies of varying severity to death [6]. Two classes of membrane proteins mediate solute movement across membranes: ion channels and transporters. Channels are proteins that function as passive devices: they form aqueous pores through which ions move downhill an electrochemical gradient at rates approaching the diffusion limit. In contrast, active transporters utilize electrochemical energy to drive movement of substrate against a gradient. Usually these two protein classes are functionally and structurally distinct. One notable exception is the CLC family which is evenly divided among $Cl^-$ channels and $H^+/Cl^-$ exchangers [7, 8].

Despite this fundamental difference ample evidence suggests that these two subgroups share the same molecular architecture, all CLC's are dimers (Fig. 1a) where each monomer forms an independent $Cl^-$ pathway, and channel gating is evolutionarily related to the exchange cycle [7, 8]. Soon after the first CLC crystal structure was solved [9, 10], several groups showed that nearly half of the known CLC's function as $H^+$-coupled anion exchangers rather than as ion channels [11-17]. The functional and structural characterization of several mutants [10, 11, 18-25] and extensive simulations [26-35] elucidated the basic characteristics of the transport cycle. All known CLC transporters share some basic features: the exchange stoichiometry is 2 anions (usually $Cl^-$) for 1 proton [11-16], two highly conserved glutamates [10, 11, 19, 36], $Glu_{in}$ and $Glu_{ex}$, define the proton pathway and movement of the $Glu_{ex}$ side chain regulates $Cl^-$ transport [10]. Additionally, two conserved side chains of a Tyrosine and of a Serine, called $Tyr_{cen}$ and $Ser_{cen}$, form the central ion binding site and constrict the pathway towards the intracellular side (Fig. 1b). Rotation of the $Glu_{ex}$ side chain after protonation opens the $Cl^-$

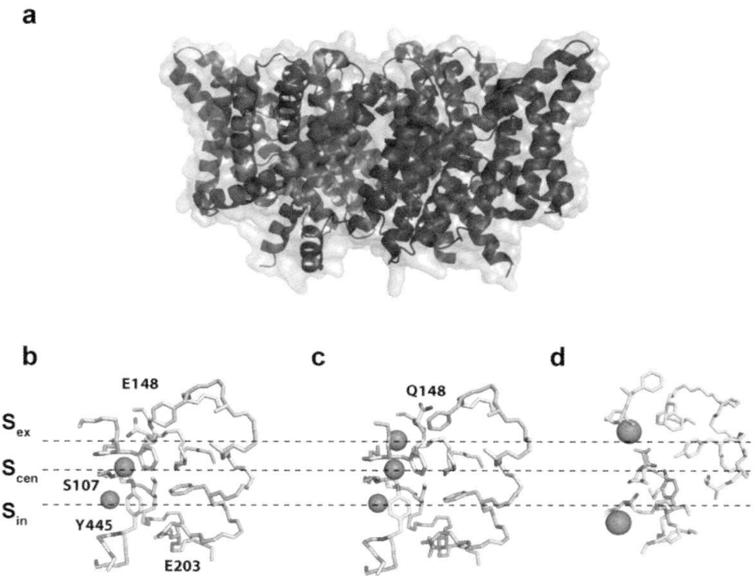

**Figure 1.** Structure of CLC transmembrane region. a) Ribbon representation of the CLC-ec1 homodimer viewed from the plane of the membrane. Chloride ions are represented as spheres. b) The chloride binding region for the WT CLC-ec1 and c) the E148Q mutant, d) The chloride binding region for WT cnCLC. (PDB accession codes WT: 1OTS, E148Q: 1OTU, cnCLC: 3ORG). Residues are numbered according to the sequence of CLC-ec1.

pathway towards the extracellular side [10] (Fig. 1c). In contrast, how protons transfer between $Glu_{ex}$ and $Glu_{in}$, remains a fundamental and unanswered question. A network of fixed protonatable sites could connect $Glu_{in}$ and $Glu_{ex}$ through the protein with minimal rearrangements. However, the paucity of suitable side chains argues against this possibility [19]. It has been recently proposed that a rotation of $Glu_{in}$ is sufficient to open a water filled crevice that connects $Glu_{in}$ and $Glu_{ex}$ [29, 35]. Protons could then move along a transiently ordered chain of water molecules connecting $Glu_{in}$ and $Glu_{ex}$ via a Grotthuss mechanism. Alternatively, the water cavity could be sufficiently large to allow the hydrodynamic diffusion of an $H_3O^+$ cation. One intriguing possibility is that protons could cover part of the distance separating $Glu_{ex}$ and $Glu_{in}$ by quantum tunneling. This would allow $H^+$ particles to cross an energy barrier even though its energy would be insufficient to go over it [37]. While not commonplace, tunneling of protons has been observed in several biological systems. For example, it has been shown that during the catalytic cycle of some enzymes a proton-transfer reaction is mediated by quantum tunneling [38-44]. The recently identified intermediate state of the CLC transport cycle in which the $Glu_{ex}$ side chain occupies $S_{cen}$ (Fig. 1d) greatly shortens the distance that a $H^+$ would need to cover when moving between $Glu_{in}$ and $Glu_{ex}$ [45] from ~13 Å to ~6 Å, a distance which thermal fluctuations could further reduce to the 1-2 Å range that is consistent with the distances allowed for proton tunneling.

Since the reaction rates of these proton transfer mechanisms are differentially affected by the replacement of hydrogen with the heavy isotope deuterium, we reasoned that we could distinguish among these three possible $H^+$ transport mechanisms by investigating the deuterium kinetic isotope effect on the transport rates of CLC-ec1 and CLC-5. The kinetic isotope effect or a reaction is defined as the ratio of the rates in hydrogen and in deuterium: $KIE=\gamma_H/\gamma_D$, where $\gamma_H$ and $\gamma_D$ are the reaction rates in hydrogen and deuterium. Hydrodynamic diffusion of hydronium ions is nearly independent from the isotope, as the diffusion coefficients of $H_3O^+$ and $D_3O^+$ are almost identical [46, 47], so we expect KIE~1. In contrast, the Grotthuss hopping rate of deuterons is ~30% slower than that of protons, so that KIE~1.5 [46, 47]. Finally, a high value of KIE would suggest that $H^+$ transfer is rate-limited by a step exquisitely sensitive to the mass of the isotope, such as tunneling [46, 48].

## 2. Materials and Methods

*Protein purification:* Expression and purification of CLC-ec1 was performed following published protocols [18, 19, 49]. In the final purification step the protein was run on a Superdex 200 column (GE Healthcare) pre-equilibrated in 100 mM NaCl, 20 mM Tris– $H_2SO_4$, 5 mM DM, pH 7.5 (buffer R).

*Molecular Biology.* All mutants were introduced with Quikchange (Stratagene) and the gene fully sequenced. RNA was synthesized with the SP6 kit (Ambion), stored at -80 °C and injected in Xenopus oocytes which were stored at 18 °C. Currents were measured 3-4 days after injection.

*Electrophysiology.* An OC-725C (Warner Inst.) amplifier was used for voltage clamp, and data was acquired using Gepulse (developed by Dr. M. Pusch). Currents were evoked by stepping voltage from -140 to 110 mV in 10 mV increments for 100 ms. The extracellular recording solution was 100 mM NaCl, 5 mM MgCl$_2$, 10 mM buffer to the appropriate pH or pD (Glycine at pH 9, Hepes from pH 8 to pH 6.5, Mes at pH 6 and 5.5 and glutamate at pH 5 and 4.5). To determine the pD of solutions prepared with D$_2$O we added 0.4 pH units to the pH meter reading [46]. Recording electrodes were filled with 3 M KCl and their resistance was 0.3-1 M$\Omega$.

*Cl-flux measurements* were carried out as previously described [20]. Temperature control was achieved by immersing the recording chamber in a custom built flow-cell connected to an E-100 water bath circulator (Lauda) which was kept at 25 ºC.

*Data analysis:* To quantify the dependence of CLC-5 currents from D$^+_{ex}$ the normalized current, $I_{norm}=I(pD)/I(pD=7.5)$, was fitted to a simple 1:1 binding curve of the form

$$I_{norm}= I_{min} + (1-I_{min})/(1+K/[D]_{ex}) \qquad (Eq.\ 1)$$

where $[D]_{ex}$ is the free extracellular deuteron concentration, K is the apparent binding constant (pK=-logK), $I_{min}$ is the minimal value of the normalized current recorded at a given voltage [8].

## 3. Results

We chose to measure the deuterium kinetic isotope effect of two CLC transporters, the bacterial CLC-ec1 and the mammalian CLC-5 because these two proteins mediate the exchange of 2 Cl$^-$ for 1 H$^+$ according to the the same basic transport principles but do so at very different rates, $\sim 10^3$ ion s$^{-1}$ for CLC-ec1 and $\sim 10^5$ ion s$^{-1}$ for CLC-5. Furthermore, transport mediated by CLC-5 is strongly dependent of voltage while CLC-ec1 displays little to no voltage dependence. These kinetic differences suggest that different steps might rate-limit transport in the two systems so that they could be differentially affected by deuterium.

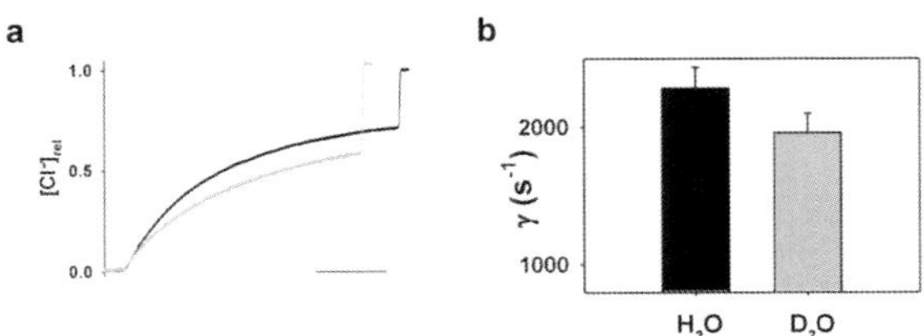

**Figure 2. Deuterium kinetic isotope effect on the transport rate of CLC-ec1.** a) Cl$^-$ efflux mediated by WT CLC-ec1 in H$_2$O (black line) and D$_2$O (gray line). Scale bar represents 20 sec. ^ denotes the addition of 1 μl of Val/FCCP and * denotes the addition of 40 μl of 1.5 M β-Octyl Glucosyde. b) Transport rate of WT CLC-ec1 in H$_2$O (black bar) and D$_2$O (gray bar).

276

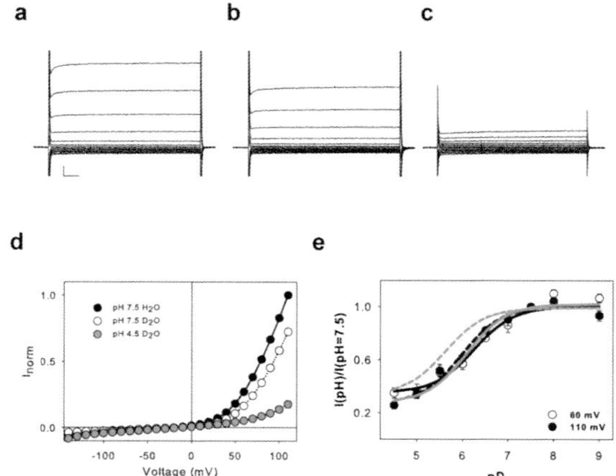



**Figure 3. Deuterium kinetic isotope effect on the transport rate of CLC-5.** a-c) Currents mediated by WT CLC-5 in $H_2O$ at pH 7.5 (a), $D_2O$ at pH 7.5 (b) and $D_2O$ at pH 4.5 (c). Vertical scale bar indicates 1 μA and horizontal scale bar indicates 10 ms. d) Normalized average I-V relationship for WT CLC-5, $I_{norm}=I(pH\ 7.5, V, X_2O)/I(pH\ 7.5, 110\ mV, H_2O)$ where $X_2O$ can be $H_2O$ or $D_2O$. I-V's at pH 7.5 and $H_2O$ (black), at pH 7.5 and $D_2O$ (white) and at pH 4.5 and $D_2O$ (gray) are shown. Connecting lines hold no theoretical meaning. e) Inhibition of WT CLC-5 by extracellular deuterium at +60 mV (white circles) and +110 mV (black circles). Solid lines represent the fits to Eq. 1 with pK (+60 mV) = 6.2±0.4 and $I_{min}$(+60 mV)=0.35±0.05 (gray line) and pK (+110 mV)= 6.0±0.3 and $I_{min}$(+110 mV)=0.27±0.04 (black line). Dashed gray and black lines represent the fits at +60 and +110 mV in $H_2O$ and are taken from [50]. The errors on the pK and $I_{min}$ are the uncertainty of the fits rather than S.E.M.

## 3.1. *Effects of deuterium on the transport rate of CLC-ec1 and CLC-5*

To determine the deuterium kinetic isotope effect on the transport rate of CLC-ec1 we measured the speed of Cl⁻ efflux from proteoliposomes [20] reconstituted with purified CLC-ec1 at 25 °C in $H_2O$ and $D_2O$ (Fig. 2a). The transport rate measured in water, $\gamma_H = (2290\pm150)$ ion s⁻¹, is consistent with published data [20]. Transport in $D_2O$ is slowed by ~17%, $\gamma_D = (1960+140)$ ion s⁻¹, so that KIE ~1.2 (Fig. 2a, b). This small KIE suggests either that H⁺ movement is not is rate limiting for transport mediated by CLC-ec1 or that it occurs through a chain of hydrogen bonds via a Grotthus mechanism.

We then measured the KIE of CLC-5 using two electrode voltage-clamp. CLC-5 currents in heavy water maintain the same characteristics as in normal water: outward rectification, kinetics and inhibition by low pH are all qualitatively preserved (Fig. 3a-d). The current in deuterium is 70% of that in water (Fig. 3a-d), so that

KIE~1.4, a value similar to that seen for CLC-ec1. Therefore, $H^+$ movement through CLC-ec1 and CLC-5 likely occurs via similar mechanisms.

### 3.2. *Kinetic isotope effect on the Voltage dependent $H^+$ block of CLC-5*

One surprising effect of deuterium on CLC-5 currents is that the voltage dependence of proton block [50] is drastically reduced (Fig. 3e). CLC-5 inhibition by $D^+_{ex}$ is nearly identical at +60 and +110 mV (Fig. 1E, solid lines) while inhibition by $H^+_{ex}$ is reduced as the applied voltage increases ([50]; Fig. 3e, dashed lines). Thus, two particles differing only by their mass are differentially affected by voltage. This suggests that protons inhibit CLC-5 through two distinct mechanisms: one that is independent of voltage and isotope mass, and another that is voltage dependent and displays a strong isotope effect.

## 4. Discussion

We set out to investigate the nature of the $H^+$ pathway in two CLC exchangers, CLC-ec1 and CLC-5. To this end we measured the deuterium kinetic isotope effect on the transport rate of CLC-ec1 and CLC-5. Both proteins are similarly affected by this substitution: their transport rate slows down only slightly when hydrogen is replaced by deuterium (Fig. 2, 3). This result shows that the rate limiting $H^+$-transfer step of the transport cycle of a CLC transporter does not involve proton tunneling. In contrast, it suggests that proton transfer takes place along a pathway formed by a series of hydrogen bonds. The distance separating $Glu_{ex}$ and $Glu_{in}$ is mostly lined by hydrophobic residues [19] which lack the hydrogen bonding capabilities necessary to form such a pathway. Our data is consistent with the recent proposal that a rotation of $Glu_{in}$ can open a small crevice which connects $Glu_{in}$ and $Glu_{ex}$ through the entry of water molecules [29, 35]. The recent structure of a eukaryotic CLC homologue [45] shows that the $Glu_{ex}$ side chain can also occupy the central $Cl^-$ binding site, reducing the total distance that $H^+$ need to cross and thus the number of water molecules needed to span this gap. These water molecules can transiently become ordered forming a Grotthuss chain which would allow fast $H^+$ movement.

It is worth noting that the substitution of hydrogen with deuterium has a profound effect on the voltage dependence of $H^+$ inhibition of CLC-5. We have recently shown that extracellular protons travel through ~50% of the electric field and bind to $Glu_{ex}$ to inhibit transport mediated by CLC-5 [50]. We expected that deuterons should behave in a similar manner since the two particles differ only by their mass. Surprisingly, we found that the fraction of the electric field traversed by $D^+_{ex}$ in reaching $Glu_{ex}$ is ~0 (Fig. 3e) while $H^+_{ex}$ is cross nearly half of electric field [50]. One possibility is that the shortened length of the hydrogen bonds in deuterium causes a structural rearrangement of the outer mouth of the CLC-5 transport pathway so that $Glu_{ex}$ becomes more exposed. However, the characteristics of the CLC-5 currents in $H_2O$ and $D_2O$ are very similar –the voltage dependence, outward rectification and kinetics are indistinguishable (Fig. 1c, d)– suggesting that the transport pathway remains intact. Furthermore, the prompt reversibility of the effects of deuterium upon washout in $H_2O$ also argues against a structural rearrangement. Another possibility is that there are two inhibitory proton binding sites in CLC-5, one closer to the surface and a second one located deep in the electric field. Protons can freely diffuse and bind to the first site while binding to the second site requires crossing a voltage-dependent barrier. This second step could take place via a mechanism that is sensitive to particle mass, such as proton tunneling [46]. A final alternative is that there are two possible pathways to reach a single binding site: free diffusion and a voltage- and isotope-sensitive pathway. Presently, we cannot distinguish between these possibilities and further work will be needed to elucidate the mechanism and site of action of extracellular protons on CLC-5.

In conclusion, our results suggest that $H^+$ movement in a CLC transporter is not rate-limited by a quantum mechanical step, but rather that it takes place in more conventional mechanisms, such as Grotthus diffusion. However, we show that the voltage dependent component of the block of CLC-5 by extracellular $H^+$ is exquisitely sensitive to the mass of the isotope suggesting that non-classical events might play a key role in this process.

## Acknowledgments

This work was supported by grant 1R01GM085232 from the National Institutes of Health (NIH) to A.A.

## References

1.  A. Kohen and J. P. Klinman, Chemistry & Biology **6**, R191 (1999)
2.  W. J. Bruno and W. Bialek, Biophys J **63**, 689 (1992)
3.  M. H. M. Olsson, P. E. M. Siegbahn and A. Warshel, J Am Chem Soc **126**, 2820 (2004)

4.  H. M. Wiseman and J. Eisert, D. Abbott, P. C. W. Davies and A. K. Pati, Imperial College Press, (2007)

5.  B. Hille, Sinauer, Pages (2001)

6.  F. M. Ashcroft, Academic Press, Pages (1999)

7.  C. Miller, Nature **440**, 484 (2006)

8.  A. Accardi and A. Picollo, Biochim Biophys Acta **1798**, 1457 (2010)

9.  R. Dutzler, E. B. Campbell, M. Cadene, B. T. Chait and R. MacKinnon, Nature **415**, 287 (2002)

10.  R. Dutzler, E. B. Campbell and R. MacKinnon, Science **300**, 108 (2003)

11.  A. Accardi and C. Miller, Nature **427**, 803 (2004)

12.  A. Picollo and M. Pusch, Nature **436**, 420 (2005)

13.  O. Scheel, A. A. Zdebik, S. Lourdel and T. J. Jentsch, Nature **436**, 424 (2005)

14.  A. De Angeli, D. Monachello, G. Ephritikhine, J. M. Frachisse, S. Thomine, F. Gambale and H. Barbier-Brygoo, Nature. **442**, 939 (2006)

15.  A. Graves, P. Curran, C. Smith and J. Mindell, Nature **453**, 5 (2008)

16.  J. Matsuda, M. Filali, K. Volk, M. Collins, J. Moreland and F. Lamb, Am J Physiol Cell Physiol **294**, 251 (2008)

17.  I. Neagoe, T. Stauber, P. Fidzinski, E. Y. Bergsdorf and T. J. Jentsch, J Biol Chem **285**, 21689 (2010)

18.  A. Accardi, S. Lobet, C. Williams, C. Miller and R. Dutzler, J. Mol. Biol. **362**, 691 (2006)

19.  A. Accardi, M. Walden, W. Nguitragool, H. Jayaram, C. Williams and C. Miller, J. Gen. Physiol. **126**, 563 (2005)

20.  M. Walden, A. Accardi, F. Wu, C. Xu, C. Williams and C. Miller, J. Gen. Physiol. **129**, 317 (2007)

21.  W. Nguitragool and C. Miller, J. Mol. Biol. **362**, 682 (2006)

22.  S. Lobet and R. Dutzler, Embo J. **25**, 24 (2006)

23.  H. Jayaram, A. Accardi, F. Wu, C. Williams and C. Miller, Proc Natl Acad Sci U S A. **105**, 6 (2008)

24.  H. H. Lim and C. Miller, J Gen Physiol. **133**, 8 (2009)

25.  A. Picollo, M. Malvezzi, J. Houtman and A. Accardi, Nat Struct Mol Biol **16**, 1294 (2009)

26.  G. V. Miloshevsky, A. Hassanein and P. C. Jordan, Biophys. J. **98**, 999 (2010)

27.  G. V. Miloshevsky and P. C. Jordan, Biophys. J. **86**, 825 (2004)

28.  D. L. Bostick and M. L. Berkowitz, Biophys. J. **87**, 1686 (2004)

29.  D. Wang and G. Voth, Biophysical journal **97**, 121 (2009)

30.  Z. Kuang, U. Mahankali and T. L. Beck, Proteins **68**, 26 (2007)

31.  J. Yin, Z. Kuang, U. Mahankali and T. L. Beck, Proteins **57**, 414 (2004)

32.  J. D. Faraldo-Gomez and B. Roux, J. Mol. Biol. **339**, 981 (2004)

33.  J. Cohen and K. Schulten, Biophys. J. **86**, 836 (2004)

34.  B. Corry, M. O'Mara and S. H. Chung, Biophys. J. **86**, 846 (2004)

35.  Y. J. Ko and W. H. Jo, Biophys. J. **98**, 2163 (2010)

36.  A. A. Zdebik, G. Zifarelli, E. Y. Bergsdorf, P. Soliani, O. Scheel, T. J. Jentsch and M. Pusch, J. Biol. Chem. **283**, 4219 (2008)

37.  J. Sakurai, Addison Wesley, Pages (1994)

38.  P. L. Dutton, A. W. Munro, N. S. Scrutton and M. J. Sutcliffe, Philos Trans R Soc Lond B Biol Sc **361**, 1293 (2006)

39.  M. H. Olsson, P. E. Siegbahn and A. Warshel, J Am Chem Soc. **126**, 9 (2004)

40.  L. Masgrau, A. Roujeinikova, L. O. Johannissen, P. Hothi, J. Basran, K. E. Ranaghan, A. J. Mulholland, M. J. Sutcliffe, N. S. Scrutton and D. Leys, Science **312**, 237 (2006)

41.  V. L. Schramm, Curr Opin Chem Biol **11**, 529 (2007)

42.  A. Ouchi, S. Nagaoka and K. Mukai, J Phys Chem B **114**, 6601 (2010)

43.  S. Hay, L. O. Johannissen, M. J. Sutcliffe and N. S. Scrutton, Biophys. J. **98**, 121 (2010)

44.  R. Banerjee, A. Dybala-Defratyka and P. Paneth, Philos Trans R Soc Lond B Biol Sci. **361**, 1333 (2006)

45.  L. Feng, E. B. Campbell, Y. Hsiung and R. MacKinnon, Science **330**, 635 (2010)

46.  T. E. DeCoursey and V. V. Cherny, J. Gen. Physiol. **109**, 415 (1997)

47.  E. Beitz, B. Wu, L. M. Holm, J. E. Schultz and T. Zeuthen, PNAS **103**, 269 (2006)

48.  G. Zifarelli, A. R. Murgia, P. Soliani and M. Pusch, J. Gen. Physiol. **132**, 185 (2008)

49.  A. Accardi, L. Kolmakova-Partensky, C. Williams and C. Miller, J. Gen. Physiol. **123**, 109 (2004)

50.  A. Picollo, M. Malvezzi and A. Accardi, J Gen Physiol **135**, 653 (2010)

Quantum Bio-Informatics V
© 2013 World Scientific Publishing Co. Pte. Ltd.
pp. 281–290

# ON POSITIVE MAPS; FINITE DIMENSIONAL CASE

WLADYSLAW A. MAJEWSKI

*Institute of Theoretical Physics and Astrophysics,*
*Gdańsk University*
*Wita Stwosza 57 80-952 Gdańsk, Poland,*
*E-mail: fizwam@univ.gda.pl*

We outline a new approach to the characterization as well as to the classification
of positive maps. This approach is based on the concept of exposed points, links
between tensor products and mapping spaces and on the convex analysis characterization of the geometry of a unit ball in Banach spaces.

## 1. Introduction

Let us begin with a brief sketch of the motivation of our work. In operator
theory, some of the most important families of linear maps are positive,
k-positive, completely positive, and decomposable maps. Furthermore, in
quantum physics dynamical maps should be positive and unital - to have the
well defined probabilistic scheme. Hence, a characterization of the structure of positive unital maps is not only an interesting **OLD** mathematical
problem but it is also a necessary step in analysis of many problems in
Quantum Mechanics and Quantum Information. In particular, let us emphasize, the theory of positive maps constitutes the essential ingredient in
the study of quantum correlations [11], [9] (so, also, this topic belongs to
quantum probability!)

For sake of convenience, we provide all necessary preliminaries and set
up the notation. Let $\mathcal{B}(\mathcal{H})$ be the set of all linear bounded operators on
a Hilbert space $\mathcal{H}$. From now on, for simplicity, we make the assumption:
$dim\mathcal{H} < \infty$. We denote the set of all positive elements of $\mathcal{B}(\mathcal{H})$ by $\mathcal{B}(\mathcal{H})^+$.
Further, we denote $\mathcal{B}(\mathcal{H})_h = \{a \in \mathcal{B}(\mathcal{H}); a = a^*\}$. The following pair

$$(\mathcal{B}(\mathcal{H})_h, \mathcal{B}(\mathcal{H})^+) \tag{1}$$

being an ordered Banach space, will play an important role in our lecture.

A *state* on $\mathcal{B}(\mathcal{H})$ is a linear functional $\phi : \mathcal{B}(\mathcal{H}) \longrightarrow \mathbb{C}$ such that $\phi(A) \geq 0$ for every $A \in \mathcal{B}(\mathcal{H})^+$ and $\phi(\mathbb{1}) = 1$, where $\mathbb{1}$ is the unit of $\mathcal{B}(\mathcal{H})$. The set of all states on $\mathcal{B}(\mathcal{H})$ is denoted by $\mathcal{S}_{\mathcal{B}(\mathcal{H})}$.

A linear map $\psi : \mathcal{B}(\mathcal{H}_1) \longrightarrow \mathcal{B}(\mathcal{H}_2)$ is called *positive* if $\psi(\mathcal{B}(\mathcal{H}_1)^+) \subset \mathcal{B}(\mathcal{H}_2)^+$. For $k \in \mathbb{N}$ we consider a map $\psi_k : M_k \otimes \mathcal{B}(\mathcal{H}_1) \longrightarrow M_k \otimes \mathcal{B}(\mathcal{H}_2)$ where $M_k \equiv M_k(\mathbb{C})$ denotes the algebra of $k \times k$-matrices with complex entries and $\psi_k = \mathrm{id}_{M_k} \otimes \psi$. We say that $\psi$ is *k-positive* if the map $\psi_k$ is positive. The map $\psi$ is said *completely positive*(CP for short) when $\psi$ is $k$-positive for every $k \in \mathbb{N}$. Due to the Stinespring theorem any CP map $T : \mathcal{B}(\mathcal{H}_1) \to \mathcal{B}(\mathcal{H}_2)$ is of the form

$$T(A) = W^*\pi(A)W, \tag{2}$$

where $\pi : \mathcal{B}(\mathcal{H}_1) \to \mathcal{B}(\mathcal{L})$ is a *-representation, and $W : \mathcal{H}_2 \to \mathcal{L}$ is a linear bounded operator. Thus, the structure of CP maps is well known.

The set of all (linear, bounded) positive maps $\alpha : \mathcal{B}(\mathcal{H}_1) \to \mathcal{B}(\mathcal{H}_2)$ will be denoted by $\mathcal{L}^+(\mathcal{B})$, while the set of extreme points of $\mathcal{L}^+(\mathcal{B})$ will be denoted by $Ext\{\mathcal{L}^+(\mathcal{B})\}$.

To characterize the structure of $\{\mathcal{L}^+(\mathcal{B})\}$ it was tempting to use Krein Milman theorem:

**Theorem 1.1.** *A compact convex set $C$ is a (closed) convex hull of its extreme points.*

It is an easy task to provide examples of positive extremal maps so elements of $Ext\{\mathcal{L}^+(\mathcal{B})\}$. Namely (see [16]):

(1) Jordan homomorphisms $\alpha : \mathcal{A} \to \mathcal{B}$,
(2) Let $\mathcal{A} \subset \mathcal{B}(\mathcal{K})$, $\mathcal{B} \subset \mathcal{B}(\mathcal{H})$. Define a map $\alpha : \mathcal{A} \to \mathcal{B}$ to be of the form $\alpha(A) = V^*AV$, where $V$ is a linear isometry of the Hilbert space $\mathcal{H}$ into $\mathcal{K}$.

However, it is an extremely difficult task to describe the set of all extremal points of $\{\mathcal{L}^+(\mathcal{B})\}$. For two dimensional case (i.e. $dim\mathcal{H} = 2$), Størmer [17] obtained such characterization. But, even for three dimensional case, such characterization is still an open problem. Motivated by this, we turn to a special subset of extremal positive maps. We begin with (see [18], [6]):

**Definition 1.2.** (Straszewicz) Let $C$ be a convex set in a Banach space $X$. A point $x \in C$ is an exposed point of $C$ ($x \in Exp\{C\}$) if there is $f \in X^*$ (dual of $X$) such that $f$ attains its maximum on $C$ at $x$ and only at $x$.

In other words, we wish to have $\langle f, x \rangle > \langle f, y \rangle$ for $x \neq y$ (where $\langle f, x \rangle \equiv f(x)$). **Thus we invoke a kind of a variational principle!**

In general, one has $Ext\{C\} \supseteq Exp\{C\}$ but there are simple examples of 2-dimensional convex compact sets such that the inclusion $Ext\{C\} \supset Exp\{C\}$ is proper (see 6]).

Our interest in exposed points stems from the following result saying that the subset of exposed points still is quite "big" (cf [18], [6]):

**Proposition 1.3.** *Every norm-compact convex set $C$ in a Banach space $X$ is the closed convex hull of its exposed points.*

We note that the concept of an exposed point is a particular case of more general idea (see [4]):

**Definition 1.4.** Let two (real) vector spaces $X$ and $Y$ be in separating duality with respect to a bilinear form $< \cdot, \cdot >$. A face $F$ of a convex set $K \subset X$ is called exposed if there exists a $y \in Y$ and an $\alpha \in \mathbb{R}$ such that $< x, y > = \alpha$ for all $x \in F$ and $< x, y > \; > \alpha$ for all $x \in K \setminus F$.

Clearly, if $F = \{x\}$ then we arrive at the definition of exposed point. The interest of the concept of exposed faces is that in the description of (dynamical) pictures of Quantum Mechanics, exposed faces play an important role [4]. Moreover, the distinguished subset of density matrices will be an exposed face of the unit ball (cf Proposition 2.3). Further, the nice property in convex analysis, so called "the transition of extremality" (if $x \in C' \subset C$ is an extreme point of $C$, then it is a fortiori an extreme point of the smaller set $C'$ ) is enhanced for faces.

To proceed with an analysis of exposed points of $\mathcal{L}^+(\mathcal{B})$, firstly we should describe the corresponding linear duality in an effective way. To this end, we will follow Grothendieck, [7], who observed the links between tensor products and mapping spaces. To describe these links, we will select certain results from the theory of tensor products of Banach spaces.

Let $\mathfrak{A}$ be a norm closed self-adjoint subspace of bounded operators on a Hilbert space $\mathcal{K}$ containing identity operator on $\mathcal{K}$. Further, $\mathfrak{T}$ will denote the set of trace class operators on a Hilbert space $\mathcal{H}$ equipped with the trace norm. $\mathcal{B}(\mathcal{H}) \ni x \mapsto x^t \equiv \tau(x) \in \mathcal{B}(\mathcal{H})$ will stand for the transpose map $\tau$ of $\mathcal{B}(\mathcal{H})$ with respect to some orthonormal basis.

Finally, we denote by $\mathfrak{A} \odot \mathfrak{T}$ the algebraic tensor product of $\mathfrak{A}$ and $\mathfrak{T}$ while $\mathfrak{A} \otimes_\pi \mathfrak{T}$ means its Banach space closure under the projective norm defined by

$$\pi(x) = \inf\{\sum_{i=1}^{n} \|a_i\|\|b_i\|_1 : x = \sum_{i=1}^{n} a_i \otimes b_i, \ a_i \in \mathfrak{A}, \ b_i \in \mathfrak{T}\}, \qquad (3)$$

where $\|\cdot\|_1$ stands for the trace norm. The promised relations between the tensor product structure and positive maps is contained in (see [15], 19]):

**Lemma 1.5.** *(Basic Lemma) There is an isometric isomorphism $\phi \mapsto \tilde{\phi}$ between $\mathcal{L}(\mathfrak{A}, \mathcal{B}(\mathcal{H}))$ and $(\mathfrak{A}\otimes_\pi \mathfrak{T})^*$ given by*

$$(\tilde{\phi})(\sum_{i=1}^{n} a_i \otimes b_i) = \sum_{i=1}^{n} \mathrm{Tr}\,(\phi(a_i)b_i^t), \qquad (4)$$

*where $\sum_{i=1}^{n} a_i \otimes b_i \in \mathfrak{A} \odot \mathfrak{T}$.*
*Furthermore, $\phi \in \mathcal{L}^+(\mathfrak{A}, \mathcal{B}(\mathcal{H}))$ if and only if $\tilde{\phi}$ is positive on $\mathfrak{A}^+\otimes_\pi \mathfrak{T}^+$.*

**Remark 1.6.**

(1) There is no restriction on the dimension of Hilbert spaces. In other words, this result can be applied to the true quantum systems.

It is an easy observation that the isomorphism given by the above lemma sends the set $Exp\{\mathcal{L}^+(\mathcal{B}(\mathcal{H}))\}$ onto the set $Exp\{(\mathcal{B}(\mathcal{H})\otimes_\pi \mathcal{B}(\mathcal{H}))^{*,+}\}$, where $(\mathcal{B}(\mathcal{H})\otimes_\pi \mathcal{B}(\mathcal{H}))^{*,+}$ stands for continuos functionals on $\mathcal{B}(\mathcal{H})\otimes_\pi \mathcal{B}(\mathcal{H})$ which are positive on $\mathcal{B}(\mathcal{H})^+ \otimes_\pi \mathcal{B}(\mathcal{H}))^+$. Consequently, we reduced our task to an analysis of exposed points of the set

$$\{(\mathcal{B}(\mathcal{H}) \otimes_\pi \mathcal{B}(\mathcal{H}))^{*,+}\}. \qquad (5)$$

## 2. The set $\mathfrak{D}$ and its properties.

As it was argued in the previous Section, cf. Lemma 1.5, to describe the structure of positive maps it is enough to carry out an analysis of exposed points of the corresponding set of functionals . Therefore, we wish to provide a rigorous description of this set . We begin with the observation: any (linear, bounded) functional in $(\mathcal{B}(\mathcal{H}) \otimes_\pi \mathcal{B}(\mathcal{H}))^{*,+}$ is of the form

$$\varphi(x \otimes y) = \mathrm{Tr}\,\varrho_\varphi\, x \otimes y, \qquad (6)$$

(the finite dimension case is assumed!)

In other words, there is one-to-one correspondence between functionals described in Lemma 1.5 and a subset of "density " matrices. We wish to describe this subset of "density" matrices in details. To this end we note that the order isomorphism of Lemma 1.5 leads the so called "block positivity" which is defined as follows: $\varrho_\varphi$ satisfies "block-positivity" (denoted "bp" for short) if

$$\varrho_\varphi \geq_{bp} 0 \quad \Leftrightarrow \quad (f \otimes g, \varrho_\varphi f \otimes g) \geq 0 \tag{7}$$

for any $f, g \in \mathcal{H}$.

To use the fact that this isomorphism is isometrical we define (see [10])

**Definition 2.1.**

$$\alpha(\varrho_\varphi) = \sup_{0 \neq a \in \mathcal{B}(\mathcal{H}) \otimes_\pi \mathcal{B}(\mathcal{H})} \frac{|\mathrm{Tr}\, \varrho_\varphi a|}{\pi(a)} \tag{8}$$

Further, to describe the above mentioned set of functionals in terms of bp density matrices we need (see [10]):

**Definition 2.2.** The set of bp normalized density matrices is defined as

$$\mathfrak{D} = \{\varrho_\phi : \alpha(\varrho_\phi) = 1,\ \varrho_\phi = \varrho_\phi^*,\ \varrho_\phi \geq_{bp} 0,\ \mathrm{Tr}\, \varrho_\phi = n\} \tag{9}$$

Now we are in position to present our first result [10]:

**Proposition 2.3.**

*(1) Lemma 1.5 gives an isometric isomorphism between the convex set of unital positive maps $\mathfrak{C}$ and the subset $\mathfrak{D}$ of bp normalized density matrices.*

*(2) $\mathfrak{D}$ is contained in the ball of $(\mathcal{B}(\mathcal{H}) \otimes \mathcal{B}(\mathcal{H}))_h$ of radius $n$ (with respect to the operator norm.)*

*(3) $\mathfrak{D}$ is an exposed face in the set of positive elements of the unit ball (with respect to the norm $\alpha$).*

As $\mathfrak{D}$ is a convex and compact subset, applying Proposition 1.3 we arrived at

**Corollary 2.4.** *$\mathfrak{D}$ is the closure of the convex hull of its exposed points!*

**Remark 2.5.** Geometrically speaking, we are using the correspondence between two "flat" subsets of balls in $\mathcal{L}(\mathcal{B}(\mathcal{H}))$ and in the set of all self-adjoint density matrices $(\mathcal{B}(\mathcal{H}) \otimes_\pi \mathcal{B}(\mathcal{H}))_h$, respectively. The interest of this remark follows from the fact that the considered balls are not so nicely shaped as the closed ball of real Euclidean 2– or 3 space.

Other useful properties of $\mathfrak{D}$ are collected in, [10],

**Proposition 2.6.** $\mathfrak{D}$ *is globally invariant with respect to the following operations:*

> *(1) local operations, LO for short, i.e. maps implemented by unitary operators $U : \mathcal{H} \otimes \mathcal{H} \to \mathcal{H} \otimes \mathcal{H}$ of the form $U = U_1 \otimes U_2$ where $U_i : \mathcal{H} \to \mathcal{H}$ is unitary, $i = 1, 2$;*
> *(2) partial transpositions $\tau_p = id_{\mathcal{H}} \otimes \tau : \mathcal{B}(\mathcal{H}) \otimes \mathcal{B}(\mathcal{H}) \to \mathcal{B}(\mathcal{H}) \otimes \mathcal{B}(\mathcal{H})$ where $\tau$ stands for transposition.*

To provide a reader with interesting examples of elements of $\mathfrak{D}$ we will need

**Definition 2.7.** A self-adjoint unitary operator $s$ is called a symmetry, i.e. $s = s^*$ and $s^2 = I$.

A self-adjoint operator $s$ is called a partial symmetry if $s^2$ is a (orthogonal) projector $e$.

To compute the norm $\alpha(a)$, at least for some interesting examples we need (see [10]):

**Lemma 2.8.**

$$\alpha(\sigma) = \max\{|\operatorname{Tr} \sigma \cdot s \otimes p \mid : s \in \mathcal{S}(\mathcal{H}), p \in Proj^1(\mathcal{H})\}$$

*where $\mathcal{S}(\mathcal{H})$ denotes the set of all symmetries in $\mathcal{B}(\mathcal{H})_h$ while $Proj^1(\mathcal{H})$ stands for the set $\{\pm|f><f| : f \in \mathcal{H}, \|f\| = 1\}$.*

and

**Corollary 2.9.**

> *(1) Let $P$ be a projector (on $\mathcal{H} \otimes \mathcal{H}$). Then $\alpha(P) = \|\operatorname{Tr}_{\mathcal{H}} P \|$ where $\operatorname{Tr}_{\mathcal{H}}$ stands for the partial trace.*
> *(2) Let $W$ be a symmetry (on $\mathcal{H} \otimes \mathcal{H}$). Then $\alpha(W) = \max\{\|\operatorname{Tr}_{\mathcal{H}}(W s \otimes 1)\| : s \in \mathcal{S}(\mathcal{H})\}$.*

We are in position to give the promised examples (see [10]):

**Example 2.1.**

> (1) Any projector of the form $P = p \otimes I$ where p is a one dimensional projector on $\mathcal{H}$, $I$ is the identity on $\mathcal{H}$.

(2) Let $\{e_i\}$ be a basis in $\mathcal{H}$. Define $f \in \mathcal{H} \otimes \mathcal{H}$ by

$$f = \frac{1}{\sqrt{n}} |e_1 \otimes e_1 + e_2 \otimes e_2 + ... + e_n \otimes e_n > \qquad (10)$$

Such vector $f$ is frequently called: a fully entangled vector. Obviously, $n|f><f| \in \mathfrak{D}$.

(3) Define

$$W = \sum_{i,j} E_{ij} \otimes E_{ji} \qquad (11)$$

where $E_{ij} \equiv |e_i><e_j|$. $W$ is a bp symmetry with $\operatorname{Tr} W = n$. Moreover, $\tau_p(W) = n|f><f|$. Hence, $W \in \mathfrak{D}$.

To appreciate the above examples we make

**Remark 2.10.**

(1) In Quantum Information, $W$ is called the swapping operation.
(2) $W g_1 \otimes g_2 = g_2 \otimes g_1$ (so it is also a "flip" operator )
(3) If we apply $W$ to the correspondence (4) and take into account (6) we get that $\phi$ is the transposition.
(4) The partial transposition of $n|f><f|$ gives $W$.

To describe one of the important consequences of our approach we need some more preliminaries. We recall that for 2D (two dimensional) case - all positive maps are decomposable, [17], i.e. they are in the set

$$conv\{CP, \quad \tau \circ CP\}.$$

However, even for 3D case, there are also non-decomposable maps ([5], [20])) To show that, within the presented approach, we can understand this difference we define the following subset of $\mathfrak{D}$

$$\mathfrak{D}_a = \{\text{symmetries}, n\mathrm{P_x}, |\mathrm{f}><\mathrm{f}| \otimes 1\} \subset \mathfrak{D}$$

where x is a fully entangled vector. An analysis of extreme positive for 2D case leads to (see [12])

**Proposition 2.11.** *A unital positive map (2D case) is a convex combination of maps from* $\mathfrak{D}_a$.

However, this Proposition is not longer true for 3D case. Namely, the proof of this Proposition is based on an analysis of extreme positive maps restricted to diagonal matrices. But, in 3D case, one can find (explicitly) examples of extreme positive maps such that their restriction to diagonal matrices are not multiplicative (contrary to 2D case!) This show that the set of maps corresponding to $\mathfrak{D}_a$ is too small (for 3D case).

## 3. A characterization of exposed points of $\mathfrak{D}$

In previous section we have seen that the set of functional $\mathfrak{D}$ distinguished by Lemma 1.5 exhibits a complicated geometrical features. Therefore, looking for a characterization of exposed points of $\mathfrak{D}$, we turn to more analytic approach to this problem with some emphasize on the underlying geometry. We wish to provide a characterization of exposed points of $\mathfrak{B}_1^{(+)} (\equiv \{\varrho_\varphi : \alpha(\varrho_\varphi) \leq 1, \ \varrho_\varphi = \varrho_\varphi^*, \ \varrho_\varphi \geq_{bp} 0\}$.

We begin with recalling selected definitions appearing in the convex analysis of real Banach space $X$ (see [1], [3], [13], and [14]).

We denote by $S_X$ $(X_1)$ the unit sphere (ball) of $X$.

**Definition 3.1.** A point $x$ of $S_X$ is said to be

    (1) an exposed point of $X_1$ if $\{x\}$ is an exposed face of $X_1$,
    (2) a rotund point of $X_1$ if every $y \in S_X$ with $\|\frac{x+y}{2}\| = 1$ satisfies $x = y$.
    (3) a smooth point of $X_1$ if there is exactly one element $f$ of $S_{X^*}$ such that $f(x) = 1$.

The sets of rotund points (smooth points) of $X_1$ will be denoted by $rot(X_1)$ $(smo(X_1))$.

We need also (see [2])

**Definition 3.2.** $x \in S_X$ is defined to be a strongly non-smooth point of $X_1$ if for every $y \in S_{X \setminus \{x\}}$ with $[x, y] \subseteq S_X$, $x$ is not smooth point of $Y_1$, where $Y = span\{x, y\}$ and $[x, y] = \{v = \lambda x + (1 - \lambda)y, \quad \lambda \in [0, 1]\}$.

and

**Theorem 3.1.** *Let $X$ be a real, separable Banach space. Then one has*

$$Exp(X_1) = rot(X_1) \cup nsmo(X_1) \tag{12}$$

Using these general scheme we were able to prove, [10]

**Theorem 3.2.** *An exposed point of $\mathfrak{B}_1^{(+)}$ is also an exposed point of $\mathfrak{B}_1$.*
    *Conversely, a bp positive, strongly non-smooth, no rotund $\varrho \in \mathfrak{B}_1$ is an exposed point of $\mathfrak{B}_1^{(+)}$.*

and

**Corollary 3.1.** *Strongly non-smooth, no rotund , bp-positive points $\sigma$ of unit sphere (with respect to the norm $\alpha$) having the normalization $\operatorname{Tr}\sigma = n$ are exposed points of $\mathfrak{D}$.*

To sum up, as we are interested in a characterization of the structure of $\mathcal{L}^+(\mathcal{B})$, i.e. linear and *continuous* maps on $\mathcal{B}(\mathcal{H})$ we are forced to use the correspondence given by Lemma 1.5. This makes the important difference with respect to the standard approach based on the Jamiolkowski-Choi isomorphism (cf [8]). Further, again using Lemma 1.5, we introduced and described the set $\mathfrak{D}$. The principal significance of the set $\mathfrak{D}$ is that it has the same structure as $\mathcal{L}^+(\mathcal{B})$ has. Using the Straszewicz concept of exposed points we are able to provide a description of the structure of the set $\mathfrak{D}$. Finally, a characterization of exposed points in terms of convex analysis was given.

## Acknowledgments

The author would like to express a deep gratitude to Masanori Ohya for kind hospitality. He would like also to acknowledge the partial supports of the grant number N N202 208238 of Polish Ministry of Science and Higher Education and QBIC grant which make his stay in Quantum Bioinformatics Center of Tokyo University of Science possible.

## References


1. A. Aizupuru, F. J. Garcia-Pacheco, Some questions about rotundity and renorming in Banach spaces, *J. Aust. Math. Soc.* **79**, 131-140, 2005.
2. A. Aizupuru, F. J. Garcia-Pacheco, A short note about exposed points in real Banach spaces, *Acta Math. Scientia* **28B**, 797-800 (2008)
3. E. M. Alfsen, *Compact convex sets and boundary integrals*, Erg. der Math. 57, Springer Verlag, 1971.
4. E. M. Alfsen and F. W. Shultz, *State spaces of operator algebras*, Birkhauser, Boston 2001.
5. M. -D. Choi, Positive semidefinite biquadratic forms, *Linear Alg. Appl.* **12** (1975), 95-100
6. V. P. Fonf, J. Lindenstrauss, R. R. Phelps, Infinite Dimensional Convexity, Chapter 15 in *Handbook of Geometry of Banach Spaces*, vol. **1**, Edited by William B. Jonson and Joram Lindenstrauss, Elsevier Science, 2001
7. A. Grothendieck, Products tensoriels topologiques et espaces nuclearies, *Memoirs of the American Mathematical Society*, **16**, Providence, Rhode Island, 1955.

8. A. Jamiolkowski, Linear transformations which preserves trace and positive semidefiniteness of operators, *Rep. Math. Phys.* **3**, 275 (1972)

9. W. A. Majewski, "On quantum correlations and positive maps", *Lett. Math. Physics,* **67**, 125-132 (2004); e-print, LANL math-ph/0403024

10. W. A. Majewski, On the structure of positive maps; finite dimensional case, submitted

11. W. A. Majewski, T. Matsuoka, M. Ohya, Characterization of partial positive transposition states and measures of entanglement, *J. Math. Phys.,* **50**, 113509 (2009)

12. W. A. Majewski, T. I. Tylec, On positive maps on low dimensional matrix algebras, in preparation

13. R. E. Megginson, *An introduction to Banach space theory,* Springer Verlag, 1998

14. R. R. Phelps, *Convex functions, monotone operators and differentiability,* Lecture Notes in Mathematics, vol. 1364, Springer Verlag, 1989

15. E. Størmer, Extension of positive maps, *J. Funct. Analysis,* **66** (1986), 235–254.

16. E. Størmer, Positive linear maps on $C^*$-algebras, in *Foundations of Quantum Mechanics and Ordered Linear spaces,* Lecture Notes in Physics, vol **29**, Springer Verlag, 1973 .

17. E. Størmer, Positive linear maps on operator algebras, *Acta Math.* **110** (1963) 233–278

18. S. Straszewicz, Über exponierte Punkte abgeschlossener Punktmengen, *Fund. Math.,* **24**, 139-143 (1935)

19. A. W. Wickstead, *Linear operators between partially ordered Banach space and some related topics,* PhD Thesis, Chelsea College, University of London, 1973

20. S. L. Woronowicz, Positive maps of low dimensional algebras, *Rep. Math. Phys.* **10** (1976) 165–183

Quantum Bio-Informatics V

© 2013 World Scientific Publishing Co. Pte. Ltd.

pp. 291–299

# A NEW NOISE DEPENDING ON A SPACE PARAMETER AND ITS APPLICATION

SI SI

*Faculty of Information Science and Technology Aichi Prefectural University,
JapanA
E-mail: sisi@ist.aichi-pu.ac.jp*


WIN WIN HTAY

*Department of Engineering Mathematics Yangon Technological University
Yangon, Myanmar*

The purpose of this paper is to introduce a *new noise* depending on the space
parameter and the construction of a space of generalized functional of the new
noise. One of the significant applications is to give a concrete answer to the jump
finding problem raised in [15] where the first step to the approach has been made.
It is noted that the new noise, denoted by $P'(u)$, is quite different from the time
derivative $\dot{P}(t)$ of a Poisson process $P(t)$.

2000 AMS Subject Classification 60H40 White Noise Theory

## 1. Introduction

Poisson noise $\dot{P}(t)$ has been studied by many people and the theory has
been widely established. Our purpose in this talk is to introduce a new
noise which will be denoted by $P'(u)$. It is a type of Poisson noise, however
the parameter is taken to be a space parameter u, which is actually height of
jump. Unlike noises depending on time, we can show significant properties
different from those of (Gaussian) white noise or Poisson noise which is the
time derivative of Poisson process $P(t)$.

As an application, we can discuss decomposition of a compound Poisson
process into infinitesimal Poisson type random variables which are independent. Those independent variables form a system of idealized random
variables, often called simply a noise.

We can roughly classify various noises according to the parameters on
which noises depend.

1) Depending on time $t \in R^1$ or $R^d$

    i) White noise (Gaussian) $\dot{B}(t)$, time derivative of a Brownian motion, defined on $(E^*, \mathcal{B}, \mu)$, where the sigma-field $\mathcal{B}$ is generated by cylinder subsets of $E^*$.

    ii) Poisson noise $\dot{P}(t)$, time derivative of a Poisson process $P(t)$, defined the same as white noise.

          Invariance : time shift.

They are independent elemental random variables (i.e.r.v) and are idealized variables, however rigorously discussed in Mathematics. They form a base of Hilbert space and form a total set.

2) Depending on space variable

    New noise $P'(u), u$ is height of jump

    Invariance : dilation of the space variable $u$.

We are going to discuss this noise $P'(\bar{u})$ with the parameter $\bar{u}$ is taken to be in $R_+^d = (0, \infty)^d, d \geq 1$.

## 2. New noise $P'(\bar{u}), \bar{u} \in R_+^d$

We first take the space parameter set to be $R_+^d = (0, \infty)^d$, $d \geq 1$, since we mean the space variable as a displacement of position.

The idea is to choose another kind of parameter, say space variable, instead of time. The decomposition of a compound Poisson process has given a hint to have this idea.

We start with a functional

$$C^P(\xi) = \exp\left[\int_{R_+^d} (e^{i(\bar{u}, \xi(\bar{u}))} - 1) dn(\bar{u})\right], \tag{1}$$

where $\bar{u} \in R_+^d, dn(\bar{u})$ is a measure on $R_+^d$.

To fix the idea we assume that the measure

$$dn(\bar{u}) = \lambda(\bar{u}) d\bar{u}, \tag{2}$$

where $\lambda(\bar{u})$ is positive almost everywhere and satisfies

$$\int_1^\infty \cdots \int_1^\infty \lambda(\bar{u}) d\bar{u} < \infty,$$

$$\int_0^1 \cdots \int_0^1 |\bar{u}|\lambda(\bar{u})d\bar{u} < \infty,$$

**Theorem 2.1.** *Under these assumptions, the functional $C^P(\xi)$ is a characteristic functional. That is*

*i) $C^P(\xi)$ is continuous in $\xi$, with a suitable choice of a nuclear space $E$,*

*ii) $C^P(0) = 1$ and*

*iii) $C^P(\xi)$ is positive definite.*

Applying the Bochner-Minlos theorem (see e.g. [5]), we can see that there exists a probability measure $\nu^P$ on $E^*$ such that

$$C^P(\xi) = \int_{E^*} e^{i\langle x, \xi \rangle} d\nu^P(x). \tag{3}$$

We introduce a notation $P'(\bar{u}, \lambda(\bar{u}))$ or write it simply $P'(\bar{u})$. We understand that $\nu^P$-almost all $x \in E^*, x = x(\bar{u}), \bar{u} \in R_+^d$ is a sample function of a generalized stochastic process $P'(\bar{u}), \bar{u} \in R_+^d$.

**Theorem 2.2.** *The $\{P'(\bar{u}), \bar{u} \in R_+^d\}$ has independent values at each $\bar{u}$ (in the sense of Gel'fand).*

Proof. Suppose $\xi_1$ and $\xi_2$ have disjoint support; let them be denoted by $A_1$ and $A_2$. Then, the integral in the expression (2.1) for $C^P(\xi_1 + \xi_2)$ can be expressed as a sum

$$\int_{A_1} (e^{i\langle \bar{u}, \xi_1(\bar{u}) \rangle} - 1)dn(\bar{u}) + \int_{A_2} (e^{i\langle \bar{u}, \xi_2(\bar{u}) \rangle} - 1)dn(\bar{u}).$$

This proves the equality

$$C^P(\xi_1 + \xi_2) = C^P(\xi_1)C^P(\xi_2).$$

This shows that $P'(\bar{u})$ has independent value at every point $\bar{u}$. Also follow the atomic property and others.

The bilinear form $\langle P', \xi \rangle, \xi \in E$, $E$ being a nuclear space, is a random variable with mean

$$\int (\bar{u}, \xi(\bar{u}))\lambda(\bar{u})d\bar{u}.$$

The variance of $\langle P', \xi \rangle$ is

$$\int (\bar{u}, \xi(\bar{u}))^2 \lambda(\bar{u}) d\bar{u}.$$

Hence $\langle P', \xi \rangle$ extends to $\langle P', f \rangle$, for $f$ such that $(\bar{u}, f(\bar{u})) \in L^2((0, \infty)^d, \lambda(\bar{u}) d\bar{u})$.

If $(\bar{u}, f(\bar{u}))$ and $(\bar{u}, g(\bar{u}))$ are orthogonal in $L^2(\lambda(\bar{u}) d\bar{u})$, then $\langle P', f \rangle$ and $\langle P', g \rangle$ are uncorrelated.

Thus, we can form a random measure and hence, we can define the space $\mathcal{H}_1(P)$ like $\mathcal{H}_1$ in the case of Gaussian white noise.

The space $H_1(P)$ can also be extended to a space $H_1^{(-1)}(P)$ of generalized linear functionals of $P'(u)$'s. Note that there we can give an *identity* to $P'(u)$ for any $u$.

Our conclusion is that single noise $P'(u)$ with the parameter $u$ can be found in $\mathcal{H}_1^{(-1)}(P)$.

**Remark 2.1.** The case where the parameter $\bar{u}$ runs through the negative interval $(-\infty, 0)^d$ can be discussed in the similar manner.

We will focus our attention to the intensity $\lambda(u)$ in the next section.

## 3. Lévy measure

Before we come to the next problem concerned with the Lévy measure we wish to make a note on the type of Poisson distribution with the following proposition.

**Proposition 3.1.** *Let $X(\lambda)$ be subject to a Poisson distribution with intensity $\lambda$. Then the probability distribution of $X(\lambda)$ and $aX(\lambda), a > 0$ are the same type, while $X(\lambda)$ and $X(\lambda')$ have distribution of different type for $\lambda \neq \lambda'$).*

Proof comes from the formulas of characteristic functions.

Consider the isotropic dilation of $\bar{u} \in R_+^d$ for $d \geq n$,

$$\bar{u} \to a\bar{u},$$

where $a$ is a positive constant.

The $P'(\bar{u})$ is said to be self-similar if for any $a > 0$ there exists $d(a)$ such that

$$P'(a\bar{u}) \sim d(a)P'(\bar{u}),$$

where $\sim$ means the same distribution. It can be also seen that

$$d(ab) = d(a)d(b).$$

**Theorem 3.1.** *If $P'(\bar{u})$ is self similar, then*

$$\lambda(\bar{u}) = \frac{c}{|\bar{u}|^{1+\alpha}},$$

*where $c$ is a positive constant and $d - 1 < \alpha < d$.*

Proof is easy and is omitted.

The above case corresponds to the stable noise and $\alpha$ is the exponent of the noise.

## 4. Invariance

Like white noise, we can see that the noise $P'(\bar{u})$ is invariant with respect to dilation, rotation and reflection.

1) Isotropic dilation

Since the parameter $\bar{u}$ runs through $R_+^d$, the dilation is the basic transformation.

The transformation

$$g_a : \bar{u} \mapsto a\bar{u}$$

is an isotropic dilation. Let

$$\mathcal{G} = \{g_a, g_a\bar{u} = a\bar{u}\}.$$

There, we have

$$g_a g_b = g_{a+b},$$

so $\mathcal{G}$ is the group of dilations.

**Remark 4.1.**

The group $G$ characterizes the class of the stable noises.

2) Rotation

Use the polar coordinate such that $\bar{u} = (r, \bar{\phi})$.
Let $g_\theta$ be a rotation, defined by

$$g_\theta \bar{u} = (r, (\bar{\phi} + \bar{\theta})\mathrm{mod}\frac{\pi}{2}),$$

it is seen that $|\bar{u}|$ is invariant under $g_\theta$. Thus, Lévy measure of stable process is invariant under $g_\theta$ and the group $\mathcal{G} = \{g_\theta\}$ is a rotation group acting on $R_+^d$.

3) Reflection with respect to unit sphere $S^d$

Take the reflection

$$\bar{u} \longmapsto \frac{\bar{u}}{|\bar{u}|^2},$$

then

$$\lambda(\bar{u}) = \frac{1}{|\bar{u}|^{\alpha+1}}, \quad d-1 < \alpha < d$$

for a stable process changes to

$$\lambda(\frac{\bar{u}}{|\bar{u}|^2}) = |\bar{u}|^{\alpha+1}.$$

Thus, we take the reflection with extra multiplier :

$$\lambda(\bar{u}) = \frac{1}{|\bar{u}|^{\alpha+1}} \longmapsto dn(\frac{\bar{u}}{|\bar{u}|^2}) \cdot \frac{1}{|\bar{u}|^{2d+1}} = \frac{1}{|\bar{u}|^{2d-\alpha}}.$$

Then, the exponent $\alpha$ changes to $2d - \alpha - 1$ and the range of exponent changes as

$$d-1 < \alpha < d \quad \longleftrightarrow \quad d > 2d - \alpha - 1 > d - 1.$$

**Theorem 4.1.** *The following facts hold.*

*1) A noise of stable type $P'_\alpha(\bar{u})$ is invariant under isotropic dilation as well as rotation (mod $\frac{\pi}{2}$), for any $\alpha$.*

*2) The family $\{P'_\alpha(\bar{u}), 0 < \alpha < 1\}$ is invariant under the reflection*

$$\bar{u} \longmapsto \frac{\bar{u}}{|\bar{u}|^2} \cdot \frac{1}{|\bar{u}|^{2d+1}},$$

*where $\alpha \to 2d - \alpha - 1$.*

*For $d = 1$, $\quad \alpha \to 1 - \alpha$.*

### Generators

For the isotropic dilation $\bar{u} \mapsto a\bar{u}$, a generator is

$$\tau = \frac{d}{dt}\bigg|_{t=0} g_t = (u, \nabla), \quad (a = e^t).$$

For the rotation, a generator is

$$\kappa_{ij} = u_i \frac{\partial}{\partial u_j} - u_j \frac{\partial}{\partial u_i}.$$

## 5. Construction of $\mathcal{H}^{(-1)}(P)$ spanned by generalized linear functionals of $P'(u)$

We will construct $\mathcal{H}^{(-1)}(P)$ for $u \in R^1_+$, since we can do the same thing for $\bar{u} \in R^d_+$.

Let $\mathcal{H}_1$ be the space spanned by

$$\{\langle P', f \rangle = \int f(u) P'^2(R^1_+)\}.$$

We have known that the variance of $\langle P', f \rangle$ is

$$\int_{R^1_+} (uf(u))^2 \lambda(u) du.$$

Thus, $uf(u)$ must be in $L^2(R^1_+, \lambda(u)du)$

Assume that $\lambda(u) = 0$ for $0 < u < \epsilon$ and $\lambda(u)$ is bounded.

Then the space $L^2(R^1_+, \lambda(u)du)$ can be considered like as $L^2([\epsilon, \infty), Leb)$

Thus, we have an isomorphism.

$$\mathcal{H}_1 \cong L^2(R^1_+, \lambda(u)du) \tag{4}$$

298

We now restrict the above isomorphism to

$$\mathcal{H}_1^{(1)} \cong K_+^{(1)}$$

by taking a strong topology on $L^2(R_+^1, \lambda(u)du)$.

The space $\mathcal{H}_1^{(1)}$ is defined so as to hold the above isomorphism.

We are now able to use the same technique as in white noise case we can extend $L^2(R_+^1, \lambda(u)du)$ to the Sobolev space $K_+^{(-1)}$ of order -1 which is the dual space of the Sobolev space $K_+^{(1)}$ of order 1.

We have the inclusion

$$K_+^{(1)} \subset L^2(R_+^1) \subset K_+^{(-1)}, \tag{5}$$

where the mapping from left to right is continuous injection of Hilbert Schmidt type.

The dual space $H_1^{(-1)}$ of $H_1^{(1)}$ can be defined in parallel with (5) and then we have

$$\mathcal{H}_1^{(-1)} \subset \mathcal{H}_1 \subset \mathcal{H}_1^{(-1)}.$$

Thus we now have constructed the space $\mathcal{H}_1^{(-1)}(P) \equiv \mathcal{H}_1^{(-1)}$ of generalized linear functionals of the noise $P'(u)$, which gives the identity of $P'(u)$.

## 6. The jump finding problem

Actually, our study has been motivated by the paper [15]. We can now speak of the decomposition of a compound Poisson process, that suggests the jump finding problem.

Suppose we are given a compound Poisson process. Assume that the Lévy measure $dn(u)$ satisfies the conditions stated before. Then, we can form the space $\mathcal{H}_1^{(-1)}(P)$ where $P'(u)$ lives. Hence, we have an infinitesimal random variable at $u$, tacitly assumed that $t = 1$.

We can, therefore, say that in the world of generalized processes, we can succeed in jump finding. In reality, we can take a space of test functionals and consider the canonical bilinear form connecting $\mathcal{H}_1^{(1)}(P)$ and $\mathcal{H}_1^{(-1)}(P)$ which defines the $P'(u)$.

The basic idea is that $P'(u)$ has identity in $\mathcal{H}_1^{(-1)}(P)$. Then we come to the method of identification of $P'(u)$ for each $u$. We recall the definition of the space $\mathcal{H}_1^{(-1)}(P)$; namely it is the dual space of $\mathcal{H}_1^{(1)}(P)$ which is separable. This means that each $P'(u)$ can be identified by at most countably many members in the space $\mathcal{H}_1^{(1)}(P)$.

Each member of $\mathcal{H}_1^{(1)}(P)$ is concrete and visualized. This means that $P'(u)$ can be obtained with the help of members of $\mathcal{H}_1^{(1)}(P)$; i.e. solving the jump finding problem.

## References


1. I. M. Gelgand and N. Ya Vilenkin, Generalized functios Vol.4. Acdemic Press 1964.
2. T. Hida, Stationary stochastic processes. Princeton Univ. Press, 1970.
3. T. Hida. Analysis of Brownian functionals. Carleton Math. Notes. no.13,1975. Carleton Univ.
4. T. Hida, White noise approach to some statistical problems. (to appear in Professor Louis Chen's Volume.)
5. T. Hida and Si Si, Lectures on white noise functionals. World Sci. Pub. Co. 2008.
6. T. Hida. Si Si and Win Win Htay, A noise of new type and its applications. Ricerche di Matematica Vol. 55, No.1. 2011.
7. P. Lévy, Notice sur les Travaux Scientifiques. Herman & $C^{ie}$,1935.
8. P. Lévy, Sur les exponentialles de polynômes et sur l'arithmétique des products de lois de Poisson. Ann. Ecole Normale Sup. 54 (1937), 231-292. (in particular Chap. 2).
9. P. Lévy, Théorie de l'addition des variables aléatoires. Gauthier-Villars, 1938. 2éme éd. 1954.
10. P. Lévy, Problèmes concrets d'analyse fonctionnelle. Gauthier-Villars, 1951.
11. B. Mandelbrot, Les objects fractals. 2ém. éd. Flammarion.1984.
12. Sidney I. Resnick, Heavy-tail phenomena. Probabilistic and Statistical modeling. Springer, 2007.
13. Si Si, Multiple Markov generalized Gaussian processes and their dualities. Infinite Dimensional Analysis, Quantum Probability and Related Topics. 13 (2010) 99-110.
14. Si Si, Introduction to Hida distributions. World Sci. Pub. Co. to appear.
15. Si Si, A.H. Tsoi and Win Win Htay, Jump finding of a stable process. Quantum Information. vol. V, World Sci. Pub. Co. (2000), 193-201.
16. Si Si, Fractional power distributions and application to statistics, June 2010, Lecture at Math. Stats Dept., National Univ. of Singapore.

Quantum Bio-Informatics V

pp. 301–313


# SCHRODINGER TYPE SEMIGROUPS VIA FEYNMAN FORMULAE AND ALL THAT


O.G.SMOLYANOV

*Lomonosov Moscow State University*
*GSP-1, Leninskie Gory, Moscow, 119991, Russia*
*E-mail: smolyanov@yandex.ru*


The Schrödinger semigroup is a semigroup, of bounded operators in a complex or real Banach space $E$ of some functions, whose generator is a pseudodifferential operator $\hat{H}$ the symbol of which is a (classical Hamilton) function $H$ on the phase space $Q \times P$ of a classical Hamiltonian system. The Schrödinger group is a semigroup, of bounded operators in a similar space that in this case is assumed to be complex one, whose generator has the form $i\hat{H}$ where $\hat{H}$ is as above.

Of course if the domain of functions from $E$ is a part of an Euclidean space or of a Riemannian manifold then to define the operator $\hat{H}$ it is necessary not only to choose what type of the symbol of $\hat{H}$ is the function $H$ (see below) but also the boundary conditions. Usually one assumes that $E$ is a Hilbert space and that $\hat{H}$ is a selfadjoint operator. But we use the formulated above more general definition.

The Feynman formula is a representation of either a Schrödinger semigroup or a Schrödinger group by limits of some multiple integrals, over Cartesian products of a space $E$, whose integrands are elementary functions. The Feynman-Kac formula is a representation of the same semigroup or group by an integral over a space of functions of real variable taking values in the same space[a]$E$.

Hence the multiple integrals from Feynman's formulae approximate infinite-dimensional integrals from the Feynman-Kac formulae. Let us also

---

[a]If $E$ is an infinite-dimensional space of functions, for instant, on $\mathbb{R}^n$, taking values in a space $E_1$, then any space of functions of the real variable taking values in $E$ can be identified with a space of functions on a domain of $\mathbb{R}^{n+1}$ taking values in $E_1$ and hence in this case one can say that the Feynman-Kac formula gives a representations of the group or semigroup by an integral over a space of functions on a subset of $\mathbb{R}^{n+1}$ (taking values in $E_1$).

notice that in the case of the Schrödinger semigroup the infinite-dimensional integrals in the Feynman-Kac formulae are integrals with respect to some $\sigma$-additive measures of the diffusion type; in the case of the Schrödinger group one needs integrals with respect to the so-called Feynman pseudomeasures; actually the latter integrals are functionals on some spaces of functions.

If $E$ coincides with the domain $\Omega$ of functions from $E$ and $\Omega \subset Q$ then the corresponding Feynman formulae and Feynman-Kac formulae are called Lagrangian ones; if, under the same assumptions about $\Omega$, $E = \Omega \times P$ where $P$ is the dual to $Q$ then the Feynman-Kac and Feynman formulae are called Hamiltonian ones. This terminology can be motivated as follows. In the Lagrangian Feynman-Kac formula the integrand is generated by the action in the Lagrangian form; in the Hamiltonian Feynman-Kac formula the integrand is generated by the action in the Hamiltonian form. If $E = Q \times P$ is equipped with the natural complex structure and the functions from $E$ are holomorphic functions on $E$ then the corresponding Feynman and Feynman-Kac formulae are called holomorphic ones, or formulae in holomorphic representation (cf. [19], [20]).

It is worth mentioning that both the Feynman path integrals (which is used in the Feynman-Kac formula) and the Feynman formulae look as almost universal tools in mathematical physics and related areas. Starting from applications of the Feynman path integrals to constructing quasi-classical asymptotic and to the theory of perturbations in quantum mechanics now one can apply such integrals and Feynman formulae in the theory of Hamilton-Dirac systems, to get representations of regularized traces of operators, to investigate quantum anomalies, etc. Moreover now by Feynman formula one calls not only a representation (by suitable multiple integrals) of a Schrödinger group or semigroup but also a similar representation of any semigroup, in particular of Feller semigroups in the theory of stochastic processes.

Usually the Feynman-Kac formulae have more conceptual meaning while the Feynman formulae can be considered in more extent as giving methods of some explicit calculations (both of solutions of equations, some transition probabilities, functional integrals, measures and pseudo-measures on functional spaces). Nevertheless such calculations may have also some conceptual meaning; typical examples are Feynman formulae for initial-boundary values problems, for differential equations on manifolds and for Feller processes.

Finally, the important feature of the Feynman formulae is that quite often they are (almost) independent of the choice of the number field; in particular they can be constructed for semigroups and groups acting in spaces over p-adic numbers [8].

In the paper we consider some recent progress in applications of Feynman formulae and functional integrals.

Namely, we discuss the following three class of problems.

1. Representations, by some Lagrangian type Feynman formulae, of the semigroups describing diffusion, of particles with continuously position-dependent mass, in domains of $R^n$ (quite similar formulae are valid for some Schrödinger groups);

2. Representations, by some similar formulae, of the semigroups describing diffusion, of particles with piecewise continuously position-dependent mass, in $R^n$ (again similar formulae are valid for some Schrödinger groups);

3. Representations, by functional integrals from the Lagrangian and Hamiltonian Feynman-Kac formulae, of regularized traces of some differential operators (corresponding functional integrals are also called Lagrangian, resp. Hamiltonian ones);

In this paper we want to stress main ideas and some times we do not formulate complete technical assumptions.

Some results of the paper have been obtained in collaboration with V.A.Sadovnichii, E.T.Shavgulidze and D.S.Tolstyga.

## 1. Lagrangian Feynman formulae for semigroups generated by initial-boundary values problems.

In this section we discuss the Feynman formulae for diffusion, in the complement to an open convex domain $\Omega$ of $R^2$, of particles with continuously position dependent mass; quite similar results are valid for open convex domains of $R^n$.

We shall use, both in this and in the next sections, two Chernoff type theorems, which we remind now.

Let, for any LCS $X$, the symbol $L(X)$ denote the space of all continuous linear operators in $X$, $\tau$ — the strong operator topology in $L(X)$ and $I_X$ — the identity operator in $X$. If $X$ is a Banach space then $\|\cdot\|$ is the operator norm on $L(X)$, for any linear (in general unbounded) operator $A$ in $X$ the symbol $D(A)$ denotes the domain of $A$ and for any function $F : [0, \infty) \to L(X)$ the symbol $F'(0)$ denotes the right strong derivative of $F$ at 0. The latter means that $D(F'(0))$ is the collection of all $x \in X$ for which there exists $\lim_{t \to 0, t > 0} \frac{F(t) - F(0)}{t} x$ and that $F'(0)x$ is equal to this limit for each $x \in D(F'(0))$. Let $X$ be a Banach space, $\delta > 0$, $F : [0, \delta) \to (L(X), \tau)$

be a continuous function such that $F(0) = I_X$, $D$ be a dense in $X$ linear subspace of $D(F'(0))$.

**Theorem 1.1.** *(Chernoff,* [3]*). Let* $\|F(t)\| \leq \exp(at)$ *for some* $a \in R$ *and all* $t \in (0, \delta)$, *and the restriction of* $F'(0)$ *to* $D$ *be a closable operator whose closure is denoted by* $C$. *If* $C$ *is the generator of a continuous semigroup* $\exp(tC)$ *in* $(L(X), \tau)$, *then* $(F(t/n))^n x \to \exp(tC)x$ *as* $n \to \infty$ *in* $(L(X), \tau)$ *uniformly with respect to* $t \in [0, S]$ *for each* $S > 0$ *and each* $x \in X$.

**Theorem 1.2.** *(Engel-Nagel,* [5]*). Let there exist* $M \geq 1$ *and* $a \in R$ *such that* $\|(F(t))^k\| \leq Me^{kat}$ *for all* $k \in N$ *and* $t \in (0, \delta)$. *If* $(\lambda_0 I_X - F'(0))D$ *is dense in* $X$ *for some* $\lambda_0 > a$ *then the closure* $C$ *of* $F'(0)$ *generates a strongly continuous semigroup* $e^{tC}$ *and* $(F(t/n))^n x \to \exp(tC)x$ *as* $n \to \infty$ *in* $(L(X), \tau)$ *uniformly with respect to* $t \in [0, S]$ *for each* $S > 0$ *and each* $x \in X$.

In what follows in this section we assume that $Q = P = R^2$; moreover to simplify the formulae, we assume that $\Omega = \{(x, z) \in Q(\equiv R^2) : x < 0\}$. The general case can be treated in a completely similar way. Let $\Omega_0 = R^2 \backslash V$.

We will discuss some Feynman formulae for Schrödinger semigroups $e^{t\hat{H}}$ both in $L_1(\Omega_0)$ and in $L_2(\Omega_0)$; here $\hat{H}$ is a differential operator whose symbol $H : \Omega_0 \times P \to R$ is defined by $H(q, p) = \frac{1}{2}g(q)(p, p)$; $(p, p)$ is the scalar product in $R^2 (\equiv P)$ and $g$ is a bounded positive infinitely many times differentiable function two first derivatives of which are bounded on $\Omega_0$; the symbol $\hat{H}$ may be equipped with additional labels. Actually those formulae are the same in both cases; only in the first case $(L_1(\Omega_0))$ the proof uses Theorem 1.1 and in the second case $(L_2(\Omega_0))$ — Theorem 1.2. The same is true for Schrödinger groups. One could mention that the situation is in a sense similar to the situation, in the group representation theory, which was once noticed by Gelfand at his famous seminar at Moscow State University: identical formulae arise in different spaces.

The Hamilton function $H$ describes a particle with a position dependent mass; the corresponding Schrödinger semigroup describes the diffusion of the particle in $\Omega_0$ and in particular transition probabilities of the stochastic process governed by the associated Ito stochastic differential equation. One could also consider a more general Hamilton function defined by $H(q, p) = (G(q)p, p)$ where $G$ is a function whose values are positive operators in $P$; then one could speak about an anisotropic position dependent mass. To consider the Schrödinger group it is necessary (at least natural) to symmetrize the corresponding differential operator; it is worth stressing that the symmetrized operator defined by the given symbol $H$ is not unique. (cf.[11]). On the other hand the Schrödinger semigroup in $L_2(\Omega_0)$ generated by the symmetrized operator has only a restricted sense as a so called Euclidian version of the Schrödinger group. The generated by the same operator Schrödinger semigroup in the space $L_1(\Omega_0)$ has a more direct sense because this semigroup describes a dynamics of probability densities while the Schrödinger group has a more natural meaning in the space $L_2(\Omega_0)$ because it describes an evolution of densities of the probability amplitudes. Let us also notice that to define differential operators generated by Schrödinger semigroups and groups in both $L_1(\Omega_0)$ and $L_2(\Omega_0)$ one need to fix boundary conditions. We will use boundary conditions defined by the following way: $\alpha(z)\frac{\partial f}{\partial x}(0, z) = f(0, z)$ where $\alpha : R^1 \to (-\infty, +\infty]$ is a continuous function and $f$ is a function from an essential domain, of the corresponding operator, which is contained in $L_1(\Omega_0) \cap L_2(\Omega_0)$ (the elements of $\Omega_0$ are denoted by symbols (x,z) where $x \in [0, \infty)$, $z \in R^1$). We do not intend to give explicit definitions of those domains here.

Now for any $\alpha$ defining the boundary conditions we will construct the Feynman formulae for the Schrödinger semigroup $e^{t\hat{H}_\alpha}$ in both $L_1(\Omega_0)$ and $L_2(\Omega_0)$ where $(\hat{H}_\alpha\varphi)(q) = \frac{1}{2}g(q)(\Delta\varphi)(q)$ for each $\varphi$ from the domain $D(\hat{H}_\alpha)$ of the operator $\hat{H}_\alpha$. So $\hat{H}_\alpha$ is a non symmetrized differential operator with the symbol $H$. Let us notice that if an operator $\hat{H}$ is symmetrized then each of the just given boundary conditions defines the selfadjoint extension of the operator. In the case of the heat type equation the same boundary conditions define the so called Newton's law of radiation. As it has already been mentioned in both those spaces the Feynman formulae are identical.

To get the Feynman formulae we need to define the function $F$ from Theorems 1.1 and 1.2.

Let $F_\alpha(t) = F_{2\alpha}(t)F_{1\alpha}(t)$, where each of factors is defined as follows. Let $X = L_1(\Omega_0)$ (resp. $X = L_2(\Omega_0)$) and, for any $t > 0$, the mapping $F_{1\alpha}(t) : X \to L_1(R^2)$ (resp. $F_{1\alpha}(t) : X \to L_2(R^2)$ ) be defined by

$$(F_{1\alpha}(t)f)(x, z) = \frac{\frac{1}{2\sqrt[3]{t}}\int_0^{2\sqrt[3]{t}} f(x, y)dy(1 + \alpha(z)x)}{1 + \alpha(z)\sqrt[3]{t}}\psi_t(x) + f(x)\varphi_t(x),$$

where $\varphi_t, \psi_t \colon R^1 \to [0, 1]$ are smooth functions, such that $\psi(x) = 0$ if $x \leq -3\sqrt[3]{t}$ or if $x \geq 2\sqrt[3]{t} + t^3$ and $\psi(x) = 1$ if $x \in [-2\sqrt[3]{t}, 2\sqrt[3]{t}]$; $\varphi_t(x) = 1$ if $x \geq 2\sqrt[3]{t} + t^3$ and $\varphi_t(x) = 0$ if $x \leq 2\sqrt[3]{t}$; $\varphi_t(x) + \psi_t(x) = 1$ for $x \geq -2\sqrt[3]{t}$. If $\alpha(z) = \infty$ then by definition $\frac{1+\alpha(z)q}{1+\alpha(z)r} = \frac{q}{r}$ for any $q, r$.

Let the mapping $F_{2\alpha}(t) : L_1(R^2) \to X$ (resp. $L_2(R^2) \to X$) be defined by

$$(F_{2\alpha}(t)f)(q) = \frac{1}{\sqrt{2\pi t g(q)}} \int_{\mathbb{R}^2} e^{-\frac{(q-q_1)^2}{2g(q)t}} f(q_1) dq_1$$

(actually the mappings $F_{2\alpha}(t)$ do not depend on $\alpha$).

**Theorem 1.3.** *For any* $\alpha : R^1 \to (-\infty, +\infty]$, $t > 0$ *and* $f \in X$ *the following Lagrangian Feynman formula is valid:*

$$e^{t\hat{H}_\alpha} f = \lim_{n\to\infty} (F_\alpha(\frac{t}{n}))^n f$$

*in* $X$.

If $X = L_1(\Omega_0)$ (resp., if $X = L_2(\Omega_0)$) then to prove the Lagrangian Feynman formula it is sufficient to check that assumptions of Theorem 1.1 (resp., of Theorem 1.2) are fulfilled.

To construct Feynman formulae for the Schrödinger semigroups generated by symmetrized differential operators with symbol $H$ it is sufficient to construct the Feynman formulae for Schrödinger semigroups generated by general (elliptic) differential operators of the second order because any symmetrized differential operator with the symbol $H$ is the sum of the defined above operator $\hat{H}$ and two additional terms with the derivative of the unknown function and with this function itself. To get such Feynman formulae one can use the representation $F_\alpha(t) = F_{4\alpha}(t)F_{3\alpha}(t)F_{2\alpha}(t)F_{1\alpha}(t)$ where the factors with indexes 3 and 4 are related to coefficients before the terms, in the elliptic operator, containing the unknown function and its derivative of the first order (cf. next section where one discusses the diffusions in $R^2$ governed by selfadjoint operators with the same symbol). To get after that the Feynman formulae for Feynman groups (generated by symmetrized operators) it is sufficient to apply an analytical continuations to the just discussed Feynman formulae for semigroups. It is curious that the Feynman formulae for semigroups generated by symmetrized operators (with the same symbol $H$) have more complicated structure than the corresponding formulae for semigroups generated by the (nonsymmetrized) operator $\hat{H}$. The reason is that the backward Kolmogorov equation for the Ito stochastic equation describing the diffusion with the position dependent diffusion coefficient (without drift or creation-annihilation new particles) contains just the (nonsymmetrized) operator $\hat{H}$.

## 2. Lagrangian Feynman formulae for Cauchy problems for equations describing multidimensional dynamics of particle with piecewise continuously position-dependent mass.

In this section we discuss the Feynman formulae for diffusion and quantum dynamics, in $R^2$, of particles with piecewise continuously position dependent mass[b]. So the difference between problems which are considered in the preceding and in the current section is that now we consider processes in the whole space but, on the other hand, we admit that the dependence of the mass on the position need not be everywhere continuous.

In this section we again assume that $Q = P = R^2$ and that $\hat{H}$ is a differential operator whose symbol $H : Q \times P \to R$ is defined by $H(q, p) = \frac{1}{2} g(q)(p, p)$ where $g$ is a bounded positive function. In addition we assume that the restriction of $g$ to $W = \{(x, z) \in R^2 : x \neq 0\}$ is an infinitely many times differentiable function three first derivatives of which are bounded on $W$; as in the preceding section the symbol $H$ may be equipped with additional indexes. We consider this simple set of discontinuity of $g$ only for brevity but by the same method one could treat and more general sets of discontinuity.

We will discuss some Feynman formulae for Schrödinger semigroups $e^{t\hat{H}}$ both in $L_1(Q)$ and in $L_2(Q)$ under the assumption that $\hat{H}$ is a selfadjoint operator in $L_2(Q)$ (whose essential domain is contained in $L_1(Q) \cap L_2(Q)$); again those formulae are the same in both cases; only in the first case the proof uses Theorem 1.1 and in the second case Theorem 1.2.

We will consider selfadjoint operators that are similar to the operators defined in [11]. Each of them is determined by a triplet (a,b,T) where $(a, b) \in [0, \frac{1}{2}] \times [0, \frac{1}{2}]$ and $T : R^1 \to L(C^2)$ the values of $T$ being invertible operators. The defined by this way operator is denoted by $\hat{H}_{(a,b,T)}$ and acts (on functions from a natural domain) by (below $(x, z) \equiv q \in Q \equiv R^2$):
$f \mapsto \frac{1}{2} g^a \frac{\partial}{\partial x}(g^{1-2a} \frac{\partial}{\partial x}(g^a f)) + \frac{1}{2} g^b \frac{\partial}{\partial z}(g^{1-2b} \frac{\partial}{\partial z}(g^b f))$. One assumes that any function $h$ from the domain of $\hat{H}_{\alpha,T}$ satisfies the following conditions: if $h^+(z) = (h^+(0, z), h'^+(0, z))$ and $h^-(z) = (h^-(0, z), h'^-(0, z))$, then $h^+(z) = T(z)h^-(z)$; here $h^+(0, z), h'^+(0, z)$ (resp., $h^-(0, z), h'^-(0, z)$) are right limits (resp., left limits) of $h$ and of $h'$ at $q = (0, z)$.

---

[b]Similar results are valid for $\mathbb{R}^n$ as well.

The definition of $\hat{H}_{(a,b,T)}$ implies that

$$(\hat{H}_{(a,b,T)}(f)(q) = \frac{1}{2}g(q)\Delta f(q) + (l(q), f'(q)) + V(q)f(q),$$

where the functions $l : R^2 \to R^2$ and $V : R^2 \to R^1$ depend on $a, b$ and $g$.

Let $X = L_1(R^2)$ (resp., $X = L_2(R^2)$), $T(z) = (t_{ij}(z))$ and $t > 0$. For any triplet $(a, b, T)$ let $F_{(a,b,T)}(t) = F_8(t)F_7(t)F_6(t)F_5(t)$, where the factors (which of course may depend on the triplet $(a, b, T)$) are defined by:

$F_5(t) : X \to X \oplus X$, $F_5(t)(f) = (h, k) \in X \oplus X$, where

$$h(q) \equiv h(x, z) = \begin{cases} (a_+(z)x + z_+(z))\psi_t(x) + f(q)\varphi_t(x), & \text{if } x > 0; \\ 0, & \text{if } x < 0; \end{cases}$$

$$k(q) \equiv k(x, z) = \begin{cases} 0, & \text{if } x > 0; \\ (a_-(z)x + z_-(z))\psi_t(x) + \varphi_t(-x)f(q), & \text{if } x < 0; \end{cases}$$

the functions $\varphi_t$ and $\psi_t$ are defined in the preceding section and $a_+(z), a_-(z), z_+(z), z_-(z)$ are determined by:

$$\begin{cases} a_+(z)t + z_+(z) = \dfrac{1}{2\sqrt[3]{t}} \displaystyle\int_0^{2\sqrt[3]{t}} f(x, z)dx, \\[2ex] -a_-(z)t + z_+(z) = \dfrac{1}{2\sqrt[3]{t}} \displaystyle\int_{-2\sqrt[3]{t}}^{0} f(x, z)dx, \\[2ex] z_+(z) = t_{11}(z)z_-(z) + t_{12}(z)a_-(z), \\[1ex] a_+(z) = t_{21}z_-(z) + t_{22}(z)a_-(z). \end{cases}$$

Let $F_6(t) : X \oplus X \to X \oplus X$, $F_6(t)((h, k))(q) = (h(q + tl(q)), k(q + tl(q)))$; $F_7(t) : X \oplus X \to X$, $(F_7(t)(h, k))(q) \equiv$

$$(F_7(t)(h, k))(x, z) = \frac{1}{\sqrt{2\pi tg(q)}} \int_{-\infty}^{\infty} exp(-\frac{(q - z)^2}{2tg(q)})h(z)dz,$$

if $x > 0$;

$$(F_7(t)(h, k))(x, z) = \frac{1}{\sqrt{2\pi tg(q)}} \int_{-\infty}^{\infty} exp(-\frac{(q - z)^2}{2tg(q)})k(z)dz,$$

if $x < 0$; $F_8(t) : X \to X$, $F_8(t)(f)(q) = e^{tV(q)}f(q)$.

**Theorem 2.1.** *For any triplet $(a, b, T)$, $t > 0$ and $f \in X$ the following Lagrangian Feynman formula is valid:* $e^{t\hat{H}_{(a,b,T)}}f = \lim_{n\to\infty}(F_{(a,b,T)}(\frac{t}{n}))^n f$ *in $X$.*

For $X = L_1(R^2)$ the theorem is implied by Theorem 1.1; for $X = L_2(R^2))$ this theorem follows from Theorem 1.2.

**Remark 2.1.** The Feynman formulae approximate the transition probabilities from the corresponding Feynman-Kac formulae.

## 3. Functional integral representations of regularized traces of some differential operators.

One calls a regularized trace (more precisely, $B$-regularized trace) of an operator $A$, which is not a trace class operator, the usual trace of the operator $A - B$, where $B$ is another operator. This definition is usually applied to differential operators that are even unbounded in natural spaces.

Let $\Omega$ be a domain in $Q$, $\dim Q = d$ and $H : \Omega \times P \to R^1$ be defined by $H(q,p) = \frac{p^2}{2} + V(q)$ and let $\hat{H}$ be the selfadjoint operator in $L_2(\Omega)$ whose symbol is $H$ and whose essential domain is the collection of all smooth functions vanishing at the boundary of $\Omega$; so $\hat{H} = -\frac{\Delta}{2} + V$.

**Proposition 3.1.** *If $K_t$ is the integral kernel of the operator $e^{-t\hat{H}}$ in $L_2(\Omega)$ then $K_t(q_1, q_2) = \int_{C_{q_1,q_2}([0,t],\Omega)} \exp\{-\int_0^t V(\xi(\tau))d\tau\} w_{q_1,q_2}^{t,\Omega}(d\xi)$ where $w_{q_1,q_2}^{t,\Omega}$ is the measure associated with the restriction, of the Wiener measure on $C_{q_1}((0,t),Q)$, the collection of all continuous functions on $[0,t]$ taking values in $Q$ , to $C_{q_1}((0,t),\Omega)$.*

Let $a \geq 0$, $f$ be a function on $\Omega$, $H_{\alpha,f}$ be the function on $\Omega \times P$, determined by $H_{a,f}(q,p) = \frac{p^2}{2} + V + af$ and let $\hat{H}_{\alpha,f}$ be the operator definition of which is similar to the given above definition of $\hat{H}$ ($\hat{H}_{0,f} = \hat{H}$).

We assume that the spectra of $\hat{H}_{\alpha,f}$ and $\hat{H}$ are discrete and positive; the corresponding eigenvalues of $\hat{H}_{\alpha,f}$ and $\hat{H}$ are denoted by $\lambda_n(a,f)$ and $\lambda_n$ ($= \lambda_n(0,f)$), $\lambda_1 \leq \lambda_2 \leq ...; \lambda_1(a,f) \leq \lambda_2(a,f) \leq ....$ Instead of the symbols $\hat{H}_{1,f}$ and $\lambda_n(1,f)$ we shall write $\hat{H}_f$ and $\lambda_n(f)$. It is known that there exists a number $c_f$ such that the series , $\sum_{n=1}^{\infty}(\lambda_n(f) - \lambda_n - c_f)$ converges; the sum of the series is called a $f$-regularized trace of $\hat{H}$ (the $f$-regularized trace is a particular case of the $B$-regularized trace).

**Theorem 3.1.** *If $d = 1$, then*

$$\sum_{n=1}^{\infty} (\lambda_n(f) - \lambda_n - c_f) e^{-\lambda_n t}$$

$$= \int_{\Omega} \int_{C_{q,q}([0,t],\Omega)} \left[ \int_0^1 f(\sqrt{t}\gamma(\tau) + q) d\tau - c_f \right]$$

$$\cdot \exp\left\{ -t \int_0^1 V(\sqrt{t}\gamma(\tau) + q) d\tau \right\} w_{q,q}^{t,\Omega}(d\gamma) dq$$

This is implied by Proposition 1 via the identity

$$\frac{d}{d\alpha}(\operatorname{tr} e^{-t\hat{H}_{\alpha,f}})|_{\alpha=0} = t \sum_{n=1}^{\infty} (\lambda_n(f) - \lambda_n(0)) e^{-\lambda_n t}.$$

Taking limit corresponding to $t \to 0$ one can get, after some calculations, an explicit formula for the f-regularized trace. Some details can be found in [22]. The obtained functional integrals can be represented by a sort of Feynman formulae.

The formula in Theorem 3.1 contains a Lagrangian functional integral; now we will briefly discuss how a Hamiltonian functional integral arises.

**Proposition 3.2.** *Let $H : \Omega \times P \to R^1$ be defined by $H(q,p) = h(p) + V(q)$ and let the definition of $\hat{H}$ is similar to the definition given at the beginning of this section. Then*

$$tr e^{-t\hat{H}} = \int_{\Omega} \int_{\xi : \forall \tau (\xi(\tau) + q) \in \Omega} \int_{p(\cdot)} e^{-\int_0^t V(\xi(\tau)+q) d\tau} e^{-\int_0^t h(p(\tau)) d\tau} \Phi_q^{t,\Omega}(d\xi dp) dq,$$

*where $\Phi_q^{t,\Omega}$ is a pseudomeasure which is as related to the Feynman pseudomeasure, whose Fourier transform is the function $(\xi, p) \mapsto e^{\int_0^t p'(\tau)\xi(\tau) d\tau}$, as the measure $w_{q,q}^{t,\Omega}$ is related to the standard Wiener measure (cf.[22], [4],[6] ).*

From Proposition 3.2 one can deduce a formula similar to the formula from the preceding theorem but containing a Hamiltonian functional integral.

## 4. Additional remarks.

1. In [7] a Hamiltonian Feynman formula is obtained for semigroups generated by some pseudodifferential operators. The main idea can be formulated

as follows. Let $Q = P = R^N$, $H : Q \times P \to C$ and $\tau \in [0, 1]$. A pseudodifferential operator with $\tau$-symbol $H$ is the operator $\widehat{H}_\tau$ in $L_2(Q)$ defined by $\left(\widehat{H}_\tau \phi\right)(q) = (2\pi)^{-N} \int_{\mathbb{R}^N} \int_{\mathbb{R}^N} H((1 - \tau)q + \tau q', p)e^{ip(q-q')}\phi(q')dq'dp;$ one assumes that the domain $D(\widehat{H}_\tau)$ of $H_\tau$ is the collection of functions $\phi \in L^2(R^N)$ for which $(\widehat{H}_\tau \phi)(\cdot)$ exists. The mapping $H \mapsto \hat{H}_\tau$ can be called $\tau$-quantization.

In general $e^{\widehat{H}_\tau} \neq \widehat{e^H_\tau}$ (only for linear $H$ one has $e^{\widehat{H}_\tau} = \widehat{e^H_\tau}$); but "infinitesimally" it is true; the following statement can be interpreted just in this sense.


**Proposition 4.1.** *Let* $X = L^2(R^N)$ *and for any* $t > 0$ *and* $\alpha = 1$, *i let* $F_{\alpha,\tau} = \left(\widehat{e^{-\alpha \frac{t}{n} H}}\right)_\tau$. *Then the following Hamiltonian Feynman formula holds:*
$e^{\alpha t \hat{H}_\tau} = \lim_{n \to \infty}(F_{\alpha,\tau}(\frac{t}{n}))^n$ *in the strong operator topology.*


Some similar holomorphic Feynman formulae are also valid.

2. The formulae which are quite similar to the formulae obtained in the first two sections can be developed for equation with respect to functions defined on Riemannian manifolds. Those formulae contain some geometric characteristics of the manifold and, in the case when one consider a submanifold of an Euclidian space, of the embedding (cf. [10]). Also some similar formulae can be obtained for manifolds, of the classical phase space, defined by the so called constrains, which arise in the theory of Hamilton-Dirac systems ([2],[21]).

3. As it has been already mentioned, the statements, which are similar to statements from Theorems 1.3 and 2.1, could be formulated and proved for the Schrödinger groups.

4. The results of the first two sections can be extended to the case of Laplace type operator in spaces of functions and measures defined on infinite dimensional Hilbert spaces.


This work was supported by the Russian foundation for basic research, (project 10-01-00724), and by the Government grant of the Russian Federation under the Resolution No. 220 "On measures designed to attract leading scientists to Russian institutions of higher learning" according the agreement No. 11.G34.31.0054 between the Ministry of education and science of the Russian Federation, , and Lomonosov Moscow State University.

# References


1. R.P. Feynman, "Space-time approach to nonrelativistic quantum mechanics," Rev.Mod.Phys. 20 367–387 (1948).

2. V.I.Arnold, V.V.Kozlov, A.I.Neishtadt. Mathematical aspects of classical and celestial mechanics.Moscow, 2002.

3. P.R. Chernoff, "Note on Product Formulas for Operator Semigroups," J. Funct. Ana. 84, 238–242 (1968).

4. O.G. Smolyanov, E.T. Shavgulidze, Functional integrals Moscow State University Press, Moscow, 1990 (in Russian).

5. Klaus-Jochen Engel, Reiner Nagel. One-Parameter Semigroups for Linear Evolution Equations. Springer, 2000.

6. O. G. Smolyanov, M. O. Smolyanova, "Transformations of the Feynman integral under nonlinear transformations of the phase space," Theoret. and Math. Phys. 100 (1994), no. 1, 803–810 (1995).

7. O.G.Smolyanov, A.G.Tokarev, A.Truman. Hamiltonian Feynman path integrals via the Chernoff formula. J.Math.Phys. 43, 10, (2002) 5161-5171.

8. O,G,Smolyanov, N.N.Shamarov. Hamiltonian Feynman integrals for equation containing the Vladimirov operator with variable coefficients. Doklady Mathematics, 84, 2 (2011), 689-694.

9. O. G. Smolyanov, D. S. Tolstyga and H. von Weizsacker Feynman description of the one-dimensional dynamics of particles whose masses piecewise continuously depend on coordinates. Doklady Mathematics, 84, 3 (2011), 804-807.

10. O.G.Smolyanov. Feynman formulae for evolutionary equations. in:Trends in Stochastic Analysis. Edited by J.Blath, P.Mörters, M.Scheutzow, 284-303, Cambridge University Press, 2009.

11. M. Gadella, S. Kuru, and J. Negro, Phys. Lett. A 362, 265-68 (2007).

12. J.Gough, O.O.Obrezkov, O.G.Smolyanov. *Randomized Hamiltonian Feynman integrals and stochastic Schrödinger-Ito equations Izvestia Mathematics* 69, 6 (2005), 3-20.

13. S. Albeverio, R. Hoegh-Krohn, Mathematical theory of Feynman path integrals Lecture notes in math 523. Berlin: Springer, 1976; the second edition, 2008.

14. O.G.Smolyanov. Feynman type formulae for quantum evolutions and diffusions on manifolds and graphs.Quantum Probability and White Noise Analysis. Vol.26. Quantum Bio-Informatics 3, eds. L.Accardi, W.Freudenberg and M.Ohya (2010), 337-347.

15. O.G.Smolyanov, H. von Weizsaecker, O.Wittich. Constructon of diffusions on sets of Mappings from an Interval to Compact Riemannian Manifolds. Doklady Mathematics, 71, 3 (2005), 390 - 395.

16. L.Accardi, O.G.Smolyanov. Feynman Formulae for Evolution Equations with Levy Laplacians on Infinite-Dimensional Manifolds. Doklady Mathematics, 73, 2 (2006), 252 - 257.

17. E. Nelson, "Feynman integrals and the Schrödinger equation," J. Math. Phys. 5 N 3 332–343 (1964).

18. V. P. Maslov, (Russian) Complex Markov chains and the Feynman path

integral for nonlinear equations (Moscow: Nauka, 1976).

19. A.A.Slavnov, L.D.Faddeev. Introduction to quantum theory of gauge fields. Moscow, 1988.

20. F.A.Berezin. Functional integral over trajectories in the phase space. Russian Phys. Surveys, 132, 3 (1980), 497-548.

21. P.A.M.Dirac. Generalized Hamiltonian dynamics. Canad. J. Math.,2, 129 (1950).

22. V.A.Sadovnichii, O.G.Smolyanov, E.T.Shavgulidze. Representations of regularized traces of differential operators by functional integrals. Doklady Mathematics (2012), to appear.

Quantum Bio-Informatics V
© 2013 World Scientific Publishing Co. Pte. Ltd.
pp. 315–325

# ENTROPY PRODUCTION AND NON-EQUILIBRIUM STEADY STATES

MASUO SUZUKI

*Tokyo University of Science*
*Kagurazaka 1-3, Shinjuku, Tokyo, 162-8601*
*E-mail: msuzuki@rs.kagu.tus.ac.jp*

The long-term issue of entropy production in transport phenomena is solved by separating the symmetry of the non-equilibrium density matrix $\rho(t)$ in the von Neumann equation, as $\rho(t) = \rho_s(t) + \rho_a(t)$ with the symmetric part $\rho_s(t)$ and antisymmetric part $\rho_a(t)$. The irreversible entropy production $(dS/dt)_{irr}$ is given in M.Suzuki, Physica A 390(2011)1904 by $(dS/dt)_{irr} = Tr(\mathcal{H}(d\rho_s(t)/dt))/T$ for the Hamiltonian $\mathcal{H}$ of the relevant system. The general formulation of the extended von Neumann equation with energy supply and heat extraction is reviewed from the author's paper (M.S.,Physica A391(2012)1074). irreversibility; entropy production; transport phenomena; electric conduction; thermal conduction; linear response; Kubo formula; steady state; non-equilibrium density matrix; energy supply; symmetry-separated von Neumann equation; unboundedness.

## 1. Introduction - duality of physical systems and life systems with respect to order formations and fluctuations

The purpose of the present paper is to give a brief review of my work on the entropy production and steady states in transport phenomena[1,2,3]. Before this review, we discuss the duality between physical phenomena and life phenomena in connection with the main theme of the present conference QBIC. In 1976, the present author [4] proposed the scaling theory of formation of macroscopic orders near the instability point using the nonlinear Langevin equation for a single classical macro-variable $x$. The order parameter $\langle |x| \rangle_t$ and the fluctuation $\langle (|x| - \langle |x| \rangle_t)^2 \rangle_t = \langle x^2 \rangle_t - \langle |x| \rangle_t^2$ obtained by this scaling theory are shown schematically in Fig.1(a). The remarkable feature of our conclusion is that fluctuation is enhanced anomalously around the onset time $t_o$ at which the order formation occurs. On the other hand, the order (or life activity) in living systems becomes maximum around some time region, while the entropy increases always as is shown in

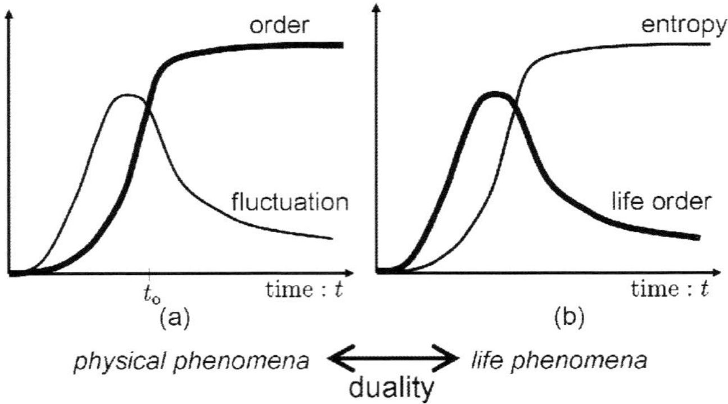



Fig. 1.   (a) The time change of the order $\langle |x| \rangle_t$ and the fluctuation $\left\langle \left( |x| - \langle |x| \rangle_t \right)^2 \right\rangle_t$ in physical systems. (b) A schematic time change of the life order and entropy in living systems.

Fig.1(b). Thus, we may say that there exists a duality between the above two phenomena.

Concerning Brownian motions of an infinite degree of freedom, the Hida calculus can be used conveniently[5]. The present scaling theory on a single macro-variable was extended to field variables by Kawasaki et al.[6]

In order to discuss the irreversibility and entropy production in quantum systems, we have to treat the von Neumann equation of the form

$$\frac{\partial \rho(t)}{\partial t} = \frac{1}{i\hbar} [\mathcal{H}(t), \rho(t)] \tag{1}$$

for the Hamiltonian $\mathcal{H}(t)$ of the relevant system, as will be discussed in Section 2. The key idea of the present theory on the entropy production is to separate the symmetry of the non-equilibrium density matrix $\rho(t)$ as $\rho(t) = \rho_s(t) + \rho_a(t)$. In Section 3, general formulations on nonlinear transport phenomena are presented, by extending Kubo's and Zubarev's schemes. In Section 4, a new scheme[2] of steady states with energy supply and heat extraction is reviewed using the two-component vector notation $\rho^\dagger(t) = (\rho_s(t), \rho_a(t))$ of the density matrix and the super-matrix $\mathcal{L}$, whose elements are hyper-operators such as the inner derivation $\delta_{\mathcal{H}}$. Summary and discussion will be given in Section 5.

## 2. Entropy production and symmetry separation of the density matrix

For long years, it has been a puzzle to derive the irreversible entropy production from the von Neumann equation (1) even in the linear response scheme[7,8,9,10,11]. This is now solved[1] to yield

$$\left(\frac{dS}{dt}\right)_{\mathrm{irr}} = \frac{1}{T}\frac{d}{dt}\mathrm{Tr}\mathcal{H}_0\rho_{\mathrm{s}}(t) = \frac{\sigma_F F^2}{T}; \quad F = |\boldsymbol{F}|, \tag{2}$$

where, $\mathcal{H}_0$ denotes the Hamiltonian of internal energy, $\boldsymbol{F}$ a static field, $\sigma_F$ the nonlinear transport coefficient and $T$ the temperature. Here, the density matrix $\rho(t)$ is separated into the two parts $\rho_{\mathrm{s}}(t)$ and $\rho_{\mathrm{a}}(t)$ defined by

$$\rho_{\mathrm{s}}(t) = \rho_0 + \rho_2(t) + \rho_4(t) + \cdots, \tag{3a}$$

and

$$\rho_{\mathrm{a}}(t) = \rho_1 + \rho_3(t) + \rho_5(t) + \cdots, \tag{3b}$$

as the power-series expansion with respect to the field $\boldsymbol{F}(t)$ defined in the Hamiltonian

$$\mathcal{H}(t) = \mathcal{H}_0 + \mathcal{H}_1(t); \quad \mathcal{H}_1(t) = -\boldsymbol{A} \cdot \boldsymbol{F}(t). \tag{4}$$

The above treatment has been shown[2] to be valid even in the thermal conductance[10,11,12,13] which is formulated[2] using a "mechanical perturbation" $\mathcal{H}_1 = -\boldsymbol{A} \cdot \boldsymbol{F}$.

## 3. General formulations on nonlinear responses – Kubo-type and Zubarev-type schemes

In the present section, we review general schemes of nonlinear responses based on the Kubo-type and Zubarev-type formulations being unified by the present author[2].

Following Kubo[7], we may put $\rho(t) = \rho_0 + \Delta\rho(t)$ with

$$\rho_0 = e^{-\beta\mathcal{H}_0}/Z_0(\beta); \quad Z_0(\beta) = \mathrm{Tr}e^{-\beta\mathcal{H}_0}. \tag{5}$$

Then, $\Delta\rho(t)$ is given formally[2] by

$$\Delta\rho(t) = \frac{1}{i\hbar}\int_{t_0}^{t} \mathcal{U}(t,t')[\mathcal{H}_1(t'),\rho_0]\mathcal{U}^{-1}(t,t')dt', \tag{6}$$

where $\mathcal{U}(t,t')$ is the ordered exponential defined[14,15,16,17,18] by

$$\mathcal{U}(t,t') = \exp_+ \left( \frac{1}{i\hbar} \int_{t'}^{t} \mathcal{H}(s)ds \right)$$

$$= 1 + \frac{1}{i\hbar} \int_{t'}^{t} \mathcal{H}(t_1)dt_1 + \left( \frac{1}{i\hbar} \right)^2 \int_{t'}^{t} dt_1 \int_{t'}^{t_1} dt_2 \mathcal{H}(t_1)\mathcal{H}(t_2) + \cdots .$$

$$(7)$$

Thus, the density matrix is written[2] as $\Delta\rho(t)$ in the Kubo-type (generalized) form

$$(\Delta\rho(t))_{\text{K-type}} = \int_{t_0}^{t} dt' \left( \int_0^{\beta} d\lambda \rho_0 \boldsymbol{j}(-i\hbar\lambda) \right) (t,t') \cdot \boldsymbol{F}(t'), \qquad (8)$$

where $\boldsymbol{j} = \dot{\boldsymbol{A}} = [\boldsymbol{A}, \mathcal{H}_0]/i\hbar$,

$$Q(t,t') = \mathcal{U}(t,t')Q\mathcal{U}^{-1}(t,t') \quad \text{and} \quad \boldsymbol{j}(-i\hbar\lambda) = e^{\lambda\mathcal{H}_0}\boldsymbol{j}e^{-\lambda\mathcal{H}_0}, \qquad (9)$$

for any operator $Q$ including the density matrix. Thus, the nonlinear current $\boldsymbol{J}_F$ is given[2] in the form

$$\boldsymbol{J}_F = \text{Tr}\boldsymbol{j}\Delta\rho(t) = \int_{t_0}^{t} dt' \int_0^{\beta} d\lambda \left\langle \boldsymbol{j}(-i\hbar\lambda)\mathcal{U}^{-1}(t,t')\boldsymbol{j}\mathcal{U}(t,t') \right\rangle_0 \cdot \boldsymbol{F}(t'), \quad (10)$$

using the nonlinear current-current correlation function in equilibrium.

If the force $\boldsymbol{F}(t)$ contains only the adiabatic factor $e^{\epsilon t}(t \to +0)$, then the above expression (8) is rewritten in the following more compact form[2]

$$(\Delta\rho(t))_{\text{K-type}} = \int_{t_0}^{t} dt' \left( \int_0^{\beta} d\lambda \rho_0 \boldsymbol{j}(-i\hbar\lambda) \right) (t' - t) \cdot \boldsymbol{F}e^{\epsilon t'}. \qquad (11)$$

Here, $Q(t)$ except the density matrix $\rho(t)$ is defined by

$$Q(t) = \exp\left( -\frac{t}{i\hbar}\mathcal{H} \right) Q \exp\left( \frac{t}{i\hbar}\mathcal{H} \right) \qquad (12)$$

for $\mathcal{H} = \mathcal{H}_0 + \mathcal{H}_1$. Thus, in the limit $t_0 \to -\infty$, we obtain[2] the relation $\boldsymbol{J}_F = \sigma_F \boldsymbol{F}$, where

$$\sigma_F = \int_0^{\infty} dt e^{-\epsilon t} \int_0^{\beta} d\lambda \left\langle \boldsymbol{j}(-i\hbar\lambda)\boldsymbol{j}_F(t) \right\rangle_0 \; ; \; \boldsymbol{j}_F(t) = e^{-\frac{t\mathcal{H}}{i\hbar}} \boldsymbol{j} e^{\frac{t\mathcal{H}}{i\hbar}}, \qquad (13)$$

with $\boldsymbol{j}(-i\hbar\lambda)$ defined in Eq.(9). This is a straight-forward generalization of the Kubo formula[7] to the nonlinear regime[1,2].

Next, we discuss another type of formulation on nonlinear responses by extending Zubarev's theory on the non-equilibrium statistical operator[19,20].

The starting point of our general formulation is the following expression of the non-equilibrium density matrix

$$\rho(t) = \mathcal{U}(t, t_0)\rho_0 \mathcal{U}^{-1}(t, t_0) = \frac{1}{Z_0(\beta)} \exp\left(-\beta \mathcal{U}(t, t_0)\mathcal{H}_0 \mathcal{U}^{-1}(t, t_0)\right)$$

$$\equiv \frac{1}{Z_0(\beta)} \exp\left(-\beta(\mathcal{H}_0 + \mathcal{R}(t, t_0))\right), \tag{14}$$

where $\mathcal{R}$ is given[2] by

$$\mathcal{R}(t, t_0) = -\int_{t_0}^{t} dt' \boldsymbol{j}(t, t') \cdot \boldsymbol{F}(t'); \quad \boldsymbol{j}(t, t') = \mathcal{U}(t, t')\boldsymbol{j}\mathcal{U}^{-1}(t, t'). \tag{15}$$

Then, the nonlinear current $\boldsymbol{J}_F$ is expressed[2] by

$$\boldsymbol{J}_F = \mathrm{Tr}\boldsymbol{j}\rho(t) = \mathrm{Tr}\boldsymbol{j}\left(\Delta\rho(t)\right)_{\text{Z-type}}, \tag{16}$$

where the deviation of $\rho(t)$ from $\rho_0$ is given by

$$\left(\Delta\rho(t)\right)_{\text{Z-type}} \equiv \rho(t) - \rho_0 = -\rho_0 \int_0^{\beta} d\lambda e^{\lambda \mathcal{H}_0} \mathcal{R}(t, t_0) e^{-\lambda(\mathcal{H}_0 + \mathcal{R}(t, t_0))}$$

$$= -\int_0^{\beta} d\lambda e^{-\lambda(\mathcal{H}_0 + \mathcal{R}(t, t_0))} \mathcal{R}(t, t_0) e^{\lambda \mathcal{H}_0} \rho_0. \tag{17}$$

Thus, the nonlinear current $\boldsymbol{J}_F$ is expressed[2] in the following two ways:

$$\boldsymbol{J}_F = \int_{t_0}^{t} dt' \int_0^{\beta} d\lambda \left\langle \boldsymbol{j} e^{-\lambda(\mathcal{H}_0 + \mathcal{R}(t, t_0))} \boldsymbol{j}(t, t'^{\lambda \mathcal{H}_0}) \right\rangle_0 \boldsymbol{F}(t')$$

$$= \int_{t_0}^{t} dt' \int_0^{\beta} d\lambda \left\langle e^{\lambda \mathcal{H}_0} \boldsymbol{j}(t, t'^{-\lambda(\mathcal{H}_0 + \mathcal{R}(t, t_0))} \boldsymbol{j} \right\rangle_0 \boldsymbol{F}(t'), \tag{18}$$

or

$$\boldsymbol{J}_F = \int_{t_0}^{t} dt' \int_0^{\beta} d\lambda \int_0^1 d\mu \left\langle e^{-(\beta-\lambda)\mathcal{H}_\mu(t, t_0)} \boldsymbol{j}(t, t'^{-\lambda \mathcal{H}_\mu(t, t_0)} \boldsymbol{j} e^{\beta \mathcal{H}_0} \right\rangle_0 \boldsymbol{F}(t') \tag{19}$$

with the notation $\mathcal{H}_\mu(t, t_0) = \mathcal{H}_0 + \mu \mathcal{R}(t, t_0)$. The above triple integral expression (19) is interesting from the view point of symmetry[2]. The average nonlinear transport coefficient is thus expressed by the "double canonical" current-current correlation function in equilibrium.

If the force $\boldsymbol{F}(t)$ depends on time $t$ only through the adiabatic factor $e^{\epsilon t}(\epsilon \to +0)$, then we have

$$\mathcal{R}(t, t_0) = -\int_{t_0}^{t} dt' e^{\epsilon t'} e^{\frac{(t-t')}{i\hbar}\mathcal{H}} \boldsymbol{j} e^{-\frac{(t-t')}{i\hbar}\mathcal{H}} \cdot \boldsymbol{F}. \tag{20}$$

Now, the Zubarev-type non-equilibrium density-matrix $\rho_{\text{Z-type}}$ is given by

$$\rho_{\text{Z-type}} = \rho(0) = \exp\left(-\beta(\mathcal{H}_0 + \mathcal{R}_F)\right)/Z_0(\beta), \tag{21}$$

in the limit $t_0 \to -\infty$ using $\mathcal{R}_F = \mathcal{R}_F(0, -\infty) = \mathcal{R}_F(t, -\infty)$. This shows

$$\mathcal{U}(t,0)(\mathcal{H}_0 + \mathcal{R}_F)\mathcal{U}^{-1}(t,0) = \mathcal{H}_0 + \mathcal{R}_F = \text{time-independent} \tag{22}$$

and consequently we have $[\mathcal{H}_0 + \mathcal{R}_F, \mathcal{H}] = 0$. Here, $\mathcal{R}_F$ is expressed[2] by the integral

$$\mathcal{R}_F = -\int_0^\infty e^{-\epsilon t}\boldsymbol{j}_F(-t)\cdot\boldsymbol{F}\,dt; \quad \boldsymbol{j}_F(-t) = e^{\frac{t\mathcal{H}}{i\hbar}}\boldsymbol{j}\,e^{-\frac{t\mathcal{H}}{i\hbar}}. \tag{23}$$

Now we make here a linear approximation of $\mathcal{R}_F$ with respect to the force $\boldsymbol{F}$:

$$\mathcal{R}_1 = -\int_0^\infty e^{-\epsilon t}\boldsymbol{j}(-t)\cdot\boldsymbol{F}\,dt \quad \text{and} \quad \boldsymbol{j}(t) = \boldsymbol{j}_0(t) = e^{-\frac{t}{i\hbar}\mathcal{H}_0}\boldsymbol{j}\,e^{\frac{t}{i\hbar}\mathcal{H}_0}. \tag{24}$$

Then we obtain Zubarev's non-equilibrium statistical operator of the form[2]

$$\rho_{\text{Zub}} = \exp\left(-\beta(\mathcal{H}_0 + \mathcal{R}_1)\right)/Z_0(\beta)$$

$$= \exp\left(-\beta\mathcal{H}_0 + \beta\int_0^\infty e^{-\epsilon t}\boldsymbol{j}(-t)\cdot\boldsymbol{F}\,dt\right)/Z_0(\beta). \tag{25}$$

It should be remarked here that the operators $\mathcal{H}_1$ and $\mathcal{R}_1$ (or more generally $\mathcal{R}_F$) are unbounded even below and consequently that we have to be careful[2] in calculating explicitly the average of a physical quantity using $\rho_{\text{Zub}}$ (or more generally $\rho_{\text{Z-type}}$ in Eq.(21)).

## 4. Density-matrix scheme of steady states with energy supply and heat extraction

In the previous papers[1,2], the present author has introduced the basic equation of the form

$$\frac{d}{dt}\boldsymbol{\rho}(t) = -\mathcal{L}\boldsymbol{\rho}(t) + \mathcal{L}_{\text{s}}(t)\boldsymbol{\rho}_0, \tag{26}$$

which describes non-equilibrium phenomena or transport phenomena with energy supply and extraction of generated heat outside the system. Here, the two-component vector $\boldsymbol{\rho}(t)$ and the super-matrices $\mathcal{L}$ and $\mathcal{L}_{\text{s}}(t)$ are defined[1,2] by

$$\boldsymbol{\rho}(t) = \begin{pmatrix} \rho_{\text{s}}(t) \\ \rho_{\text{a}}(t) \end{pmatrix}, \quad \mathcal{L} = \begin{pmatrix} \epsilon_{\text{r}}\mathcal{A} & -\epsilon_{\text{r}}\mathcal{B} \\ -\epsilon\mathcal{D} & \epsilon\mathcal{C} \end{pmatrix}, \quad \mathcal{L}_{\text{s}}(t) = \begin{pmatrix} \epsilon_{\text{r}} & 0 \\ 0 & \mathcal{L}_{\text{s}}^{(22)}(t) \end{pmatrix}, \tag{27}$$

where the super-operators $\mathcal{A}, \mathcal{B}, \mathcal{C}$ and $\mathcal{D}$ are defined [1] by

$$\mathcal{A} = 1 - \omega_{\mathrm{r}} \delta_{\mathcal{H}_0}, \quad \mathcal{B} = \omega_{\mathrm{r}} \delta_{\mathcal{H}_1}, \quad \mathcal{C} = 1 - \omega_{\epsilon} \delta_{\mathcal{H}_0}, \quad \mathcal{D} = \omega_{\epsilon} \delta_{\mathcal{H}_1},$$

$$\omega_{\mathrm{r}} = \frac{1}{i\hbar\epsilon_{\mathrm{r}}}, \quad \omega_{\epsilon} = \frac{1}{i\hbar\epsilon}, \quad \delta_A Q \equiv [A, Q] = AQ - QA. \tag{28}$$

The parameter $\epsilon_{\mathrm{r}}$ denotes[1,2] the rate of extracting generated heat outside the system. Furthermore, the two-component vector $\boldsymbol{\rho}_0$ and the super-operator $\mathcal{L}_{\mathrm{s}}^{(22)}(t)$ are defined[2] as

$$\boldsymbol{\rho}_0 = \begin{pmatrix} \rho_0 \\ \rho_0 \end{pmatrix} \quad \text{and} \quad \mathcal{L}_{\mathrm{s}}^{(22)}(t)\rho_0 = \eta_{\mathrm{es}}(t). \tag{29}$$

Here, $\eta_{\mathrm{es}}(t)$ is given[2] by

$$\eta_{\mathrm{es}}(t) = \epsilon'_{\mathrm{es}}(t)\{\mathcal{H}_1 \rho_0\} - \frac{1}{i\hbar}\epsilon_{\mathrm{es}}(t)[\mathcal{H}_0, \{\mathcal{H}_1 \rho_0\}] - \frac{1}{i\hbar}[\mathcal{H}_1, \tilde{\rho}_{\mathrm{s}}(t)], \tag{30}$$

where $\tilde{\rho}_{\mathrm{s}}(t)$ is given[2] by

$$\tilde{\rho}_{\mathrm{s}}(t) = \frac{1}{i\hbar}\int_{t_0}^{t} e^{\frac{(t-s)}{i\hbar}\mathcal{H}_0}[\mathcal{H}_1, \tilde{\rho}_{\mathrm{a}}(s)]e^{-\frac{(t-s)}{i\hbar}\mathcal{H}_0}ds, \quad \text{and} \quad \tilde{\rho}_{\mathrm{a}}(t) = \epsilon_{\mathrm{es}}(t)\{\mathcal{H}_1 \rho_0\}. \tag{31}$$

and the parameter $\epsilon_{\mathrm{es}}(t)$ is given in the form

$$\epsilon_{\mathrm{es}}(t) = (t + a)e^{\epsilon t}\epsilon_{\mathrm{es}} \tag{32}$$

for $\epsilon \to +0$ and for an appropriate constant $a$. The notation $\{AB\}$ denotes the symmetrized product $(AB + BA)/2$. The energy source operator $\eta_{\mathrm{es}}(t)$ given by Eq.(30) with Eq.(31) is linear in $\rho_0$, and consequently the super operator $\mathcal{L}_{\mathrm{s}}^{(22)}(t)$ can be defined as above. The formal solution of Eq.(26) is given by

$$\boldsymbol{\rho}(t) = \int_{t_0}^{t} e^{-(t-t')\mathcal{L}}\mathcal{L}_{\mathrm{s}}(t')\boldsymbol{\rho}_0 dt'^{-(t-t_0)\mathcal{L}} \begin{pmatrix} \rho_0 \\ 0 \end{pmatrix}. \tag{33}$$

In the limit $t_0 \to -\infty$, Eq.(33) is transformed into the "stationary" solution

$$\boldsymbol{\rho}^{(\mathrm{st})}(t) = \int_{0}^{\infty} e^{-s\mathcal{L}}\mathcal{L}_{\mathrm{s}}(t - s)\boldsymbol{\rho}_0 ds = \boldsymbol{\rho}(0) + t\boldsymbol{\rho}'(0) \tag{34}$$

using the formula

$$\mathcal{L}_{\mathrm{s}}(t)\boldsymbol{\rho}_0 = \mathcal{L}_{\mathrm{s}}(0)\boldsymbol{\rho}_0 + t\mathcal{L}'_{\mathrm{s}}(0)\boldsymbol{\rho}_0. \tag{35}$$

which are correct for any $t$ as is seen from Eq.(32) and Eq.(29) with Eq.(30) for $t_0 \to -\infty$ and $\epsilon \to +0$. Here, the vector matrices $\boldsymbol{\rho}(0)$ and $\boldsymbol{\rho}'(0)$ in Eq.(34) are given by[2]

$$\boldsymbol{\rho}(0) = \int_0^\infty e^{-t\mathcal{L}} \mathcal{L}_{\mathrm{s}}(-t)\boldsymbol{\rho}_0 dt = \mathcal{L}^{-1}\left(\mathcal{L}_{\mathrm{s}}(0)\boldsymbol{\rho}_0 - \mathcal{L}^{-1}\mathcal{L}_{\mathrm{s}}'(0)\boldsymbol{\rho}_0\right), \qquad (36)$$

and

$$\boldsymbol{\rho}'(0) = \int_0^\infty e^{-t\mathcal{L}} \mathcal{L}_{\mathrm{s}}'(0)\boldsymbol{\rho}_0 dt = \mathcal{L}^{-1}\mathcal{L}_{\mathrm{s}}'(0)\boldsymbol{\rho}_0, \qquad (37)$$

respectively. Note that $\mathrm{Tr}\boldsymbol{\rho}'(0) = 0$ and consequently that $\mathrm{Tr}\rho(t) = \mathrm{Tr}\rho_{\mathrm{s}}(t) = 1$, as it should be. (See Appendix in Ref.2 for the inverse $\mathcal{L}^{-1}$ of the super-matrix $\mathcal{L}$ whose elements are super-operators.)

As is seen from the above arguments, the concept of a "stationary density matrix" $\rho^{(\mathrm{st})}(t)$ is extended as[2]

$$\frac{d}{dt}\rho^{(\mathrm{st})}(t) = \text{constant}, \quad \text{namely} \quad \rho^{(\mathrm{st})}(t) = \rho(0) + t\rho'(0) \qquad (38)$$

in the limit $t_0 \to -\infty$.

In general, the time derivative of the average $\langle Q \rangle_t$ of any physical operator $Q = Q_{\mathrm{s}} + Q_{\mathrm{a}}$ for the symmetric part $Q_{\mathrm{s}}$ and antisymmetric part $Q_{\mathrm{a}}$ can be calculated by the formula[2]:

$$\frac{d}{dt}\langle Q \rangle_t = \frac{d}{dt}\left(\langle Q_{\mathrm{s}}\rangle_t + \langle Q_{\mathrm{a}}\rangle_t\right) = \frac{d}{dt}\left(\mathrm{Tr}Q_{\mathrm{s}}\rho_{\mathrm{s}}(t) + \mathrm{Tr}Q_{\mathrm{a}}\rho_{\mathrm{a}}(t)\right)$$

$$= \frac{d}{dt}\mathrm{Tr}\left(\boldsymbol{Q}\cdot\boldsymbol{\rho}(t)\right) = -\mathrm{Tr}\left(\boldsymbol{Q}\cdot\mathcal{L}\boldsymbol{\rho}(t)\right) + \mathrm{Tr}\left(\boldsymbol{Q}\cdot\mathcal{L}_{\mathrm{s}}(t)\boldsymbol{\rho}_0\right), \qquad (39)$$

using the basic equation (26). Here, we have used the two-component vector notation $\boldsymbol{Q}$ defined by[2]

$$\boldsymbol{Q} = \begin{pmatrix} Q_{\mathrm{s}} \\ Q_{\mathrm{a}} \end{pmatrix}. \qquad (40\mathrm{a})$$

For example, we have

$$\boldsymbol{\mathcal{H}} = \begin{pmatrix} \mathcal{H}_{\mathrm{s}} \\ \mathcal{H}_{\mathrm{a}} \end{pmatrix} = \begin{pmatrix} \mathcal{H}_0 \\ \mathcal{H}_1 \end{pmatrix} \quad \text{and} \quad \boldsymbol{\rho}(t) = \begin{pmatrix} \rho_{\mathrm{s}}(t) \\ \rho_{\mathrm{a}}(t) \end{pmatrix} \qquad (40\mathrm{b})$$

with the inner product[2]

$$\left(\boldsymbol{\mathcal{H}}\cdot\boldsymbol{\rho}(t)\right) = \mathcal{H}_{\mathrm{s}}\rho_{\mathrm{s}}(t) + \mathcal{H}_{\mathrm{a}}\rho_{\mathrm{a}}(t) = \mathcal{H}_0\rho_{\mathrm{s}}(t) + \mathcal{H}_1\rho_{\mathrm{a}}(t). \qquad (41)$$

The above formula (39) is convenient for studying the average of dynamical derivatives[1] such as the entropy production in steady states, because partial contributions to the time-derivative $d\langle Q\rangle_t/dt$ can be calculated separately with use of the last expression in Eq.(39)

The time-derivative of the energy $\langle \mathcal{H}_1\rangle_1$ is thus given[2] by

$$\frac{d}{dt}\langle \mathcal{H}_1\rangle_t = \mathrm{Tr}\left(\mathcal{H}_1\eta_{\mathrm{es}}(t)\right) - \boldsymbol{J}_F\cdot\boldsymbol{F} = \epsilon'_{\mathrm{es}}(t)\mathrm{Tr}\left(\mathcal{H}_1\{\mathcal{H}_1\rho_0\}\right) - \boldsymbol{J}_F\cdot\boldsymbol{F}$$

$$= \epsilon_{\mathrm{es}}\mathrm{Tr}\left(\mathcal{H}_1^2\rho_0\right) - \boldsymbol{J}_F\cdot\boldsymbol{F}, \tag{42}$$

using Eq.(39). Therefore, by imposing the condition[2]

$$\epsilon_{\mathrm{es}} = \frac{\boldsymbol{J}_F\cdot\boldsymbol{F}}{\mathrm{Tr}\left(\mathcal{H}_1^2\rho_0\right)}, \tag{43}$$

we arrive at the conservation relation of $\langle \mathcal{H}_1\rangle_t$:$\langle \mathcal{H}_1\rangle_t$ = constant as well as the conservation of $\langle \mathcal{H}_0\rangle_t$, while the entropy production $(dS/dt)_{\mathrm{irr}} = \boldsymbol{J}_F\cdot\boldsymbol{F}/T$ is still positive. Thus, we have obtained[2] the true steady state with the irreversible (positive) entropy production in this scheme.

## 5. Summary and discussion

The present paper has reviewed the author's recent work on the irreversibility and entropy production in transport phenomena[1,2]. One of the main ideas is to separate the density matrix $\rho(t)$ into the two parts $\rho_{\mathrm{s}}(t)$ and $\rho_{\mathrm{a}}(t)$ by symmetry. The derivation of this entropy production using the symmetric part $\rho_{\mathrm{s}}(t)$ is one of the most important results in our theory. This unifies the two research trends of Prigogine[31,32,33] and Kubo[7].

Quite recently, the author[2] has found a new principle of minimum "integrated" entropy production (or energy dissipation) in steady transport phenomena including nonlinear responses. This theory is shown[2] to be applicable to electric circuits, thermal conduction and chemical reaction.

The relation between the information entropy and thermodynamic entropy was discussed by the present author[36] with the emphasis of their difference that the latter is related to the energy of the relevant system, or a physical force, while Shannon's "entropy" is not necessarily related to such quantities.

Originally the concept of entropy was introduced to express the thermal state. That is, the entropy change $dS$ is calculated as $dS = dQ/T$ using the heat energy $dQ$ and the temperature $T$. After Prigogine, many people have been using the terminology of entropy flow and some people misunderstand that there occurs irreversibility at the boundary of the relevant system where the entropy flow is counted. This is only a manipulative interpretation of irreversibility. The irreversibility is a bulk effect in essence, as is seen in the derivation of the irreversible entropy production given in the present author's series of papers[1,2,3]. One may say "Back to Clausius".

The unboundedness of the non-equilibrium density operator has been discussed in detail in Ref.2. The relation of the present general theory to other works[21,22,23,24,25,26,27,28,29,30,31,32,33,34,35] on non-equilibrium fluctuations and other types of steady states will be discussed elsewhere.

## Acknowledgments

The author would like to thank Prof.B.K.Chakrabarti for stimulating comments and Dr.Y.Hashizume for discussion and digitization of the present manuscript. The author also thanks his late wife Noriko for her moral support.

## References


1. M.Suzuki, *Physica* **A 390**,1904 (2011).
2. M.Suzuki, *Physica***A 391**,1074(2012) and to be submitted.
3. M.Suzuki, *J.Phys.Conf.Ser.* **297**, 021029 (2011) (IOP Publishing).
4. M.Suzuki, *Prog.Theor.Phys.* **56**, 77, 477 and 380 (1976).
   See also M.Suzuki, *Int.J.Mod.Phys.B*, (2012) in press.
5. T.Hida, in this book.
6. K.Kawasaki, M.C.Yalabik and J.D.Gunton, *Phys.Rev.* **17** 455 (1978)
7. R.Kubo, *J.Phys.Soc.Jpn.* **12** 570 (1957).
8. H.Nakano, *Prog.Theor.Phys.* **15** 77 (1955).
   H.Nakano, *Int.J.Mod.Phys.B* **7** 2397 (1993).
   See also K.Tani, *Prog.Theor.Phys.* **32** 167 (1964).
   and M. Ichiyanagi, *Prog.Theor.Phys.* **76** 37 (1986).
9. M.Suzuki, *Prog.Theor.Phys.* **53** 1657 (1975);
   ibid **55** 1064 (1976); *J.Phys.Soc.Jpn.* **69** (2000) Suppl.A.
10. R.Kubo,M.Yokota and S.Nakajima, *J.Phys.Soc.Jpn.* **12** 1203 (1957).
11. J.M. Luttinger, *Phys.Rev.***135A** 1505 (1964).
12. H.Mori, *Phys.Rev.* **112** 1829 (1958).
13. K.Kawasaki and J.D.Gunton, *Phys.Rev.A* **8** 2048 (1973).
14. R.Kubo, *J.Phys.Soc.Jpn.* **17** 1100 (1962).
15. M.Suzuki, *Prog.Theor.Phys.* **69** 160 (1980).
16. M.Suzuki, *Commun.Math.* **183** 339 (1997).
17. M.Suzuki, *J.Math.Phys.* **38** 1183 (1997).
18. M.Suzuki, *Rev.Math.Phys.* **11** 243 (1999).
19. D.N.Zubarev, in *Nonequilibrium Statistical Mechanics* (Nauka, Moscow, 1988). See also A.L.Kuzemsky, *Int.J.Mod.Phys.B* **19** 1029 (2005).
20. M.Suzuki, *Prog.Theor.Phys.* **100** 475 (1998).
    See also S.Hershfield, *Phys.Rev.Lett.* **70** 2134 (1993).
21. D.J.Evans, E.G.D.Cohen and G.P.Morriss, *Phys.Rev.Lett.* **71** 2401 (1993).

22. J.Kurchan, *J.Phys.A:Math.Gen.* **31** 3719 (1998).

23. J.L.Lebowitz and H.Spohn, *J.Stat.Phys.* **95** 333 (1999).

24. M.Esposito, U.Harbola and S.Mukamel, *Rev.Mod.Phys.* **81** 1665 (2009).

25. M.Esposito and C.Van den Broeck, arXive:0911.2666v2[Cond-mat. stat-mech]

26. M.Suzuki, *Int.J.Mod.Phys.* **B5** 1821 (1991).

27. Y.Oono and M.Paniconi, *Prog.Theor.Phys.Suppl.* **130** 29 (1998).

28. D.Ruelle, *Proc.Natl.Acad.U.S.A.* **100** 3054 (2003).

29. S.Sasa and H.Tasaki, *J.Stat.Phys.* **125** 125 (2006).

30. T.S.Komatsu, N.Nakagawa, S.Sasa and H.Tasaki, *Phys.Rev.Lett.* **100** 230602 (2008).

31. I.Prigogine, *Int.J.Quantum Chemistry* **53** 105 (1995).

32. I.Antoniou and S.Tasaki, *Int.J.Quantum Chemistry* **46** 425 (1993).

33. T.Petrosky and I.Prigogine, *Proc.Natl.Acad.Sci.U.S.A.* **90** 9393 (1993).

34. H.Tasaki, *Phys.Rev.Lett.* **80** 1373 (1998).

35. A.V.Ponomarev,S.Denisov and P.Hänggi, *Phys.Rev.Lett.* **106** 010405 (2011).

36. M.Suzuki, in QP-PQ Quantum Probability and White Noise Analysis vol.XIX, Quantum Information and Computing, eds. L.Accardi, M.Ohya and N.Watanabe (World Scientific, Singapore, 2006)

Quantum Bio-Informatics V

© 2013 World Scientific Publishing Co. Pte. Ltd.

pp. 327–338

# TEST AND MEASURE ON DIFFERENCE OF ASYMMETRY BETWEEN SEVERAL SQUARE TABLES AND APPLICATION TO MEDICAL DATA

KOUJI TAHATA[1], KOUJI YAMAMOTO[2], NOBUKO MIYAMOTO[3], AND SADAO TOMIZAWA[4]

[1,3,4] *Department of Information Sciences, Tokyo University of Science,*
*Noda City, Chiba 278-8510, Japan*
[2] *Center for Clinical Investigation and Research, Osaka University Hospital,*
*2-15, Yamadaoka, Suita, Osaka, 565-0871, Japan*
[4] *E-mail: tomizawa@is.noda.tus.ac.jp*

For the analysis of square contingency tables with ordered categories, Tahata, Yamamoto, Nagatani and Tomizawa (2009) considered an average symmetry model and a measure of degree of departure from the model. The present paper proposes (1) the statistic to test the equality of degree of asymmetry and (2) the measure to represent the difference of degree of asymmetry between several square contingency tables. They are applied to unaided distance vision data in Britain and in Japan, and to decayed teeth data in Japan.

*Keywords*: Asymmetry; Average symmetry; Conditional symmetry; Linear column-parameter symmetry; Measure; Test.

## 1. Introduction

For the analysis of an $r \times r$ square contingency table with the same ordered row and column categories, let $p_{ij}$ denote the probability that an observation will fall in the $i$th row and $j$th column of the table ($i = 1, \ldots, r; j = 1, \ldots, r$). Bowker (1948) considered the symmetry model defined by

$$p_{ij} = \psi_{ij} \quad (i \neq j),$$

where

$$\psi_{ij} = \psi_{ji}.$$

Also, see, for example, Bishop, Fienberg and Holland (1975, p.282), Caussinus (1965), and Tomizawa and Tahata (2007). When the symmetry model does not hold, some extended symmetry models and asymmetry models had been proposed by many statisticians, for example, Stuart (1955), Caussinus (1965), McCullagh (1978), Goodman (1979), Agresti (1983), Tomizawa (1993), Tomizawa, Miyamoto, Yamamoto and Sugiyama (2007), Tomizawa, Miyamoto and Yamamoto (2006b), Tahata and Tomizawa (2008), and Tahata and Tomizawa (2010); although the details are omitted.

When the symmetry model does not hold, we are also interested in measuring the degree of departure from the symmetry. The measures to represent the degree of departure from the symmetry had been proposed by Tomizawa, Seo and Yamamoto (1998), and Tomizawa, Miyamoto and Hatanaka (2001).

Tahata, Yamamoto, Nagatani and Tomizawa (2009) considered the average symmetry model and proposed the measure to represent the degree of departure from the model. Some measures to represent the degree of departure from other models had been proposed by, for example, Tomizawa (1995), Tomizawa, Miyamoto, and Ashihara (2003), Yamamoto and Tomizawa (2007), and Tahata and Tomizawa (2011).

Consider the several square contingency tables. Tomizawa et al. (2001) also considered the estimate of the difference of degree of departure from the symmetry between several square tables. We are now interested in inferring the difference of degree of departure from the average symmetry between several square tables.

In the present paper, Section 2 reviews the measure to represent the degree of departure from the average symmetry, namely, the degree of asymmetry. Section 3.1 proposes the test statistic for the hypothesis of equality of degree of asymmetry between several square tables. Section 3.2 proposes the measure to represent the difference of degree of asymmetry between several square tables, and gives the asymptotic variance of the estimated measure. Section 4 analyzes two kinds of medical data using the proposed test statistic and measure.

## 2. Review of measure of average symmetry

We shall review briefly the average symmetry model and the measure which represents the degree of departure from average symmetry, considered by Tahata et al. (2009). Assuming $\{p_{ij} + p_{ji} \neq 0\}$, the measure is given by

$$\psi = \frac{4}{\pi\delta}\sum_{i<j}\sum(p_{ij} + p_{ji})(\theta_{ij} - \frac{\pi}{4}),$$

where

$$\delta = \sum_{s \neq t} \sum p_{st},$$

$$\theta_{ij} = \cos^{-1}\left(\frac{p_{ij}}{\sqrt{p_{ij}^2 + p_{ji}^2}}\right).$$

If the symmetry model holds, then $\psi$ equals zero. However, the converse does not hold. The average symmetry model is defined by the structure of probabilities with $\psi = 0$. Thus the symmetry model implies the average symmetry model. Note that (1) the symmetry model is equivalent to $\{\theta_{ij} - \frac{\pi}{4} = 0\}$, $i < j$; and (2) the average symmetry model indicates that the average of $\{\theta_{ij} - \frac{\pi}{4}\}$, $i < j$, equals zero.

We see that (1) $-1 \leq \psi \leq 1$, (2) $\psi = -1$ if and only if $p_{ji} = 0$ for all $i < j$ (then $p_{ij} > 0$ for all $i < j$), (called the complete-upper-asymmetry), and (3) $\psi = 1$ if and only if $p_{ij} = 0$ for all $i < j$ (then $p_{ji} > 0$ for all $i < j$), (called the complete-lower-asymmetry).

## 3. Test and measure of difference of asymmetry

We shall propose the test statistic and the measure of difference of degree of asymmetry between several square tables.

Consider two $r \times r$ contingency tables, say, $T_1$ and $T_2$. For each table $T_k$ $(k = 1, 2)$, let $p_{ij}^{(k)}$ denote the probability that an observation will fall in the $i$th row and $j$th column of $T_k$ $(i = 1, \ldots, r; j = 1, \ldots, r)$, with $\sum_i \sum_j p_{ij}^{(k)} = 1$. Also for table $T_k$ $(k = 1, 2)$, let $n_{ij}^{(k)}$ denote the observed frequency in the $i$th row and $j$th column of $T_k$ $(i = 1, \ldots, r; j = 1, \ldots, r)$ with $n^{(k)} = \sum_i \sum_j n_{ij}^{(k)}$.

Assume that $\{n_{ij}^{(1)}\}$ are independent of $\{n_{ij}^{(2)}\}$, and for each table $T_k$ $(k = 1, 2)$, $\{n_{ij}^{(k)}\}$ have a multinomial distribution. For table $T_k$ $(k = 1, 2)$, denote the measure to represent the degree of departure from average symmetry by $\psi^{(k)}$, which is defined by $\psi$ with $\{p_{ij}\}$ replaced by $\{p_{ij}^{(k)}\}$.

### 3.1. *Test*

We shall propose the statistic to test the hypothesis of equality of degrees of asymmetry between tables $T_1$ and $T_2$, i.e., the hypothesis of $\psi^{(1)} = \psi^{(2)}$. For table $T_k$ $(k = 1, 2)$, the sample version of $\psi^{(k)}$, i.e., $\hat{\psi}^{(k)}$, is given by $\psi^{(k)}$ with $\{p_{ij}^{(k)}\}$ replaced by $\{\hat{p}_{ij}^{(k)}\}$, where $\hat{p}_{ij}^{(k)} = n_{ij}^{(k)}/n^{(k)}$. Although the detail is omitted, when $n^{(1)}$ and $n^{(2)}$ are large, we obtain that $\hat{\psi}^{(1)} - \hat{\psi}^{(2)}$ has asymptotically a normal distribution with mean $\psi^{(1)} - \psi^{(2)}$ and standard error

$$\sqrt{\frac{\sigma^2[p_{st}^{(1)}]}{n^{(1)}} + \frac{\sigma^2[p_{st}^{(2)}]}{n^{(2)}}},$$

where

$$\sigma^2[p_{st}^{(k)}] = \sum\sum_{i<j} \left[ p_{ij}^{(k)} \left( D_{ij}^{(k)} \right)^2 + p_{ji}^{(k)} \left( D_{ji}^{(k)} \right)^2 \right],$$

with for $i < j$,

$$D_{ij}^{(k)} = \frac{4}{\pi\delta^{(k)}} \left[ \theta_{ij}^{(k)} - (\sin\theta_{ij}^{(k)})(\sin\theta_{ij}^{(k)} + \cos\theta_{ij}^{(k)}) \right] - \frac{\psi^{(k)} + 1}{\delta^{(k)}},$$

$$D_{ji}^{(k)} = \frac{4}{\pi\delta^{(k)}} \left[ \theta_{ij}^{(k)} + (\cos\theta_{ij}^{(k)})(\sin\theta_{ij}^{(k)} + \cos\theta_{ij}^{(k)}) \right] - \frac{\psi^{(k)} + 1}{\delta^{(k)}},$$

$$\theta_{ij}^{(k)} = \cos^{-1} \left( \frac{p_{ij}^{(k)}}{\sqrt{(p_{ij}^{(k)})^2 + (p_{ji}^{(k)})^2}} \right),$$

$$\delta^{(k)} = \sum\sum_{s\neq t} p_{st}^{(k)}.$$

Let $\sigma^2[\hat{p}_{st}^{(k)}]$ denote $\sigma^2[p_{st}^{(k)}]$ with $\{p_{st}^{(k)}\}$ replaced by $\{\hat{p}_{st}^{(k)}\}$ ($k = 1, 2$). Therefore, for testing the null hypothesis of $\psi^{(1)} = \psi^{(2)}$, we propose the test statistic $Z^{(1,2)}$, where

$$Z^{(1,2)} = \frac{\hat{\psi}^{(1)} - \hat{\psi}^{(2)}}{\sqrt{\frac{\sigma^2[\hat{p}_{st}^{(1)}]}{n^{(1)}} + \frac{\sigma^2[\hat{p}_{st}^{(2)}]}{n^{(2)}}}}.$$

Under the null hypothesis, $Z^{(1,2)}$ has asymptotically a standard normal distribution.

### 3.2. *Measure*

We shall propose the measure to represent the difference of degrees of asymmetry between tables $T_1$ and $T_2$. We propose the measure $\psi^{(1,2)}$ defined by

$$\psi^{(1,2)} = \psi^{(1)} - \psi^{(2)}.$$

Although the detail is omitted, we can see that (1) $-2 \leq \psi^{(1,2)} \leq 2$, (2) $\psi^{(1,2)} = 0$ if and only if the degree of asymmetry in table $T_1$ is equal to that in table $T_2$, (3) $\psi^{(1,2)} = 2$ if and only if there is the complete-lower-asymmetry in table $T_1$ and the complete-upper-asymmetry in table $T_2$, and (4) $\psi^{(1,2)} = -2$ if and only if there is the complete-upper-asymmetry in table $T_1$ and the complete-lower-asymmetry in table $T_2$.

An approximate $100(1 - \alpha)\%$ confidence interval for $\psi^{(1,2)}$ is given by

$$\hat{\psi}^{(1)} - \hat{\psi}^{(2)} \pm z_{\frac{\alpha}{2}} \sqrt{\frac{\sigma^2[\hat{p}_{st}^{(1)}]}{n^{(1)}} + \frac{\sigma^2[\hat{p}_{st}^{(2)}]}{n^{(2)}}},$$

where $z_{\frac{\alpha}{2}}$ is the percentage point of the standard normal distribution corresponding to a two-tail probability of $\alpha$.

## 4. Analysis of data

### 4.1. *Analysis of unaided vision*

Consider two kinds of unaided vision data. Table 1a taken from Stuart (1955) is data on unaided distance vision data of 7477 women aged 30 to 39 employed in Royal Ordnance factories in Britain from 1943 to 1946. Table 1b taken from Tomizawa (1984) is data on unaided distance vision of 4746 students aged 18 to 25 including about 10% women in Faculty of Science and Technology, Science University of Tokyo in Japan examined in April 1982.

We see from Table 3 that (1) for the data in Table 1a, the average symmetry for the women's right and left eyes departs toward the complete-upper-asymmetry, and (2) for the data in Table 1b, the average symmetry for the students' right and left eyes departs toward the complete-lower-asymmetry (Tahata et al., 2009). Using test statistic and measure proposed in Section 3, we shall analyze the difference of degrees of asymmetry between women's vision data in Table 1a and students' vision data in Table 1b. According to the two-sided test for the null hypothesis $\psi^{(1a)} = \psi^{(1b)}$, there is a significant difference at a significant level 0.05 ($Z^{(1a,1b)} = -4.866, p < 0.001$). Therefore, the degree of asymmetry of the women's right eye and left eye vision is not equal to the degree of asymmetry of the students' right eye and left eye vision. Also the estimated value of measure on difference of degrees of asymmetry between women's vision data in Table 1a and students' vision data in Table 1b, is $\hat{\psi}^{(1a,1b)}(= \hat{\psi}^{(1a)} - \hat{\psi}^{(1b)}) = -0.222$. An approximate 95% confidence interval for $\psi^{(1a,1b)}(= \psi^{(1a)} - \psi^{(1b)})$ is $(-0.312, -0.133)$. Since the values of this interval are negative, $\psi^{(1b)}$ is estimated to be greater than $\psi^{(1a)}$. Therefore the students' vision data in Table 1b rather than the women's vision data in Table 1e tend to depart toward the complete-lower-asymmetry.

## 4.2. *Analysis of decayed teeth*

Consider the data in Table 2, taken from Tomizawa, Miyamoto and Iwamoto (2006a). Table 2a (Table 2b) is constructed from the data of the decayed teeth of 349 men (363 women) aged 18-39, for the patients visiting a dental clinic in Sapporo City, Japan, from 2001 to 2005. Tables 2a and 2b are classified by the numbers of decayed teeth on the left side of the mouth of a patient and those on the right side. Note that each of these patients has at least one decayed tooth. Tables 2c and 2d are re-classified by the numbers of decayed teeth on the lower side of the mouth of a patient and those on the upper side.

We see from Table 3 that since the estimated values of the measure are negative, (1) for men (for women) classified by the left and right decayed teeth, the average symmetry departs toward the complete-upper-asymmetry, and (2) for men (for women) classified by the lower and upper decayed teeth, the average symmetry also departs toward the complete-upper-asymmetry. In addition, we see from Table 3 that the degree of asymmetry is estimated to be greater for men (for women) classified by the lower and upper decayed teeth than for men (for women) classified by the left and right decayed teeth.

Using test statistic and measure proposed in Section 3, we shall analyze the difference of degrees of asymmetry between men and women. According to the two-sided test of $\psi^{(2a)} = \psi^{(2b)}$, there is not a significant difference at $\alpha$ level 0.05 ($Z^{(2a,2b)} = -0.883, p = 0.377$). Also, for the test of $\psi^{(2c)} = \psi^{(2d)}$, there is not a significant difference ($Z^{(2c,2d)} = -0.256, p = 0.798$). Therefore (1) the degree of asymmetry for men classified by the left and right decayed teeth is equal to that for women classified by the left and right decayed teeth, and (2) the degree of asymmetry for men classified by the lower and upper decayed teeth is equal to that for women classified by the lower and upper decayed teeth. Also, the estimated values of measure on difference of degrees of asymmetry between men's and women's decayed teeth are $\hat{\psi}^{(2a,2b)}(= \hat{\psi}^{(2a)} - \hat{\psi}^{(2b)}) = -0.136$ and $\hat{\psi}^{(2c,2d)}(= \hat{\psi}^{(2c)} - \hat{\psi}^{(2d)}) = -0.018$. An approximate 95% confidence interval for $\psi^{(2a,2b)}$ is $(-0.438, 0.166)$, and that for $\psi^{(2c,2d)}$ is $(-0.157, 0.121)$. Since these intervals include zero, the degree of asymmetry for men and that for women would be equal or the difference would be slight.

## 5. Concluding remarks

For the $r \times r$ contingency tables with the probabilities $\{p_{ij}\}$, some asymmetry models had been proposed. For example, the conditional symmetry model (McCullagh, 1978) is defined by

$$p_{ij} = \Delta p_{ji} \quad (i < j).$$

A special case of this model with $\Delta = 1$ is the symmetry model. The linear column-parameter symmetry model, proposed by Tomizawa et al. (2006a), is defined by

$$p_{ij} = \Gamma^{j-1} p_{ji} \quad (i < j).$$

A special case of this model with $\Gamma = 1$ is the symmetry model. When the conditional symmetry model holds, the measure $\psi$ can be expressed as

$$\psi = \frac{4}{\pi} \cos^{-1}\left(\frac{\Delta}{\sqrt{\Delta^2 + 1}}\right) - 1.$$

Thus $\psi = 0$ if and only if the symmetry model holds, i.e., $\Delta = 1$, thus $\{p_{ij} = p_{ji}\}$. Also, when the linear column-parameter symmetry model holds, the measure $\psi$ can be expressed as

$$\psi = \frac{4}{\pi\delta} \sum_{i<j} \sum (p_{ij} + p_{ji})(\theta_{ij} - \frac{\pi}{4}),$$

where

$$\theta_{ij} = \cos^{-1}\left(\frac{\Gamma^{j-1}}{\sqrt{\Gamma^{2(j-1)} + 1}}\right).$$

Thus $\psi = 0$ if and only if the symmetry model holds, i.e., $\Gamma = 1$, thus $\{p_{ij} = p_{ji}\}$. Namely, when the conditional symmetry model or the linear column-parameter symmetry model holds, the average symmetry is equivalent to the symmetry. Therefore, for tables $T_1$ and $T_2$, the test statistic $Z^{(1,2)}$ proposed in Section 3 would be adequate to test the hypothesis of equality of degrees of departure from *symmetry* between tables $T_1$ and $T_2$ when the conditional symmetry model or the linear column-parameter symmetry model holds. By the similar reason, the measure $\psi^{(1,2)}$ proposed in Section 3 would be adequate to represent the difference of degrees of departure from *symmetry* between tables $T_1$ and $T_2$ when the conditional symmetry model or the linear column-parameter symmetry model holds.

Although the details are omitted, the values of the likelihood ratio chi-squared statistics $G^2$ for testing goodness-of-fit of the conditional symmetry model, are 7.35 for the data in Table 1a, and 4.98 for the data in Table 1b, with both 5 degrees of freedom (Tomizawa, 1985). Thus the conditional symmetry model fits the data in Tables 1a and 1b well. Also, the values of $G^2$ for testing goodness-of-fit of the linear column-parameter symmetry model, are 1.03, 2.59, 1.22, and 0.39 for the data in Tables 2a, 2b, 2c, and 2d, respectively, with each 2 degrees of freedom (Tomizawa et al., 2006a). Thus the linear column-parameter symmetry model fits the data in Tables 2a, 2b, 2c, and 2d well. Therefore for the vision data and for the decayed teeth data, it would be adequate to use the test statistic and the measure proposed in Section 3, for inferring the difference of degrees of departure from symmetry between several tables.

## References


1. Agresti, A. (1983). A simple diagonals-parameter symmetry and quasi-symmetry model. *Statistics and Probability Letters*, **1**, 313-316.
2. Bishop, Y. M. M., Fienberg, S. E. and Holland, P. W. (1975). *Discrete Multivariate Analysis: Theory and Practice*. The MIT Press, Cambridge, Massachusetts.
3. Bowker, A. H. (1948). A test for symmetry in contingency tables. *Journal of the American Statistical Association*, **43**, 572-574.
4. Caussinus, H. (1965). Contribution à l'analyse statistique des tableaux de corrélation. *Annales de la Faculté des Sciences de l'Université de Toulouse*, **29**, 77-182.
5. Goodman, L. A. (1979). Multiplicative models for square contingency tables with ordered categories. *Biometrika*, **66**, 413-418.
6. McCullagh, P. (1978). A class of parametric models for the analysis of square contingency tables with ordered categories. *Biometrika*, **65**, 413-418.
7. Stuart, A. (1955). A test for homogeneity of the marginal distributions in a two-way classification. *Biometrika*, **42**, 412-416.
8. Tahata, K. and Tomizawa, S. (2008). Orthogonal decomposition of point-symmetry for multiway tables. *Advances in Statistical Analysis*, **92**, 255-269.
9. Tahata, K. and Tomizawa, S. (2010). Double linear diagonals-parameter symmetry and decomposition of double symmetry for square tables. *Statistical Methods and Applications*, **19**, 307-318.
10. Tahata, K. and Tomizawa, S. (2011). Measure for uniform association based on concordant and discordant pairs for cross-classifications. *Journal of Statistical Theory and Practice*, to appear.
11. Tahata, K., Yamamoto, K., Nagatani, N. and Tomizawa, S. (2009). A measure of departure from average symmetry for square contingency tables with ordered categories. *Austrian Journal of Statistics*, **38**, 101-108.
12. Tomizawa, S. (1984). Three kinds of decompositions for the conditional symmetry model in a square contingency table. *Journal of the Japan Statistical Society*, **14**, 35-42.
13. Tomizawa, S. (1985). Analysis of data in square contingency tables with

ordered categories using the conditional symmetry model and its decomposed models. *Environmental Health Perspectives*, **63**, 235-239.

14. Tomizawa, S. (1993). Diagonals-parameter symmetry model for cumulative probabilities in square contingency tables with ordered categories. *Biometrics*, **49**, 883-887.

15. Tomizawa, S. (1995). Measures of departure from marginal homogeneity for contingency tables with nominal categories. *The Statistician*, **44**, 425-439.

16. Tomizawa, S. and Tahata, K. (2007). The analysis of symmetry and asymmetry: Orthogonality of decomposition of symmetry into quasi-symmetry and marginal symmetry for multi-way tables. *Journal de la Société Française de Statistique*, **148**, 3-36.

17. Tomizawa, S., Miyamoto, N. and Ashihara, N. (2003). Measure of departure from marginal homogeneity for square contingency tables having ordered categories. *Behaviormetrika*, **30**, 173-193.

18. Tomizawa, S., Miyamoto, N. and Hatanaka, Y. (2001). Measure of asymmetry for square contingency tables having ordered categories. *Australian and New Zealand Journal of Statistics*, **43**, 335-349.

19. Tomizawa, S., Miyamoto, N. and Iwamoto, M. (2006a). Linear column-parameter symmetry model for square contingency tables: application to decayed teeth data. *Biometrical Letters*, **43**, 91-98.

20. Tomizawa, S., Miyamoto, N. and Yamamoto, K. (2006b). Decomposition for polynomial cumulative symmetry model in square contingency tables with ordered categories. *Metron*, **64**, 303-314.

21. Tomizawa, S., Seo, T. and Yamamoto, H. (1998). Power-divergence-type measure of departure from symmetry for square contingency tables that have nominal categories. *Journal of Applied Statistics*, **25**, 387-398.

22. Tomizawa, S., Miyamoto, N., Yamamoto, K. and Sugiyama, A. (2007). Extensions of linear diagonal-parameter symmetry and quasi-symmetry models for cumulative probabilities in square contingency tables. *Statistica Neerlandica*, **61**, 273-283.

23. Yamamoto, K. and Tomizawa, S. (2007). Decomposition of measure for marginal homogeneity in square contingency tables with ordered categories. *Austrian Journal of Statistics*, **36**, 105-114.

Table 1.   Unaided vision data of (a) 7477 women in Britain (Stuart, 1955), and (b) 4746 students in Japan (Tomizawa, 1984).

(a) Women in Britain

| Right eye grade | Best (1) | Second (2) | Third (3) | Worst (4) | Total |
|---|---|---|---|---|---|
| Best (1)   | 1520 | 266  | 124  | 66  | 1976 |
| Second (2) | 234  | 1512 | 432  | 78  | 2256 |
| Third (3)  | 117  | 362  | 1772 | 205 | 2456 |
| Worst (4)  | 36   | 82   | 179  | 492 | 789  |
| Total      | 1907 | 2222 | 2507 | 841 | 7477 |

(b) Students in Japan

| Right eye grade | Best (1) | Second (2) | Third (3) | Worst (4) | Total |
|---|---|---|---|---|---|
| Best (1)   | 1291 | 130 | 40   | 22   | 1483 |
| Second (2) | 149  | 221 | 114  | 23   | 507  |
| Third (3)  | 64   | 124 | 660  | 185  | 1033 |
| Worst (4)  | 20   | 25  | 249  | 1429 | 1723 |
| Total      | 1524 | 500 | 1063 | 1659 | 4746 |

Table 2.  Decayed teeth data of 349 men and 363 women aged 18-39, for patients visiting a dental clinic in Sapporo City, Japan, from 2001 to 2005 (Tomizawa et al., 2006a).

(a) For men with left and right decayed teeth

| Left (numbers of decayed teeth) | Right (numbers of decayed teeth) | | | |
| --- | --- | --- | --- | --- |
| | 0-4 (1) | 5-8 (2) | 9+ (3) | Total |
| 0-4 (1) | 118 | 37 | 2 | 157 |
| 5-8 (2) | 21 | 87 | 23 | 131 |
| 9+ (3) | 2 | 11 | 48 | 61 |
| Total | 141 | 135 | 73 | 349 |

(b) For women with left and right decayed teeth

| Left (numbers of decayed teeth) | Right (numbers of decayed teeth) | | | |
| --- | --- | --- | --- | --- |
| | 0-4 (1) | 5-8 (2) | 9+ (3) | Total |
| 0-4 (1) | 103 | 45 | 1 | 149 |
| 5-8 (2) | 35 | 84 | 33 | 152 |
| 9+ (3) | 3 | 17 | 42 | 62 |
| Total | 141 | 146 | 76 | 363 |

(c) For men with lower and upper decayed teeth

| Lower (numbers of decayed teeth) | Upper (numbers of decayed teeth) | | | |
| --- | --- | --- | --- | --- |
| | 0-4 (1) | 5-8 (2) | 9+ (3) | Total |
| 0-4 (1) | 115 | 55 | 25 | 195 |
| 5-8 (2) | 16 | 49 | 60 | 125 |
| 9+ (3) | 1 | 7 | 21 | 29 |
| Total | 132 | 111 | 106 | 349 |

(d) For women with lower and upper decayed teeth

| Lower (numbers of decayed teeth) | Upper (numbers of decayed teeth) | | | |
| --- | --- | --- | --- | --- |
| | 0-4 (1) | 5-8 (2) | 9+ (3) | Total |
| 0-4 (1) | 97 | 62 | 15 | 174 |
| 5-8 (2) | 20 | 63 | 75 | 158 |
| 9+ (3) | 2 | 6 | 23 | 31 |
| Total | 119 | 131 | 113 | 363 |

Table 3. Estimated measure $\hat{\psi}$, estimated approximate standard error of $\hat{\psi}$, and approximate 95% confidence interval for $\psi$, applied to the data in Tables 1 and 2.

| Applied data | Estimated measure | Standard error | Confidence interval |
| --- | --- | --- | --- |
| Table 1a | $-0.093$ | 0.027 | $(-0.146, -0.040)$ |
| Table 1b | 0.129 | 0.037 | $(0.057, 0.201)$ |
| Table 2a | $-0.360$ | 0.114 | $(-0.584, -0.136)$ |
| Table 2b | $-0.224$ | 0.103 | $(-0.426, -0.022)$ |
| Table 2c | $-0.775$ | 0.050 | $(-0.873, -0.678)$ |
| Table 2d | $-0.757$ | 0.050 | $(-0.856, -0.659)$ |

Quantum Bio-Informatics V
© 2013 World Scientific Publishing Co. Pte. Ltd.
pp. 339–350

# FUNCTIONAL MECHANICS AND KINETIC EQUATIONS

ANTON S. TRUSHECHKIN

*Steklov Mathematical Institute of the Russian Academy of Sciences,
Gubkina Str. 8, 119991 Moscow, Russia;
National Research Nuclear University "MEPhI",
Kashirskoe Highway 31, 115409 Moscow, Russia
E-mail: trushechkin@mi.ras.ru*

The Boltzmann–Enskog kinetic equation is known to describe the irreversible dynamics of a dilute hard sphere gas. From the other side, the Boltzmann–Enskog equation is known to have reversible microscopic solutions. We show that the reversibility or irreversibility property of the Boltzmann–Enskog equation depends on the considered class of solutions. If the considered class solutions is sums of delta-functions, then the equation is reversible. If the considered class of solutions is continuous functions, then the equation is irreversible. We suggest a new way of the reconciliation of the irreversible kinetic equation with the reversible particle dynamics. We postulate the Boltzmann–Enskog equation and regard its microscopic solutions as a kind of derivation of the microscopic equations of motion from the Boltzmann–Enskog equation. From this point of view, the "fundamental" laws of motion are not the laws for individual particles (reductionism), but the laws for the single-particle density function, which describes the system as a whole (holism). Laws of motion for the individual particles are approximations. This work can be regarded as a development of recently proposed functional mechanics, which is aimed to make mechanics more appropriate for applications to complex and biological systems.

*Keywords* : reversibility paradox, functional mechanics, Liouville equation, kinetic equations, Boltzmann–Enskog equation, microscopic solutions.

## 1. Introduction

An important problem in interdisciplinary researches is a problem of a consistent description of different levels of the nature. A particular question within this scope is: what description of the particle (atomic and molecular) dynamics is the most appropriate for the applications to complex and biological systems? See Ref. 1 for applications of fundamental physics and mathematics to complex systems and life sciences. The fundamental reversibility paradox shows that the microscopic and macroscopic descriptions of the same physical system are not consistent with each other.

Particle dynamics described by the Newton, Hamilton, Schrödinger equations, etc. is known to be time-reversible (any physical process can pass in both directions) and recurrent (the system return to states very close to the initial state infinitely many times). From the other side, we observe that macroscopic processes pass in a distinguished direction (the direction of the entropy growth if the system is isolated). This fact is expressed by the macroscopic equations (kinetic, hydrodynamic equations, etc.). So, microscopic and macroscopic descriptions for the same system lead to different conclusions.

One can point out the extremely small probability of the inverse process and the extremely long recurrence time for the systems with a large number of particles as a possible solution of this paradox (both arguments originated with Boltzmann). However, rigorous estimations of the corresponding probabilities and recurrence times based on the Hamilton equations have not been obtained up to now.

Besides, a further question arises: how to derive the macroscopic properties of a system (e.g., the rate of the relaxation to the equilibrium) from the particle dynamics. Even more complicated problems arise in the context of complex and biological systems: how to derive the properties of such systems from the particle dynamics.

Recently, functional mechanics was proposed to make progress in these problems, see Refs. 2 and 3, see also Refs. 4, 5, 6, and 7.

According to the traditional (Newtonian) paradigm, the state of a system is specified by a point in a phase space. The evolution of the system is specified by a trajectory of the phase point. The microscopic laws of motion are assumed to be known. One tries to derive macroscopic equations of motion (Boltzmann, Navier–Stocks etc.) from the microscopic ones.

The fundamental notion of functional mechanics is not a material point, but an (integrable) probability density function, even if only a single particle is considered. Hence, the fundamental equation of mechanics is not the Newton equation (or Hamilton equations), but the Liouville equation (or even the Focker–Planck–Kolmogorov equation).

The Liouville equation is time-symmetric. But the integrable solutions of the Liouville equation have the property of delocalization. For example, suppose we have a particle in a box with hard walls (or, equivalently, in a torus) and its state is described by an integrable probability density function. Then the spatial probability distribution weekly tends to the uniform distribution in the box as time increases indefinitely (this is the assertion of the diffusion theorems for the Liouville equation[8]). So, initially spatially localized density function diffuses and fill the whole of the available volume.

A mechanics based on the notion of a function, rather than on the notion of a material point, seems to be more appropriate to describe the behaviour of complex and biological systems. The following research program is to derive the macroscopic properties from the new particle mechanics or to reconcile them in another way.

In this report we propose a reconciliation of the Boltzmann–Enskog kinetic equation with the particle dynamics. The Boltzmann–Enskog equation describes the dynamics of a hard sphere gas. In the traditional approach, one tries to derive the kinetic equation from particle dynamics. On the contrary, we postulate the Boltzmann–Enskog equation for the system of hard spheres. From this point of view, the "fundamental" laws of motion are not the laws of motion for individual particles (reductionism), but the laws of motion for the function $f$, which describes the system as a whole (holism). The Boltzmann–Enskog equation is known to have microscopic solutions[9,10]. We treat these solutions as the microdynamics. So, the laws of motions for the individual particles are approximations.

The establishment of the existence of microscopic solutions for the kinetic equation can be regarded as a derivation of the particle dynamics from it. Thus, instead of the derivation of the kinetic equations from the microdynamics we suggest a kind of derivation of the microdynamics from the kinetic equations.

## 2. Microscopic solutions of the Boltzmann–Enskog equation

Let us consider a gas of hard spheres with diameter $a > 0$ in some 3-dimensional domain. In order to avoid difficulties with taking a boundary of the domain into account, let us assume that the domain is the 3-dimensional torus $\mathbb{T}^3$. The dynamics in a rectangular box is known to be reduced to the dynamics on a torus. The Boltzmann–Enskog kinetic equation for this gas takes the form

$$\frac{\partial f(r_1, v_1, t)}{\partial t} = -v_1 \frac{\partial f(r_1, v_1, t)}{\partial r_1} + na^2 \int_{(v_{21},\sigma)\geq 0} (v_{21}, \sigma)$$

$$\times [f(r_1, v_1', t)f(r_1 + a\sigma, v_2', t) - f(r_1, v_1, t)f(r_1 - a\sigma, v_1, t)]d\sigma dv_2. \quad (1)$$

Here $f(r_1, v_1, t)$ is the density of particles with positions near $r_1 \in \mathbb{T}^3$ and velocities near $v_1 \in \mathbb{R}^3$ at the moment of time $t$. Let $f$ be normalized on

342

the volume $V$ of the torus $\mathbb{T}^3$:

$$\int_{\mathbb{T}^3 \times \mathbb{R}^3} f(r_1, v_1, t)\, dr_1 dv_1 = V.$$

The integral term in the right-hand side of the equation (1) is called the collision integral, we will denote it by $\mathrm{St}\, f$. $n > 0$ is a constant (the mean concentration of particles), $v_1$ and $v_2$ are the velocities of the colliding particles just before the collision, $v_1'$ and $v_2'$ are the velocities of the particles just after the collision:

$$v_1' = v_1 + \sigma(v_{21}, \sigma),$$
$$v_1' = v_2 - \sigma(v_{21}, \sigma). \tag{2}$$

Here $v_{21} = v_2 - v_1$, the vector $\sigma$ is a unit vector from the center of the second particle to the center of the first particle (so, $\sigma$ belongs to the 2-dimensional unit sphere: $\sigma \in S^2$), $(\cdot, \cdot)$ is a scalar product.

The Boltzmann–Enskog equation describes the irreversible dynamics of a hard sphere gas and the entropy production, a rigorous $H$-theorem for this equation (and more general Enskog equation) can be found in [11]. The $H$-function for the Boltzmann–Enskog equation slightly differs from the $H$-function for the Boltzmann equation.

Now let us consider a function $f$ which has the form of a sum of delta-functions:

$$f(r_1, v_1, t; \Gamma) = n^{-1} \sum_{i=1}^{N} \delta(r_1 - q_i(t, \Gamma), v_1 - w_i(t, \Gamma)), \tag{3}$$

where $q_1(t, \Gamma), w_1(t, \Gamma), \ldots, q_N(t, \Gamma), w_N(t, \Gamma)$ are the positions and velocities of $N$ hard spheres at the moment $t$. Here $n = \frac{N}{V}$. The time dependence is defined by the initial positions and velocities

$$\Gamma = (q_1^0, w_1^0, \ldots, q_N^0, w_N^0)$$

and the laws of motion for hard spheres (i.e., law (2) for pairwise collisions and the free motion between the collisions, we neglect the collisions of three and more spheres simultaneously, because they have zero measure).

Denote $G = \mathbb{T}^3 \times \mathbb{R}^3$ (the single-particle phase space), $\Omega$ is the set of points $(r_1, v_1, \ldots, r_N, v_N) \in G^N$ such that $|r_i - r_j| \geq a$ for all $i \neq j$. Obviously, $\Gamma \in \Omega$, since hard sphere centers cannot be closer than $a$ to each other.

$f$ is a generalized function. Let it be defined on test functions from $C^1(\Omega)$ (continuously differentiable functions on $\Omega$) that satisfy the following additional condition:

$$D(\ldots, r_1, v_1, \ldots, r_1 - a\sigma, v_2, \ldots) = D(\ldots, r_1, v_1', \ldots, r_1 - a\sigma, v_2', \ldots), \quad (4)$$

for all $r_1, v_1, v_2, \sigma$ and other arguments (denoted as "$\ldots$"). Here $v_1'$ and $v_2'$ are defined by (2). The arguments $(r_1, v_1)$ and $(r_1 - a\sigma, v_2)$ stand on arbitrary positions in $D$, the other arguments in both sides of the equality are the same. Every function $D \in C^1(\Omega)$ can be regarded as a function on $G^N$ if we put $D(r_1, v_1, \ldots, r_N, v_N) = 0$ in the case $|q_i^0 - q_j^0| \le a$ for some $i \ne j$. These test functions constitute a subspace of $C^1(\Omega)$ denoted by $CC^1(\Omega)$.

Denote

$$\int f(x_1, t; \Gamma) D(\Gamma) \, d\Gamma = F_1(x_1, t),$$

$$\int f(x_1, t; \Gamma) f(x_2, t; \Gamma) D(\Gamma)) \, d\Gamma = F_2(x_1, x_2, t). \tag{5}$$

$F_1$ and $F_2$ are the single-particle and two-particle density functions correspondingly.

Condition (4) can be rewritten in terms of the two-particle density function $F_2$:

$$F_2(r_1, v_1, r_1 - a\sigma, v_2, t) = F_2(r_1, v_1', r_1 - a\sigma, v_2', t) \tag{6}$$

for all $r_1, v_1, v_2, \sigma$ and $t$.

The meaning of condition (4) (or (6)) is easy to understand if $D$ is a probability density function. It is a continuity condition: since the collision of hard spheres is instantaneous, the two-particle density just after the collision should be equal to the density just before the collision.

In fact, we can drop condition (4). But this case is more sophisticated to analyze. For simplicity we will assume condition (4).

**Proposition 2.1.** *The (generalized) function $f$ given by (3) satisfies the Boltzmann–Enskog equation (1). The both sides of the equation are understood as functionals over $C^1(\Omega)$ with additional condition (4).*

**Proof.** The single-particle and two-particle density functions for the system of $N$ hard spheres are known to satisfy the following equation (the first equation of the BBGKY hierarchy for hard spheres, see [12]; condition (4) is used in the derivation of this equation):

$$\frac{\partial F_1(r_1, v_1, t)}{\partial t} = -v_1 \frac{\partial F_1(r_1, v_1, t)}{\partial r_1} + na^2 \int_{(v_{21}, \sigma) \geq 0} (v_{21}, \sigma)$$

$$\times [F_2(r_1, v_1', r_1 + a\sigma, v_2', t) - F_2(r_1, v_1, r_1 - a\sigma, v_1, t)] d\sigma dv_2, \quad (7)$$

where $n = \frac{N}{V}$. Here the argument $r_1 \pm a\sigma$ means $r_1 \pm a\sigma \pm 0$, since $F_2(r_1, v_1, r_2, v_2, t) = 0$ as $|r_1 - r_2| < a$. Due to (5) this equation can be written as

$$\int \{\frac{\partial f(r_1, v_1, t)}{\partial t} + v_1 \frac{\partial f(r_1, v_1, t)}{\partial r_1} - \mathrm{St}\, f\} D(\Gamma)\, d\Gamma = 0.$$

Since $D(\Gamma)$ is an arbitrary function from $CC^1(\Omega)$, this equality means that $f$ satisfies the Boltzmann–Enskog equation in the stated sense. $\qquad\square$

This proposition has been proved by Bogolyubov[9]. Now we have formulated it rigorously (namely, we have specified the space of test functions and condition (4)).

A solution of form (3) is called a microscopic solution of the considered kinetic equation. This result is surprising for two reasons. The first reason is that the Boltzmann–Enskog is regarded as an approximation to the $N$-particle dynamics in the case of a dilute gas ($n$ is small). But this result shows that this equation has exact microscopic solutions as well. The Vlasov kinetic equation is also known to have microscopic solutions[13].

But, in contrast to the Vlasov equation, the Boltzmann–Enskog equation is time-irreversible and describes the entropy production. This proposition tells that it has solutions $f$, which are defined by the time-reversible dynamics of $N$ hard spheres. In other words, both $f(r_1, v_1, t; \Gamma)$ and $\widetilde{f}(r_1, v_1, t; \Gamma) = f(r_1, -v_1, -t; \Gamma)$ are solutions of the Boltzmann–Enskog equation, despite of the fact that this equation is time-irreversible!

Indeed,

$$f(r_1, -v_1, -t; \Gamma) = f(r_1, v_1, t; \widetilde{\Gamma}), \quad \widetilde{\Gamma} = (q_1^0, -w_1^0, \ldots, q_N^0, -w_N^0).$$

So, the function $\widetilde{f}$ also has form (3) and, hence, is a solution of (1).

## 3. Reversible and irreversible classes of solutions of the Boltzmann–Enskog equation

Now, we are able to answer the question about the relation of the irreversible Boltzmann–Enskog equation (1) and its reversible solution (3).

For generality, let us denote $\Delta$ the generalized functions in the form (3) with arbitrary functions $q_i(t, \Gamma)$, $w_i(t, \Gamma)$, $i = 1, \ldots, N$.

**Theorem 3.1.** *1) Let us treat the Boltzmann–Enskog equation (1) as an equation for functions $f \in C^1(\mathbb{R}^3 \times \mathbb{R}^3 \times [0, T))$, $T > 0$ (the classical solution of the Boltzmann–Enskog equation is defined in this class [14]). In this case the equation is irreversible, i.e., if some function $f(r_1, v_1, t)$ is a solution of this equation, then, in general, the function $\widetilde{f}(r_1, v_1, t) = f(r_1, -v_1, -t)$ is not a solution;*

*2) Let us treat the Boltzmann–Enskog equation as an equation for generalized functions $f \in \Delta$ over test functions from $CC^1(\Omega)$. In this case the equation is reversible, i.e., if some function $f(r_1, v_1, t; \Gamma)$ is a solution of this equation, then the function $\widetilde{f}(r_1, v_1, t; \Gamma) = f(r_1, -v_1, -t; \Gamma)$ is also a solution.*

A proof can be found in Ref. 15. The main idea of the proof is the following: the condition

$$f(r_1, v_1, t)f(r_1 - a\sigma, v_2, t) = f(r_1, v_1', t)f(r_1 - a\sigma, v_2', t) \tag{8}$$

is crucial for the reversibility of the Boltzmann–Enskog equation. If $f$ is a continuous function, this condition is, in general, not satisfied. Hence, the Boltzmann–Enskog equation is irreversible in this case. But if $f \in \Delta$, then condition (8) is satisfied. In this case condition (8) is equivalent to (6) by definition, since the both sides of (8) are understood as functionals over the space of probability density functions:

$$\int f(r_1, v_1, t)f(r_1 - a\sigma, v_2, t)D(\Gamma)\,d\Gamma = \int f(r_1, v_1', t)f(r_1 - a\sigma, v_2', t)D(\Gamma)\,d\Gamma \tag{9}$$

Of course, this is a direct consequence of (5) and (6). Hence, in this case the Boltzmann–Enskog equation is reversible.

So, if we consider continuous solutions of the Boltzmann–Enskog equation, then the solution is irreversible, whereas if we consider the solutions in the form of a sum of delta-functions, then the solutions are reversible.

This result can be compared with the Kozlov's diffusion theorems for a collisionless continuous medium in a rectangular box [8]. The dynamics is described by the Liouville equation. If we consider a solution in the form of a delta-function, then the solution is periodic or almost periodic (since this case is reduced to the single-particle dynamics). But if we consider an integrable solution, the spatial density weakly converges to the uniform distribution, as time indefinitely increases. So, the asymptotic dynamical properties of solutions depend on the considered class of functions.

Here we show that in the case of the Boltzmann–Enskog equation not only the asymptotic properties of solutions, but also the reversibility or irreversibility property of them depend on the considered class of functions.

## 4. Reconciliation of the kinetic equations with the microscopic dynamics

Of course, the above arguments cannot be regarded as a derivation of the kinetic Boltzmann–Enskog equation from the microdynamics. According to the traditional point of view, we must start with some initial $N$-particle distribution function $D$ (a randomization of the initial $N$-particle phase point) and analyse the dynamics of $F_1(x_1, t)$ (see (5)). By doing this, we get equation (7). Then, we must obtain the Boltzmann–Enskog (or Boltzmann) equation for $F_1$ using some limiting procedure. Bogolyubov[16] applies the thermodynamic limit $N \to \infty$, $V \to \infty$, $\frac{N}{V} = const$ to get the Boltzmann–Enskog equation. Lanford[17] applies the so called Boltzmann–Grad limit $N \to \infty$, $a \to 0$, $Na^2 = const$ to get the Boltzmann equation.

But both derivations have significant drawbacks. Bogolyubov's derivation uses some additional assumptions (which do not follow from the Liouville equation or its initial data). Besides, this derivation leads to the divergences in high order corrections to the Boltzmann equation[18]. The Lanford's derivation can be applied only to small times. On the contrary, the Boltzmann equation is interesting from the viewpoint of the large time asymptotics (the relaxation to the Maxwell distribution)[19].

However, the microscopic solutions of the Boltzmann–Enskog equation and the investigations of the reversibility or irreversibility property of it suggest us another way of the reconciliation of the kinetic equations with the microdynamics.

Let us postulate not the microdynamics, but the Boltzmann–Enskog equation. That is, the dynamics of a system that we call "a system of hard spheres" is postulated to be described by the Boltzmann–Enskog equation (1) with a *continuous* function $f$. Then this equation is noticed to have also special (noncontinuous) solutions in form (3). The dynamics of $(q_i(t, \Gamma), w_i(t, \Gamma))$, $i = 1, \ldots, N$, in these solutions are interpreted as the microscopic dynamics.

So, the establishment of the existence of microscopic solutions for a kinetic equation can be regarded as a derivation of the particle dynamics from it. Instead of the derivation of the kinetic equations from the microdynamics we suggest a kind of derivation of the microdynamics from the kinetic equations.

According to this point of view, the "fundamental" laws of motion are not the laws of motion for individual particles (reductionism), but the laws of motion for the function $f$, which describes the system as a whole (holism). Laws of motions for the individual particles are approximations. This approximation is suitable if the density function has a special (delta-like) form (see, e.g., theorem 1).

Functional mechanics [2,3,4,5,6] also claims that the Newton equation (or Hamilton equations) is not fundamental even in the classical nonrelativistic world.

Also it should be noted that, in principle, the function $f$ in the Boltzmann–Enskog equation, as well as other kinetic equations, can be interpreted without any reference to the notion of a particle. Let $f$ be normalized not on the volume $V$, but on the total mass $M = Nm$ ($m$ is the mass of a particle): $\int f(r, v, t) dr dv = M$. Then the Boltzmann–Enskog equation has the same form, but the constant $n = \frac{N}{V}$ in the collision integral is replaced by $\frac{1}{m}$. $f(r, v, t)\Delta r \Delta v$ has the meaning of the mass of the matter which is contained in the volume element $\Delta r$ and moves with the velocity in the range $\Delta v$. So, $f(r, v, t)$ is the density function of the matter in the phase point $(r, v) \in \mathbb{T}^3 \times \mathbb{R}^3$.

We have a kind of continuous medium. In contrast to the usual continuum mechanics, there is no definite velocity in a fixed point of physical space: in the same volume element $\Delta r$ different parts of the matter move with different velocities. The terms $f(r_1, v_1, t)f(r_1 \pm a\sigma, v_2, t)$ in the collision integral are intensities of interaction of different parts of the matter.

Initially, atomic theory of matter was an attempt to reduce the properties of the matter to the properties of its elementary blocks. But the problems like the reversibility paradox show that this program and the notion of an atom should be revised.

Thus, we may still interpret the function $f$ as the single-particle density function, or we may interpret it as the density function of a continuous matter. In both cases, we postulate the Boltzmann–Enskog equation for the system as a whole rather than the equations for the individual particles.


**Remark 4.1.** A question concerns the notion of the derivation can arise. Mathematically we can postulate a kinetic equations and speak about the derivation of the particle dynamics from it. But actually, kinetic equations are known to be obtained heuristically with the use of the notion of a particle and the laws of motion for particles.

It is more preferable to speak not about a derivation, but about a reconciliation of the kinetic dynamics with the microdynamics. Usually, the reconciliation is understood as a derivation of the kinetic equations from the microdynamics. We claim that neither kinetic dynamics can be derived from the microdynamics, nor microdynamics can be derived from the kinetic dynamics in the "pure" sense. Every level of nature has its own laws, but one can speak about some agreement between them.

We do not demand a kinetic equation to be derived from the microdynamics, by we say that a kinetic equation is in agreement with the microdynamics if it has microscopic solution or if it is an approximation of another kinetic equation which has microscopic solutions (like the Boltzmann equation, see Ref. 15).

## 5. Conclusions

The reversibility or irreversibility property of the Boltzmann–Enskog kinetic equation is shown to be dependent on the considered class of solutions. The equation is irreversible if the considered class of solutions is continuous functions. It is reversible if the considered class of solutions is sums of the delta-functions. Such solutions are called the microscopic solutions.

Functional mechanics suggests to describe $N$-particle mechanical systems in terms of the probability density function in a $6N$-dimensional phase space. The evolution of this $N$-particle density function is given by the Liouville equation. A description in terms of the density function seems to be more appropriate for the applications to complex and biological systems.

This report is a development of functional mechanics. We have considered the problem of the reconciliation of $N$ hard sphere dynamics with the Boltzmann–Enskog kinetic equation. Instead of the derivation of the Boltzmann–Enskog equation from the microdynamics (like in the traditional paradigm), we have postulated the Boltzmann–Enskog equation and have claimed that it is in agreement with the microdynamics in the sense that it has microscopic solutions.

Thus, we have postulated that the fundamental description of a mechanical system is not a point in the $6N$-dimensional phase space (like in the Newtonian mechanics), not an $N$-particle probability density function (like in the original functional mechanics), but a single-particle probability density function, which describes the system as a whole. In accordance to the ideas of functional mechanics, this is one more step from the reductionist picture of the world towards the holistic one.

## Acknowledgments

The author is grateful to V. V. Kozlov, S. V. Kozyrev, A. I. Mikhailov, V. V. Vedenyapin, and I. V. Volovich for helpful discussions and remarks. This work was partially supported by the Russian Foundation of Basic Research (projects 11-01-00828-a and 11-01-12114-ofi-m-2011), the grant of the President of the Russian Federation (project NSh-2928.2012.1), and the Division of Mathematics of the Russian Academy of Sciences.

## References

1. M. Ohya and I. Volovich, *Mathematical Foundations of Quantum Information and Computation and Its Applications to Nano- and Bio-systems* (Springer, Dordrecht, 2011).
2. I. V. Volovich, in *Quantum Bio-Informatics III* (World Scientific, Singapore, 2010).
3. I. V. Volovich, *Found. Phys.* **41** 516 (2011); arXiv: 0910.5391v1 [quant-ph]
4. A. I. Mikhailov, *p-Adic Numbers, Ultrametric Analysis and Applications* **3** 205 (2011).
5. E. V. Piskovskiy and I. V. Volovich, in *Quantum Bio-Informatics IV* (World Scientific, Singapore, 2011).
6. A. S. Trushechkin and I. V. Volovich, *p-Adic Numbers, Ultrametric Analysis and Applications* **1** 361 (2009); arXiv: 0910.1502 [math-ph].
7. I. V. Volovich, *Theor. Math. Phys.* **164** 1128 (2010).
8. V. V. Kozlov, *Thermal Equilibrium in the sense of Gibbs and Poincar'e* (Institute of Computer Research, Moscow–Izhevsk, 2002) [in Russian]; V. V. Kozlov, *Dokl. Math.* **65** 125 (2002).
9. N. N. Bogolyubov, *Theor. Math. Phys.* **24** 804 (1975).
10. N. N. Bogolubov, N. N. Bogolubov Jr., *Introduction to Quantum Statistical Mechanics* (World Scientific, Singapore, 2010).
11. N. Bellomo and M. Lachowicz, *Lecture Notes in Mathematics: Mathematical Aspects of Fluid and Plasma Dynamics* **1191** 15 (1991).
12. C. Cercignani, *Theory and Application of the Boltzmann Equation* (Scottish Academic Press, Edinburgh, 1975; Elsevier, New York, 1975); C. Cercignani, *Transport Theory and Statistical Physics* **2** 211 (1972).
13. A. A. Vlasov, *Many-particle theory and its application to plasma* (Gordon and Breach, New York, 1961).
14. S.-Y. Ha and S.-E. Noh, *Nonlinearity* **19** 1219 (2006).
15. A. S. Trushechkin, "Derivation of the particle dynamics from the kinetic equations" *p-Adic Numbers, Ultrametric Analysis and Applications*, to be published; arXiv: 1201.3607 [math-ph].

16. N. N. Bogoliubov, *Problems of Dynamic Theory in Statistical Physics* (Gostekhizdat, Moscow–Leningrad, 1946; North-Hòlland, Amsterdam, 1962; Interscience, New York, 1962).
17. O. E. Lanford, *Lect. Notes Phys.* **38** 1 (1975).
18. N. N. Bogolyubov, *Kinetic Equations and Green Functions in Statistical Mechanics* (Institute of Physics of the Azerbaijan SSR Academy of Sciences, Baku, 1977), Preprint N.57 (in Russian).
19. C. Villani, *Handbook of Mathematical Fluid Dynamics* **1** 71 (2002).

Quantum Bio-Informatics V
© 2013 World Scientific Publishing Co. Pte. Ltd.
pp. 351–362

# IMPLICATIONS OF DNA-NANOSTRUCTURES BY HOOGSTEEN-DINUCLEOTIDES ON TRANSCRIPTION FACTOR BINDING

DIERK WANKE[*,1], LUISE H. BRAND[1], NINA M. FISCHER[2], FLORIAN PESCHKE[1], JOACHIM KILIAN[1] AND KENNETH W. BERENDZEN[+]

[1] *ZMBP Pflanzenphysiologie, Universität Tübingen*
*Auf der Morgenstelle 1, D-72076 Tübingen, Germany*
[*] *e-mail: dierk.wanke@zmbp.uni-tuebingen.de*
[+] *e-mail: kenneth.berendzen@zmbp.uni-tuebingen.de*
[2] *Center for Bioinformatics Tübingen (ZBIT), Universität Tübingen*
*Sand 1, D-72076 Tübingen, Germany*

Recent findings showed that non-harmonic DNA-nanostructures are formed by Hoogsteen (HG) dinucleotides *in vivo*. In contrast to Waston-Crick (WC) base pairing, the purine base component is flipped from *anti-* to *syn*-conformation. This change consequently alters the width of the DNA-helix, the sizes of minor and major groove and biophysical properties, such as the melting temperature. Three dinucleotides (CA, TG and TA) have been identified that form stable HG conformations. Functional data and structural models imply that transcription factors specifically bind DNA-motifs that consist of both HG and WC base pairs - especially at the topological transition between HG and WC dinucleotides. We could show that most know *cis*-regulatory elements contain at least one HG dinucleotide. In addition, we focused our work on human promoter sequences that encode gene regulatory information within double stranded DNA. We compared occurrences of HG dinucleotides to all 16 dinucleotides. These ratios differed most in sequences closer to gene transcripts, where the promoters are located. These findings imply that transcription factors might explicitly recognize their DNA-motifs in regulatory promoter sequences that exhibit HG nanostructure islands.

*Keywords*: DNA-nanostructures; Hoogsteen base-pairing; Watson-Crick base-pairing; transcription factor-DNA-interaction; *cis*-regulatory element (CRE); DPI-ELISA.

## 1. Introduction

Gene expression is controlled by direct binding of diverse sets of proteins to DNA, which integrate incoming developmental and environmental signals to translate them into specific responses. The places where proteins,

352

typically transcription factors, make direct contact to DNA are usually located in the non-coding double-stranded DNA of promoters upstream of the coding sequences. One turn of the B-form DNA-double helix is about 10 base pairs in length, whereas core DNA-motifs that are recognized by the DNA-binding domains of proteins are usually short and range between 4 and 6 base pairs [1,2,3,4]. How transcription factors discriminate their specific binding motif from all other short and abundant DNA-patterns is not understood to date[5,6]. Thus, it is still a challenging task for biologists and bioinformation scientists to extract meaningful information from these promoter sequences and to conclude on regulatory function based solely on sequence pattern.

Recent findings suggest that protein-DNA-interaction might not only be governed by the linear sequence of the four different DNA bases adenine, cytosine, guanine and thymine (A, C, G and T), but also by different DNA topology[5,6,7,8,9,10,11]. It is widely accepted textbook knowledge that the double helix is of harmonic B-form *in vivo*, which is structurally achieved by almost equidistant Watson-Crick base pairing between purine and pyrimidine bases (T • A and G • C)[12]. Hydrogen bonds, two in T • A and three in G • C (Fig. 1), stabilize the two antiparallel oriented DNA strands and influence the energy, which is needed to separate the two strands (i.e. melting temperature $T_m$), e.g. during transcription. So far, the B-form of DNA is thought to be the only relevant one important for organism life. Other types of DNA, such as the A- and Z-form, were only known to occur during *in vitro* experiments. In contrast, Z-like DNA-nanodomains have been identified in DNA that were bound by proteins[5,6,7,8,11,13,14,15]. Interestingly, this Z-like base pairing had a special conformational geometry that had first been reported by Karst Hoogsteen in 1963[16] just ten years after the publication of the B-form by Watson and Crick[12]. In Hoogsteen (HG) base pairs the purine base is rotated to *syn*-conformation at its glycosidic bond, while Watson-Crick (WC) base pairs are both in *anti*-conformation. The resulting *syn* / *anti*-conformation in HG base pairs has major effects on the biophysical properties of the double helix, such as a smaller diameter (~1.84 nm), greater height between each base pair (~0.38 nm) and a higher number of base pairs per turn (~12)[5,6,10,11,13,15,16]. Moreover, HG base pairs have a lower melting temperature for G • C pairs, because only two hydrogen bonds are formed instead of three (Fig. 1)[9,14,15,16,17]. In consequence the DNA topology in HG pairs is a larger and flattened major groove, while the minor groove becomes even more narrow and deeper than in WC[6,11,13,14,16,17]. Recently, it was shown that stable HG base pairing

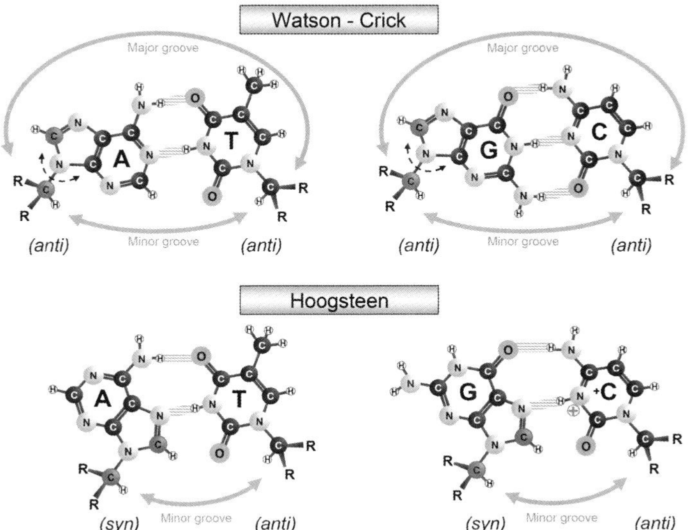



Figure 1. Scheme for Watson-Crick and Hoogsteen base pairing. In Hoogsteen base pairs the purine base is rotated at the glycosidic bond from *anti-* to *syn*-conformation, which reduces the size of the minor grove. Note, G • C base pairs form three hydrogen bonds in Watson-Crick pairing, while there are only two bonds for the Hoogsteen pair.

occurs in detectable amounts also under native conditions inside cells, particularly when two consecutive base pairs (dinucleotides) both flip into HG conformation[15]. This is especially the case for CA (antisense TG) and the palindromic TA dinucleotides[15]. HG conformation has also been found in DNA bound by proteins[7,8,9,14,15]. This implies that DNA-binding proteins might exist, which specifically recognize these nanostructures of HG dinucleotides in a WC environment or even induce the transition of WC to HG conformation[6,11,14,15,13]. Witnessing these processes is not possible with the present analysis methods and is a challenging task for future scientists. However, a bioinformatics approach might shine some light on HG and WC functionality in promoter sequences, where transcription factors are known to bind to DNA and modulate gene expression.

Therefore, we here give an overview on the possible effects of HG base pairing on protein-DNA-interaction. As double stranded DNA carries the information for gene expression regulation, we investigated the occurrence of possible Hoogsteen dinucleotides in promoter sequences in comparison to Watson-Crick dinucleotides. Furthermore, we investigated the presence of HG and WC dinucleotides in known *cis*-regulatory elements (CREs). In contrast to our expectations, dinucleotides with possible HG conformation could be found more frequent in some known CREs, which implies that some transcription factors might specifically recognize these Z-like DNA-nanostructures in regulatory promoter sequences.

## 2. Material and Methods

### 2.1. *Structural information and modeling*

Molecular structure data has been retrieved from the Protein Data Bank (PDB) Resource for Studying Biological Macromolecules (http://pdb.org)[18,19]: 1QNE [TATA-binding protein with TATA-box contact][20], 3KZ8 [DNA double helix with Watson-Crick and Hoogsteen base pairing][8]. Models were conducted by using AMBER11[21] or BALLView[22,23].

Chain C of structure 3KZ8 was transformed from PDB transformation-matrix to generate the double stranded Hoogsteen-containing helix by using $C++$ and implemented BALL-methods. Chain A of structure 1QNE is imaged as solvent excluded surface for better visualization; chains C and D are stick-and-ribbon models. Non-informative base pairs have been omitted for better visibility.

### 2.2. *Occurrence of Hoogsteen dinucleotides in sequence datasets*

The genome sequence of the chromosome pseudomolecules for the human (*Homo sapiens*) were retrieved from GenBanks Genomes Central database (http://www.ncbi.nlm.nih.gov/genomes/); Assembly GRCh37(Patch Release 4).

Promoters were extracted either as -1500 bp and -600 bp sequence sets or as -1500 bp to +1500 bp sets 5′ of the annotated transcription (TSS) or translation (ATG) start site for each gene (without redundancy) using Motif Mapper (5.1.1.39) or Motif Mapper for Python (v1.3) (http://www.zmbp.uni-tuebingen.de/PlantPhysiology/ ResearchGroups/harter/berendzen/programs.html)[1].

Datasets for known functional *cis*-regulatory elements have been retrieved from The JASPAR database (http://jaspar.genereg.net/)[24]. JASPAR PWM-matrices were read with the Motif package from biopython[25]. The PWM was then converted to all possible combinations to measure the HG/dinuclotide content using the default Motif function.

Occurrence and frequencies for HG/dinuclotide were counted in Python with scripts from Motif Mapper. The data was visualized with Kernel-density plots called from the standard R libraries (http://cran.r-project.org/).

## 2.3. *DNA-Protein-Interaction-ELISA (DPI-ELISA)*

DPI-ELISA was performed with crude extracts from *E.coli*, expressing the respective proteins from suitable expression-vectors. Preparation of double stranded DNA-oligonucleotide probes, testing and readout were performed as published previously in Brand *et al.* (2010)[26].

## 2.4. *Statistical analysis*

For multiple comparison procedure and statistical testing the Tukey HSD had been applied to the different sequence datasets set to the median containing the HG/dinuclotide occurrences, by calling the multcompBoxplot of the multcompView package implemented in R (http://cran.r-project.org/).

## 3. Results and Discussion

Previous analyses suggested that the formation of Hoogsteen (HG) dinucleotides harbors different information content than its Watson-Crick (WC) surrounding[8,9,13,15]. The formation of HG base pairs increases the DNAs entropy by shaping non-harmonic nanostructure domains that differ from harmonic WC double helices (Fig. 2).

DNA that is solely composed of WC base pairs is harmonic and one has to assume that DNA-binding proteins have to read the linear sequence of bases until they recognize their specific DNA-motif consensus for binding (Fig. 2). In contrast, a DNA-sequence that is supplemented with HG dinucleotide islands is structurally diverse. Here, the HG dinucleotides form short stretches of Z-like DNA structures that interrupt the tube like B-type DNA of WC base pairs with their flat surface topology (Fig. 2).

To gain a deeper insight into the proteins that bind to HG dinucleotides, we retrieved the atomic coordinates of a TATA-binding protein (TBP) that is bound to the TATA-box DNA motif[20]. It has been reported that TBPs preferentially target TATA-motifs in HG conformation[8,14,15,20] and, indeed, it became evident that the protein makes contact to DNA sequences that contain both WC (white arrows) and HG (black arrow) base pairs (Fig. 3A and B). Figure 3C schematically illustrates that a total of six base pairs

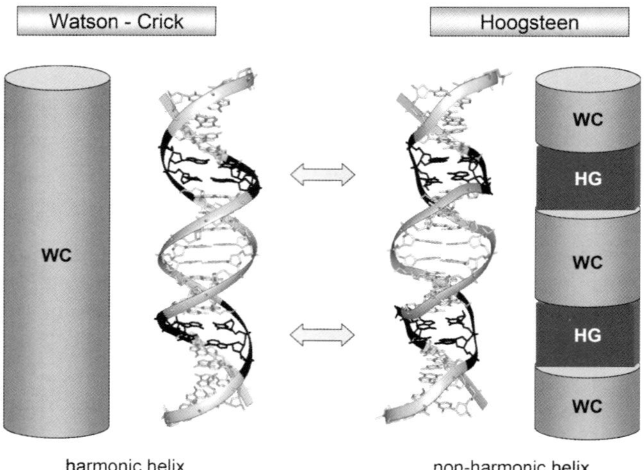



Figure 2. Harmonic Watson-Crick and non-harmonic Hoogsteen DNA. Watson-Crick (WC) and Hoogsteen (HG) base pairing of identical DNA sequences (PDB: 3KZ8). Harmonic B-form of DNA is characterized by Watson-Crick base pairs. Hoogsteen dinucleotides form a Z-like DNA topology, in which the major groove composes a flat surface that. Harmonic B-form DNA is interrupted by Hoogsteen dinucleotides that forms non-harmonic regions by Z-like nanodomains.

are important for the thermodynamically stable binding and positioning of TBP, four of them are two consecutive TA dinucleotides that form HG base pairs. The AA dinucleotides at $5^{th}$ and $6^{th}$ position of the TATAAA-consensus and all dinucleotides adjacent to it are regular WC base pairs (Fig. 3C). The TATA-box motif is an evolutionary conserved *cis*-regulatory element in the core-promoters of all organisms[1,27]. Therefore, it is not a far-fetched assumption that the TBPs from all organisms will target the TATA-boxes at the transition between HG and WC base pairs. At present, the formation of stable HG base pairs could only be shown for dinucleotides *in vivo*[15]. However, longer stretches of HG base pairs possibly exist, but their identification will eventually cause problems: In between each of the three HG dinucleotides (CA, TG and TA) principal WC dinucleotides are positioned (Fig. 3D). In addition, it is not well understood whether HG conformation might also have an influence on the binding of other transcription factors that bind to more complex DNA signatures than the TBP.

Next we investigated the binding signatures of previously published data on *At*bZIP63 and *At*WRKY11[26]. In DPI-ELISA experiments *At*bZIP63

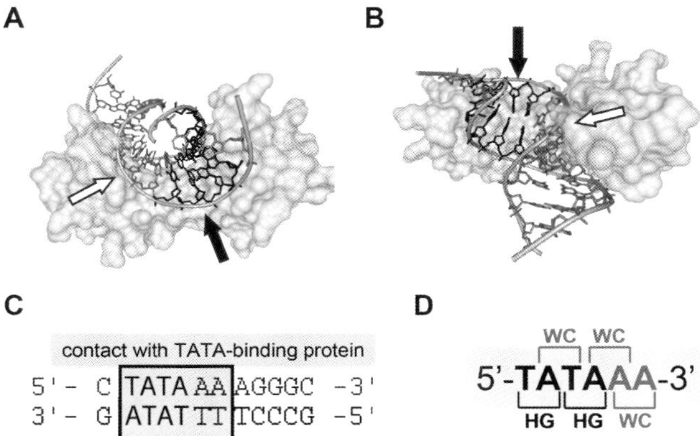



Figure 3. TATA-binding protein (TBP) binds to Hoogsteen dinucleotides. Three-dimensional structural model of the TBP (PDB: 1QNE) reveals the surfaces that interact with both HG (black) and WC (grey) dinucleotides (A) and (B). Arrows indicate where HG (black) and WC (white) base pairs make contact to TBP. (C) DNA-sequence that was used in the structural model. Bases that contact the TBP surface are surrounded by a box. Hoogsteen and Watson-Crick dinucleotides are displayed in black and grey, respectively. (D) DNA-pattern analysis of HG or WC dinucleotides is difficult as both structures can possibly overlap.

has been shown to specifically bind to the so called C-box DNA-motif (Fig. 3A). No binding was detectable for a mutated C-box version, where the palindromic motif has been deleted by the inversion of a dinucleotide (Fig. 4A and C). The C-box contains the highly conserved bZIP binding core ACGT, which is solely composed of non-HG dinucleotides. This WC core-motif is flanked by HG dinucleotides on either side (Fig. 4C). In the mutant C-box the dinucleotide inversion not only affected the conserved ACGT sequence, but also generated an additional HG dinucleotide. In contrast, $At$WRKY11 might be an example for the necessity of HG base pairs within the W2-box consensus. The $2^{nd}$ and $3^{rd}$ positions of the W2-box are essentially needed for the strong and specific binding of $At$WRKY11 and other WRKY transcription factors (Fig. 4B)[28]. This TG position is possibly forming a Hoogsteen dinucleotide, embedded in a WC environment (Fig. 4D). Mutations that affect this dinucleotide completely abolishes the binding of WRKY transcription factors to W-boxes (Fig. 4B)(Ciolkowski et al., 2008). Here again, the HG dinucleotide is surrounded by WC base

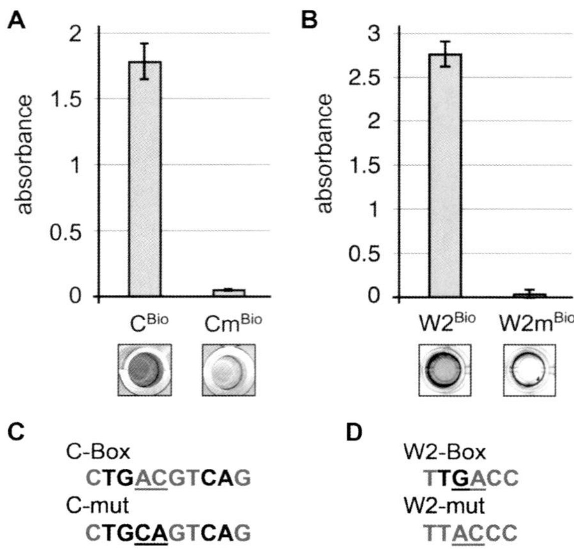



Figure 4. In Vitro DNA binding studies with DPI-ELISA. Binding of *At*bZIP63 to positive C-box and negative C-mut DNA-probe (A) and of *At*WRKY11 to positive W2-box and negative W2-mut DNA-probe (B) by DPI-ELISA. Binding to DNA-probes was quantified by colorimetric detection[26]. (C) and (D) show the core DNA-binding consensi. The *At*bZIP63 binding core is composed of Watson-Crick (grey) and flanked by Hoogsteen (black) dinucleotides (C). *At*WRKY11 binds to the W2-box hexanucleotide that contains a Hoogsteen dinucleotide (black) in its core binding consensus (D).

pairs, like for the TATA-box binding motif. The data presented here and previous data suggested that the presence of HG dinucleotides in functional CREs might be essential for their recognition by DNA-binding proteins.

To investigate the significance of Hoogsteen *versus* Watson-Crick dinucleotides in known transcription factor DNA-binding sites, we retrieved sequence information for all known CREs from the JASPAR-database[24]. As there are 16 possible dinucleotides, we assumed that the three HG dinucleotides should be statistically more frequent in CREs, if they were essentially needed for the recognition by the proteins. However, due to the problem of overlapping dinucleotides in any given sequence (Fig. 3D), the pattern matching algorithms will always yield a maximal HG content of only 60% ($p = 0.6$).

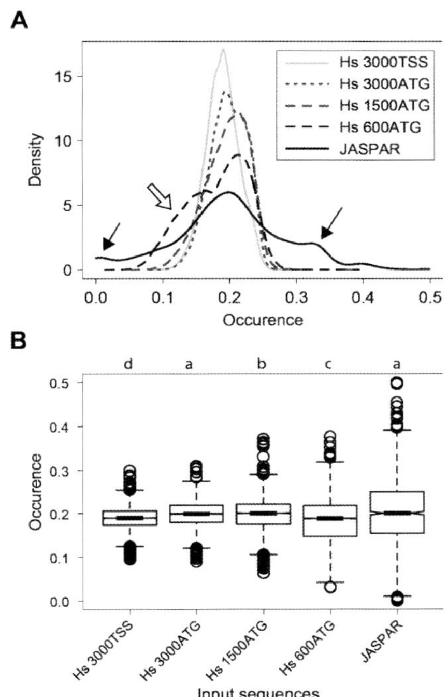

Figure 5. Assessing the HG-dinucleotide in various human promoter datasets. (A) Kernel density distribution plots of the HG-dinucleotide content eukaryotic *cis*-elements available from JASPAR and in the promoter datasets rooted to the translation start site (ATG) or the transcription start site (TSS). WHITE arrow: The 600ATG dataset has a secondary distribution that is depleted in HG-dinucleotides compared to the other promoter datasets. BLACK arrows: regions of HG-dinucleotide depleted or enriched content that are drastically different compared to the promoter contents. (B) Box plot representations of HG-dinucleotide content in the 5 datasets (median centered). Letters indicate independent statistical classes based on TukeyHSD.

In accordance with our assumption, several CREs exist that exhibit a significantly higher number of HG dinucleotides, nearly up to the maximal content of 60%. Conversely, there are also a considerable number of motifs that do not contain any HG dinucleotides and are composed of WC base pairs only (Fig. 5A). The average and the median of HG dinucleotide occurrence in known CREs was $p = 0.2$.

We next examined the occurrence of HG dinucleotides in four promoter sequence sets of the human genome. The sets displayed significant differ-

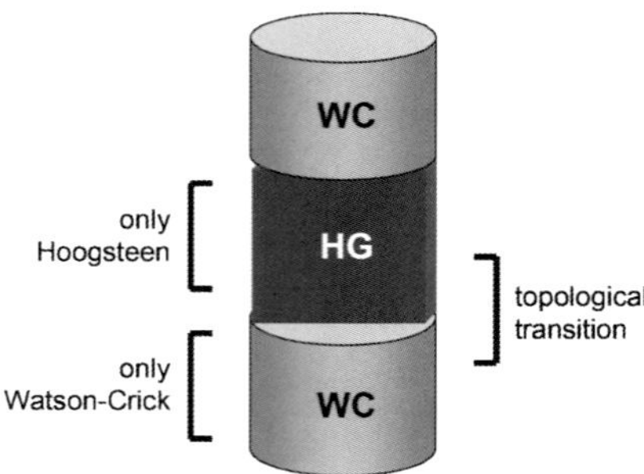

Figure 6. Scheme of nanostructural islands of HG domains in WC B-form DNA Transcription factors recognize a combinatorial set of features, such as the linear DNA sequence (WC or HG), DNA topology (WC or HG) and possibly HG nanodomains in CREs

ences in their HG composition (Fig. 5B), although they contain the same promoters in varying lengths. The variability of HG dinucleotide content within the datasets increases from 1500 bp to 600 bp upstream of the ATG and, thus, HG occurrence had a higher variability closer to the translation start site (ATG) (Fig. 5A and B). The dataset containing 600 bp upstream of the ATG has two populations of promoters, those with average HG content and those that are depleted in HG dinucleotides (Fig. 5A). This is especially noteworthy, as known CREs are also highly diverse in HG content (Fig. 5A and B) and, hence, the reason for the HG differences in the promoters might be due to the varying numbers and types of CREs.

Our findings suggest that some core-promoter sequences close to the ATG display reduced occurrences of possible HG dinucleotides, while their frequency in several known functional CREs was enriched. Combined with our knowledge on the structural properties of the DNA, we can propose a model in which islands of nanostructural HG domains close to the ATG are formed, which direct transcription factors to their specific CREs (Fig. 6). In principal, four different types of DNA-binding proteins can exist: First, those which follow the classical image and read the linear DNA sequence literally. Second, that class of proteins, which recognize any of the possible HG conformations. Third, proteins that bind at the transition between HG and WC topology of the B- and Z-like form of the DNA (Fig. 6). The fourth type of proteins will combine every of the above pattern recognition mechanisms.

These preliminary results suggest that DNA-binding proteins recognize a combinatorial set of features[13,14,15], such as the linear DNA sequence, DNA shape and possibly HG nanodomains in CREs that form islands in WC dominated promoter sequence.

## 4. Conclusions

In the present work we discuss the possible impact of HG dinucleotides on DNA-protein-interaction by shaping distinct Z-like DNA-nanostructures in harmonic WC helices. We could show that possible HG dinucleotides are present in almost all known functional CREs, while they are depleted in some human promoters close to the ATG. Thus, nanostructural islands of HG domains might occur close to the ATG that enable transcription factors to distinguish their CREs from WC dominated DNA. The synergy between both computational and experimental methods will be needed for future analyses of how transcription factors recognize their specific DNA binding pattern. This will require more than just the evaluation of linear DNA sequence, but has to take genomic contexts, topological features and biophysical properties into account.

## Acknowledgements

We thank Klaus Harter and Oliver Kohlbacher for continuous support and encouragement. We thank Andreas Hecker and Friederike Ladwig for their friendship. JK and KWB acknowledge Rothaus for isotonic drinks. We are grateful to acknowledge financial support from the Landesgraduierten-frderung Baden-Wrttemberg to LHB. We are especially thankful to the organizers of the QBIC conference for five years of invaluable support and the opportunity for vivid exchange of interdisciplinary information.

## References

1. K. Berendzen, K. Stuber, K. Harter and D. Wanke, BMC Bioinformatics 7, p. 522 (2006).
2. M. Berger and M. Bulyk, Nat Protoc 4, 393 (2009).
3. S. Brohee, R. Janky, F. Abdel-Sater, G. Vanderstocken, B. Andre and J. van Helden, Nucleic Acids Res (May 2011).
4. M. Godoy, J. Franco-Zorrilla, J. Perez-Perez, J. Oliveros, O. Lorenzo and R. Solano, Plant J 66, 700(May 2011).
5. R. Rohs, S. West, A. Sosinsky, P. Liu, R. Mann and B. Honig, Nature 461, 1248 (Oct 2009). July 13, 2011 13:20 Proceedings Trim Size: 9in x 6in swp000012

6. R. Rohs, X. Jin, S.West, R. Joshi, B. Honig and R. Mann, Annu Rev Biochem 79, 233 (2010).

7. Y. Chen, R. Dey and L. Chen, Structure 18, 246(Feb 2010).

8. M. Kitayner, H. Rozenberg, R. Rohs, O. Suad, D. Rabinovich, B. Honig and Z. Shakked, Nat Struct Mol Biol 17, 423(Apr 2010).

9. D. Nair, R. Johnson, S. Prakash, L. Prakash and A. Aggarwal, Nature 430, 377(Jul 2004).

10. A. Palmer and F. Massi, Chem Rev 106, 1700(May 2006).

11. S. Parker, L. Hansen, H. Abaan, T. Tullius and E. Margulies, Science 324, 389(Apr 2009).

12. J. Watson and F. Crick, Nature 171, 737(Apr 1953).

13. J. Aishima, R. Gitti, J. Noah, H. Gan, T. Schlick and C. Wolberger, Nucleic Acids Res 30, 5244(Dec 2002).

14. B. Honig and R. Rohs, Nature 470, 472(Feb 2011).

15. E. Nikolova, E. Kim, A. Wise, P. O'Brien, I. Andricioaei and H. Al-Hashimi, Nature 470, 498(Feb 2011).

16. K. Hoogsteen, Acta Crystallographica 16, 907 (1963).

17. K. Liu, H. Miles, J. Frazier and V. Sasisekharan, Biochemistry 32, 11802(Nov 1993), Mf820 Times Cited:66 Cited References Count:45.

18. H. Berman, K. Henrick, H. Nakamura and J. Markley, Nucleic Acids Res 35, D301(Jan 2007).

19. S. Dutta, C. Zardecki, D. Goodsell and H. Berman, J Appl Crystallogr 43, 1224(Oct 2010).

20. G. Patikoglou, J. Kim, L. Sun, S. Yang, T. Kodadek and S. Burley, Genes Dev 13, 3217(Dec 1999).

21. D. Case, T. Cheatham, T. Darden, H. Gohlke, R. Luo, K. Merz, A. Onufriev, C. Simmerling, B. Wang and R. Woods, J Comput Chem 26, 1668(Dec 2005).

22. A. Hildebrandt, A. Dehof, A. Rurainski, A. Bertsch, M. Schumann, N. Toussaint, A. Moll, D. Stockel, S. Nickels, S. Mueller, H. Lenhof and O. Kohlbacher, BMC Bioinformatics 11, p. 531 (2010).

23. A. Moll, A. Hildebrandt, H. Lenhof and O. Kohlbacher, Bioinformatics 22, 365(Feb 2006).

24. J. Bryne, E. Valen, M. Tang, T. Marstrand, O. Winther, I. da Piedade, A. Krogh, B. Lenhard and A. Sandelin, Nucleic Acids Res 36, D102(Jan 2008).

25. P. Cock, T. Antao, J. Chang, B. Chapman, C. Cox, A. Dalke, I. Friedberg, T. Hamelryck, F. Kauff, B. Wilczynski and M. de Hoon, Bioinformatics 25, 1422(Jun 2009).

26. L. Brand, T. Kirchler, S. Hummel, C. Chaban and D. Wanke, Plant Methods 6, p. 25 (2010).

27. K. Feldmeier, J. Kilian, K. Harter, D. Wanke and K. Berendzen, Quantile-Quantile Plots: An approach for the inter-species comparison of promoter architecture in eukaryotes. 2011, pp. 373-386.

28. I. Ciolkowski, D. Wanke, R. Birkenbihl and I. Somssich, Plant Mol Biol 68, 81(Sep 2008).

Quantum Bio-Informatics V
© 2013 World Scientific Publishing Co. Pte. Ltd.
pp. 363–373

# ON TREATMENT OF GAUSSIAN COMMUNICATION PROCESS BY QUANTUM ENTROPIES

NOBORU WATANABE

*Department of Information Sciences,*
*Tokyo University of Science,*
*Noda City, Chiba 278-8510, Japan*
*E-mail: watanabe@is.noda.tus.ac.jp*

In order to discuss the efficiency of information transmission of the Gaussian communication processes consistently, we introduced an entropy type functional and a mutual entropy type functional in reference [1]. In that study, we used the set of Gaussian input measures with covariance operators of trace one.
In this paper, we introduce a concept of structure equivalent class to study more general input Gaussan measures and Gaussian channels.

## 1. Introduction

In Shannon's theory, one can discuss the efficiency of the communication processes by using two entropies. One is the Shannon entropy $S(p)$ and another is the mutual entropy (information) $I(p; \Lambda^*)$. Two entropies satisfy the following inequalities

$$0 \leq I(p; \Lambda^*) \leq \min\{S(p), S(\Lambda^* p)\},$$

which is called the Shannon's type fundamental inequalities. Based on the Shannon's type inequalities, one can understand that the mutual entropy means the amount of information exactly sending through the channel from the input system to the output system. In Gaussian communication process, the quantity of the information transmitted correctly from the input Gaussian measure through the Gaussian communication channel is given by the Kullback - Leibler information (mutual entropy) [5]. To measure the amount of information of the input Gaussian measure, one can use the entropy of the discrete probability distribution associated with the set of all finite partitions of the initial space. Since it is infinite for all input Gaussian measures, it is impossible to compare an input Gaussian state with other Gaussian states. If the differential entropy is used to measure the input Gaussian measure, the Kullback - Leibler information and the differential entropy in the Gaussian communication process do not satisfy the Shannon's type fundamental inequalities. In order solve an unsuitability of the Gaussian transmission processes, we defined the entropy functional and the mutual entropy functional in [14] under the condition of trace one based on the quantum communication theory.

In this paper, we introduced a concept of the structure equivalent class to extend our entropy type functionals to more general Gaussian communication processes.

## 2. Gaussian Communication Process

In this section, based on [14], we briefly explain the Gaussian communication process.

Let $\mathcal{H}_1$ and $\mathcal{H}_2$ be real separable Hilbert spaces of an input and output systems and $\mathbf{B}(\mathcal{H}_k)$ and $\mathbf{T}(\mathcal{H}_k)_+$ be the sets of all bounded linear operators and positive self-adjoint trace class operators (i.e., $\mathbf{T}(\mathcal{H}_k)_+ \equiv \{\rho \in \mathbf{B}(\mathcal{H}_k);\ \rho \geq 0,\ \rho = \rho^*,\ tr\rho < \infty\}$) on $\mathcal{H}_k$, respectively ($k = 1, 2$). $\mu$ is called a Gaussian measure in $\mathcal{H}_k$ if $\mu$ is a Borel measure such that for $x \in \mathcal{H}_k$, there exist real numbers $m_x$ and $\sigma_x$ $(> 0)$ satisfying

$$\mu\left(\{y \in \mathcal{H}_k;\ < y, x > \leq a\}\right) = \int_{-\infty}^{a} \frac{1}{\sqrt{2\pi\sigma_x}} \exp\left\{\frac{-(t - m_x)^2}{2\sigma_x}\right\} dt.$$

$\hat{\mu}$ is the characteristic function of $\mu$ given by

$$\hat{\mu}(x) = \exp\left\{i \langle x, m_x \rangle - \frac{1}{2} \langle x, R_\mu x \rangle\right\}, R_\mu \in \mathbf{T}(\mathcal{H}_k)_+ .$$

We denote a Gaussian measure $\mu$ with a mean vector $m$ and a covariant operator $R$ by $\mu = [m, R]$. $\mu$ is a Borel probability measure on $\mathcal{B}_k$ satisfying

$$\int_{\mathcal{H}_k} \|x\|^2 d\mu(x) < \infty,$$

where $\mathcal{B}_k$ is the Borel $\sigma-$field of $\mathcal{H}_k$ ($k = 1, 2$). Let $\mathbf{P}(\mathcal{H}_k)$ be the set of all Borel probability measures on $\mathcal{H}_k$. For given $\mu$, the mean vector $m_\mu \in \mathcal{H}_k$ and the covariance operator $R_\mu$ of $\mu$ are defined by

$$\langle x_1, m_\mu \rangle = \int_{\mathcal{H}_k} \langle x_1, y \rangle \mu(dy)$$

$$\langle x_1, R_\mu x_2 \rangle = \int_{\mathcal{H}_k} \langle x_1, y - m_\mu \rangle \langle y - m_\mu, x_2 \rangle \mu(dy)$$

for any $x_1, x_2, y \in \mathcal{H}_k \ (k = 1, 2)$.

Now we explain a mathematical treatment of Gaussian communication process according to the reference [14]. Let $(\mathcal{H}_1, \mathcal{B}_1)$ and $(\mathcal{H}_2, \mathcal{B}_2)$ be an input and output spaces, respectively. We represent the set of all Gaussian probability measures on $(\mathcal{H}_k, \mathcal{B}_k) \ (k = 1, 2)$. $\mu_1 \in \mathbf{P}_G^{(1)}$ is a Gaussian probability measure of the input space and $\mu_0 \in \mathbf{P}_G^{(2)}$ is a Gaussian probability measure showing an additive noise of the channel. A mapping $\Gamma^*$ from $\mathbf{P}_G^{(1)}$ to $\mathbf{P}_G^{(2)}$ is defined by the Gaussian channel $\lambda : \mathcal{H}_1 \times \mathcal{B}_2 \to [0, 1]$ such as

$$\Gamma^* (\mu_1)(Q) \equiv \int_{\mathcal{H}_1} \lambda(x, Q) d\mu_1(x)$$

$$\lambda(x, Q) \equiv \mu_0 \left( \{y \in \mathcal{H}_2; \ Ax + y \in Q\} \right), \ x \in \mathcal{H}_1, \ Q \in \mathcal{B}_2,$$

where $A$ is a linear transformation from $\mathcal{H}_1$ to $\mathcal{H}_2$, $\lambda$ holds (1) $\lambda(x, \bullet) \in \mathbf{P}_G^{(2)}$ for each fixed $x \in \mathcal{H}_1$ and (2) $\lambda(\bullet, Q)$ is a measurable function on $(\mathcal{H}_1, \mathcal{B}_1)$ for each fixed $Q \in \mathcal{B}_2$. The compound measure $\mu_{12}$ derived from the input measure $\mu_1$ and the output measure $\mu_2$ is obtained by

$$\mu_{12}(Q_1 \times Q_2) = \int_{Q_1} \lambda(x, Q_2) d\mu_1(x)$$

for any $Q_1 \in \mathcal{B}_1$ and $Q_2 \in \mathcal{B}_2$. Then the mutual entropy with respect to the input Gaussian measure $\mu_1$ and the mapping $\Gamma^*$ is given by the Kullback - Leibler information as follows:

$$I(\mu_1; \lambda) = S(\mu_{12} | \mu_1 \otimes \mu_2)$$
$$= \begin{cases} \int_{\mathcal{H}_1 \times \mathcal{H}_2} \frac{d\mu_{12}}{d\mu_1 \otimes \mu_2} \log \frac{d\mu_{12}}{d\mu_1 \otimes \mu_2} d\mu_1 \otimes \mu_2 & (\mu_{12} \ll \mu_1 \otimes \mu_2) \\ \infty & \text{else} \end{cases}$$

where $\frac{d\mu_{12}}{d\mu_1 \otimes \mu_2}$ is the Radon - Nikodym derivative of $\mu_{12}$ in respect to $\mu_1 \otimes \mu_2$. In [14], we showed that (1) if we use the differential entropy $S(\mu_1)$

$$S(\mu_1) = -\int_{\mathbf{R}^2} \frac{d\mu_1}{dm} \log \frac{d\mu_1}{dm} dm$$

of the input Gaussian measure $\mu_1$, the Shannon's inequalities do not satisfy

$$S(\mu_1) < I(\mu_1; \lambda)$$

for a simple model of the Gaussian communication process, and (2) if we use the entropy $S(\mu_1)$

$$S(\mu_1) = \sup \left\{ -\sum_{A_k \in \tilde{\mathcal{A}}} \mu_1(A_k) \log \mu_1(A_k); \ \ \tilde{\mathcal{A}} \in \mathcal{P}(\mathcal{B}_2) \right\}$$

$(\mathcal{P}(\mathcal{B}_2)$ is the set of all finite partitions of $\mathcal{B}_2)$

of the discrete probability distribution associated with the set of all finite partitions of the initial space, then it is always infinite. It might be difficult to compare the input Gaussian measure with other Gaussian measures.

### 2.1. *Quantum Channels*

Let $\mathcal{H}_k$ $(k = 1, 2)$ be complex separable Hilbert spaces. We denote the set of all bounded linear operators on $\mathcal{H}_k$ by $\mathbf{B}(\mathcal{H}_k)$ $(k = 1, 2)$ and we express the set of all density operators on $\mathcal{H}_k$ by $\mathfrak{S}(\mathcal{H}_k)$ $(k = 1, 2)$. Let $(\mathbf{B}(\mathcal{H}_k), \mathfrak{S}(\mathcal{H}_k))$ $(k = 1, 2)$ be input $(k = 1)$ and output $(k = 2)$ quantum systems, respectively. A mapping from $\mathfrak{S}(\mathcal{H}_1)$ to $\mathfrak{S}(\mathcal{H}_2)$ is called a quantum channel $\Lambda^*$.

(1) $\Lambda^*$ is called a linear channel if $\Lambda^*$ satisfies the affine property such as $\Lambda^*(\sum_k \lambda_k \rho_k) = \sum_k \lambda_k \Lambda^*(\rho_k)$ for any $\rho_k \in \mathfrak{S}(\mathcal{H}_1)$ and any non-negative number $\lambda_k \in [0, 1]$ with $\sum_k \lambda_k = 1$.

For the quantum channel $\Lambda^*$, the dual map $\Lambda$ of $\Lambda^*$ is defined by

$$tr\Lambda^*(\rho)B = tr\rho\Lambda(B), \quad \forall \rho \in \mathfrak{S}(\mathcal{H}_1), \quad \forall B \in \mathbf{B}(\mathcal{H}_2).$$

(2) $\Lambda^*$ is called a completely positive (CP) channel if $\Lambda^*$ is linear channel and its dual map $\Lambda : \mathbf{B}(\mathcal{H}_2) \to \mathbf{B}(\mathcal{H}_1)$ of $\Lambda^*$ holds

$$\left\langle x, \sum_{i,j=1}^{n} A_i^* \Lambda(\overline{A}_i^* \overline{A}_j) A_j\, x \right\rangle \geq 0 \quad (\forall x \in \mathcal{H}_1)$$

for any $n \in \mathbf{N}$, any $\{\overline{A}_i\} \subset \mathbf{B}(\mathcal{H}_2)$ and any $\{A_i\} \subset \mathbf{B}(\mathcal{H}_1)$.

One can discuss the efficiency of the information transmission of the quantum communication processes described by the CP channels [7], [4], [11], [12], [15,19].

### 2.1.1. *Quantum Communication Channel*

Here we explain about an example of the quantum communication channels.

Let $\mathcal{K}_1$ and $\mathcal{K}_2$ be complex separable Hilbert spaces of noise and loss systems, respectively. Quantum channel of quantum communication process with noise and loss was discussed by [8,13].

**Noisy quantum channel and Generalized Beam Splitter** For an input state $\rho$ in $\mathfrak{S}(\mathcal{H}_1)$ and a noise state $\xi \in \mathfrak{S}(\mathcal{K}_1)$, Ohya and Watanabe defined in [13] a generalized beam splitting $\Pi^*$ by

$$\Pi^*(\rho \otimes \xi) \equiv V(\rho \otimes \xi) V^*,$$

where $V$ is a linear mapping from $\mathcal{H}_1 \otimes \mathcal{K}_1$ to $\mathcal{H}_2 \otimes \mathcal{K}_2$ given by for the $n_1, m_1, j, (n_1 + m_1 - j)$ photon number state vectors $|n_1\rangle \in \mathcal{H}_1, |m_1\rangle \in \mathcal{K}_1, |j\rangle \in \mathcal{H}_2, |n_1 + m_1 - j\rangle \in \mathcal{K}_2$

$$V(|n_1\rangle \otimes |m_1\rangle) = \sum_{j}^{n_1+m_1} C_j^{n_1,m_1} |j\rangle \otimes |n_1 + m_1 - j\rangle$$

and

$$
\begin{aligned}
C_j^{n_1,m_1} \\
= \sum_{r=L}^{K} (-1)^{n_1+j-r} \frac{\sqrt{n_1! m_1! j! (n_1 + m_1 - j)!}}{r!(n_1 - j)!(j - r)!(m_1 - j + r)!} \\
\times \alpha^{m_1 - j + 2r} \left(-\bar{\beta}\right)^{n_1 + j - 2r}
\end{aligned}
\tag{1}
$$

$\alpha$ and $\beta$ are complex numbers satisfying $|\alpha|^2 + |\beta|^2 = 1$. $K$ and $L$ are constants given by $K = \min\{n_1, j\}$, $L = \max\{m_1 - j, 0\}$. For the coherent input state $\rho = |\theta\rangle\langle\theta| \otimes |\kappa\rangle\langle\kappa| \in \mathfrak{S}(\mathcal{H}_1 \otimes \mathcal{K}_1)$, the output state of $\Pi^*$ is obtained by

$$
\begin{aligned}
\Pi^*\left(|\theta\rangle\langle\theta| \otimes |\kappa\rangle\langle\kappa|\right) &= |\alpha\theta + \beta\kappa\rangle\langle\alpha\theta + \beta\kappa| \\
&\otimes \left|-\bar{\beta}\theta + \bar{\alpha}\kappa\right\rangle\left\langle-\bar{\beta}\theta + \bar{\alpha}\kappa\right|.
\end{aligned}
$$

By using $\Pi^*$, Ohya and Watanabe introduced in [13] the noisy quantum channel $\Lambda^*$ with a fixed noise state $\xi \in \mathfrak{S}(\mathcal{K}_1)$ defined by

$$\Lambda^*(\rho) \equiv tr_{\mathcal{K}_2} \Pi^*(\rho \otimes \xi) = tr_{\mathcal{K}_2} V(\rho \otimes \xi) V^*. \tag{2}$$

The generalized beam splitting $\Pi^*$ with the vacuum noise state $\xi_0 = |0\rangle\langle0|$ is called the beam splitter $\Pi_0^*$ given by

$$\Pi_0^*\left(|\theta\rangle\langle\theta| \otimes \xi_0\right) = |\alpha\theta\rangle\langle\alpha\theta| \otimes \left|-\bar{\beta}\theta\right\rangle\left\langle-\bar{\beta}\theta\right|$$

for the coherent input state $\rho \otimes \xi_0 = |\theta\rangle\langle\theta| \otimes |0\rangle\langle0| \in \mathfrak{S}(\mathcal{H}_1 \otimes \mathcal{K}_1)$. The beam splitter $\Pi_0^*$ was described by means of the lifting $\mathcal{E}_0^*$ from $\mathfrak{S}(\mathcal{H})$ to $\mathfrak{S}(\mathcal{H} \otimes \mathcal{K})$ in the sense of Accardi and Ohya [1] as follows

$$\mathcal{E}_0^*\left(|\theta\rangle\langle\theta|\right) = |\alpha\theta\rangle\langle\alpha\theta| \otimes |\beta\theta\rangle\langle\beta\theta|.$$

Based on the liftings, the beam splitting was studied by Accardi - Ohya and Fichtner - Freudenberg - Libsher [3]. Moreover, the noisy quantum channel $\Lambda_0^*$ with the vacuum noise state $\xi_0 = |0\rangle \langle 0|$ is called the attenuation channel given by Ohya [8] as

$$\Lambda_0^*(\rho) \equiv tr_{\mathcal{K}_2} \Pi_0^*(\rho \otimes \xi_0) = tr_{\mathcal{K}_2} V_0(\rho \otimes |0\rangle \langle 0|) V_0^*, \tag{3}$$

which plays an important role for investigating the quantum communication processes.

## 3. Quantum Entropies

### 3.0.2. *von Neumann Entropy and Ohya mutual entropy*

We here review the von Neumann entropy and the Ohya mutual entropy. The von Neumann entropy $S(\rho)$ [6] is defined by

$$S(\rho) = -tr\rho \log \rho$$

for any density operators $\rho \in \mathfrak{S}(\mathcal{H}_1)$, which satisfies (1) $S(\rho) \geq 0$, (2) if $\Lambda^* = id$ ($id$ is the identity channel), then $S(\Lambda^* \rho) = S(\rho)$, (3) $S(\rho_1 \otimes \rho_2) = S(\rho_1) + S(\rho_2)$. The quantum mutual entropy with respect to the initial state $\rho$ and the quantum channel $\Lambda^*$ is introduced by Ohya [8] defined such as

$$I(\rho; \Lambda^*) \equiv \sup \left\{ \sum_n S(\sigma_E, \rho \otimes \Lambda^* \rho), \rho = \sum_n \lambda_n E_n \right\}.$$

where $\sigma_E$ is the compound state given by $\sigma_E = \sum_n \lambda_n E_n \otimes \Lambda^* E_n$ associated with the Schatten-von Neumann (one dimensional spectral) decomposition [16] $\rho = \sum_n \lambda_n E_n$ of the input state $\rho$, and $S(\cdot, \cdot)$ is the Umegaki's relative entropy [17] denoted by

$$S(\rho, \sigma) \equiv \begin{cases} tr\rho (\log \rho - \log \sigma) & (\overline{ran\rho} \subset \overline{ran\sigma}) \\ \infty & \text{otherwise} \end{cases} \tag{4}$$

which was extended to more general quantum systems by Araki and Uhlmann [2,9,10,11,18]. The quantum mutual entropy $I(\rho, \Lambda^*)$ and the von Neumann entropy hold the Shannon's fundamental inequalities such as

$$0 \leq I(\rho, \Lambda^*) \leq \min \{S(\rho), S(\Lambda^* \rho)\}.$$

If $\Lambda^* = id$, then $I(\rho, id) = S(\rho)$ is satisfied.

## 4. A treatment of Gaussian communication process by using quantum entropies

Let us review a mathematical treatment of Gaussian communication process corresponding to the reference [14]. We restrict the Gaussian measure $\mu = [0, R]$ into the subset $\mathbf{P}_{G,1}^{(k)}$ with $tr R = 1$ of $\mathbf{P}_G^{(k)}$

$$\mathbf{P}_{G,1}^{(k)} = \left\{ \mu = [0, R] \in \mathbf{P}_G^{(k)}; \ tr R = 1 \right\} \quad (k = 1, 2).$$

For the covariant operator $R_0$ of $\mu_0$, we suppose that $A^* A = (1 - tr R_0) I_1$ holds. For any $\mu_1 \in \mathbf{P}_{G,1}^{(1)}$ and any $Q \in \mathcal{B}_2$, a transformation $\Gamma^* : \mathbf{P}_{G,1}^{(1)} \to \mathbf{P}_{G,1}^{(2)}$ associated with the Gaussian channel $\lambda$ is given by

$$(\Gamma^* \mu_1)(Q) = \int_{\mathcal{H}_1} \lambda(x, Q) \mu_1(dx).$$

For any $\mu_1 = [0, \rho_1] \in \mathbf{P}_{G,1}^{(1)}$, $\Gamma^*(\mu_1)$ is obtained by

$$\Gamma^*(\mu_1) = [0, \ A\rho_1 A^* + R_0].$$

For any $A_k \in \mathbf{B}(\mathcal{H}_k)$ and any $\mu_k \in \mathbf{P}_{G,1}^{(k)}$ $(k = 1, 2)$, there exists a bijection $\Xi_k^*$ from $\mathbf{P}_{G,1}^{(k)}$ to $\mathfrak{S}(\mathcal{H}_k)$ given by

$$tr \Xi_k^*(\mu_k) A_k = \int_{H_k} \langle \xi, \ A_k \xi \rangle \mu_k(d\xi).$$

Thus the quantum channel $\Lambda^* : \mathfrak{S}(\mathcal{H}_1) \to \mathfrak{S}(\mathcal{H}_2)$ consisted of $\Xi_1^*$, $\Xi_2^*$, $\Gamma^*$ is defined by

$$\Lambda^* \rho_1 = \Xi_2^* \circ \Gamma^* \circ (\Xi_1^*)^{-1} \rho_1, \quad (\forall \rho_1 \in \mathfrak{S}(\mathcal{H}_1)),$$

$$
\begin{array}{ccc}
\mu_1 \in \mathbf{P}_{G,1}^{(1)} \left( \subset \mathbf{P}_G^{(1)} \right) & \xrightarrow{\ \Gamma^* \ } & \Gamma^* \mu_1 \in \mathbf{P}_{G,1}^{(2)} \left( \subset \mathbf{P}_G^{(2)} \right) \\
\Xi_1^* \downarrow & & \downarrow \Xi_2^* \\
\rho_1 \in \mathfrak{S}(\mathcal{H}_1) \left( \subset \mathbf{T}(\mathcal{H}_1)_+ \right) & \xrightarrow{\ \Lambda^* \ } & \Lambda^* \rho_1 \in \mathfrak{S}(\mathcal{H}_2) \left( \subset \mathbf{T}(\mathcal{H}_2)_+ \right)
\end{array}
$$

which is described by

$$\Lambda^* \rho_1 = A\rho_1 A^* + R_0, \quad (\forall \rho_1 \in \mathfrak{S}(\mathcal{H}_1)).$$

We had the following theorems in [14].

**Theorem 4.1.** $\Lambda^*$ *is a completely positive channel from* $\mathfrak{S}(\mathcal{H}_1)$ *to* $\mathfrak{S}(\mathcal{H}_2)$

**Theorem 4.2.** *The Gaussian measure $\mu = [0, \sigma_E]$ is a compound state (measure) derived from the input measure $\mu_1 = [0, \Xi_1^* (\mu_1)]$ on $\mathcal{H}_1$ and the output measure $\mu_2 = [0, \Lambda^* \circ \Xi_1^* (\mu_1)]$ on $\mathcal{H}_2$ in the sense that*

$$\bar{\mu}_1 (A) = \bar{\mu} (A \otimes \mathcal{H}_2) \quad \textit{for any subspace } A \textit{ in } \mathcal{B}_1,$$
$$\bar{\mu}_2 (B) = \bar{\mu} (\mathcal{H}_1 \otimes B) \quad \textit{for any subspace } B \textit{ in } \mathcal{B}_2,$$

*where $\bar{\mu}_k (A) = \int_A \|\xi\|^2 d\mu_k (\xi)$ for any $A \in \mathcal{B}_k$ and any $\mu_k \in \mathbf{P}_{G,1}^{(k)}$ $(k = 1, 2)$.*

One can define (1) the entropy type functional $\tilde{S} (\mu_1)$ of the input Gaussian measure $\mu_1 = [0, \Xi_1^* (\mu_1)]$ and (2) the mutual entropy type functional $\tilde{I} (\mu_1; \lambda)$ with respect to the input Gaussian measure $\mu_1$ and the Gaussian channel $\lambda$ such as

$$\tilde{S} (\mu_1) = -tr\Xi_1^* (\mu_1) \log \Xi_1^* (\mu_1),$$
$$\tilde{I} (\mu_1; \lambda) = \sup_E S (\sigma_E, \Xi_1^* (\mu_1) \otimes \Lambda^* \circ \Xi_1^* (\mu_1)),$$

where $\sigma_E$ is the Ohya compound state with respect to $\Xi_1^* (\mu_1)$ and $\Lambda^*$. Therefore one can show the following theorem [14].

**Theorem 4.3.** *[14] For any $\mu_1 \in \mathbf{P}_{G,1}^{(k)}$ and for some Gaussian channel $\lambda$, one obtain the Shannon's type fundamental inequalities:*

$$0 \leq \tilde{I} (\mu_1; \lambda) \leq \tilde{S} (\mu_1).$$

In the above discussion [14], we assume the following three conditions:

- **Our previous approach**
  - (1) Linearity condition (linear approximation)
  - (2) Trace preserving condition
  - (3) Normality condition (trace of covariance operator is equal to one.)

In this paper, we suppose two conditions without (3) in the above three conditions.

- **A treatment I**
  - (1) Linearity condition (linear approximation)
  - (2) Trace preserving condition : $tr\Pi^* (R) = trR$ is hold for any $R \in \mathbf{T}(\mathcal{H}_1)_+$
- **A treatment II**
  - (1) Linearity condition (linear approximation)

(2) Weak Trace preserving condition : $tr\Pi^*(R) = tr\Pi^*(R')$ is hold for any $R, R' \in \mathbf{T}(\mathcal{H}_1)_+$ satisfying $trR = trR'$

**Theorem 4.4.** *If $\Pi^*$ satisfies the trace preserving condition, then $\Pi^*$ holds the weak trace preserving condition.*

**Proof.** For any $R, R' \in \mathbf{T}(\mathcal{H}_1)_+$ satisfying $trR = trR'$, one can obtain the weak trace preserving condition $tr\Pi^*(R) = trR = trR' = tr\Pi^*(R')$. $\quad\square$

We consider a mapping $\Pi^* : \mathbf{T}(\mathcal{H}_1)_+ \to \mathbf{T}(\mathcal{H}_2)_+$ consisted of $\Xi_1^*$, $\Xi_2^*$, $\Gamma^*$ defined by

$$\Pi^* R_1 = \Xi_2^* \circ \Gamma^* \circ (\Xi_1^*)^{-1} R_1, \quad \left(\forall R_1 \in \mathbf{T}(\mathcal{H}_1)_+\right),$$

$$\begin{array}{ccc}
\mu_1 \in \mathbf{P}_G^{(1)} & \xrightarrow{\Gamma^*} & \Gamma^* \mu_1 \in \mathbf{P}_G^{(2)} \\
\Xi_1^* \downarrow & & \downarrow \Xi_2^* \\
R_1 \in \mathbf{T}(\mathcal{H}_1)_+ & \xrightarrow{\Pi^*} & \Pi^* R_1 \in \mathbf{T}(\mathcal{H}_2)_+
\end{array}$$

which is denoted by

$$\Pi^* R_1 = AR_1 A^* + R_0, \quad \left(\forall R_1 \in \mathbf{T}(\mathcal{H}_1)_+\right).$$

Now we introduce a concept of the structure equivalent class in the Gaussian communication processes.

**Definition 4.1. Structure equivalent of $\mathbf{T}(\mathcal{H}_1)_+$ and $\mathbf{P}_G^{(1)}$**

(1) $R_1$ and $R_2$ are structure equivalent (i.e., $R_1 \overset{s}{\sim} R_2$) if there exists a positive number $\lambda > 0$ such that $R_1 = \lambda R_2$ holds,
(2) $\mu_1 = [0, R_1]$ and $\mu_2 = [0, R_2]$ are structure equivalent (i.e., $\mu_1 \overset{s}{\sim} \mu_2$) if $R_1 \overset{s}{\sim} R_2$ is satisfied.

**Definition 4.2. Structure equivalent class of $\mathbf{T}(\mathcal{H}_1)_+$ and $\mathbf{P}_G^{(1)}$**

(1) $\widetilde{R_1} \equiv \left\{R \in \mathbf{T}(\mathcal{H}_1)_+; \quad R_1 \overset{s}{\sim} R\right\}$,
(2) $\widetilde{\mu_1} \equiv \left\{\mu \in \mathbf{P}_G^{(1)}; \quad \mu_1 \overset{s}{\sim} \mu\right\}$.

We have the following theorems.

**Theorem 4.5.** *$\Pi^*$ is a completely positive channel from $\mathbf{T}(\mathcal{H}_1)_+$ to $\mathbf{T}(\mathcal{H}_2)_+$.*

**Theorem 4.6.** *The Gaussian measure* $\bar{\mu} = [0, \sigma_E]$ *is a compound state (measure) given by the input measure* $\mu_1 = [0, \Xi_1^*(\mu_1)]$ *on* $\mathcal{H}_1$ *and the output measure* $\mu_2 = [0, \Pi^* \circ \Xi_1^*(\mu_1)]$ *on* $\mathcal{H}_2$ *in the sense that*

$$\bar{\mu}_1(A) = \bar{\mu}(A \otimes \mathcal{H}_2) \quad \text{for any subspace } A \text{ in } \mathcal{B}_1,$$

$$\bar{\mu}_2(B) = \bar{\mu}(\mathcal{H}_1 \otimes B) \quad \text{for any subspace } B \text{ in } \mathcal{B}_2,$$

*where* $\bar{\mu}_k(A) = \int_A \|\xi\|^2 d\mu_k(\xi)$ *for any* $A \in \mathcal{B}_k$ *and any* $\mu_k \in \mathbf{P}_G^{(k)}$ *(k = 1, 2).*

We introduce (1) the entropy type functional of structure equivalent $\tilde{S}_{SE}(\mu_1)$ of the input Gaussian measure $\mu_1 = [0, \Xi_1^*(\mu_1)]$ and (2) the mutual entropy type functional of structure equivalent $\tilde{I}_{SE}(\mu_1; \lambda)$ with respect to the input Gaussian measure $\mu_1$ and the Gaussian channel $\lambda$ such as

$$\tilde{S}_{SE}(\mu_1) \equiv -tr \frac{\Xi_1^*(\mu_1)}{tr\,[\Xi_1^*(\mu_1)]} \log \frac{\Xi_1^*(\mu_1)}{tr\,[\Xi_1^*(\mu_1)]},$$

$$\tilde{I}_{SE}(\mu_1; \lambda) \equiv \sup_E S\left(\sigma_E, \frac{\Xi_1^*(\mu_1) \otimes \Pi^* \circ \Xi_1^*(\mu_1)}{tr\,[\Xi_1^*(\mu_1) \otimes \Pi^* \circ \Xi_1^*(\mu_1)]}\right),$$

where $\sigma_E$ is the Ohya compound state in respect to $\Xi_1^*(\mu_1)$ and $\Pi^*$. One can obtain the following structure equivalent of compound states and input states.

- **Structure Equivalent of Compound states**

  (1) $\Xi_1^*(\mu_1) \otimes \Pi^* \circ \Xi_1^*(\mu_1) \overset{s}{\sim} \sigma_0 = \frac{\Xi_1^*(\mu_1) \otimes \Pi^* \circ \Xi_1^*(\mu_1)}{tr\left[\Xi_1^*(\mu_1) \otimes \Pi^* \circ \Xi_1^*(\mu_1)\right]}$

  (2) $\sum_n \tau_n E_n \otimes \Pi^*(E_n) \overset{s}{\sim} \sigma_E = \frac{\sum_n \tau_n E_n}{tr\left[\Xi_1^*(\mu_1)\right]} \otimes \frac{\Pi^*(E_n)}{tr[\Pi^*(E_n)]}$

- **Structure Equivalent of the input state**

  (3) $\Xi_1^*(\mu_1) \overset{s}{\sim} \frac{\Xi_1^*(\mu_1)}{tr\left[\Xi_1^*(\mu_1)\right]}$ Therefore one obtain the following

  result.

**Theorem 4.7.** *For any* $\mu_1 \in \mathbf{P}_G^{(k)}$ *and for some Gaussian channel* $\lambda$, *one obtain the Shannon's type fundamental inequalities as*

$$0 \leq \tilde{I}_{SE}(\mu_1; \lambda) \leq \tilde{S}_{SE}(\mu_1).$$

## References


1. L. Accardi and M. Ohya, Compound channels, transition expectation and liftings, Appl. Math, Optim., **39**, 33-59 (1999).
2. H. Araki, Relative entropy for states of von Neumann algebras, Publ. RIMS Kyoto Univ. **11**, 809–833, (1976).

3. K.H. Fichtner, W. Freudenberg and V. Liebscher, Beam splittings and time evolutions of Boson systems, Fakultat fur Mathematik und Informatik, Math/ Inf/96/ **39**, Jena, 105 (1996).

4. R.S. Ingarden, A. Kossakowski and M. Ohya, *Information Dynamics and Open Systems*, Kluwer, (1997).

5. Kullback, S. and Leibler, R., On information and sufficiency, **22**, 79-86 (1951).

6. J. von Neumann, *Die Mathematischen Grundlagen der Quantenmechanik*, Springer-Berlin, (1932).

7. M. Ohya, Quantum ergodic channels in operator algebras , J. Math. Anal. Appl., **84**, 318-328, (1981).

8. M. Ohya, On compound state and mutual information in quantum information theory, IEEE Trans. Information Theory, **29**, 770-774 (1983).

9. M. Ohya, Note on quantum probability, L. Nuovo Cimento, **38**, 402-404, (1983).

10. M. Ohya, Some aspects of quantum information theory and their applications to irreversible processes, Rep. Math. Phys., **27**, 19-47 (1989).

11. M. Ohya and D. Petz, *Quantum Entropy and its Use*, Springer, Berlin, 1993.

12. M. Ohya and I. Volovich, *Mathematical Foundations of Quantum Information. and Computation and Its Applications to Nano- and Bio-systems*, Springer, (2011).

13. M. Ohya and N. Watanabe, Construction and analysis of a mathematical model in quantum communication processes. Electronics and Communications in Japan, Part 1, **68**, No.2, 29-34 (1985).

14. M. Ohya and N. Watanabe, A new treatment of communication processes with Gaussian channels, Japan Journal on Applied Mathematics, **3**, 197-206 (1986).

15. M. Ohya and N. Watanabe, *Foundation of Quantum Communication Theory (in Japanese)*, Makino Pub. Co., (1998).

16. R. Schatten, *Norm Ideals of Completely Continuous Operators*, Springer - Verlag, (1970).

17. H. Umegaki, Conditional expectations in an operator algebra IV (entropy and information), Kodai Math. Sem. Rep., **14**, 59-85 (1962).

18. A. Uhlmann, Relative entropy and the Wigner-Yanase-Dyson-Lieb concavity in interpolation theory, Commun. Math. Phys., **54**, 21–32, (1977).

19. N. Watanabe, Some aspects of complexities for quantum processes, Open Systems and Information Dynamics, **16**, 293-304 (2009).

Quantum Bio-Informatics V
© 2013 World Scientific Publishing Co. Pte. Ltd.
pp. 375–387

# IMPORTANCE OF EXCLUDED VOLUME AND HYDRODYNAMIC INTERACTIONS ON MACROMOLECULAR DIFFUSION *IN VIVO*

TADASHI ANDO

*Center for the Study of Systems Biology, School of Biology, Georgia Institute of Technology*
*250 14th Street NW, Atlanta, GA 30318-5304, USA*


JEFFREY SKOLNICK

*Center for the Study of Systems Biology, School of Biology, Georgia Institute of Technology*
*250 14th Street NW, Atlanta, GA 30318-5304, USA*

The interiors of all living cells are highly crowded with macromolecules, which results in a considerable difference between the thermodynamics and kinetics of biological reactions *in vivo* from that *in vitro*. To begin to elucidate the principles of intermolecular dynamics in the crowded environment of cells, employing Brownian dynamics (BD) simulations, we examined possible mechanism(s) responsible for the great reduction in diffusion constants of macromolecules *in vivo* from that at infinite dilution. In an *E. coli* cytoplasm model comprised of 15 different macromolecule types at physiological concentrations, where macromolecules were represented by spheres with their Stokes radii, BD simulations were performed with and without hydrodynamic interactions (HI). Without HI, the calculated diffusion constant of green fluorescent protein (GFP) is much larger than experiment. On the other hand, when HI were considered, the *in vivo* experimental GFP diffusion constant is almost reproduced without adjustable parameters. In addition, HI give rise to significant, size independent intermolecular dynamic correlations. These results suggest that HI play an important role on macromolecular dynamics *in vivo*.

## 1.  Introduction

One of the most characteristic features of the interiors of cells is the high total concentration of biological macromolecules. Typically, 20-40% of the cytoplasmic volume is occupied by proteins, nucleic acids and other macromolecules [1-3]. Under these conditions, although the molar concentration of each protein ranges from nM to μM, the distance between neighboring proteins is comparable to the size of the proteins. Therefore, simulating the crowded intracellular environment is crucial to understanding the nature of living systems.

Diffusion is one of the most important physical parameters that describe motions of molecules in a fluid. Recently, Elowitz et al. [4] and Konopka et al. [5] applied fluorescence recovery after photobleaching to measure the diffusion coefficient of green fluorescent protein (GFP) in the *E. coli* cytoplasm. Both groups reported that the diffusion coefficient of GFP *in vivo* is about 10 times less than that at infinite dilution in water. What is responsible for this reduction? In this study, as a necessary first step towards whole cell modeling, we performed Brownian dynamics (BD) simulations of the *E. coli* cytoplasm to address the importance of hydrodynamic interactions on diffusion. While intermolecular HI play an important role in determining the dynamics of concentrated particles [6, 7], it has been neglected in the most simulations of biological macromolecules due to its long-range nature and high computational cost. Here, we apply Stokesian dynamics to simulate the diffusion of a polydisperse collection of macromolecules in crowded, heterogeneous intracellular 3D environments.

## 2.  Methods

### 2.1.  *Hydrodynamic Interactions*

The Rotne-Prager-Yamakawa (RPY) tensor is commonly used to take into account the HI in dynamic simulations of biomolecules because this tensor retains positive definiteness even when particles overlap [8-10]. Actually, the RPY tensor contains the two-body long-range or far-field contributions to particle mobility. However, in concentrated systems, e.g. the inside of cells, far-field many-body HI as well as near-field HI, so-called "lubrication forces", play important roles in determining mobility (see Fig. 4 in Ref. [11]). In order to include not only the far-field HI but also the many-body and near-field HI in simulations, the Durlofsky-Brady-Bossis approach to "Stokesian dynamics" was used [6, 12]. Their approach can reproduce the properties of dense monodisperse suspensions whose volume fraction is up to 0.50 [13, 14]. Here, we briefly outline this method. For details, please refer to the original reference [12].

For $N$ particles system in a Newtonian fluid and in absence of an external shear flow, the hydrodynamic forces acting on particles, $\mathbf{F}$, are related to the particle velocities, $\mathbf{U}$, through the Stoke equation:

$$\mathbf{F} = \mathbf{R} \cdot \mathbf{U} \, , \tag{1}$$

where $\mathbf{R}$ is the resistance matrix and is the inverse of the mobility matrix, $\mathbf{M}$:

$$\mathbf{R} = \mathbf{M}^{-1} . \tag{2}$$

When torque-angular velocity is not considered, the so-called "F version" in Ref. [12], $\mathbf{F}$ and $\mathbf{U}$ are $3N \times 1$ vectors and $\mathbf{R}$ and $\mathbf{M}$ are $3N \times 3N$ matrices. Then, the diffusion matrix of the system is simply given by

$$\mathbf{D} = k_{\mathrm{B}}T\mathbf{M} . \tag{3}$$

Here, $k_{\mathrm{B}}$ is Boltzmann constant and $T$ is the temperature.

The resistance tensor $\mathbf{R}$, which contains both near-field lubrication effects and far-field many-body interactions, is calculated as

$$\mathbf{R} = \left(\mathbf{M}^{\infty}\right)^{-1} + \mathbf{R}_{2\mathrm{B}} - \mathbf{R}_{2\mathrm{B}}^{\infty} . \tag{4}$$

The first term, $(\mathbf{M}^{\infty})^{-1}$, represents the contribution of many-body, far-field interactions. The second term, $\mathbf{R}_{2\mathrm{B}}$, represents the exact two-body HI, which includes both near-field and far-field interactions. The third term $\mathbf{R}_{2\mathrm{B}}^{\infty}$ is the resistance tensor that represents two-body far-field interactions. The far-field part has already been included on $(\mathbf{M}^{\infty})^{-1}$. Thus, in order not to count these interactions twice, we must subtract off the two-body interactions. This is the standard method to correct for the lubrication effects in the resistance tensor.

$\mathbf{M}^{\infty}$ can be estimated by the Ewald summation of the RPY tensor developed by Beenakker for the case of periodic boundary conditions [15]. Due to the long range nature of HI (which decays as $1/r$, with $r$ the intermolecular distance between particles), use of the Ewald summation technique is necessary not only for accuracy but also for obtaining positive definite matrices in the calculation of the mobility tensor under periodic boundary conditions [14]. In addition, inverting $\mathbf{M}^{\infty}$ corresponds to including many-body interactions [12]. $\mathbf{R}_{2\mathrm{B}}$ is calculated by the exact two-body solution of Jeffrey and Onishi [16]. $\mathbf{R}_{2\mathrm{B}}^{\infty}$ is obtained by simply inverting a two-body mobility matrix containing terms to the same order in $1/r$ as $\mathbf{M}^{\infty}$.

When lubrication effects were added to the resistance tensor, the correction method developed by Cichocki et al. [17] is used to prevent a divergence of the translational self-diffusion coefficient for certain configurations. Collective motions are separated out from the standard lubrication correction as follows [17]:

$$\mathbf{R} = \left(\mathbf{M}^{\infty}\right)^{-1} + \mathbf{s}_{2\mathrm{B}} - \mathbf{s}_{2\mathrm{B}}^{\infty} , \tag{5}$$

where

$$\mathbf{s}_{2\mathrm{B}} = \sum_{\alpha}^{N}\sum_{\beta>\alpha}^{N}\mathbf{q}^{T} \cdot \mathbf{R}_{2\mathrm{B}}\left(\alpha,\beta\right) \cdot \mathbf{q} , \tag{6}$$

378

$$\mathbf{s}_{2B}^{\infty} = \sum_{\alpha}^{N} \sum_{\beta>\alpha}^{N} \mathbf{q}^{T} \cdot \mathbf{R}_{2B}^{\infty}(\alpha,\beta) \cdot \mathbf{q} . \tag{7}$$

Here, $\alpha$ and $\beta$ represents indices of particles, and the matrix $\mathbf{q}$ is given by

$$\mathbf{q} = \frac{1}{2} \begin{pmatrix} \mathbf{I} & -\mathbf{I} \\ -\mathbf{I} & \mathbf{I} \end{pmatrix} . \tag{8}$$

This modified correction method was essential to prevent a divergence of the translational self-diffusion coefficient for our crowded and heterogeneous system.

### 2.2. *Brownian Dynamics Algorithm*

When HI are considered, the diffusion tensor macromolecule depends in principle on the configuration of the entire system and varies over time. We can write the propagation equation for such Brownian particles as

$$\begin{aligned} \mathbf{r} &= \mathbf{r}^{0} + (\nabla \cdot \mathbf{D})\Delta t + \frac{\mathbf{D} \cdot \mathbf{F}^{\mathrm{p}}}{k_{\mathrm{B}}T}\Delta t + \mathbf{G}(\Delta t) \\ &= \mathbf{r}^{0} + k_{\mathrm{B}}T(\nabla \cdot \mathbf{M})\Delta t + (\mathbf{M} \cdot \mathbf{F}^{\mathrm{p}})\Delta t + \mathbf{G}(\Delta t), \end{aligned} \tag{9}$$

where $\mathbf{r}$ is the particle's position vector and $\mathbf{F}^{\mathrm{p}}$ is the deterministic conservative force acting on the particle. $\mathbf{G}(\Delta t)$ is the random displacement due to Brownian motion, which has the following properties:

$$\langle \mathbf{G}(\Delta t) \rangle = 0, \langle \mathbf{G}(\Delta t)\mathbf{G}(\Delta t) \rangle = 2k_{\mathrm{B}}T\mathbf{M}\Delta t . \tag{10}$$

In contrast to a BD algorithm with constant diffusion tensors, we need to evaluate the spatial gradient of the mobility tensor in the BD simulation with HI, in which the explicit computation of $\nabla \cdot \mathbf{M}$ is a $O(N^3)$ task. To avoid this expensive calculation, we used a method introduced by Banchio and Brady [18], which is based on Fixman's idea [19], the so-called "mid-point scheme". In this method, Eqs. 9 and 10 are re-written as follows:

$$\mathbf{r} = \mathbf{r}^{0} + k_{\mathrm{B}}T(\nabla \cdot \mathbf{M})\Delta t + \mathbf{M} \cdot (\mathbf{F}^{\mathrm{p}} + \mathbf{F}^{\mathrm{B}})\Delta t , \tag{11}$$

$$\langle \mathbf{F}^{\mathrm{B}} \rangle = 0, \langle \mathbf{F}^{\mathrm{B}}(0)\mathbf{F}^{\mathrm{B}}(t) \rangle = 2k_{\mathrm{B}}T\mathbf{R}/\Delta t . \tag{12}$$

Here, $\mathbf{F}^{\mathrm{B}}$ is the Brownian force, obtained through the Cholesky decomposition [20]. The procedure for this mid-point algorithm is the following:
1.　　　Compute the velocity $\mathbf{U}^{0}$ using an initial configuration $\mathbf{r}^{0}$

$$\mathbf{U}^{0} = (\mathbf{R}^{0})^{-1} \cdot (\mathbf{F}^{\mathrm{p},0} + \mathbf{F}^{\mathrm{B},0}) . \tag{13}$$

Here, superscript 0 represents the value evaluated at position $\mathbf{r}^0$.

2. Move the particles to intermediate positions $\mathbf{r}'$ by a small fraction of a time step, $\Delta t / m$

$$\mathbf{r}' = \mathbf{r}^0 + \frac{\Delta t}{m} \mathbf{U}^0 . \tag{14}$$

In this study, an $m$ of 100 was used.

3. Calculate a new velocity $\mathbf{U}'$ at the intermediate positions using the forces evaluated at $\mathbf{r}^0$

$$\mathbf{U}' = \left(\mathbf{R}'\right)^{-1} \cdot \left(\mathbf{F}^{p,0} + \mathbf{F}^{B,0}\right). \tag{15}$$

4. Calculate the drift velocity, $\mathbf{U}^{\mathrm{drift}}$,

$$\mathbf{U}^{\mathrm{drift}} = \frac{m}{2}\left(\mathbf{U}' - \mathbf{U}^0\right). \tag{16}$$

5. Finally, update the positions of the particles for time step $\Delta t$.

$$\mathbf{r} = \mathbf{r}^0 + \left(\mathbf{U}^0 + \mathbf{U}^{\mathrm{drift}}\right)\Delta t . \tag{17}$$

For the simulations without HI, Ermak and McCammon algorithm described by Eqs. 9 and 10 was used [20, 21].

### 2.3. Simulation System

A simulation system having over 1,000 macromolecules consisting of 15 different kinds of proteins and tRNA was constructed in a 100 nm × 100 nm × 100 nm box based on the data reported by Ridgway et al [22] and the CyberCell database of the physical properties of *E. coli* [23], where each macromolecule was represented by an equivalent sphere with their Stokes radii (Table 1, Fig. 1). Stokes radii were estimated by using rigid-particle theory [21, 24]. The macromolecular concentration was set to 300 mg/ml, which is a reasonable estimate of the macromolecular concentration in *E. coli* (300–340 mg/ml) [3]. Volume occupancies calculated with the Stokes radii and radii of gyration of the macromolecules are 51% and 22%, respectively.

Table 1. Simulated macromolecules, their physical properties, and number in 100 nm × 100 nm × 100 nm simulation box.

| Name | PDB ID | Molecular weight (kDa) | Molecular weight range (kDa)[*] | Radius of gyration (Å) | Stokes radius (Å)[†] | $D_0$ (Å$^2$/ns)[‡] | Number of molecules in the box |
|---|---|---|---|---|---|---|---|
| Ribonuclease HI | 1JL1 | 17.5 | 0-20 | 14.9 | 21.4 | 11.4 | 70 |
| GFP | 1W7S | 26.9 | 20-40 | 16.9 | 24.0 | 10.2 | 113 |
| Malonyl CoA-acyl carrier protein transacylase | 1MLA | 32.4 | 20-40 | 18.5 | 25.7 | 9.52 | 94 |
| Triosephosphate isomerase | 1TRE | 54.0 | 40-60 | 24.5 | 31.7 | 7.73 | 126 |
| Fructose 1-6 bisphosphate aldolase | 1DOS | 78.2 | 60-80 | 28.5 | 36.8 | 6.67 | 91 |
| Enolase | 1E9I | 91.3 | 80-100 | 26.5 | 35.9 | 6.82 | 92 |
| 6-phosphogluconate dehydrogenase | 2ZYA | 106.6 | 100-120 | 29.4 | 39.2 | 6.26 | 31 |
| Phosphoenolpyruvate-protein phosphotransferase | 2HWG | 129.3 | 120-140 | 31.8 | 42.5 | 5.77 | 34 |
| Glyceraldehyde-3-P dehydrogenase | 1S7C | 145.0 | 140-160 | 31.6 | 43.1 | 5.69 | 42 |
| Cystathionine gamma-synthase | 1CS1 | 167.5 | 160-180 | 33.2 | 45.4 | 5.40 | 6 |
| Phosphoglycerate dehydrogenase | 1YBA | 182.1 | 180-200 | 40.1 | 49.2 | 4.99 | 12 |
| RNA polymerase | 1IW7 | 432.9 | 200+ | 50.3 | 66.5 | 3.69 | 76 |
| GroEL/ES | 1AON | 877.6 | 200+ | 68.4 | 85.2 | 2.88 | 37 |
| 70S Ribosome | 3I1Q, 3I1R | 2,155.2 | Ribosomes | 84.3 | 115.2 | 2.13 | 29 |
| Initial tRNA | 3CW5 | 24.8 | tRNAs | 22.7 | 27.8 | 8.83 | 299 |

[*]This molecular weight range corresponds to the class used in Ref. [22].

[†]Stokes radii were calculated by $6\pi\eta a = k_B T/D_0$, where $\eta$ is the viscosity of water, $D_0$ is the translational diffusion coefficient in dilute solution estimated by the rigid-particle theory [21, 24], and $a$ is the Stokes radius.

[‡]The diffusion constants are at 298 K.

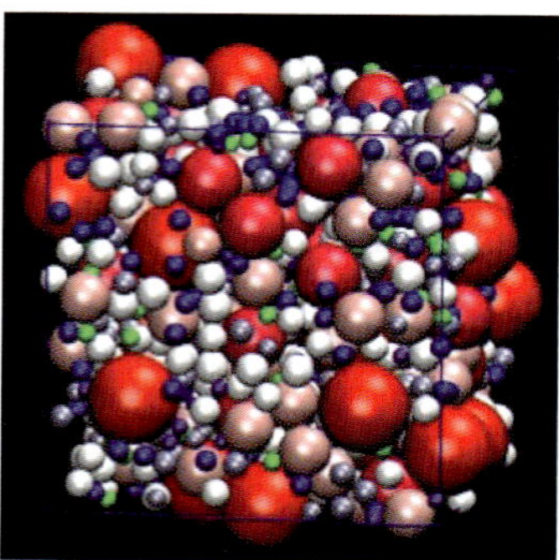

Figure 1. Simulation system in 100 nm × 100 nm × 100 nm box. Macromolecules are represented in different colors. Radii of molecules correspond to their Stokes radii.

## 2.4.  *Repulsive Interaction Model*

Repulsive interactions between intermolecular particles in BD simulations without HI were represented by a soft-sphere potential described by

$$V_{ss}\left(r_{ij}\right)=\begin{cases}k_{ss}\left(r_{ij}-r_{m}\right)^{2} & \text{if } r_{ij}\leq r_{m}\\ 0 & \text{if } r_{ij}>r_{m}\,,\end{cases} \tag{18}$$

where $r_{ij}$ is the distance between particles $i$ and $j$, and $k_{ss}$ is a force constant. $r_{m}$ is $r_{c}+\Delta_{ss}$, in which $r_{c}$ is sum of radii of particles $i$ and $j$, $a_{i}$ and $a_{j}$, and $\Delta_{ss}$ is an arbitrary parameter representing a buffer distance between particles. In this study, a $\Delta_{ss}$ of 2 Å and $k_{ss}$ of $5k_{B}T/\Delta_{ss}^{2}$ were used, which means $V_{ss}=5k_{B}T$ at the distance $r_{c}$. In BD simulations with lubrication forces, we do not use any repulsive forces between particles, since the lubrication forces prevent particles from overlapping.

## 2.5.  *Simulation Conditions and Analysis*

All simulations were performed at 298 K with periodic boundary conditions. For all simulation systems, ten different initial configurations for each system were randomly generated without significant overlaps to insure representative conformational sampling. For BD simulations of the repulsive model without HI, 30 μs simulations were performed with a time step of 0.5 ps. For simulations with HI, 12 μs simulations were performed with a time step of 2 ps. Trajectories for the first 5 μs of simulations were discarded for analysis. In calculating lubrication forces, particle pairs having $s < 4$ were evaluated every time step, where $s = 2r_{ij}/(a_{i} + a_{j})$. On the other hand, since far-field HI are insensitive to small configuration changes, $\mathbf{M}^{\infty}$ was computed every 500 steps.

In simulations that include HI, the short-time self diffusion coefficient for translational motion of particle type $i$, $D_{i}^{S}$, is defined by [11]

$$D_{i}^{S}=\frac{1}{3N_{i}}\sum_{\alpha\in i}^{N_{i}}\text{tr}\left(\mathbf{D}^{\alpha\alpha}\right). \tag{19}$$

Here, $N_{i}$ is the number of type $i$ particles in the system and $\mathbf{D}^{\alpha\alpha}$ is the 3 × 3 matrix of the self part of the diffusion tensor. The long-time self diffusion coefficient is defined as

$$D^{L}=\lim_{t\to\infty}\frac{1}{6t}\left\langle\left|\mathbf{r}(t)-\mathbf{r}(0)\right|^{2}\right\rangle. \tag{20}$$

Finally, to analyze the correlations between particles in time and space, we calculate the normalized pair correlation function, $C_{ij}$, given by

$$C_{ij}(d_0, \tau) = \frac{\sum\left[\left(\Delta\mathbf{r}_i(\tau)\cdot\Delta\mathbf{r}_j(\tau)\right)\delta(d_0 - d_{ij})\right]}{\sqrt{\sum\left|\Delta\mathbf{r}_i(\tau)\delta(d_0 - d_{ij})\right|^2}\sqrt{\sum\left|\Delta\mathbf{r}_j(\tau)\delta(d_0 - d_{ij})\right|^2}}, \qquad (21)$$

where $d_0$ is a specified the surface distance between particles $i$ and $j$, and $\tau$ is the time interval. $\delta(d_0 - d_{ij})$ is the Dirac delta function. $d_{ij}$ is the surface distance between particles $i$ and $j$ at time $t$ given by

$$d_{ij} = \left|\mathbf{r}_i(t) - \mathbf{r}_j(t)\right| - a_i - a_j. \qquad (22)$$

The summation in Eq. 21 is over all time points $t$ and all independent simulations.

## 3. Results

### 3.1. *Effect of Excluded Volume and Hydrodynamic Interactions on Diffusion*

We performed BD simulations with and without HI to evaluate the effects of HI on diffusion in crowded environments. In addition to the RPY [25, 26] interaction tensor, (widely used in biomolecular simulations to incorporate the long-range effects of HI), we also account for the lubrication forces that play a crucial role in the short-range interactions that are especially important in dense systems [6, 11].

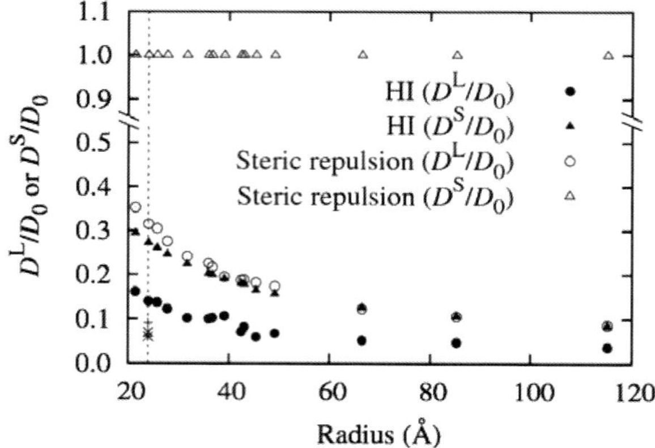


Figure 2. Reduction in diffusivity as a function of Stokes radius. Plus, cross, and asterisk are diffusivity of GFP measured *in vivo* of DH5α [4], BL21(DE3) [5], and K-12 [5] *E. coli*, respectively.

Fig. 2 shows $D^S/D_0$ and $D^L/D_0$ values at three different concentrations as a function of Stokes radius, where $D_0$ is the diffusion coefficient in infinite dilution.

$D^{\mathrm{L}}/D_0$ obtained from the BD simulations of sphere systems with just steric repulsions are also shown. Similar to the simulations without HI, $D^{\mathrm{S}}/D_0$ and $D^{\mathrm{L}}/D_0$ decrease with increasing radius. For $D^{\mathrm{S}}$, HI greatly reduces the diffusion constants of all particles; in contrast, $D^{\mathrm{S}}$ is always equal to $D_0$ when HI are ignored; the reduction in short-time diffusion coefficient is a purely hydrodynamic property; $D^{\mathrm{S}}$ equals $D_0$ when HI are absent [13].

In *in vivo* experiments, $D^{\mathrm{L}}/D_0$ of GFP is 0.06–0.09 [4, 5] (see Fig. 2). On the other hand, $D^{\mathrm{L}}/D_0$ value of GFP in the simulated equivalent sphere system without HI is 0.31; this is more than 3 times larger than experiment. This result indicates that although excluded volume effects reduce macromolecular diffusion in intracellular environments, they cannot explain the factor of ~10-16 reduction observed *in vivo*. However, when HI are considered, BD simulations give $D^{\mathrm{L}}/D_0$ of 0.14 for at 300 mg/ml, which is close to the observed experimental values (Fig. 2). Because the concentration of macromolecules in the *E. coli* cytoplasm is estimated to be in range of 300–340 mg/ml [3], the concentration of 300 mg/ml used in this study is a lower limit of physiological conditions. Therefore, diffusivity of GFP obtained from the simulation at higher concentrations get even closer to experiment. These results indicate that steric crowding and HI are two major factors responsible for the reduction in diffusion of macromolecules in intracellular environments. Indeed, without any other assumptions, these two effects well reproduce the experimentally observed diffusion constant of GFP *in vivo*.

### 3.2. *Large-distance and Long Time Intermolecular Correlations*

Next, the dynamical correlations in space and time between macromolecules were examined. Such effects are expected to be present when HI are included. To analyze the correlation between particles, we calculated a normalized pair correlation function, $C_{ij}$. $C_{ij}$ ranges from $-1$ to 1. When two particles are positively correlated, $C_{ij} > 0$, and when they are negatively correlated, $C_{ij} < 0$.

Representative $C_{ij}$ of large and small particle pairs up to 100 ns in time and 10 Å in space for the sphere system with and without HI as well as with the non-specific attractive interaction model are shown in Fig. 3. In the model without HI, where $D^{\mathrm{L}}$ ~3 times larger than in the HI model, (Fig 3 top) $C_{ij} < 0.1$ even at short times ($< 30$ ns) for both pairs. In contrast, for both pairs of molecules, a significant positive intermolecular dynamic correlation for the simulation with HI is evident, though these are on average weak, $C_{ij} < 0.3$. These results clearly show that HI give rise to quite long distance and time dynamical correlations between particles of all sizes even in the high crowded environment.

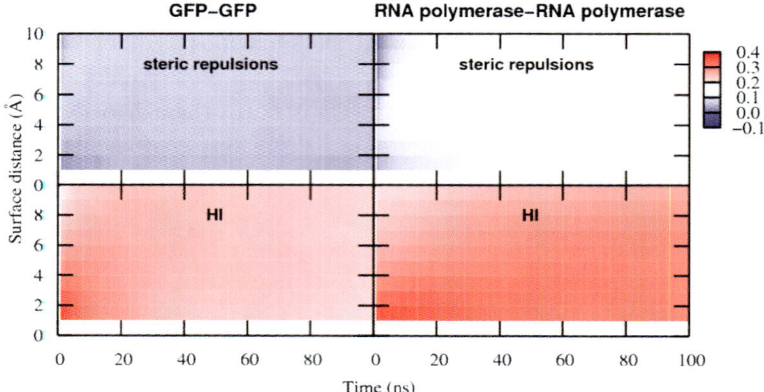



Figure 3. Normalized pair correlation function, $C_{ij}$, averaged over GFP-GFP (left) and RNA polymerase-RNA polymerase (right) pairs. The Stokes radii of GFP and RNA polymerase are 24.0 and 66.5 Å, respectively.

## 4.  Discussion

The goal of this study is to evaluate the role of HI in the reduction of macromolecular diffusion in intracellular environments. To ascertain the importance of HI, we performed BD simulations that take into account not only far-field HI but also near-field lubrication effects in three dimensions. The key finding of this work is that the two factors, excluded volume effects and HI, are sufficient to explain the large reduction in diffusion of macromolecules observed *in vivo*. Indeed, the diffusion constant of GFP *in vivo* can be almost quantitatively predicted by the inclusion of HI *without any adjustable parameters or other ad hoc assumptions.*

A number of other factors can also affect intracellular diffusion: 1) *Electrostatic interactions between molecules.* In principle, electrostatic interactions are long-ranged. However, the salt concentration inside cells is ~150 mM, so that they are well screened with a Debye length of ~8 Å. McGuffee and Elcock recently simulated a bacterial cytoplasm model where electrostatic interactions were treated by using Poisson-Boltzmann equations [27]. However, the diffusion coefficient of GFP is just slightly smaller than that obtained without electrostatic interactions; both values were 3-4 times larger than experiment. Heterogeneous charge distributions on molecular surface, like in real biomolecules, may affect macromolecular motions. However, since electrostatic interactions are highly screened due to the short Debye length found in physiological conditions, we believe that our conclusions would not qualitatively change. 2) *Viscosity of the cytoplasm.* In our simulations, the viscosity of the cytoplasm equals the value in water. The *in vivo* cytoplasm viscosity has been measured and is not significantly larger than bulk water, i.e. it is less than 2 cP [2, 22, 28]. 3) *GFP dimerization.* It is well known that GFP tends to dimerize in solutions of low (< 100 mM) ionic strength [29]. All of these physical factors will decrease macromolecular diffusivity *in vivo*. However, based on other work [27] and our results, which show that the reduction in diffusion of GFP in the simulation with HI is close to experiment, we expect the contribution of these three factors to be small.

Our results have demonstrated the likely importance of HI in macromolecular diffusion *in vivo*. However, there are a few possible limitations: First, the properties of a fluid on the nanometer scale are different from the bulk [30-32]. HI determined by solving Stokes equations may not fully describe the molecular situation. In order to fully validate the continuum limit assumption, molecular dynamics simulations with explicit solvent models would be necessary. Second, HI were considered only for the equivalent sphere system where the detailed molecular shape is ignored. Without HI, we demonstrated that this is a very good approximation [21, 24], but we have not explicitly shown this for the system with HI. Recently, an analytical formula that estimates the crossover time from anisotropic to isotropic diffusion of an arbitrarily shaped object in three dimensions using its $6N \times 6N$ diffusion tensor matrix was introduced [33]. The longest crossover time of molecule in our simulation system estimated by using this formula is 1.7 μs for the ribosome. Therefore, the effect of shape and diffusion anisotropy on the analysis of long-time translational diffusion is expected to be small.

Genome-sequencing has provided a detailed "parts list" for life [34]. Recently, the proteome wide prediction of protein structure and function has also become practical [34-40]. The next frontier in biophysics is to integrate this information and construct *in silico* cells that not only can describe the behavior of living systems in terms of individual biomolecules but which also can elucidate new biological principles describing their collective behavior. Until now, little attention has been paid to the biophysical properties of the crowded, heterogeneous environments found in cells, which have a great impact on the biological processes taking place. Therefore, modeling these crowding effects is an important first step towards whole cell simulation. The simulation method developed in this study could be a good tool to analyze the *in vivo* thermodynamics and kinetics of macromolecular dynamics.

## Acknowledgements

This work was supported in part by grant No. GM-37408 of the Division of General Medical Sciences of the National Institutes of Health.

## References


1. R. J. Ellis, *Trends Biochem Sci* **26**, 597 (2001).
2. K. Luby-Phelps, *Int Rev Cytol* **192**, 189 (2000).
3. S. B. Zimmerman, S. O. Trach, *J Mol Biol* **222**, 599 (1991).
4. M. B. Elowitz, M. G. Surette, P. E. Wolf, J. B. Stock, S. Leibler, *J Bacteriol* **181**, 197 (1999).
5. M. C. Konopka, I. A. Shkel, S. Cayley, M. T. Record, J. C. Weisshaar, *J Bacteriol* **188**, 6115 (2006).
6. J. F. Brady, G. Bossis, *Annu Rev Fluid Mech* **20**, 111 (1988).
7. W. B. Russel, D. A. Saville, W. R. Schowalter, *Colloidal dispersions.* Cambridge monographs on mechanics and applied mathematics (Cambridge University Press, Cambridge ; New York, 1989), pp. xvii, 525 p., 1 leaf of plates.
8. J. Antosiewicz, J. A. McCammon, *Biophys J* **69**, 57 (1995).
9. G. Arya, Q. Zhang, T. Schlick, *Biophys J* **91**, 133 (2006).
10. H. Jian, T. Schlick, A. Vologodskii, *J Mol Biol* **284**, 287 (1998).
11. R. J. Phillips, J. F. Brady, G. Bossis, *Phys Fluids* **31**, 3462 (1988).
12. L. Durlofsky, J. F. Brady, G. Bossis, *J Fluid Mech* **180**, 21 (1987).
13. J. F. Brady, *J Fluid Mech* **272**, 109 (1994).
14. J. F. Brady, R. J. Phillips, J. C. Lester, G. Bossis, *J Fluid Mech* **195**, 257 (1988).
15. C. W. J. Beenakker, *J Chem Phys* **85**, 1581 (1986).
16. D. J. Jeffrey, Y. Onishi, *J Fluid Mech* **139**, 261 (1984).
17. B. Cichocki, M. L. Ekiel-Jezewska, E. Wajnryb, *J Chem Phys* **111**, 3265 (1999).
18. A. J. Banchio, J. F. Brady, *J Chem Phys* **118**, 10323 (2003).
19. M. Fixman, *J Chem Phys* **69**, 1527 (1978).
20. D. L. Ermak, J. A. McCammon, *J Chem Phys* **69**, 1352 (1978).
21. T. Ando, J. Skolnick, in *QUANTUM BIO-INFORMATICS IV,* L. Accardi, W. Freudenberg, M. Ohya, Eds. (World Scientific Publishing Co. Pte. Ltd., Singapore, 2011), vol. 28, pp. 413-426.
22. D. Ridgway *et al.*, *Biophysical Journal* **94**, 3748 (2008).
23. S. Sundararaj *et al.*, *Nucleic Acids Res* **32**, D293 (2004).
24. T. Ando, J. Skolnick, *Proc Natl Acad Sci USA* **107**, 18457 (2010).
25. J. Rotne, S. Prager, *J Chem Phys* **50**, 4831 (1969).
26. H. Yamakawa, *J Chem Phys* **53**, 436 (1970).
27. S. R. McGuffee, A. H. Elcock, *PLoS Comput Biol* **6**, e1000694 (2010).
28. A. S. Verkman, *Trends Biochem Sci* **27**, 27 (2002).
29. F. Yang, L. G. Moss, G. N. Phillips, *Nat Biotechnol* **14**, 1246 (1996).

30. M. Benz, N. H. Chen, G. Jay, J. I. Israelachvili, *Ann Biomed Eng* **33**, 39 (2005).

31. B. Bhushan, J. N. Israelachvili, U. Landman, *Nature* **374**, 607 (1995).

32. A. Maali, B. Bhushan, *J Phys-Condens Mat* **20**, 315201 (2008).

33. J. Schluttig, C. B. Korn, U. S. Schwarz, *Phys Rev E* **81**, 030902 (2010).

34. J. Skolnick, J. S. Fetrow, *Trends Biotechnol* **18**, 34 (2000).

35. A. K. Arakaki, Y. Huang, J. Skolnick, *BMC Bioinformatics* **10**, 107 (2009).

36. D. Baker, A. Sali, *Science* **294**, 93 (2001).

37. M. Brylinski, J. Skolnick, *Proteins* **78**, 118 (2010).

38. M. Gao, J. Skolnick, *Nucleic Acids Res* **36**, 3978 (2008).

39. S. B. Pandit *et al.*, *Bioinformatics* **26**, 687 (2010).

40. Y. Zhang, J. Skolnick, *Proc Natl Acad Sci U S A* **101**, 7594 (2004).

Quantum Bio-Informatics V
© 2013 World Scientific Publishing Co. Pte. Ltd.
pp. 389–401

# SELF-REPELLING FRACTIONAL BROWNIAN MOTION - A GENERALIZED EDWARDS MODEL FOR CHAIN POLYMERS*

JINKY BORNALES

*Physics Dept., MSU-IIT, Iligan, The Philippines*
*E-mail: jinky.bornales@g.msuiit.edu.ph*

MARIA JOÃO OLIVEIRA

*Univ. Aberta and CMAF, University of Lisbon*
*E-mail: oliveira@cii.fc.ul.pt*

LUDWIG STREIT

*BiBoS, Univ. Bielefeld and CCM, Univ. da Madeira*
*E-mail: streit@uma.pt*

We present an extension of the Edwards model for conformations of individual chain molecules in solvents in terms of fractional Brownian motion, and discuss the excluded volume effect on the end-to-end length of such trajectories or molecules.

## 1. Introduction

Individual chain polymers in good solvents are typically modelled by trajectories of random walks, or - in the continuum limit - by Brownian paths. Such models by themselves however do not take into account that self-crossings of these paths should be suppressed, this "the excluded volume" effect will make the trajectories less curly and more extended. Fractional Brownian paths have been suggested as a heuristic model for such swelling, or on the other hand for polymers in a collapsed state[2], but a more proper model would be based on self-avoiding random walks, or on a weight factor which penalizes self-crossings, such as in the continuum Edwards[3] [6] [20] [21] [22] [23] or the discrete Domb-Joyce[5] model.

---

*This work is supported PTDC/MAT/100983/2008, ISFL-1-209

The ensuing swelling of the molecular conformations is given by the Flory index[8] [9] which describes the scaling of the end-to-end distance as a function the number of monomers. It has been extensively studied both in the (chemical) physics and the mathematics community. The physics literature is characterized by structural intuition and far-reaching predictions, the mathematical results are less far-reaching but provide the high reliability characteristic of the mathematical approach. Both are too vast to be quoted here, we refer for this to recent reviews[12] [18].

In the present paper, after a few words on fractional Brownian motion fBm, we shall see that one can extend to fBm the Edwards model of Brownian paths with exponentially suppressed self-intersections, a mathematical existence proof has been established recently[11]. In the third part of the paper we generalize some by now classical arguments from the physics literature to explore what the Flory index might be in the fBm case.

## 2. The fBm Edwards Model

### 2.1. *Fractional Brownian Motion*

Fractional Brownian motion on $\mathbb{R}^d$, $d \geq 1$, with "Hurst parameter" $H \in (0, 1)$ is a $d$-dimensional centered Gaussian process $B^H = \{B^H(t) : t \geq 0\}$ with covariance function

$$\mathbb{E}(B_i^H(t) B_j^H(s)) = \frac{\delta_{ij}}{2} \left( t^{2H} + s^{2H} - |t - s|^{2H} \right), \quad i, j = 1, \ldots, d, \ s, t \geq 0.$$

For $H = 1/2$ it is ordinary $d$-dimensional Brownian motion $B$. We refer to the recent monographs by Biagini et al.[1] and by Y. Mishura[16]; for self-intersection local times of fBm see Hu and Nualart[13].

### 2.2. *The Edwards Model*

Self-repelling Brownian paths for a time interval $0 \leq t \leq l$ can be modelled via a "Gibbs factor" to suppress self-intersections:

$$G = \frac{1}{Z} \exp \left( -g \int_0^l ds \int_0^l dt \delta \left( B(s) - B(t) \right) \right).$$

Technically one defines this expression as a limit, using

$$\delta_\varepsilon(x) := \frac{1}{(2\pi\varepsilon)^{d/2}} e^{-\frac{|x|^2}{2\varepsilon}}, \quad \varepsilon > 0,$$

in particular

$$Z = \lim_{\varepsilon \to +0} \mathbb{E}\left(\exp\left(-g \int_0^l ds \int_0^l dt \delta_\varepsilon\left(B(s) - B(t)\right)\right)\right)$$

if this quantity is well defined; otherwise a renormalization is required, as, more generally, in Theorem 2.2 below.

Recently, generalizing an argument of Varadhan[20], this was extended in [11] to

$$G = \frac{1}{Z} \exp\left(-g \int_0^l ds \int_0^l dt \delta\left(B^H(s) - B^H(t)\right)\right),$$

as follows.

**Theorem 2.1.** *The Edwards model is well defined for all $H < 1/d$, with*

$$G = \frac{1}{Z} \exp\left(-g \int_0^l ds \int_0^l dt \delta\left(B^H(s) - B^H(t)\right)\right).$$

**Theorem 2.2.** *For $H = 1/d$ and $g$ sufficiently small*

$$G = \lim_{\varepsilon \searrow 0} \frac{1}{Z_\varepsilon} \exp\left(-g \int_0^l ds \int_0^l dt \delta_\varepsilon\left(B^H(s) - B^H(t)\right)\right),$$

*with*

$$Z_\varepsilon \equiv \mathbb{E}\left(\exp\left(-g \int_0^l ds \int_0^l dt \delta_\varepsilon\left(B^H(s) - B^H(t)\right)\right)\right)$$

*is well-defined.*

## 3. The Flory Index

When the number $N$ of monomers of a polymer becomes large one expects its end-to-end length $R$ to scale[10]

$$R(N) \sim N^\upsilon.$$

For (fractional) Brownian motion the root-mean-square length

$$R = \sqrt{\mathbb{E}\left(B^H(N)^2\right)}$$

is scaling with

$$\upsilon = H.$$

But the excluded volume effect makes the paths and polymers swell: the end-to-end length increases. For the Brownian motion case there is the famous Flory formula

$$v = v\,(d) = \frac{3}{d+2}$$

based originally on a mean field argument. Since its proposal by Flory[7][9], numerous methods were invoked to put it on a more solid mathematical basis, a process which has up to now been fully successful in the case $d = 1$ [12].

To obtain what may be considered as a first guess of a similar formula for fBm we shall return to the modest beginnings, generalizing Fisher's original argument[7][8] (see e.g. the review given in McKenzie[15]) to the case at hand.

### 3.1. *The Fisher Argument*

A partition function $Z(R)$ for a freely jointed chain of $N$ segments for which the end-to-end length has fixed modulus $R$ is given by

$$Z(R) = aR^{d-1}\exp(-\frac{dR^2}{2N}),$$

and leads to a free energy

$$F_1 = -\ln Z \sim \frac{dR^2}{2N} - (d-1)\ln R.$$

Instead of such a chain a continuous model is that of a Brownian trajectory from time zero to time $N$, for which one computes

$$\mathbb{E}\left(\delta\left(B\,(N) - \vec{R}\right)\right) = (2\pi N)^{-d/2}\exp\left(-\frac{R^2}{2N}\right). \tag{1}$$

For the fBm case this formula generalizes to

$$\mathbb{E}\left(\delta\left(B^H\,(N) - \vec{R}\right)\right) = (2\pi N^{2H})^{-d/2}\exp\left(-\frac{R^2}{2N^{2H}}\right) \tag{2}$$

from which we see that $N \to N^{2H}$, and hence we should consider

$$Z(R) = aR^{d-1}\exp(-\frac{dR^2}{2N^{2H}})$$

i.e.

$$F_1 = -\ln Z \sim \frac{dR^2}{2N^{2H}} - (d-1)\ln R.$$

For the repulsive excluded volume energy of fBm paths $x$ with $x(N) = \vec{R}$,

$$F_2 = -\ln \mathbb{E}_{x(N)=\vec{R}} \left( \exp \left( -g \int_0^N ds \int_0^N dt \delta \left( x(s) - x(t) \right) \right) \right)$$

dimensional considerations and mean field arguments[15] suggest

$$F_2 \sim const.\frac{N^2}{R^d}$$

Maximizing

$$F(N, R) = F_1(N, R) + F_2(N, R)$$

with regard to $R$ leads to

$$0 = \frac{dR}{N^{2H}} - \frac{d-1}{R} - const.N^2 R^{-d-1}.$$

Assuming that the 2nd term is negligible one finds

$$R^{d+2} \sim N^{2H+2}$$

i.e.

$$R \sim N^{\upsilon_H}$$

with

$$\upsilon_H(d) = \frac{2H + 2}{d + 2}. \tag{3}$$

**Remark 3.1.** A polymer model with $B(t^{2H})$ instead of $B^H(t)$ would produce the same expression as in (2), hence also the same Flory index, but would not share the homogeneity implied by the stationary increments of $B^H(\cdot)$.

## 3.2. *The Critical Dimension*

The derivation of $\upsilon_H$ is evidently heuristic and needs validation. For this it is worth noting that for Brownian motion there is a critical dimension $d_c = 4$ defined by the fact that for $d \geq d_c$ there is no excluded volume effect, so that $R$ scales like the unperturbed Brownian motion:

$$R \sim N^{1/2}$$

i.e.

$$\upsilon_{1/2}(4) = 1/2.$$

We can ask for which dimension, more generally, the fBm Flory index will show no excluded volume effect from self-crossings, i.e.

$$v_H(d_c) = H.$$

Inserting our ansatz (3) one finds

$$Hd_c = 2.$$

and indeed it is known (Theorem 1.1 of Talagrand[19]) that $d$-dimensional fBm has no double points iff

$$Hd \geq 2,$$

in other words, our $v_H$ predicts $d_c$ correctly.

**Remark 3.2.** As a consequence, any Flory formula should be considered only up to the critical dimension, i.e. as long as there are double points and an excluded volume effect. Similarly, any prediction of $v_H > 1$ would be unphysical: the end-to-end distance cannot grow faster than the number N of monomers. In the case at hand this suggests for the one-dimensional case

$$v_H(1) = \begin{cases} \frac{2H+2}{3} & \text{if } H \leq \frac{1}{2} \\ 1 & \text{if } H > \frac{1}{2} \end{cases}.$$

Note that for small $H$ the scaling exponent as predicted would be strictly less than one while for the Brownian motion case

$$v_{1/2}(1) = 1$$

has been proven[12][23]. The infimum

$$\lim_{H \to 0} v_H(1) = 2/3$$

happens to be the scaling exponent of the myopic random walk[12] .

**Remark 3.3.** In the attached figure 1 the two red lines correspond to $v_H(d) = 1$ and to the critical dimension as a function of the Hurst index $H$, respectively. Above these the Flory index is unphysical. On the green lines $v_H$ is validated. (The existence proof of the fBm Edwards model in [11] works below the dashed line.)

**Remark 3.4.** For fixed dimension $d$, any extension $F(H)$ of the Flory formula to general Hurst indices $H$ will have to obey

$$F(\frac{1}{2}) = \frac{3}{d+2} \tag{4}$$

for the usual Brownian motion (Flory-Fisher), and

$$F(\tfrac{2}{d}) = \frac{2}{d} \tag{5}$$

at the critical point (Talagrand). Note that our ansatz (3)

$$F(H) = \upsilon_H(d) = \frac{2H + 2}{d + 2}$$

is just the unique linear interpolation between those two values.

### 3.3.  *A Recursion Formula*

For $H = 1/2$ Kosmas and Freed[14] derive a recursion formula

$$2 - \frac{1}{\upsilon(d)} = \frac{4 - d}{3} \left( 2 - \frac{1}{\upsilon(1)} \right) \tag{(4.13)}$$

(Here and in the following we label formulas from - or analogous to those in - the paper [14] by their numbers in that article, in double brackets.)

The derivation of this formula is specific to the Brownian motion case and does not hold for general $\upsilon_H$. Hence in what follows we shall generalize their arguments which led to ((4.13)) to first obtain a valid recursion formula and then check whether it is satisfied by $\upsilon_H$ as given in (3).

We begin by considering

$$Z(g, N) \equiv \mathbb{E} \left( \exp \left( -g \int_0^N ds \int_0^N dt\delta \left( B^H(s) - B^H(t) \right) \right) \right).$$

From the defining relation

$$\mathbb{E} \left( B^H(s) B^H(t) \right) = \frac{1}{2} \left( s^{2H} + t^{2H} - |s - t|^{2H} \right)$$

we see that for $a > 0$ the processes $\{ B^H(t) : t > 0 \}$ and $\{ a^{-H} B^H(at) : t > 0 \}$ obey the same law. Making this substitution and a change of integration variables $as = \sigma$, $at = \tau$ we obtain

$$Z(g, N) = \mathbb{E} \left( \exp \left( -ga^{Hd-2} \int_0^{aN} ds \int_0^{aN} dt\delta \left( B^H(s) - B^H(t) \right) \right) \right) \tag{6}$$

$$= Z \left( a^{Hd-2} g, aN \right). \tag{(2.14)}$$

Likewise we find for the mean-square end-to-end distance

$$
\langle R^2 \rangle
$$
$$
\equiv \frac{1}{Z(g,N)} \mathbb{E}\left( \left(B^H(N)\right)^2 \times \right.
$$
$$
\left. \times \exp\left(-g \int_0^N ds \int_0^N dt\, \delta\left(B^H(s) - B^H(t)\right)\right)\right) \tag{7}
$$

$$
\langle R^2 \rangle
$$
$$
= a^{-2H} \mathbb{E}\left( \left(B^H(aN)\right)^2 \right.
$$
$$
\left. \times \exp\left(-ga^{Hd-2} \int_0^{aN} ds \int_0^{aN} dt\, \delta\left(B^H(s) - B^H(t)\right)\right)\right)
$$
$$
= a^{-2H} f\left(a^{Hd-2}g, aN\right) \tag{(2.18)}
$$
$$
= N^{2H} f\left(N^{2-Hd}g, 1\right). \tag{(2.19)}
$$

(Note the critical dimension $Hd = 2$ where $\langle R^2 \rangle^{1/2} \sim N^H$.) For large $N$ one expects a power law behavior for the unknown function $f$, i.e.

$$
\langle R^2 \rangle \sim N^{2H} \left(N^{2-Hd}g\right)^x \tag{(2.22)}
$$

with an exponent $x$ to be determined.

As a next step we restrict one coordinate of the positions, $x_i(t) = B_i^H(t)$ to the interval $[0, D]$ by inserting

$$
1_{[0,D]}(B_i^H) = \begin{cases} 1 \text{ if } B_i^H(t) \in [0, D] \text{ for all } t \\ 0 \qquad\qquad \text{otherwise} \end{cases}
$$

into (7). One obtains

$$\langle R^2 \rangle_D$$

$$= \frac{1}{Z_D(g,N)} \mathbb{E}\left( 1_{[0,D]}(B_i^H)\left(B^H(N)\right)^2 \times \right.$$

$$\left. \times \exp\left(-g \int_0^N ds \int_0^N dt\, \delta\left(B^H(s) - B^H(t)\right)\right)\right)$$

$$= \frac{a^{-2H}}{Z_D(g,N)} \mathbb{E}\left( 1_{[0,a^H D]}(B_i^H)\left(B^H(aN)\right)^2 \times \right.$$

$$\left. \times \exp\left(-a^{Hd-2}g \int_0^{aN} d\sigma \int_0^{aN} d\tau\, \delta\left(B^H(\sigma) - B^H(\tau)\right)\right)\right)$$

$$= a^{-2H} F(a^H D,\ a^{Hd-2}g, aN) = N^{2H} F(N^{-H}D,\ N^{2-Hd}g, 1). \tag{4.3}$$

Now assume that asymptotically there is a dimensionless correction factor $h$ for

$$\langle R^2 \rangle_D \approx \langle R^2 \rangle\, h\left(\frac{D}{\sqrt{\langle R^2 \rangle}}\right).$$

It should grow as $D$ becomes small which suggests a power law behavior for the function $h$:

$$\langle R^2 \rangle_D \approx \langle R^2 \rangle \left(\frac{D}{\sqrt{\langle R^2 \rangle}}\right)^{-y} \tag{4.5}$$

with $y$ to be determined. As $D$ approaches a minimal value $D_0$ - approximately the extension of a monomer ("Kuhn length") - the polymer becomes effectively $(d-1)$-dimensional:

$$\langle R_{d-1}^2 \rangle \approx \langle R_d^2 \rangle_D \approx D_0^{-y} \langle R_d^2 \rangle^{1+\frac{y}{2}}. \tag{4.6}$$

This provides a relation between the end-to-end length for dimensions $d$ and $d-1$. To obtain from this a recursion relation, recall equation $(2.22)$:

$$\langle R_d^2 \rangle \sim N^{2H+x(2-Hd)} g^x$$

and introduce instead of $x$ the (unknown)

$$2\upsilon_H(d) \equiv 2H + x(2 - Hd)$$

so that

$$\langle R_d^2 \rangle = c_d N^{2\upsilon_H(d)} g^{\frac{2\upsilon_H(d)-2H}{2-Hd}} \tag{4.8}$$

and

$$\left\langle R_{d-1}^2 \right\rangle = c_{d-1} N^{2 v_H (d-1)} g^{\frac{2 v_H (d-1) - 2H}{2 - H \cdot (d-1)}} . \qquad ((4.9))$$

On the other hand from ((4.6)) we have

$$\left\langle R_{d-1}^2 \right\rangle \approx D_0^{-y} \left\langle R_d^2 \right\rangle^{1+\frac{y}{2}} = const. N^{2 v_H (d)\left(1+\frac{y}{2}\right)} g^{\frac{2 v_H (d) - 2H}{2 - Hd}\left(1+\frac{y}{2}\right)} . \qquad (8)$$

Comparing exponents in these two expressions we find

$$v_H (d-1) = v_H (d) \left(1 + \frac{y}{2}\right)$$

$$\frac{v_H (d-1) - H}{2 - H \cdot (d-1)} = \frac{v_H (d) - H}{2 - Hd} \left(1 + \frac{y}{2}\right) .$$

The first of these equations gives

$$1 + \frac{y}{2} = \frac{v_H (d-1)}{v_H (d)} ,$$

with this the second one becomes

$$\frac{1}{v_H (d-1)} \frac{v_H (d-1) - H}{2 - H \cdot (d-1)} = \frac{1}{v_H (d)} \frac{v_H (d) - H}{2 - Hd}$$

i.e. this expression does not depend on the dimension $d$ so that all the $v_H (d)$ are given in terms of e.g. $v_H (1)$ :

$$\frac{1}{v_H (d)} \frac{v_H (d) - H}{2 - Hd} = \frac{1}{v_H (1)} \frac{v_H (1) - H}{2 - H} ,$$

$$\implies v_H (d) = \frac{(2 - H) \, v_H (1)}{(d - 1) \, v_H (1) + 2 - dH} . \qquad (9)$$

**Proposition 3.1.**

$$v_H (d) = \frac{2H + 2}{d + 2}$$

*satisfies this recursion equation, with*

$$\frac{1}{v_H (d)} \frac{v_H (d) - H}{2 - Hd} = \frac{1}{2H + 2} .$$

**Remark 3.5.** The standard Flory index (3) for $H = 1/2$ obeys the recursion formula.

**Remark 3.6.** The recursion formula (9) implies the correct critical behavior, i.e. any solution will obey $v_H (d) = H$ for $d = 2/H \equiv d_c$, whatever the choice of $v_H (1)$. To see this explicitly, insert $d = 2/H$ and find

$$v_H \left(\frac{2}{H}\right) = \frac{(2 - H) \, v_H (1)}{(2/H - 1) \, v_H (1) + 0} = H .$$

**Remark 3.7.** If $\upsilon_H(1)$ turned out to be equal to one for all $H$, the recursion formula would suggest

$$\upsilon_H(d) = \frac{2 - H}{d + 1 - dH}, \tag{10}$$

an expression which then also produces the standard Flory formula for $H = 1/2$, as well as the critical dimension $d = 2/H$.

## 4. Summary

The Edwards type model for self-repelling fBm now at hand will raise the question of how the end-to-end length of trajectories scales as a function of time (or "number of monomers"). The original Fisher argument, while criticized regarding its assumptions[17][4], provides a simple heuristic "derivation" of the Flory formula which allows an extension to fBm. The obtained scaling law needs further verification; we note that it correctly predicts the critical dimension for which the excluded volume become negligible and obeys a recursion formula based on dimension reduction. The latter would provide a useful constraint on any alternate scaling laws.

$$* \ * \ *$$

J. B. and L. S. would like to express their gratitude for the kind hospitality of the Centro de Matemática e Aplicações Fundamentais CMAF in Lisbon which made our collaboration possible and pleasant.

## References


1. F. Biagini, Y. Hu, B. Oksendal: *Stochastic Calculus for Fractional Brownian Motion and Applications*. Springer, Berlin, 2007.
2. P. Biswas, B. J. Cherayil: Dynamics of Fractional Brownian Walks. *J. Phys. Chem.* **99**, 816-821 (1995).
3. E. Bolthausen: On the construction of the three dimensional polymer measure. *Prob. Theory. Rel. Fields* **97**, 81-101 (1993).
4. J. Des Cloizeaux: On the absence of flory terms in the energy and in the entropy of a polymer chain. *J. Phys. France* **37**, 431-434 (1976).
5. C. Domb, G. S. Joyce: Cluster expansion for a polymer chain. *J. Physics C* **5**, 956-976 (1972).
6. S. F. Edwards: The statistical mechanics of polymers with excluded volume. *Proc. Phys. Sci.* **85** 613-624 (1965).
7. M. E. Fisher: Shape of a self-avoiding walk or polymer chain, *J. Chem. Phys.* **44,** 616-622 (1966).
8. M. E. Fisher: *J. Phys. Soc. Japan* **26** Suppl. 44 (1969).

9. P.J. Flory: *Principles of Polymer Chemistry*. Cornell University Press, 1953.

10. P. G. de Gennes: Scaling Concepts in Polymer Physics. Cornell University Press, Ithaca, NY, 1979.

11. M. Grothaus, M.J. Oliveira, J.-L. Silva, L.Streit: Self-avoiding fractional Brownian motion - The Edwards model. *J. Stat. Phys.* **145**, 1513–1523 (2011).

12. R. van der Hofstad, W. König: A survey of one-dimensional random polymers. *J. Stat. Phys.* **103**, 915-944 (2001).

13. Y. Hu and D. Nualart: Renormalized self-intersection local time for fractional Brownian motion. *Ann. Probab.* **33**, 948–983 (2005).

14. M. K. Kosmas, K. F. Freed: On scaling theories of polymer solutions. *J. Chem. Phys.* **69**, 3647-3659 (1978).

15. D. S. McKenzie: Polymers and Scaling. *Phys. Rep.* **27**, 35-88 (1976).

16. Y. Mishura: *Stochastic Calculus for Fractional Brownian Motion and Related Processes*. Springer LNM 1929, 2008.

17. M. A. Moore, A. J. Bray: On the Flory formula for the polymer size exponent v. ,*J. Phys. A: Math. Gen.* **11**, 1353-1359 (1978).

18. A. Pelissetto, E. Vicari: Critical phenomena and renormalization-group theory. *Phys. Rep.* **368**, 549–727 (2002), Chapter 9.

19. M. Talagrand: Multiple points of trajectories of multiparameter fractional Brownian motion. *Probab. Theory Related Fields* **112**, 545–563 (1998).

20. S. R. S. Varadhan: Appendix to "*Euclidean quantum field theory*" by K. Symanzik, in: R. Jost, ed., Local Quantum Theory, Academic Press, New York, p. 285, 1970.

21. J. Westwater: On Edwards' model for polymer chains. *Comm. Math. Phys.* **72**, 131-174 (1980).

22. J. Westwater: On Edwards' model for polymer chains. III. Borel summability. *Comm. Math. Phys.* **84**, 459-470 (1982).

23. J. Westwater: On Edwards model for polymer chains, in Trends and Developments in the Eighties, S. Albeverio and P. Blanchard, eds., Bielefeld Encounters in Math. Phys. 4/5 (World Scientific, Singapore, 1984).

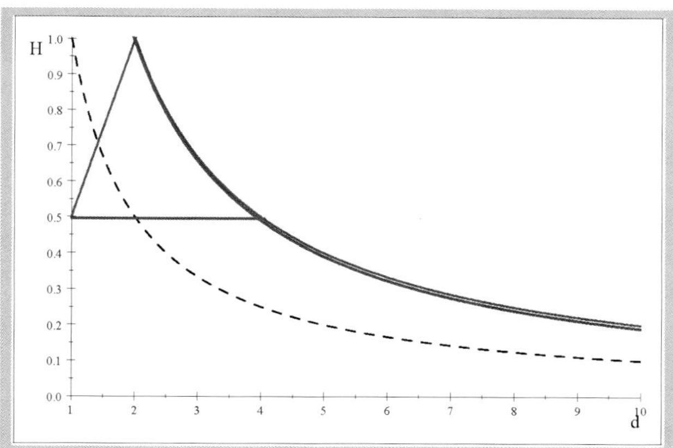

Figure 1.    The domain of the Flory index.

Quantum Bio-Informatics V
© 2013 World Scientific Publishing Co. Pte. Ltd.
pp. 403–407

# SIGNALING NETWORKS INVOLVING REACTIVE OXYGEN SPECIES AND Ca$^{2+}$ IN PLANTS

KAZUYUKI KUCHITSU[†]

*Department of Applied Biological Science, Tokyo University of Science, Noda,Chiba 278-8510 Japan*

Although plants never evolved central information processing organs such as brains, plants have evolved distributed information processing systems and are able to sense various environmental changes and reorganize their body plan coordinately without moving. Recent molecular biological studies revealed molecular bases for elementary processes of signal transduction in plants. Though reactive oxygen species (ROS) are highly toxic substances produced through aerobic respiration and photosynthesis, plants possess ROS-producing enzymes whose activity is highly regulated by binding of Ca$^{2+}$. In turn, Ca$^{2+}$-permeable channel proteins activated by ROS are shown to be localized to the cell membrane. These two components are proposed to constitute a positive feedback loop to amplify cellular signals. Such molecular physiological studies should be important steps to understand information processing systems in plants and future application for technology related to environmental, energy and food sciences.

## 1. Regulation of enzymes producing reactive oxygen species in plants

Plants have various scavenging systems for reactive oxygen species (ROS) such as $\cdot O_2^-$, $H_2O_2$ and $\cdot OH$ that are produced as byproducts of respiration and photosynthesis. ROS oxidizes various biomolecules including proteins, lipids, carbohydrates and nucleic acids, and therefore are very toxic to cells. In contrast, plants enzymatically produce ROS for use in cell signal transduction. Due to the high toxicity of ROS, enzymes producing ROS need to be tightly regulated spatially and temporally.

Respiratory burst oxidase homolog (rboh) proteins have been identified as NADPH oxidases (Nox) in diverse plant species. Plant Rboh are predicted to contain six conserved transmembrane helices and cytosolic

FAD- and NADPH-binding domains in the C-terminal region homologous to its mammalian homolog, NOX2. Unlike NOX2, all Rboh have an extended N-terminal region that includes two $Ca^{2+}$-binding EF-hand motifs. ROS produced by Rboh are proposed to play many crucial roles in signaling and development in plants.

Ten *rboh* genes (*AtRbohA–J*) have been identified in the genome of *Arabidopsis thaliana*. So far, mutant phenotype analyses have shown that *AtrbohB* plays a role in seed after-ripening and *AtrbohC/ROOT HAIR DEFECTIVE 2 (RHD2)* is essential for root hair tip growth. Both *AtrbohD* and *AtrbohF* function in defense responses and signaling of several plant hormones including abscisic acid (ABA), ethylene and jasmonic acid. On the other hand, phenotypes of *atrbohD* and *atrbohF* mutants exhibit differences in an avirulent bacteria-induced ROS accumulation and ion leakage, stomatal closure induced by ABA and ethylene. Rbohs have ROS-producing activity synergistically activated by protein phosphorylation and $Ca^{2+}$ [1-3].

A heterologous expression system using human embryonic kidney 293T (HEK293T) cells is a useful tool for analyzing the ROS-producing activity of Rboh [1]. AtRbohC/RHD2 shows good correlation between the ROS-producing activity of in the heterologous expression system and phenotypes *in planta* [2]. Thus the ROS-producing activity of Rboh observed in the heterologous expression system reflects their activity *in planta* at least in part.

A Ser/Thr protein phosphatase inhibitor, calyculin A, enhances the protein phosphorylation and activates the ROS-producing activity of Rbohs. On the other hand, K-252a, a protein kinase inhibitor, inhibited their $Ca^{2+}$-induced ROS-producing activity. These results suggest that $Ca^{2+}$-induced ROS production depends on protein phosphorylation [3]. The protein phosphorylation may contribute to the $Ca^{2+}$ affinity of EF-hand motif(s) to $Ca^{2+}$, leading to the synergistic activation by protein phosphorylation and $Ca^{2+}$.

## 2.   Positive feedback regulatory network involving ROS and $Ca^{2+}$

ROS produced by NADPH oxidases have been shown to activate $Ca^{2+}$ permeable channels in various cell types including stomatal guard cells and root hairs. We proposed that ROS and $Ca^{2+}$ constitute a positive feedback loop in root hair tip growth. ROS produced by AtRbohC/RHD2 activates hyperpolarization-activated $Ca^{2+}$ channels and induce $Ca^{2+}$ influx. $Ca^{2+}$ in turn activates AtRbohC/RHD2 through binding to the EF-hand motifs. ROS produced by AtRbohD and AtRbohF are involved in ABA-induced $Ca^{2+}$

influx and ABA-activation of $Ca^{2+}$ permeable $I_{Ca}$ channels in guard cells, implying that similar positive feedback regulation may also take part in ABA signal transduction.

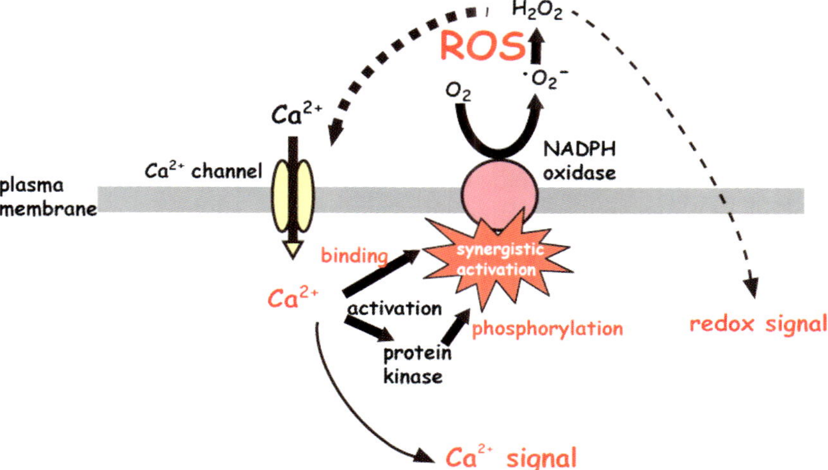


Positive feedback regulation of ROS and $Ca^{2+}$ has been proposed to play a crucial role in maintaining growth at the root hair tip; AtRbohC/RHD2 produces ROS, which in turn activate $Ca^{2+}$ channels on the plasma membrane to induce $Ca^{2+}$ influx into the cell. For the maintenance of tip growth, apical localization of the ROS-producing enzyme at the tip is a prerequisite. Once located at the tip, a positive feedback loop involving Rboh is initiated. ROS derived from the NADPH oxidase may activate the $Ca^{2+}$ channels that transport $Ca^{2+}$ into the cells that in turn activates Rboh activity through its EF hand and a $Ca^{2+}$-dependent protein kinase activity. Such a system of positive feedback, in concert with an independent mechanism for locating the Rboh protein to the tip of the cell, provides a robust mechanism that explains how cells such as root hairs maintain polarity during morphogenesis [2].

$Ca^{2+}$ directly binds the EF-hand motif of Rboh to activate ROS production. Protein phosphorylation is prerequisite for $Ca^{2+}$-induced ROS production by Rbohs. Conversely, protein phosphorylation-induced ROS production is independent of $Ca^{2+}$. Taken together, we proposed a model of how phosphorylation of Rboh may function as the initial trigger for the positive feedback regulation as well as the $Ca^{2+}$-ROS signaling network in plants: First, protein phosphorylation of Rboh produces ROS that activate $Ca^{2+}$ channels to induce $Ca^{2+}$ influx. Subsequently, $Ca^{2+}$ further activates Rboh [3].

## 3. Ca$^{2+}$-ROS signaling network in mechano/osmo-sensing in plants

Hypo-osmotic shock triggers ROS generation following a $[Ca^{2+}]_{cyt}$ increase in cytosolic $Ca^{2+}$ concentration. Extracellular $Ca^{2+}$ is required for both $Ca^{2+}$ influx and NADPH oxidase-mediated ROS generation induced by hypo-osmotic shock, suggesting that ROS generation requires $Ca^{2+}$ influx across the plasma membrane. We have recently identified a putative mechanosensitive $Ca^{2+}$-permeable channel component from rice (OsMCA1) [4] and tobacco (NtMCA1,2) [5]. Overexpression of *OsMCA1* enhances ROS generation. Binding $Ca^{2+}$ to the EF-hand regions of cytosolic regulatory domains of plant NADPH oxidases directly activates them. A functional NADPH oxidase/Rboh affects mechanical stress-induced ROS generation in a $Ca^{2+}$-dependent manner. Overproduction of the plasma membrane $Ca^{2+}$-permeable channels may induce the mobilization of excess $Ca^{2+}$ in response to mechanical stimuli, which may cause enhanced activation of NADPH oxidases. These results indicate that OsMCA1 is involved in regulation of plasma membrane $Ca^{2+}$ influx and NADPH oxidase-mediated ROS production induced by hypo-osmotic stress in cultured cells [4]. These findings shed light on our understanding of mechanical sensing pathways.

## 4. Concluding remarks

Although plants never evolved central information processing organs such as brains, plants have evolved distributed information processing systems and are able to sense various environmental changes and reorganize their body plan coordinately without moving. Recent advancement in plant molecular biology revealed molecular bases for elementary processes of signal transduction in plants. Though ROS are highly toxic substances produced through aerobic respiration and photosynthesis, plants possess ROS-producing enzymes whose activity is highly regulated by binding of $Ca^{2+}$. In turn, $Ca^{2+}$-permeable channel proteins activated by ROS are shown to be localized to the cell membrane. These two components are proposed to constitute a positive feedback loop to amplify cellular signals. Such molecular physiological studies should be important steps to understand information processing systems in plants and future application for technology related to environmental, energy and food sciences.

**References**


1. Y. Ogasawara, H. Kaya, G. Hiraoka, F. Yumoto, S. Kimura, Y.Kadota, H. Hishinuma, E. Senzaki, S. Yamagoe, K. Nagata, M. Nara, K. Suzuki, M. Tanokura and K. Kuchitsu, *J. Biol. Chem.* 283, 8885-8892 (2008).
2. S. Takeda, C. Gapper, H. Kaya, E. Bell, K. Kuchitsu and L. Dolan, *Science* 319, 1241-1244 (2008).
3. S. Kimura, H. Kaya, T. Kawarazaki, G. Hiraoka, E. Senzaki, M. Michikawa, K. Kuchitsu, *Biochim. Biophys. Acta -Molecular Cell Research* doi: 10.1016/j.bbamcr.2011.09.011 (2011).
4. T. Kurusu, D. Nishikawa, Y. Yamazaki, M. Gotoh, M. Nakano, H. Hamada, T. Yamanaka, K. Iida, Y. Nakagawa, H. Saji, K. Shinozaki, H. Iida and K. Kuchitsu K, *BMC Plant Biol in press.* (2011)
5. T. Kurusu, T. Yamanaka, M. Nakano, A. Takiguchi, Y. Ogasawara, T. Hayashi, K. Iida, S. Hanamata, K. Shinozaki, H. Iida and K. Kuchitsu, *J. Plant Res.* DOI 10.1007/s10265-011-0462-6 (2011)

Quantum Bio-Informatics V

© 2013 World Scientific Publishing Co. Pte. Ltd.

pp. 409–424

# A NOVEL MEASURE FOR FINDING DISEASE-SPECIFIC GENES FROM THE BIOMEDICAL LITERATURE

YEONDAE KWON

*Tokyo University of Science,*
*Noda, Chiba 278-8510, Japan*
*E-mail: yekwon@rs.noda.tus.ac.jp*

HIDEAKI SUGAWARA

*National Institute of Genetics,*
*Mishima, Shizuoka 411-8540, Japan*
*E-mail: hsugawar@genes.nig.ac.jp*

SHOGO SHIMIZU

*Advanced Inst. of Industrial Technology,*
*Shinagawa, Tokyo 140-0011, Japan*
*E-mail: shimizu-syogo@aiit.ac.jp*

SATORU MIYAZAKI

*Tokyo University of Science,*
*Noda, Chiba 278-8510, Japan*
*E-mail: smiyazak@rs.noda.tus.ac.jp*

We present a novel measure that identifies disease-associated genes from the biomedical literature in terms of causing less side-effects. Our aim is to extract genes that not only have strong associations with a given disease, but also have no or weak associations with other diseases. This enables identification of specific disease-associated genes, the decreased expression of which would result in a lower probability of side-effects, thus contributing to efficient drug development. Our proposed method incorporates transitive associations between the disease and genes based on the frequency of co-occurrence of gene terms.

*Keywords*: disease-associated gene; term co-occurrence; specificity; transitive association; side-effect.

## 1. Introduction

Recently, a lot of works have been done on extracting biological knowledge, especially gene-disease associations, from the biomedical literature such as the PubMed abstracts. Most of the works use the co-occurrence frequency of gene and disease terms to extract disease-associated genes[1,3,4,13]. For example, Adamic *et al.*[1] use the statistical significance of the occurrence frequency of a gene term under documents that contain a particular disease term. Cheng *et al.*[4] measures the degree of an association between terms basically by term frequencies, and refines results using other scoring strategies such as rule-based pattern matching in sentences.

On the other hand, for the purpose of supporting new drug development, it is desirable that only specifically associated genes with a particular disease are identified so that drug developers can avoid cost and time-consuming wet experiments with those genes which have associations with other diseases, in other words, may cause some side-effects. Although frequency-based measures can present enough good candidates for finding related genes with a particular disease, extracted genes may also have associations with other diseases which are not of interest. That is, those genes are probably related, but may not be good target genes for the disease. Therefore, to ensure that extracted genes can be actually used as target genes, one must verify that those genes are not extracted as candidate target genes for other diseases.

In this work, we propose another measure for extracting genes specifically associated with a given disease. This enables the identification of associated genes that are expected to have fewer side-effects, which contributes to efficient drug development. Specificity of a gene is measured by a tf-idf (term frequency-inverse document frequency) like method, where the number of diseases associated with a gene is used instead of the number of documents. Our measure is different from existing approaches in that it incorporates the number of associated diseases as a factor of specificity, while others, such as mutual information based on term occurrence probabilities[19], focus on an association between a particular pair of a disease and a gene, and do not make a distinction among the rest of associated diseases. We consider that if some gene is chosen as a target gene for some disease, then side-effects are more likely to happen as the number of associated diseases of the gene increases.

Furthermore, based on an assumption that a disease is also indirectly associated with a gene *via* an intermediate gene, we extend the notion of specificity to incorporate indirect gene-disease associations. This idea comes from a heuristics that if two genes co-occur in same part of a document, then they may belong to a same family or be on a same pathway, and thus, share same affects to diseases. This type of transitivity is also adopted for mutual information to rank inferred literature relationships using common terms.

Another approach to extract gene-disease associations is to use known disease genes[12], phenotypes[7], expression data[6], and ontologies[18]. For example, GeneSeeker[6] collects these data from multiple human and mouse

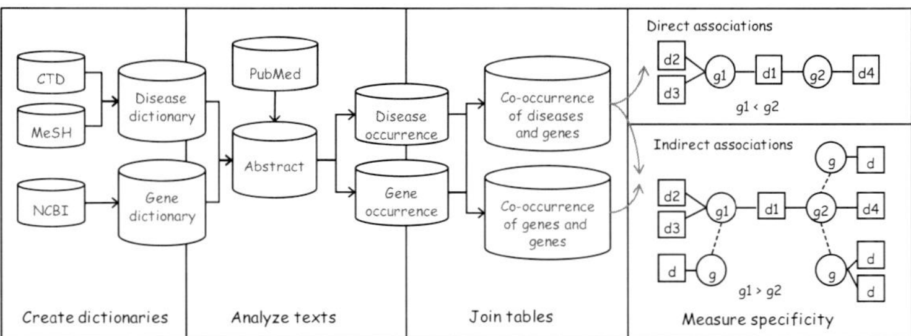



Fig. 1. The outline of our method. First, the occurrence tables of disease and gene names are created using term dictionaries. By joining these tables, the co-occurrence tables of gene-disease and gene-gene names are obtained. Next, for each disease, extract its associated genes from the co-occurrence table of gene-disease names. Then, we assign a specificity score to each associated gene using direct and indirect associations between a disease and genes. These indirect associations are extracted from the co-occurrence tables of disease-gene and gene-gene names. Finally, output genes of higher specificity scores as specific genes to the disease.

databases and prioritizes candidate genes for particular diseases based on positional, expression and model data. Tiffin *et al.*[18] uses the eVOC anatomical ontology[8] and human gene expression data, and evaluates their approach using known 17 disease genes. Protein-interaction networks can be used to predict gene-disease associations[12]. As an example of such approaches, the method[12] first constructs gene networks for a disease by literature mining based on dependency trees of sentences and support vector machines which classify a sentence as describing an interaction between genes or not. Then central nodes are identified as candidate genes under the assumption that central genes in the network are likely to be associated with the disease. Yu *et al.* compares various alternatives in gene prioritization methods such as the representation of a term vector, a ranking algorithm of associated genes, available vocabularies[22]. Our approach currently uses documents only, but can be combined as a basis with these methods where additional data such as pathways are available.

## 2. Methods

The outline of our method is shown in Figure 1. First, we create term dictionaries of disease and gene names. Using these term dictionaries, we create occurrence tables of disease and gene names in the collection of PubMed abstracts. By joining the occurrence tables on PMIDs, we obtain co-occurrence tables of disease-gene and gene-gene names. For each disease, its associated genes are extracted from the co-occurrence table of disease-gene names, and specificity score is assigned to each gene based on a tf-idf like method. Furthermore, we incorporate indirect associations between genes and diseases into specificity scores so that we do not miss the possibility of implicit side-effects. Indirect associations are extracted using the co-occurrence tables of disease-gene and gene-gene names.

We describe the detail of each step in the following.

## 2.1. *Term Dictionaries*

*Gene dictionary:* We downloaded human gene data from NCBI (National Center for Biotechnology Information) FTP site (`ftp://ftp.ncbi.nlm.gov/gene/DATA`) in March 2010. Then, we select Entrez Gene ID, gene symbol, gene synonym, and gene name fields from the data. The gene dictionary contains a total of 115,624 entries including gene synonyms.

*Disease dictionary:* We use CTD (Comparative Toxicogenomics Database) disease terms in January 2010[5] and NLM (National Library of Medicine) MeSH (Medical Subject Headings) database (`http://www.nlm.nih.gov/mesh/filelist.html`) in March 2010. CTD provides curated disease names, while MeSH provides a lot of synonyms for disease names. To receive benefit from the two databases, we adopt CTD disease names as primary diseases and MeSH thesaurus as synonyms for CTD disease names. The disease dictionary contains a total of 45,522 entries.

## 2.2. *Term Occurrences*

As a collection of documents, we downloaded MEDLINE/PubMed abstracts from NLM (`ftp://ftp.nlm.nih.gov/nlmdata/`) in January 2010 and select PubMed ID, ArticleTitle, and AbstractText fields from each abstract. We use 18,502,912 document abstracts for keyword search. First, all gene symbol occurrences are extracted from the extracted PubMed data using keyword search. The occurrence table consists of Gene ID, PubMed ID, and the sentence number in which a symbol appears in an abstract. The occurrences of synonyms of a gene are normalized into the occurrences of the corresponding single official symbol.

In addition to keyword search, additional checking, called neighbor search, is performed to reduce false positives of gene symbol occurrences. Some official symbols and synonyms of short lengths have the same spells as general words, such as CELL, which is a synonym for CEL (carboxyl ester lipase). For such a symbol that may produce many false positives, we

PubMed ID=7772531

> Recent molecular characterization of the translocation breakpoint has identified a gene fusion between NPM (nucleophosmin) and ALK (anaplastic lymphoma kinase).

PubMed ID=1522609

> There was no histopathological evidence of hepatic damage with ethanol alone, and no effect on hepatic cytochrome P-450 and glutathione levels or on serum levels of alanine aminotransferase (ALT), aspartate aminotransferase (AST), and alkaline phosphatase (ALK).

Fig. 2. Example of neighbor search. From ALK's gene names, the constitution words are generated: anaplastic, lymphoma, CD246, and 2p23. Because the first sentence (PubMed ID=7772531) contains "anaplastic", this occurrence of ALK is considered to be positive. The second sentence (PubMed ID=1522609) does not contain any of the constitution words, and thus, this occurrence is discarded.

further check whether any constitution word of the symbol appears near the symbol, in our case, in the same sentence. Constitution words of a symbol are created by splitting its gene name into a set of words delimited by special signs such as pluses, minuses, parentheses, brackets, hyphens, and spaces. General words such as body, cell, and protein, which are defined manually, are deleted from constitution words because they do not positively support the occurrence of a particular symbol in general. If any constitution word is found in the same sentence, the occurrence of the symbol is decided to be positive. Given an occurrence of a symbol, whether neighbor search is performed is determined by the character length of the symbol, and characteristic letters such as digits and hyphens.

Figure 2 shows an example of neighbor search. Assume that gene symbol ALK, Entrez Gene ID 238, occurs in an abstract. ALK's gene names are anaplastic lymphoma receptor tyrosine kinase, tyrosine kinase receptor, CD246 antigen, and 2p23. By splitting these gene names by delimiters, obtain a set of constitution words as anaplastic, lymphoma, receptor, tyrosine, kinase, CD246, antigen, and 2p23. Among these words, receptor, tyrosine, kinase, and antigen are dropped from constitution words because they are considered to be common. Next, neighbor search tries to find one of these constitution words in the same sentence where ALK appears. In the first sentence in Fig. 2 (PubMed ID=7772531), the word "anaplastic" or "lymphoma" occurs. On the other hand, the second sentence (PubMed ID=1522609) does not contain any of these words, and therefore, this occurrence of ALK is decided to be a false positive and deleted from the gene occurrence table.

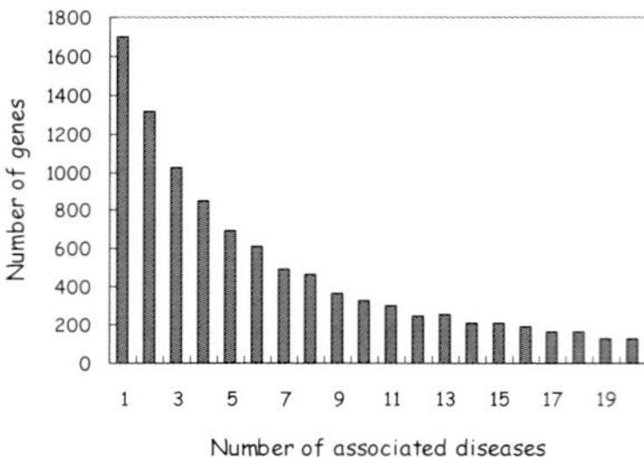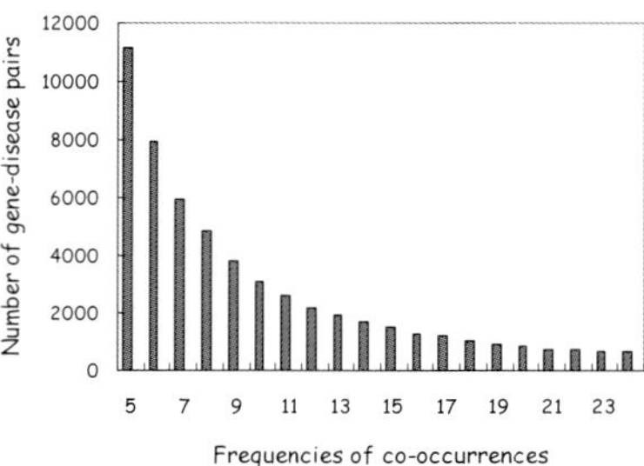

Fig. 3. (left) The number of genes associated with the given number of diseases. (right) The number of gene-disease pairs for the given frequencies of co-occurrences.

For disease terms, construct the occurrence table using keyword search. Also, synonyms are normalized into its representative disease name. If gene and disease terms co-occur in a same sentence, we consider there is an association between the two terms. There are other alternatives to the range of co-occurrence of two terms, e.g., one document, one paragraph, and a fixed length of words. In general, a broad range produces high recall and low precision results. Among these, we adopt one sentence in a same abstract because it is enough to find a small number of candidate genes that are worth being verified for new drug development. By joining the gene occurrence table and disease occurrence table on the PubMed ID and the sentence number fields, we obtain co-occurrence tables of gene-disease and gene-gene associations. Gene-gene associations are used for incorporating indirect gene-disease associations, which will be described later.

In total, 400,556 gene-disease associations are extracted. The left histogram in Fig. 3 shows the number of genes that have associations with the given number of diseases. About 1,700 genes are associated with only one disease. These genes can be expected to cause no side-effects if they are chosen as target genes for the disease. On the other hand, there are over 100 genes that have associations with 20 distinct diseases, all of which show low specificity to every disease. The right histogram in Fig. 3 shows the number of gene-disease pairs for the given frequencies of co-occurrences.

Further refinement methods to extract gene-disease associations are also applicable such as natural language processing and machine learning

techniques. However, these methods take a significant amount of time and/or need a large amount of training data and thus, are not suited for the exhaustive analysis of a large set of documents.

## 2.3. *Measuring Associations*

For the purpose of supporting new drug development, it is desirable that only specifically associated genes with a particular disease are identified so that drug developers can avoid cost and time-consuming wet experiments with those genes which have associations with other diseases, that is, possibly cause some side-effects. Our method of measuring specificity is based on a tf-idf (term frequency-inverse document frequency) method. The difference from the original definition is that the number of diseases associated with a particular gene is used in the idf definition instead of the number of documents in which a gene appears, since we want to estimate specificity in terms of gene-disease associations.

First, we define the gene term frequency (gtf). The *gtf* term evaluates the frequency of co-occurrences between a particular disease and its associated gene. Similar to the original logarithmic *tf* definition, the *gtf* term of gene $g$ with respect to disease $d$ is defined as follows:

$$gtf_d(g) = \log(1 + n(d, g)),$$

where $n(d, g)$ denotes the number of co-occurrences of $d$ and $g$. In the context of drug development, a gene of high *gtf* value is more appropriate for a target gene.

Next, we define the associated disease frequency (adf). The *adf* term evaluates the specificity of co-occurrences between a particular disease and its associated gene. Let $ad(g)$ be the number of diseases associated with gene $g$. Here, disease $d$ is said to be associated with $g$ if $n(d, g) \geq c$, where $c$ is a constant. Then, similar to the original idf definition, the *adf* term of $g$ is defined as follows:

$$adf(g) = \log \frac{m}{ad(g)},$$

where $m$ is the number of distinct diseases. In the context of drug development, a gene of high *adf* value has less possibility of side-effects and therefore, can be considered as a good target gene.

The association score of $g$ to $d$, denoted as $as_d(g)$, is defined as follows:

$$as_d(g) = gtf_d(g) \cdot adf(g).$$

## 2.4. *Indirect Associations via Intermediate Genes*

In addition to direct gene-disease associations, there may be indirect gene-disease associations *via* intermediate genes. The notion of indirect associations is based on the assumption that gene $g$ has an association with disease $d$ if there is another gene $g'$ that co-occurs with $d$ and frequently co-occurs with $g$ in literature, even if $g$ does not directly co-occur with $d$. Genes that co-occur in a same sentence often appear because of belonging to a same family or being located in a same pathway. Those genes are expected to share similar affection to a particular disease, since they have similar functions in the former case, and have multiplier effects in the latter case. For example, apolipoprotein E (APOE), which is known as a causal gene for Alzheimer disease (AD), co-occurs with lipoprotein lipase (LPL) in 74 document abstracts. This implies that diseases associated with LPL such as Cachexia may be associated also with APOE. In fact, APOE and LPL are both lipid metabolism-related genes. So, we consider APOE also has some relationship to Cachexia *via* LPL. According to this assumption, we redefine the *adf* of a gene so that the number of indirect gene-disease associations *via* intermediate genes are taken into account.

First, we define the strength of associations between genes. Among various techniques on detecting gene relationships from the PubMed abstracts have been proposed such as natural language processing, statistics[2], and multiple thesauri[15], we adopt frequencies of co-occurrences of gene symbols because of its simplicity and efficiency. Let $g_i$ and $g_j$ be genes. Then, the *similarity* of $g_i$ and $g_j$, denoted $sim(g_i, g_j)$, is defined as:

$$sim(g_i, g_j) = \frac{|PM(g_i) \cap PM(g_j)|}{|PM(g_i) \cup PM(g_j)|},$$

where $PM(g)$ denotes the set of PubMed IDs where gene $g$ appears. Then, an extended version of the $adf(g)$ definition that incorporates indirect gene-diseases associations is defined as follows:

$$adf_I(g_i) = \log \frac{m}{ad(g_i) + \sum_{j \neq i} sim(g_i, g_j)\, ad_{\neq i}(g_j)},$$

where $ad_{\neq i}$ is the number of diseases other than directly associated diseases with $g_i$. In the above expression, the left term in the denominator corresponds to the contribution of a direct association to the specificity, and the right term corresponds to that of indirect gene-disease associations. The number of diseases associated with $g_j$ is weighted by $sim(g_i, g_j)$ under the assumption that the probability that two genes share same associated diseases gets higher depending on the similarity of the two genes. Note that so-called hubgenes which have many links with other genes in a gene-gene association network are ranked lower in our algorithm because of their many indirect associations with other diseases. This is different from the existing method like using centrality on a literature mined gene-

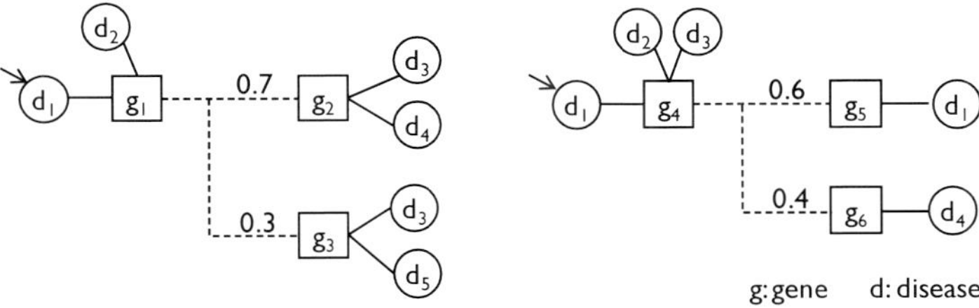



Fig. 4. Incorporation of indirectly associated genes. Indirect associations are incorporated into association scores. Numbers on arrows between genes represent similarity between genes. In the left part of the figure, $adf_I(g_1) = 2 + (0.7 * 2 + 0.3 * 2) = 4$. In the right part of the figure, $adf_I(g_4) = 3 + 0.4 = 3.4$. Since $d_1$ is already counted in the calculation of direct associations, its weight is omitted in the calculation of indirect associations. As a result, $g_4$ is more specific to $d_1$ from the perspective of less side-effects.

interaction network[12], where hubgenes are extracted as most related genes with a disease.

Figure 4 shows an example of $adf_I$ calculation when incorporating indirect gene-disease associations. Numbers on lines between genes represent similarity between genes. Assume that we want to find genes specific to disease $d_1$. A gene directly associated with $d_1$ in the left (and respectively, right) part of the figure is $g_1$ (and respectively, $g_4$). With direct associations only, $g_1$ is more specific to $d_1$ because $g_1$ has an association with only one disease $d_2$ other than $d_1$. However, with direct and indirect associations, $g_4$ is more specific to $d_1$ because $g_4$ has less indirect associations with other diseases *via* intermediate genes. Thus, from the perspective of drug development, it is effective to set $g_4$ as a target gene because $g_4$ is considered to have less possibility of side-effects than $g_1$.

The above definition is given for indirect gene-disease associations *via* one intermediate gene, but can be naturally extended for an arbitrary number of intermediate genes. It is sufficient to create an adjacency matrix, where an element of the $i$-th row and the $j$-th column is $sim(g_i, g_j)$, and obtain indirect associations *via* $n$ intermediate genes by multiplying the matrix $n$ times.

## 3. Results and Discussion

To verify the effectiveness of our novel measure, we experimented with well-studied Alzheimer disease (AD) and diabetes mellitus (DM) by checking whether drug targets and known related genes are ranked higher than by the frequency-based measure. Table 1 shows the top-30 results for AD.

The listed genes are sorted based on the association scores. APP, PSEN1, PSEN2, and APOE, which are emphasized in Table 1, are known causal genes for AD[20]. AD is characterized by a decrease in acetylcholine levels in the brain owing to a reduction in the activity of the cholinergic neurons that results in loss of cognitive functions, such as memory, communication skills, judgement and reasoning. Many current drug therapies of AD such as donepezil, rivastigmine and galantamine use acetyl cholinesterase (ACHE) inhibitors to reduce the rate at which acetylcholine is broken down[10,14,16]. PharmGKB[9], a curated database of gene-drug-disease relationships, enumerates butyrylcholine esterase (BCHE) other than ACHE as drug targets of rivastigmine. ACHE and BCHE are ranked at the 5th and 20th positions by *gtf* values, respectively. However, ACHE co-occurs with over 500 disease names including AD, brain neoplasms, breast neoplasms, colonic neoplasms, intestinal neoplasms, lung neoplasms, ovarian neoplasms, stomach neoplasms, and thyroid neoplasms, while BCHE co-occurs with only 60 disease names including AD and lung neoplasms. BCHE does not co-occur with other neoplasms other than lung neoplasms, and thus, BCHE is ranked higher than ACHE by *as* values. Actually, PharmGKB enumerates a lot of side-effects of donepezil such as severe nausea, vomiting, salivation, sweating, bradycardia, hypotension, respiratory depression, collapse and convulsions as symptoms of overdose. We also show the results for DM in Table 2. Neonatal diabetes mellitus (NDM) is classified clinically into a transient form (TNDM) and a permanent form (PNDM), and its causal genes are recently identified[17]. Among these, KCNJ11 and ABCC8 mutations are recognized as the major causes of TNDM and PNDM, respectively. The two genes are ranked at the 12th and 32th positions by *as* values, whereas ranked lower by *gtf* values. Also, WFS1, TCF7L2, and PPARG known as clinical risk factors of type 2 diabetes mellitus[11] all get into the top-30 results by *as* values.

Next, we verify the effectiveness of indirect associations *via* intermediate genes for the results of AD in Table 1. There are over one hundred genes associated with BCHE including APOE. According to an article of PubMed ID 15519745, there is a synergic association between butyrylcholinesterase-K variant (BChE-K) and apolipoproteinE-epsilon 4 (ApoE-epsilon 4) to promote risk for AD. APOE co-occurs with 451 diseases such as atherosclerosis and hypercholesterolemia. Thus, one of the reasons that the rank of BCHE gets lower by $as_I$ than by *as* is that a number of diseases associated with APOE are considered to be transitively associated with BCHE in the $as_I$ measure.

Table 3 shows a summary of the ranking results for cancers that contain "neoplasms." The 2nd and 3rd columns are related drugs and target genes to cancers, where their field values are cited from PharmGKB. The 4th and 5th columns represent the ranks of the corresponding target genes by $as$ and $as_I$ measures. The 6th and 7th columns are genes decided to be most specifically associated with the corresponding cancers by our methods. For example, anastrozole is a drug used to treat breast cancer, CYP19A1 is a target gene of anastrozole, and CYP19A1 is ranked at the 11th and 17th positions by $as$ and $as_I$. ANIB1 and PGR are top genes that are associated with breast cancer in the $as$ and $as_I$ measures in the literature. Capecitabine is a drug given as a treatment for many types of cancer, including breast cancer, colorectal cancer, gastrointestinal cancer, head and neck cancer, pancreatic cancer, and stomach cancer. DPYD, which is a target gene of capecitabine, is ranked at the 5th and the 1st positions with gastrointestinal neoplasms by $as$ and $as_I$, respectively, but do not appear in top-100 results with breast cancer in both measures. Therefore, we can expect that DPYD is the most specific gene to gastrointestinal cancer because the gene is strongly associated with gastrointestinal cancer in the literature and capecitabine that uses DPYD as a target is more effective to gastroinstinal cancer than other cancers. On the other hand, although there are many drugs used to treat lung cancer, any target gene to their drugs is not specific to lung neoplasms in the literature.

## 4. Conclusions

We proposed a novel measure for identifying disease-associated genes that incorporates direct and indirect disease-gene associations in the literature, and experimented with some diseases to confirm the effectiveness of our new measure.

Although the way of measuring the possibility of side-effects is rather simple, there is at least a case that incorporating associations with other diseases improves a result of the frequency-based method, as the experimental result of AD shows. In addition, the number of associated diseases in the literature does not directly reflect actual possibility of side-effects, but we consider that our notion is still useful for eliminating those genes that co-occur with a number of other diseases from candidates before starting wet experiments.

As a future work, we plan to perform further detailed analysis and extend our method to combine other fact data such as pathways with the literature analysis.

## Acknowledgements

This work was supported by Institute for Bioinformatics Research and Development (BIRD), Japan Science and Technology Agency (JST).

## References


1. L. A. Adamic, D. Wilkinson, B. A. Huberman, and E. Adar, A literature based method for identifying gene-disease connections, *Proc. IEEE Conf. Bioinformatics*, 109–117 (2002).
2. M. A. Andrade and A. Valencia, Automatic extraction of keywords from scientific text: application to the knowledge domain of protein families, *Bioinformatics*, **14**(7), 600–607 (1998).
3. A. R. Allende, Accelerating searches of research grants and scientific literature with novoseek, *Nature Methods*, **6**(5) (2009).
4. D. Cheng, C. Knox, N. Young, P. Stothard, S. Damaraju, and D. Wishart, PolySearch: a web-based text mining system for extracting relationships between human diseases, genes, mutations, drugs and metabolites, *Nucleic Acids Res.*, **36**, W399–W405 (2008).
5. A. P. Davis, C. G. Murphy, C. A. Saraceni-Richards, M. C. Rosenstein, T. C. Wiegers, and C. J. Mattingly, Comparative toxicogenomics database: a knowledgebase and discovery tool for chemical-gene-disease networks, *Nucleic Acids Res.*, **37**, D786–D792 (2009).
6. M. A. van Driel, K. Cuelenaere, P. P. C. W. Kemmeren, J. A. M. Leunissen, H. G. Brunner, and G. Vriend, GeneSeeker: extraction and integration of human disease-related information from web-based genetic databases, *Nucleic. Acids Res.*, **33**, W758–W761 (2005).
7. J. Freudenberg and P. Propping, A similarity-based method for genome-wide prediction of disease-relevant human genes, *Bioinformatics*, **18**, S110–S115 (2002).
8. J. Kelso, J. Visagie, G. Theiler, A. Christoffels, S. Bardien, D. Smedley, D. Otgaar, G. Greyling, C. V. Jongeneel, M. I. McCarthy, T. Hide, and W. Hide, eVOC: a controlled vocabulary for unifying gene expression data, *Genome Res.*, **13**, 1222–1230 (2003).
9. T. E. Klein, J. T. Chang, M. K. Cho, K. L. Easton, R. Fergerson, M. Hewett, Z. Lin, Y. Liu, S. Liu, D. E. Oliver, D. L. Rubin, F. Shafa, J. M. Stuart, and R. B. Altman, Integrating Genotype and Phenotype Information: An Overview of the PharmGKB Project, *The Pharmacogenomics Journal*, **1**, 167–170 (2001).
10. K. L. Lanctôt, N. Herrmann, K. K. Yau, L. R. Khan, B.A. Liu, M. M. Loulou, and T.R. Einarson, Efficacy and safety of cholinesterase inhibitors in Alzheimer's disease: a meta-analyisis, *Can. Med. Assoc. J.*, **169**(6), 557–564, (2003).
11. V. Lyssenko, A. Jonsson, P. Almgren, N. Pulizzi, B. Isomaa, T. Tuomi, G. Berglund, D. Altshuler, P. Nilsson, and L. Groop, Clinical risk factors, DNA variants, and the development of type 2 diabetes, *N. Engl. J. MED.*, **359**(21), 2220–2232 (2008).

12. A. Özgür, T. Vu, G. Erkan, and D. R. Radev, Identifying gene-disease associations using centrality on a literature mined gene-interaction network, *Bioinformatics*, **24**(13), i277–i285 (2008).
13. C. Plake, L. Royer, R. Winnenburg, J. Hakenberg, and M. Schroeder, Go-Gene: gene annotation in the fast lane, *Nucleic Acids Res.*, **37**, W300–304 (2009).
14. E. Scarpini, P. Scheltens, and H. Feldman, Treatment of Alzheimer's disease: current status and new perspectives, *Lancet Neurol.*, **2**, 539–547 (2003).
15. M. Stephens, M. Palakal, S. Mukhopadhyay, and R. Raje, Detecting gene relations from Medline abstracts, *Pac. Symp. Biocomput.*, **6**, 483–496 (2001).
16. Y. H. Suh and F. Checler, Amyloid precursor protein, preseniloins, and $\alpha$-synuclein: molecular pathogenesis and pharmacological applications in Alzheimer's disease, *Pharmocol Rev.*, **54**, 469–525 (2002).
17. S. Suzuki, Y. Makita, T. Mukai, K. Matsuo, O. Ueda, and K. Fujieda, Molecular basis of neonatal diabetes in Japanese patients, *J. Clin. Endocrinol. Metab.*, **92**(10), 3979–3985 (2007).
18. N. Tiffin, J. F. Kelso, A. R. Powell, H. Pan, V. B. Bajic, and Hide, W.A., Integration of text- and data-mining using ontologies successfully selects disease gene candidates, *Nucleic. Acids Res.*, **33**(5), 1544–1552 (2005).
19. Y. Tsuruoka, J. Tsujii, and S. Ananiadou, FACTA: a text search engine for finding associated biomedical concepts, *Bioinformatics*, **24**(21), 2559–2560 (2008).
20. S. C. Waring and R. N. Rosenberg, Genome-wide association studies in Alzheimer disease, *Arch. Neurol.*, **65**(3), 329–334 (2008).
21. J. D. Wren, Extending the mutual information measure to rank inferred literature relationships, *BMC Bioinformatics*, **5**(145) (2004).
22. S. Yu, S. V. Vooren, L. C. Tranchevent, B. D. Moor, and Y. Moreau, Comparison of vocabularies, representations and ranking algorithms for gene prioritization by text mining, *Bioinformatics*, **24**(16), i119–i125 (2008).

Table 1. Top-30 associated genes for Alzheimer disease.

| Gene | Description | $gtf$ value | $gtf$ rank | $adf$ value | $adf$ rank | $as$ rank | $adf_I$ value | $adf_I$ rank | $as_I$ rank |
|---|---|---|---|---|---|---|---|---|---|
| MAPT | microtubule-associated protein tau | 4.91 | 3 | 4.10 | 195 | 1 | 3.96 | 75 | 1 |
| PSEN1 | presenilin 1 | 4.30 | 4 | 4.64 | 124 | 2 | 4.36 | 59 | 3 |
| APP | amyloid beta (A4) precursor protein | 6.30 | 1 | 3.06 | 298 | 3 | 2.99 | 90 | 2 |
| PSEN2 | presenilin 2 (Alzheimer disease 4) | 3.47 | 8 | 5.33 | 71 | 4 | 4.77 | 50 | 5 |
| BACE1 | beta-site APP-cleaving enzyme 1 | 3.50 | 6 | 5.12 | 86 | 5 | 4.82 | 48 | 4 |
| SORL1 | sortilin-related receptor, L(DLR class) A repeats-containing | 2.71 | 11 | 5.72 | 45 | 6 | 5.60 | 28 | 6 |
| APOE | apolipoprotein E | 5.50 | 2 | 2.75 | 327 | 7 | 2.67 | 95 | 7 |
| CHAT | choline acetyltransferase | 3.09 | 10 | 4.20 | 176 | 8 | 3.90 | 76 | 8 |
| APBB1 | amyloid beta A4 precursor protein-binding family B member 1 | 2.08 | 28 | 5.97 | 34 | 9 | 5.71 | 26 | 9 |
| NCSTN | nicastrin | 2.30 | 17 | 4.82 | 109 | 11 | 4.79 | 49 | 10 |
| SLC6A3 | solute carrier family 6 member 3 | 3.50 | 7 | 3.02 | 304 | 12 | 3.00 | 88 | 12 |
| BCHE | butyrylcholinesterase | 2.20 | 20 | 4.77 | 115 | 13 | 4.41 | 58 | 17 |
| A2M | alpha-2-macroglobulin | 2.30 | 18 | 4.54 | 135 | 14 | 4.45 | 56 | 13 |
| IDE | insulin-degrading enzyme | 1.95 | 33 | 5.30 | 73 | 15 | 5.25 | 35 | 14 |
| CST3 | cystatin C | 1.61 | 41 | 6.37 | 25 | 16 | 5.98 | 23 | 18 |
| PRNP | prion protein | 2.71 | 12 | 3.68 | 239 | 17 | 3.65 | 77 | 15 |
| HTT | huntingtin | 2.71 | 13 | 3.66 | 242 | 18 | 3.64 | 78 | 16 |
| CDK5 | cyclin-dependent kinase 5 | 2.20 | 21 | 4.41 | 156 | 19 | 4.08 | 71 | 23 |
| ACHE | acetylcholinesterase (Yt blood group) | 3.66 | 5 | 2.63 | 340 | 20 | 2.60 | 96 | 19 |
| ADAM10 | ADAM metallopeptidase domain 10 | 1.95 | 34 | 4.91 | 103 | 21 | 4.86 | 45 | 20 |
| LRP1 | low density lipoprotein receptor-related protein 1 | 2.20 | 22 | 4.27 | 166 | 22 | 4.11 | 69 | 22 |
| SERPINI1 | serpin peptidase inhibitor, clade I (neuroserpin), member 1 | 1.79 | 38 | 5.22 | 79 | 23 | 5.20 | 36 | 21 |
| NGF | nerve growth factor (beta polypeptide) | 3.30 | 9 | 2.70 | 329 | 24 | 2.57 | 98 | 24 |
| TYROBP | TYRO protein tyrosine kinase binding protein | 1.61 | 42 | 5.10 | 88 | 25 | 4.98 | 40 | 25 |
| DHX40 | DEAH (Asp-Glu-Ala-His) box polypeptide 40 | 1.39 | 59 | 5.72 | 46 | 26 | 5.72 | 25 | 26 |
| CTSD | cathepsin D | 1.39 | 60 | 5.56 | 54 | 27 | 5.39 | 32 | 27 |
| EP300 | E1A binding protein p300 | 2.40 | 15 | 3.13 | 290 | 28 | 3.09 | 86 | 29 |
| RTN3 | reticulon 3 | 1.10 | 77 | 6.78 | 15 | 29 | 6.78 | 15 | 28 |
| BDNF | brain-derived neurotrophic factor | 2.40 | 16 | 3.08 | 295 | 30 | 2.91 | 92 | 32 |

Table 2. Associated genes for diabetes mellitus.

| Gene | Description | $gtf$ value | $gtf$ rank | $adf$ value | $adf$ rank | $as$ rank | $adf_I$ value | $adf_I$ rank | $as_I$ rank |
|---|---|---|---|---|---|---|---|---|---|
| IDDM2 | insulin-dependent diabetes mellitus 2 | 8.24 | 1 | 3.01 | 870 | 1 | 2.96 | 92 | 25 |
| TNDM | diabetes mellitus, transient neonatal | 3.69 | 31 | 5.60 | 161 | 2 | 5.54 | 12 | 2 |
| IAPP | islet amyloid polypeptide | 4.85 | 8 | 4.12 | 545 | 3 | 4.09 | 51 | 11 |
| GAD1 | glutamate decarboxylase 1 | 4.93 | 5 | 3.98 | 602 | 4 | 3.85 | 61 | 4 |
| ZGLP1 | zinc finger, GATA-like protein 1 | 4.71 | 10 | 4.01 | 586 | 5 | 3.77 | 68 | 98 |
| INS | insulin | 4.69 | 11 | 3.99 | 593 | 6 | 3.84 | 62 | 5 |
| GAD2 | glutamate decarboxylase 2 | 4.17 | 19 | 4.18 | 531 | 7 | 3.97 | 57 | 10 |
| DMPK | dystrophia myotonica-protein kinase | 4.98 | 2 | 3.50 | 733 | 8 | 3.49 | 76 | 12 |
| WFS1 | Wolfram syndrome 1 (wolframin) | 3.37 | 45 | 5.15 | 266 | 9 | 5.09 | 20 | 8 |
| SLC19A2 | solute carrier family 19 (thiamine transporter), member 2 | 3.14 | 60 | 5.43 | 191 | 10 | 5.41 | 13 | 9 |
| PTPRN | protein tyrosine phosphatase, receptor type, N | 3.93 | 23 | 4.27 | 497 | 11 | 4.09 | 52 | 3 |
| KCNJ11 | potassium inwardly-rectifying channel, subfamily J, member 11 | 3.58 | 37 | 4.49 | 438 | 12 | 3.94 | 58 | 14 |
| SLC2A4 | solute carrier family 2, facilitated glucose transporter member 4 | 3.66 | 32 | 4.25 | 502 | 13 | 3.90 | 59 | 13 |
| GIP | gastric inhibitory polypeptide | 3.30 | 48 | 4.72 | 373 | 14 | 3.61 | 73 | 32 |
| GNAI2 | guanine nucleotide-binding protein G(i), alpha-2 subunit | 3.85 | 24 | 3.86 | 642 | 15 | 3.48 | 78 | 16 |
| ETV3 | ets variant 3 | 4.17 | 20 | 3.52 | 728 | 16 | 3.49 | 77 | 7 |
| HNF4A | hepatocyte nuclear factor 4, alpha | 3.37 | 46 | 4.21 | 512 | 17 | 4.05 | 54 | 15 |
| PDX1 | pancreatic and duodenal homeobox 1 | 3.04 | 67 | 4.54 | 419 | 18 | 4.37 | 41 | 18 |
| CAPN10 | calpain 10 | 2.48 | 112 | 5.36 | 203 | 19 | 5.36 | 15 | 20 |
| TCF7L2 | transcription factor 7-like 2 (T-cell specific, HMG-box) | 2.94 | 72 | 4.49 | 439 | 20 | 4.34 | 43 | 22 |
| MAP4K2 | mitogen-activated protein kinase kinase kinase kinase 2 | 2.64 | 99 | 5.01 | 287 | 21 | 5.01 | 23 | 19 |
| CETP | cholesteryl ester transfer protein, plasma | 3.14 | 61 | 4.16 | 535 | 22 | 4.00 | 55 | 26 |
| ADIPOQ | adiponectin, C1Q and collagen domain containing | 3.26 | 50 | 3.99 | 594 | 23 | 3.84 | 63 | 27 |
| ACE | angiotensin I converting enzyme (peptidyl-dipeptidase A) 1 | 4.96 | 4 | 2.61 | 946 | 24 | 2.56 | 97 | 23 |
| PPARG | peroxisome proliferator-activated receptor gamma | 4.16 | 21 | 3.10 | 843 | 25 | 2.97 | 90 | 30 |
| HNF1A | HNF1 homeobox A | 2.83 | 80 | 4.54 | 420 | 26 | 4.37 | 42 | 29 |
| GCK | glucokinase (hexokinase 4) | 2.40 | 124 | 5.36 | 204 | 27 | 5.36 | 16 | 17 |
| HBA1 | hemoglobin, alpha 1 | 4.26 | 18 | 3.02 | 866 | 28 | 2.96 | 91 | 1 |
| SLC2A2 | solute carrier family 2 (facilitated glucose transporter), member 2 | 2.64 | 100 | 4.85 | 318 | 29 | 4.85 | 25 | 21 |
| ADRB3 | beta-3 adrenoreceptor | 2.40 | 125 | 5.28 | 225 | 30 | 5.28 | 19 | 24 |
| ABCC8 | ATP-binding cassette transporter sub-family C member 8 | 3.22 | 53 | 3.86 | 643 | 32 | 3.66 | 72 | 41 |

Table 3.   Related drugs and target genes to cancers.

| Disease | Drug[†] | Target[†] | Rank[‡] | | Top gene | |
|---|---|---|---|---|---|---|
| | | | $as$ | $as_I$ | $as$ | $as_I$ |
| Breast neoplasms | capecitabine | DPYD | NR | NR | ANIB1 | PGR |
| | cetuximab | EGFR | 52 | NR | | |
| | anastrozole | CYP19A1 | 11 | 17 | | |
| Colonic neoplasms | acetaminophen | PTGS2 | 28 | 50 | PMS2 | SLC5A8 |
| | cetuximab | EGFR | 37 | 51 | | |
| | lapatinib | ERBB2 | 86 | NR | | |
| | acetaminophen | PTGS2 | 28 | 50 | | |
| Colorectal neoplasms | capecitabine | DPYD | 9 | 7 | DCC | EIF4G2 |
| | cetuximab | EGFR | 37 | 49 | | |
| | bevacizumab | VEGFA | 89 | 91 | | |
| Gastrointestinal neoplasms | capecitabine | DPYD | 5 | 1 | PSG2 | DPYD |
| Head and neck neoplasms | capecitabine | DPYD | 43 | 36 | EGFR | SERPINB3 |
| | cetuximab | EGFR | 1 | 2 | | |
| | docetaxel | BCL2 | 25 | 27 | | |
| Kideney neoplasms | bevacizumab | VEGFA | 39 | 42 | VHL | VHL |
| Liver neoplasms | celecoxib | PTGS2 | 65 | NR | DLC1 | DLC1 |
| | paclitaxel | BCL2 | 79 | NR | | |
| Lung neoplasms | cetuximab | EGFR | 7 | 11 | STK11 | CALU |
| | lapatinib | ERBB2 | 18 | 24 | | |
| | gemcitabine | RRM1 | 44 | 26 | | |
| | topotecan | ABCG2 | 62 | 46 | | |
| | rifampin | ABCB1 | 65 | 93 | | |
| | docetaxel | BCL2 | 97 | NR | | |
| Ovarian neoplasms | lapatinib | ERBB2 | 6 | 10 | BRCA2 | BRCA2 |
| | rifampin | ABCB1 | 37 | 64 | | |
| Pancreatic neoplasms | capecitabine | DPYD | 88 | 69 | MIA | MIA |
| | lapatinib | ERBB2 | 22 | 29 | | |
| | cetuximab | EGFR | 25 | 32 | | |
| | celecoxib | PTGS2 | 70 | NR | | |
| Prostatic neoplasms | testosterone | AR | 12 | 8 | MSMP | MSMP |
| | docetaxel | BCL2 | 99 | NR | | |
| Stomach neoplasms | capecitabine | DPYD | 21 | 18 | DSG2 | DSG2 |
| | lapatinib | ERBB2 | 23 | 28 | | |
| | docetaxel | BCL2 | 72 | 84 | | |
| Thyroid neoplasms | sorafenib | BRAF | 4 | 4 | CCDC6 | CCDC6 |
| Uterine neoplasms | trastuzumab | EGFR | 27 | 28 | AKR1B1 | AKR1B1 |

[†] This field is cited from PharmGKB. [‡] NR denotes that the corresponding target is not ranked in top-100 results. *ABCB1* ATP-binding cassette, sub-family B, member 1; *AKR1B1* aldo-keto reductase family 1, member B1; *ANIB1* aneurysm, intracranial berry 1; *AR* androgen receptor; *BCL2* B-cell CLL/lymphoma 2; *BRAF* v-raf muring sarcoma viral oncogene homolog B1; *BRCA2* breast and ovarian cancer susceptibility gene, early onset; *CALU* calumenin; *CCDC6* coiled-coil domain containing 6; *CYP19A1* cytochrome P450, family 19, subfamily A, polypeptide 1; *DCC* deleted in colorectal carcinoma; *DLC1* deleted in liver cancer; *DPYD* dihydropyrimidine dehydrogenase; *DSG2* desmoglein 2; *EGFR* epidermal growth factor receptor; *EIF4G2* eukaryotic translation initiation factor 4 gamma, 2; *ERBB2* e-erb-b2 erythroblastic leukemia viral oncogene homolog 2; *MIA* melanoma inhibitory activity; *MSMP* prostate-associated microseminoprotein; *PGR* progesterone receptor; *PMS2* DNA mismatch repair protein PMS2; *PSG2* pregnancy specific beta-1-glycoprotein 2; *RRM1* ribonucleotide reductase M1; *SLC5A8* solute carrier family 5, member 8; *SERPINB3* serpin peptidase inhibitor, clade B, member 3; *STK11* serine/threonine kinase 11; *VEGFA* vascular endothelial growth factor A; *VHL* von Hippel-Landau tumor suppressor

Quantum Bio-Informatics V
© 2013 World Scientific Publishing Co. Pte. Ltd.
pp. 425–434

# THREE-TANGLE AND THREE-$\pi$ FOR A CLASS OF TRIPARTITE MIXED STATES

TENG MA

*School of Mathematical Sciences, Capital Normal University,
Beijing, 100048, China
E-mail: m.tengteng@163.com*

SHAO-MING FEI

*chool of Mathematical Sciences, Capital Normal University,
Beijing, 100048, China
Max-Planck-Institute for Mathematics in the Sciences,
Leipzig, 04103, Germany
E-mail: feishm@mail.cnu.edu.cn*

We study the tripartite entanglement for a class of mixed states defined by the mixture of GHZ and W states, $\rho = p|GHZ\rangle\langle GHZ| + (1-p)|W\rangle\langle W|$. Based on the Caratheodory theorem and the periodicity assumption, the possible optimal decomposition of the states has been derived, which is independent on the detailed measure of entanglement. We find that, according to $p$, there are two different decompositions. The entanglement measures three-tangle and three-$\pi$ and their relations are investigated in detailed. It is shown that the three-tangle is smaller than or equal to the three-$\pi$. Moreover, the three-$\pi$ has a minimal point in the interval $p \in [0, 1]$, while the three-tangle is a nondecreasing function of $p$.

## 1. Introduction

Quantum entangled states are the key resources in quantum computation and quantum information processing [1]. Evaluation and detection of quantum entanglement are the essential subjects in the theory of quantum entanglement. For bipartite states, there are already many well defined entanglement measures such as entanglement of formation [2,3], concurrence [4], negativity [5], relative entropy [6], PPT criterion [7] and Bell inequalities[8].

Tripartite entanglement is more complicated than bipartite entanglement. Investigation of tripartite entanglement is the basis of studying multipartite entanglement. However, although tripartite entanglement is well defined, it is formidably difficult to compute the tripartite entanglement analytically. Up to now, there is no general formulation to calculate tripartite entanglement, only a few class of tripartite entanglement can be calculated efficiently[10,?].

In this article, we study the entanglement of a class of tripartite mixed states $\rho = p|GHZ\rangle\langle GHZ| + (1-p)|W\rangle\langle W|$. We investigate possible optimal decompositions in a more general way for some entanglement measures $E$ including three-tangle [12]. It is shown that the possible optimal decompositions may contain three or four quantum pure states. When the decomposition contains three pure states, the tripartite entanglement $E$ of the mixed state is simply the entanglement of superposition states of GHZ and W. When the decomposition contains four pure states, the tripartite entanglement $E$ of the mixed state is a liner function of $p$

## 2. The general formulations

Consider the tree-qubit state

$$\rho = p|GHZ\rangle\langle GHZ| + (1-p)|W\rangle\langle W|, \tag{1}$$

where $|GHZ\rangle = 1/\sqrt{2}(|000\rangle + |111\rangle)$, $|W\rangle = 1/\sqrt{3}(|001\rangle + |010\rangle + |100\rangle)$. Let $E(|\Psi_i\rangle)$ be the entanglement of a tripartite pure state $|\Psi_i\rangle$. Then the definition is extended to general mixed states $\rho$ by the convex roof,

$$E(\rho) = \min_{\{p_i,|\psi_i\rangle\}} \left\{ \sum_i p_i E(|\psi_i\rangle) : \rho = \sum_i p_i |\psi_i\rangle\langle\psi_i| \right\}. \tag{2}$$

As the mixed state (1) has rank two, the pure states in decomposition have the form, $|q,\theta\rangle = \sqrt{q}|GHZ\rangle - \sqrt{1-q}e^{i\theta}|W\rangle$. Moreover according to the Caratheodory's theorem [13], at most four pure states are sufficient to minimize the entanglement for rank-2 states. Hence we need only to investigate the possible optimal decompositions with two to four pure states in the decomposition.

In the following we consider a class of tripartite entanglement measures $E$ such that $E(|q,\theta\rangle)$ is a periodical function of $\theta$ with period $2\pi/3$, which covers some well-known measures like three-tangle and three-$\pi$. We assume that $E(|q,\theta\rangle)$ gets the minimal value when $\theta_n = \theta_* + 2\pi n/3$, $n \in N$, for some fixed $\theta_*$. Therefore the pure states in the optimal decomposition have the form, $|q,\theta_n\rangle = \sqrt{q}|GHZ\rangle - \sqrt{1-q}e^{i\theta_n}|W\rangle$.

We first investigate the possible optimal decompositions containing three pure states,

$$\rho_3^{opt} = a|q_1,\theta_{n_1}\rangle\langle q_1,\theta_{n_1}| + b|q_2,\theta_{n_2}\rangle\langle q_2,\theta_{n_2}| + c|q_3,\theta_{n_3}\rangle\langle q_3,\theta_{n_3}|, \tag{3}$$

where $a, b, c \in [0, 1]$ and $a + b + c = 1$. $\rho_3^{opt}$ also includes the situation containing two pure states in the optimal decomposition, since $a, b, c$ can be zero. Under the bases $\{|GHZ\rangle, |W\rangle\}$, $\rho_3^{opt}$ and $\rho$ can be reexpressed as:

$$\rho_3^{opt} = \begin{pmatrix} x & z \\ \bar{z} & y \end{pmatrix},$$

where $x = aq_1 + bq_2 + cq_3$, $z = -a\sqrt{q_1(1-q_1)}e^{-i\theta_{n_1}} - b\sqrt{q_2(1-q_2)}e^{-i\theta_{n_2}} - c\sqrt{q_3(1-q_3)}e^{-i\theta_{n_3}}$, $y = 1 - (aq_1 + bq_2 + cq_3)$, and

$$\rho = \begin{pmatrix} p & 0 \\ 0 & 1-p \end{pmatrix}$$

respectively. From $\rho_3^{opt} = \rho$ we have relations

$$\begin{cases} aq_1 + bq_2 + cq_3 = p, \\ a\sqrt{q_1(1-q_1)}e^{-i\theta_{n_1}} + b\sqrt{q_2(1-q_2)}e^{-i\theta_{n_2}} + c\sqrt{q_3(1-q_3)}e^{-i\theta_{n_3}} = 0, \\ a + b + c = 1. \end{cases}$$

$$(4)$$

If $\rho_3^{opt}$ is the optimal decomposition, then $E(\rho) = aE(|q_1, \theta_{n_1}\rangle) + bE(|q_2, \theta_{n_2}\rangle) + cE(|q_3, \theta_{n_3}\rangle)$. Because $aE(|q_1, \theta_{n_1}\rangle)$, $bE(|q_2, \theta_{n_2}\rangle)$ and $cE(|q_3, \theta_{n_3}\rangle) \geq 0$, we have

$$\begin{aligned} &aE(|q_1, \theta_{n_1}\rangle) + bE(|q_2, \theta_{n_2}\rangle) + cE(|q_3, \theta_{n_3}\rangle) \\ &\geq 3\sqrt[3]{aE(|q_1, \theta_{n_1}\rangle)bE(|q_2, \theta_{n_2}\rangle)cE(|q_3, \theta_{n_3}\rangle)}. \end{aligned}$$

$$(5)$$

The definition of $E(\rho)$ demands that the equality of (5) must hold, i.e.

$$aE(|q_1, \theta_{n_1}\rangle) = bE(|q_2, \theta_{n_2}\rangle) = cE(|q_3, \theta_{n_3}\rangle).$$

$$(6)$$

Equations (4) and (6) can be satisfied if

$$\begin{cases} a = b = c, \quad q_1 = q_2 = q_3 \\ \theta_{n_1} = \theta_*, \quad \theta_{n_2} = \theta_* + \frac{2\pi}{3}, \quad \theta_{n_3} = \theta_* + \frac{4\pi}{3}. \end{cases}$$

$$(7)$$

From (4), (6) and (7), (1) becomes

$$\begin{aligned} \rho_3^{opt} &= \tfrac{1}{3}|p, \theta_*\rangle\langle p, \theta_*| + \tfrac{1}{3}|p, \theta_* + \tfrac{2\pi}{3}\rangle\langle p, \theta_* + \tfrac{2\pi}{3}| \\ &+ \tfrac{1}{3}|p, \theta_* + \tfrac{4\pi}{3}\rangle\langle p, \theta_* + \tfrac{4\pi}{3}|, \end{aligned}$$

$$(8)$$

which is the possible optimal decomposition containing three pure states.

Let us investigate the possible optimal decomposition that containing four pure states, by adding one pure state to the decomposition (8),

$$\rho_4^{opt} = a \left| q', \theta_n \right\rangle \left\langle q', \theta_n \right| + b(\left| q, \theta_* \right\rangle \left\langle q, \theta_* \right| + \left| q, \theta_* + \frac{2\pi}{3} \right\rangle \left\langle q, \theta_* + \frac{2\pi}{3} \right|$$
$$+ \left| q, \theta_* + \frac{4\pi}{3} \right\rangle \left\langle q, \theta_* + \frac{4\pi}{3} \right|), \tag{9}$$

where $a, b \in [0, 1]$ and $a + 3b = 1$. Under the bases $\{\left| GHZ \right\rangle, \left| W \right\rangle\}$, $\rho_4^{opt}$ can be expressed as

$$\rho_4^{opt} = a \begin{pmatrix} q' & -\sqrt{q'(1-q')}e^{-i\theta_n} \\ -\sqrt{q'(1-q')}e^{i\theta_n} & 1-q' \end{pmatrix} + b \begin{pmatrix} 3q & 0 \\ 0 & 3(1-q) \end{pmatrix}.$$

From $\rho_4^{opt} = \rho$ we have

$$aq' + 3bq = p, \quad a\sqrt{q'(1-q')}e^{-i\theta_n} = 0, \quad a + 3b = 1. \tag{10}$$

As $a \neq 0$, one gets $\sqrt{q'(1-q')} = 0$, that is, $q' = 0$ or $q' = 1$. Hence the added pure state is either $\left| W \right\rangle$ or $\left| GHZ \right\rangle$.

If $q' = 0$, from (10) we get $b = p/3q$, $a = (q-p)/q$, due to that $a, b \geq 0$ and $q > 0$. $p$ has a range $0 \leq p \leq q$. The corresponding possible optimal decomposition is $\rho_4^{opt}(q' = 0) = \frac{q-p}{q} \left| W \right\rangle \left\langle W \right| + \frac{p}{3q}(\left| q, \theta_* \right\rangle \left\langle q, \theta_* \right| + \left| q, \theta_* + \frac{2\pi}{3} \right\rangle \left\langle q, \theta_* + \frac{2\pi}{3} \right| + \left| q, \theta_* + \frac{4\pi}{3} \right\rangle \left\langle q, \theta_* + \frac{4\pi}{3} \right|)$. The corresponding entanglement is $E_4^{opt}(q' = 0) = \frac{q-p}{q} E(\left| W \right\rangle) + \frac{p}{q} E(\left| q, \theta_* \right\rangle)$. For given $p$, $E_4^{opt}(q' = 0)$ varies with $q$. Let $q_0^*$ be the minimal point of $E_4^{opt}(q' = 0)$. Then the final $\rho_4^{opt}(q' = 0)$ is

$$\rho_4^{opt}(q' = 0) = \frac{q_0^* - p}{q_0^*} \left| W \right\rangle \left\langle W \right| + \frac{p}{3q_0^*}(\left| q_0^*, \theta_* \right\rangle \left\langle q_0^*, \theta_* \right|$$
$$+ \left| q_0^*, \theta_* + \frac{2\pi}{3} \right\rangle \left\langle q_0^*, \theta_* + \frac{2\pi}{3} \right| + \left| q_0^*, \theta_* + \frac{4\pi}{3} \right\rangle \left\langle q_0^*, \theta_* + \frac{4\pi}{3} \right|). \tag{11}$$

If $q' = 1$, similarly we have $E_4^{opt}(q' = 1) = \frac{p-q}{1-q} E(\left| GHZ \right\rangle) + \frac{1-p}{1-q} E(\left| q, \theta_* \right\rangle)$, and

$$\rho_4^{opt}(q' = 1) = \frac{p - q_1^*}{1 - q_1^*} \left| GHZ \right\rangle \left\langle GHZ \right| + \frac{1}{3}\frac{1-p}{1-q_1^*}(\left| q_1^*, \theta_* \right\rangle \left\langle q_1^*, \theta_* \right|$$
$$+ \left| q_1^*, \theta_* + \frac{2\pi}{3} \right\rangle \left\langle q_1^*, \theta_* + \frac{2\pi}{3} \right| + \left| q_1^*, \theta_* + \frac{4\pi}{3} \right\rangle \left\langle q_1^*, \theta_* + \frac{4\pi}{3} \right|), \tag{12}$$

where $q_1^* \leq p \leq 1$ and $q_1^*$ is the minimal point of $E_4^{opt}(q' = 1)$.

We have investigated all possible optimal decompositions containing two to four pure states. With respect to decompositions (8), (11) and (12) we have the corresponding entanglement $E_3^{opt}$, $E_4^{opt}(p' = 0)$ and $E_4^{opt}(q' = 1)$. The entanglement of $\rho$ is then given by,

$$E(\rho) = \min\{E_3^{opt}, E_4^{opt}(q_4'^{opt}(q' = 1))\},\qquad(13)$$

where $E_3^{opt} = E(|p, \theta_*\rangle)$, $0 \le p \le 1$; $E_4^{opt}(q' = 0) = \frac{q_0^* - p}{q_0^*} E(|W\rangle) + \frac{p}{q_0^*} E(|q_0^*, \theta_*\rangle)$, $0 \le p \le q_0^*$; $E_4^{opt}(q' = 1) = \frac{p - q_1^*}{1 - q_1^*} E(|GHZ\rangle) + \frac{1-p}{1-q_1^*} E(|q_1^*, \theta_*\rangle)$, $q_1^* \le p \le 1$, and $q_0^*$, $q_1^*$ are the minimal points of $E_4^{opt}(q' = 0)$, $E_4^{opt}(q' = 1)$ respectively. Therefore if the entanglement $E(|q, \theta\rangle)$ of a pure state $|q, \theta\rangle$ is a periodical function of $\theta$ with period $2\pi/3$, then by equation (13) one can evaluate the entanglement of the mixed state $\rho$. If $E_4^{opt}(q' = 0)$ and $E_4^{opt}(q' = 1)$ are differentiable in $q \in (0, 1)$, the minimal points $q_0^*$ and $q_1^*$ of $E_4^{opt}(q' = 0)$ and $E_4^{opt}(q' = 1)$ can be obtained from the following equations, respectively

$$\left.\frac{\partial E_{opt40}}{\partial q}\right|_{q=q_0^*} = p\left(\frac{1}{q^2} E(|W\rangle) - \frac{1}{q^2} E(|q, \theta_*\rangle) + q\frac{dE(|q, \theta_*\rangle)}{dq}\right)\bigg|_{q=q_0^*} = 0,$$

$$\left.\frac{\partial E_{opt41}}{\partial q}\right|_{q=q_1^*} = (1 - p)\left(-\frac{1}{(1 - q)^2} E(|GHZ\rangle) + \frac{1}{(1 - q)^2} E(|q, \theta_*\rangle)\right.$$

$$\left. + \frac{1}{1 - q}\frac{dE(|q, \theta_*\rangle)}{dq}\right)\bigg|_{q=q_1^*} = 0.$$

$$\qquad(14)$$

From (14) it can be seen that $q_0^*$ and $q_1^*$ do not depend on $p$.

## 3. The study of three-tangle and three-$\pi$

Three-tangle[12] and Three-$\pi$[9] are two different important tripartite entanglement measures. In this section we study these two measures by using our results in section 2.

We first consider the case of three-tangle. The three-tangle of the state $|q, \theta\rangle = \sqrt{q}\,|GHZ\rangle - \sqrt{1 - q}\,e^{i\theta}\,|W\rangle$ is given by [10],

$$\tau(|q, \theta\rangle) = \left|q^2 - \frac{8}{9}e^{i3\theta}\sqrt{6q(1 - q)^3}\right|.\qquad(15)$$

Obviously, $\tau(|q, \theta\rangle)$ is a periodical function of $\theta$ with periodic $2\pi/3$. Moreover, when $\theta_n = \frac{2\pi}{3}n$, $n \in \mathbb{Z}$, the three-tangle of the state $|q, \theta\rangle$ gets the minimal for a fixed $q$, that is, $\theta_* = 0$. $\tau_3^{opt}$ has a minimal point $q_* = \frac{4\sqrt[3]{2}}{3+4\sqrt[3]{2}} \doteq 0.627 \in (0, 1)$, which is also the local minimal point of $\tau_4^{opt}(q' = 0)$ and $\tau_4^{opt}(q' = 1)$ from (14). According to (13) we have

$$\tau(\rho) = \min\{\tau_3^{opt}, \tau_4^{opt}(q_4'^{opt}(q' = 1))\},\qquad(16)$$

where $\tau_3^{opt} = \tau(|p,0\rangle)$ when $0 \le p \le 1$; $\tau_4^{opt}(q' = 0) = \frac{p}{q_0^*}\tau(|q_0^*,0\rangle)$ when $0 \le p \le q_0^*$; $\tau_4^{opt}(q' = 1) = \frac{p-q_1^*}{1-q_1^*} + \frac{1-p}{1-q_1^*}\tau(|q_1^*,0\rangle)$ when $q_1^* \le p \le 1$.

As $\tau_4^{opt}(q' = 0) = \frac{p}{q_*}\tau(|q_*,0\rangle) = 0$ for $q = q_*$, $q = q_*$ is the minimal point for $\tau_4^{opt}(q' = 0)$, i.e. $q_0^* = q_*$.

For $\tau_4^{opt}(q' = 1)$, we only need to consider the case $q \in (q_*, 1)$, since $0 \le p \le 1$. For $q = q_*$, $\tau_4^{opt}(q' = 1) = \frac{p-q_*}{1-q_*} + \frac{1-p}{1-q_*}\tau(|q_*,0\rangle) = \frac{p-q_*}{1-q_*}$. Using (14) and taking into account that $0 < q < 1$, we have $2q - 1 > 0$ and $155q^2 - 155q + 32 = 0$. Therefore another possible minimal point is $q_1^{*\prime} = \frac{1}{2} + \frac{3}{310}\sqrt{465}$ . $Since \tau_4^{opt}(q' = 1)(q_*) > \tau_4^{opt}(q' = 1)(q_1^{*\prime})$, the true minimal point of $\tau_4^{opt}(q' = 1)$ is $q_1^* = q_1^{*\prime}$.

For $0 \le p \le q_0^*$, $\tau_{opt40} \le \tau_{opt3}$, where the equality holds if and only if $p = 0$ or $p = q_0^*$. For $q_1^* \le p \le 1$, $\tau_{opt41} \le \tau_{opt3}$ and the equality holds if and only if $p = q_0^*$ or $p = 1$. For $q_0^* < p < q_1^*$, $\tau_3^{opt}$ is the only possible optimal value. Therefore from (16) we get

$$\tau(\rho) = \begin{cases} \frac{p}{q_0^*}\left(-(q_0^*)^2 + \frac{8}{9}\sqrt{6q_0^*(1-q_0^*)^3}\right), & 0 \le p \le q_0^*, \\ p^2 - \frac{8}{9}\sqrt{6p(1-p)^3}, & q_0^* < q < q_1^*, \\ \frac{p-q_1^*}{1-q_1^*} + \frac{1-p}{1-q_1^*}\left((q_1^*)^2 - \frac{8}{9}\sqrt{6q_1^*(1-q_1^*)^3}\right), & q_1^* \le p \le 1, \end{cases} \qquad (17)$$

where $q_0^* = 4\sqrt[3]{2}/(3 + 4\sqrt[3]{2}) = 0.627$, $q_1^* = \frac{1}{2} + \frac{3}{310}\sqrt{465} = 0.709$. The corresponding optimal decompositions are (8), (11) and (12).

Now we study the tripartite entanglement three-$\pi$. The definition of three-$\pi$ of a tripartite pure state is given by [9],

$$\pi = \frac{1}{3}(\pi_a + \pi_b + \pi_c), \qquad (18)$$

where $\pi_a = N_{a|bc}^2 - N_{ab}^2 - N_{ac}^2$, $\pi_b = N_{b|ac}^2 - N_{ba}^2 - N_{bc}^2$, and $\pi_c = N_{c|ab}^2 - N_{ca}^2 - N_{cb}^2$. $N$ stands for the bipartite entanglement measure negativity [14,5], $N_{ab} = \left\|\rho_{ab}^{T_a}\right\| - 1$, where $\|\rho\|$ is the trace norm of $\rho$, $\rho_{ab}^{T_a}$ is the partial transpose of $\rho_{ab}$ with respect to the part $a$, satisfying $(\rho_{ab}^{T_a})_{ij,kl} = (\rho_{ab})_{kj,il}$.

For the tripartite pure state $|q, \theta\rangle$, we have $\rho_{ab} = Tr_c(\rho)$,

$$\rho_{ab} = \begin{pmatrix} \frac{q}{2} + \frac{1-q}{3} & -\sqrt{\frac{q(1-q)}{6}}e^{-i\theta} & -\sqrt{\frac{q(1-q)}{6}}e^{-i\theta} & -\sqrt{\frac{q(1-q)}{6}}e^{i\theta} \\ -\sqrt{\frac{q(1-q)}{6}}e^{i\theta} & \frac{1-q}{3} & \frac{1-q}{3} & 0 \\ -\sqrt{\frac{q(1-q)}{6}}e^{i\theta} & \frac{1-q}{3} & \frac{1-q}{3} & 0 \\ -\sqrt{\frac{q(1-q)}{6}}e^{-i\theta} & 0 & 0 & \frac{q}{2} \end{pmatrix}, \qquad (19)$$

$$\rho_{ab}^{T_a} = \begin{pmatrix} \frac{q}{2} + \frac{1-q}{3} & -\sqrt{\frac{q(1-q)}{6}}e^{-i\theta} & -\sqrt{\frac{q(1-q)}{6}}e^{i\theta} & \frac{1-q}{3} \\ -\sqrt{\frac{q(1-q)}{6}}e^{i\theta} & \frac{1-q}{3} & -\sqrt{\frac{q(1-q)}{6}}e^{-i\theta} & 0 \\ -\sqrt{\frac{q(1-q)}{6}}e^{-i\theta} & -\sqrt{\frac{q(1-q)}{6}}e^{i\theta} & \frac{1-q}{3} & 0 \\ \frac{1-q}{3} & 0 & 0 & \frac{q}{2} \end{pmatrix} \tag{20}$$

and

$$\rho_a = Tr_{bc}(\rho) = \begin{pmatrix} \frac{q}{2} + \frac{2(1-q)}{3} & \sqrt{\frac{q(1-q)}{6}}e^{-i\theta} \\ \sqrt{\frac{q(1-q)}{6}}e^{i\theta} & \frac{q}{2} + \frac{(1-q)}{3} \end{pmatrix}. \tag{21}$$

It can be verified that due to the symmetry of $|GHZ\rangle$ and $|W\rangle$, $\rho_{ab} = \rho_{bc} = \rho_{ac}$, $\rho_a = \rho_b = \rho_c$. Hence $N_{ab} = N_{ac} = N_{bc}$, $\pi_a = \pi_b = \pi_c$, $\pi = \pi_a$. For three-qubit pure states, $N_{a|bc} = C_{a|bc}$, where $C_{a|bc}$ is the concurrence [4] between the subsystems $a$ and $b, c$. From (19) we get $N_{ab} = C_{a|bc} = \sqrt{2(1 - Tr\rho_a^2)} = \sqrt{\frac{5}{9}q^2 - \frac{4}{9}q + \frac{8}{9}}$. By (18) we get the characteristic equation of the matrix $\rho_{ab}^{T_a}$,

$$\begin{aligned} &\lambda^4 - \lambda^3 + \left(\tfrac{5}{36}q^2 - \tfrac{q}{9} + \tfrac{2}{9}\right)\lambda^2 \\ &+ \left[\tfrac{(q(1-q))^{3/2}}{3\sqrt{6}}\cos 3\theta - \tfrac{7}{27}q^3 + \tfrac{7}{18}q^2 - \tfrac{q}{6} + \tfrac{1}{27}\right]\lambda \\ &+ \left[-\tfrac{q(q(1-q))^{3/2}}{6\sqrt{6}}\cos 3\theta - \tfrac{41}{648}q^4 + \tfrac{149}{648}q^3 - \tfrac{13}{54}q^2 + \tfrac{7}{81}q - \tfrac{1}{81}\right] = 0. \end{aligned} \tag{22}$$

The three-$\pi$ of the pure state $|q, \theta\rangle$ can be expressed as

$$\begin{aligned} \pi = \pi_a &= N_{a(bc)}^2 - N_{ab}^2 - N_{ac}^2 = C_{a|bc}^2 - 2N_{ab}^2 \\ &= \tfrac{5}{9}q^2 - \tfrac{4}{9}q + \tfrac{8}{9} - 2\left(\sum_{i=1}^{4}|\lambda_i\,(q, Cos3\theta)| - 1\right)^2, \end{aligned} \tag{23}$$

where $i = 1, 2, 3, 4$ and $\lambda_i\,(q, cos3\theta)$ is the solutions of (22).

From Fig. 1 and (23) we can see that the three-$\pi$ is a periodical function of $\theta$ with period $2\pi/3$. When $\theta = 2\pi n/3$, $n \in \mathbb{Z}$, the three-$\pi$ gets minimal value, that is, $\theta_* = 0$. Hence we can use our formulations in section 2 to deal with the possible optimal decomposition of three-$\pi$ for the state $\rho$.

Using the formula (13) and the method used for the case of three-tangle,

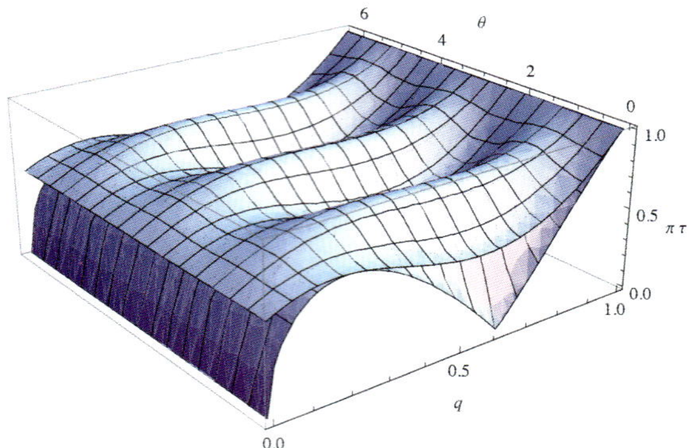

Figure 1. The three-$\pi$ and three-tangle for pure state $|q,\theta\rangle = \sqrt{q}\,|GHZ\rangle - \sqrt{1-q}\,e^{i\theta}\,|W\rangle$ as a function of $\theta$ and $q$. The top surface is the three-$\pi$, the bottom surface is the three-tangle. The three-$\pi$ is always greater than or equal to three-tangle. For a fixed $q$, three-tangle and three-$\pi$ are all periodical functions of $\theta$. They get minimal when $\theta = 0, 2\pi/3, 4\pi/3, \dots$

we can obtain

$$\pi(\rho) = \begin{cases} \dfrac{q_0^*-p}{q_0^*}\dfrac{4}{9}(\sqrt{5}-1) + \dfrac{p}{q_0^*}[\dfrac{5}{9}(q_0^*)^2 - \dfrac{4}{9}q_0^* \\ \qquad\qquad + \dfrac{8}{9} - 2(\sum\limits_{i=1}^{4}|\lambda_i(q_0^*,1)|-1)^2], & 0 \le p \le q_0^*, \\[2ex] \dfrac{5}{9}p^2 - \dfrac{4}{9}p + \dfrac{8}{9} - 2(\sum\limits_{i=1}^{4}|\lambda_i(p,1)|-1)^2, & q_0^* \le q \le q_1^*, \\[2ex] \dfrac{p-q_1^*}{1-q_1^*} + \dfrac{1-p}{1-q_1^*}[\dfrac{5}{9}(q_1^*)^2 - \dfrac{4}{9}q_1^* + \dfrac{8}{9} \\ \qquad\qquad -2(\sum\limits_{i=1}^{4}|\lambda_i(q_1^*,1)|-1)^2], & q_1^* \le p \le 1, \end{cases} \qquad (24)$$

where $q_0^* = 0.564$, $q_1^* = 0.963$, $i = 1,2,3,4$, $\lambda_i(q,1)$ is the solutions of (22). (8), (11) and (12) are the corresponding optimal decompositions.

Let us make a comparison between the three-tangle and the three-$\pi$. For $m \otimes n$, $m \le n$, bipartite mixed states, we have $\sqrt{\dfrac{2}{m(m-1)}}(\|\rho_{ab}^{Ta}\|-1) \le C(\rho_{ab})$ [15], and for $2 \otimes 2$ systems, $N(\rho_{ab}) \le C(\rho_{ab})$. For three-qubit pure

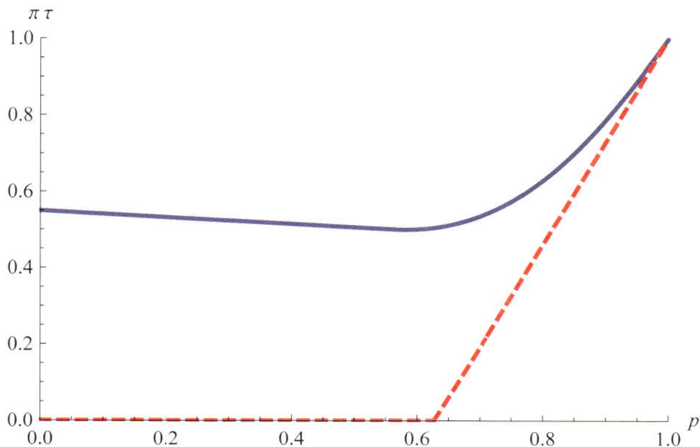

Figure 2. The three-$\pi$ and three-tangle for mixed states $\rho = p\,|GHZ\rangle\,\langle GHZ| + (1-p)\,|W\rangle\,\langle W|$ as a function of $p$. The top curve is for the three-$\pi$, the bottom curve is for the three-tangle. They are all liner functions of $p$ at $(0, q_0^*)$ and $(q_1^*, 1)$. For three-tangle, $q_0^* = 0.627$, $q_1^* = 0.709$, and for three-$\pi$, $q_0^* = 0.564$, $q_1^* = 0.963$.

states we have $N_{a|bc} = C_{a|bc}$. Hence

$$
\begin{aligned}
\pi &= \tfrac{1}{3}\big(\pi_a + \pi_b + \pi_c\big) \\
&= \tfrac{1}{3}\big(N^2_{a(bc)} - N^2_{ab} - N^2_{ac} + N^2_{b(ac)} - N^2_{ba} - N^2_{bc} + N^2_{c(ab)} - N^2_{ca} - N^2_{cb}\big) \\
&\geq \tfrac{1}{3}\big(C^2_{a(bc)} - C^2_{ab} - C^2_{ac} + C^2_{b(ac)} - C^2_{ba} - C^2_{bc} + C^2_{c(ab)} - C^2_{ca} - C^2_{cb}\big) \\
&= \tfrac{1}{3}\big(\tau_a + \tau_b + \tau_c\big) = \tau.
\end{aligned}
$$

$$(25)$$

Therefore for three qubit pure states, three-$\pi$ is great than or equal to the three-tangle, as shown in Fig. 1.

For rank-2 mixed states $\rho = p\,|GHZ\rangle\,\langle GHZ| + (1-p)\,|W\rangle\,\langle W|$, from (17) and (24) one sees, Fig. 2, that the three-tangle is smaller or equal to the three-$\pi$. The two measures show different behavior when $p$ increases. At $p = q_0^* = 0.564$ the three-$\pi$ gets minimal value $0.50103$, which implies that the increase of the weight of the maximally entangled state $|GHZ\rangle$ does not means the increase of entanglement [16]. While the three-tangle is a nondecreasing function of $p$.

## 4. Conclusions

We have studied the entanglement for a class of rank-2 mixed states $\rho = p\,|GHZ\rangle\,\langle GHZ| + (1-p)\,|W\rangle\,\langle W|$. Base on the the periodicity assumption, the possible optimal decomposition has been derived. Our optimal decomposition does not depend on the detailed entanglement measure and works for all entanglement measures satisfying our assumptions. By applying our formulations, two important tripartite entanglement measures, three-tangle and three-$\pi$, have been investigated. It has been shown that the three-tangle is always smaller than or equal to three-$\pi$. Moreover the three-tangle is a nondecreasing function of $p$, while the three-$\pi$ has a minimal point at $p \in (0,1)$, showing that different entanglement measures may have different characters for the same state. Our approach may be also useful for studying other tripartite and multipartite quantum states.

## References


1. M. A. Nielsen, and I. L. Chuang, *Quantum Computation and Quantum Information* (Cambridge University Press, Cambridge, 2000).
2. C. H. Bennett, D. P. DiVincenzo, J .A. Smolin, and W. K. Wootters, Phys. Rev. A **54**, 3824(1996).
3. M. Horodecki, Quantum Inf. Comp. **1**, 3(2001);
   M. B. Plenio, and S. Virmani, Quantum Inf. Comp. **7**, 1(2007).
4. A. Uhlmann, Phys. Rev. A **62**, 032307(2000);
   P. Rungta, V. Bužek, C. M. Caves, M. Hillery, and G. J. Milburn, Phys. Rev. A **64**, 042315(2001);
   S. Albeverio, and S. M. Fei, J. Opt. B: Quantum Semiclass. Opt. **3**, 223(2001).
5. G. Vidal, and R. F. Werner, Phys. Rev. A **65**, 032314(2002).
6. V. Vedral, and M. B. Plenio, Phys Rev A **57**, 1619(1998).
7. A. Peres, Phys. Rev. Lett. **77**, 1413(1996).
8. J. S. Bell, Physics (Long Island, N. Y.) **1**, 195(1964).
9. Y.C. Ou and H. Fan, Phys. Rev. A **75**, 062308(2007).
10. R. Lohmayer, A. Osterloh, J. Siewert, A. Uhlmann, Phys. Rev. Lett. **97**, 260502(2006).
11. C. Eltschka, A. Osterloh, J. Siewert and A. Uhlmann, New J. Phys. **10**, 043014 (2008).
12. V. Coffman, J. Kundu, and W. K. Wootters, Phys. Rev. A **61**, 052306(2000)
13. R. Horodecki, P. Horodecki, M. Horodecki, K. Horodecki, Rev. Mod. Phys. **81** (2009).
14. M. Horodecki, P. Horodecki, and R. Horodecki, Phys. Rev. Lett. **80**, 5239 (1998).
15. K. Chen, S. Albeverio, and S.M. Fei, Phys. Rev. Lett. **95**, 040504 (2005).
16. N. Gisin and H.B. Pasquinucci, Phys. Lett. A **246**, 1(1998).

Quantum Bio-Informatics V
© 2013 World Scientific Publishing Co. Pte. Ltd.
pp. 435–446

# ENERGY FLOW AND INFORMATION FLOW IN SUPERCONDUCTING QUBIT MEASUREMENT PROCESS

HAYATO NAKANO

*NTT Basic Research Laboratories, NTT Corporation,*
*Atsugi-shi, Kanagawa 243-0198, Japan*
**E-mail: nakano.hayato@lab.ntt.co.jp*

A quantum mechanical measurement is information conversion from quantum one to classical one. As an example, we analyzed the Josephson bifurcation amplifier (JBA) measurement process of a superconducting qubit. We theoretically examined the dynamics of the density operator of a detector (a driven nonlinear oscillator) and a qubit coupled system during the measurement process. When the initial state of the qubit is a superposition state, in other words, the qubit contains quantum information initially, we observe the formation of a qubit-detector (JBA) entangled state and it is divided into two separable states at the moment the JBA transition begins. The resulting JBA state shows the measurement result, which is classical information. We also discuss the process from the viewpoint of time variations in information quantities, such as mutual information between the qubit and detector. This clearly shows that the measurement process corresponds to an information conversion from quantum to classical one.

Keywords : superconducting qubit measurement; detector; entanglement; decoherence; quantum correlation.

## 1. Introduction

In the last century, quantum mechanics became well established and it is now used to develop new technologies. In particular, in the field of quantum information science and related engineering research, quantum mechanics is the core concept that guides us to the new world of information technology [1]. Among many unconventional properties of physical systems that quantum mechanics reveals, measurement is the most subtle and plays one of the most important roles in quantum information technology. A (projective) measurement on a quantum mechanically superposed state selects one of the eigenstates of the measured physical quantity (observable) and the rest of possibilities disappear after the measurement [2]. A projective measurement is also very important for quantum information processing applications, for example, measurement-based quantum computing [3].

A perfect understanding of the strange properties of quantum measurement is currently beyond us [4]. However, we can calculate and design a measurement process quantitatively with our knowledge of quantum open systems if we have sufficient information on the measured physical system, the measurement apparatus (detector) and the interaction between them. In particular, the concept of 'decoherence' [5,6,7] is very useful for analyzing an actual measurement process, when it is applied to a system-detector composite system [11,9,10,8].

Qubit readout is one of simple quantum measurements. According to a simple explanation of quantum measurements, it is described as if we can directly obtain an observable eigenvalue of the quantum system by a measurement. In a real measurement, however, we must always use a detector that interacts with the quantum system. A detector is often a macroscopic object. We actually measure a macroscopic quantity in the detector, and merely *postulate* the physical quantity of the quantum system we want to measure. To discuss what we really measure when we measure a qubit, we have to take account of the detector system. A measurement process is a quantum dynamics of the total system consisting of a qubit and a detector.

A good qubit measurement gives a probabilistic projection of a qubit state $|\psi_{\mathrm{q}}\rangle = a| \uparrow \rangle + b| \downarrow \rangle$ into $| \uparrow \rangle$ or $| \downarrow \rangle$ with a probability $|a|^2$ or $|b|^2$, respectively. Here, $| \uparrow \rangle$ and $| \downarrow \rangle$ are two of the eigenstates of the *observable* in the qubit. The observable is mainly determined by the operator of the qubit appearing in the interaction Hamiltonian between the qubit and the detector. However, the observable is sometimes different from the operator because it is determined by the entire dynamics of the total system during the measurement process.

Below, we discuss the measurement process for a superconducting qubit. We use a Josephson Bifurcation Amplifier (JBA) as the detector in the measurement [12]. The JBA is an AC-driven SQUID that behaves as a nonlinear oscillator. We find many important quantum characters by investigating the dynamics during the measurement process [13].

Usually, the total system is expected to evolve as follows, when the qubit is in a superposition of two of observable eigenstates $|\psi(0)_{\mathrm{q}}\rangle = a| \uparrow \rangle + b| \downarrow \rangle$. When the qubit-detector interaction becomes on, an entanglement starts to be formed: $|\psi(0)_{\mathrm{q}}\rangle \otimes |\mathrm{C}_0\rangle \to a| \uparrow \rangle \otimes |\mathrm{A}_1\rangle + b| \downarrow \rangle \otimes |\mathrm{B}_1\rangle$. Since the detector is a macroscopic system where a finite decoherence is inevitable, dephasing occurs. Then, the total system becomes a classical mixture like $|a|^2|\uparrow\rangle|\mathrm{A}_1\rangle\langle\uparrow|\langle\mathrm{A}_1| + |b|^2|\downarrow\rangle|\mathrm{B}_1\rangle\langle\downarrow|\langle\mathrm{B}_1|$. Here, $|\mathrm{C}_0\rangle$, $|\mathrm{A}_1\rangle$, $|\mathrm{B}_1\rangle$ are macroscopic states of the detector, and we can easily distinguish $|\mathrm{A}_1\rangle$ or $|\mathrm{B}_1\rangle$. Therefore, when we measure the detector state, we can postulate whether the qubit state is $|\uparrow\rangle$ or $|\downarrow\rangle$.

## 2. Superconducting Qubit Measurement with a Detector (SQUID)

A superconducting flux qubit is a superconducting ring interrupted by a few Josephson junctions [14,15,16,17,18,19]. When the external magnetic field piercing the ring is in the vicinity of the half flux quantum $\Phi_0/2$, the ring becomes a superposition of two macroscopically distinctive states. Supercurrent flows clockwise around the ring in one state $|\mathrm{R}\rangle$, and it flows counterclockwise in the other state $|\mathrm{L}\rangle$. The qubit Hamiltonian is given by [20,21,22]

$$H_{\mathrm{q}} = \frac{1}{2}(\varepsilon\sigma_z + \Delta\sigma_x),$$

where $\sigma_z$ and $\sigma_x$ are two Pauli matrices, and

$$\sigma_z = |\mathrm{L}\rangle\langle\mathrm{L}| - |\mathrm{R}\rangle\langle\mathrm{R}|.$$

We can control $\varepsilon$ by changing the externally applied magnetic flux, but $\Delta$ is fixed. The energy-eigenstates of the qubit are given by

$$|g\rangle = \cos(\theta/2)|\mathrm{L}\rangle + \sin(\theta/2)|\mathrm{R}\rangle,$$

and

$$|e\rangle = \sin(\theta/2)|\mathrm{L}\rangle - \cos(\theta/2)|\mathrm{R}\rangle$$

where $\tan\theta = \Delta/\varepsilon$.

In order to measure superconducting flux qubit states, a Superconducting Quantum Interference Device (SQUID) is used as the detector. The SQUID detects a very small magnetic flux induced by supercurrent flowing along the qubit ring, which is only one observable in a flux qubit we can measure from the outside (see, Fig. 1). A SQUID can be approximated as a single Josephson junction whose Josephson energy depends on the magnetic flux $\Phi_{\mathrm{SQ}}$ piercing the SQUID ring. The Hamiltonian for the SQUID phase $\gamma$(=the Josephson phase of the effective single junction) is given by [23],

$$H_{\mathrm{SQUID}} = (\hbar/2e)^2\dot{\gamma}^2/(2C) - E_{\mathrm{J}}(\Phi_{\mathrm{SQ}})\cos\gamma - (\hbar/2e)I_{\mathrm{b}}\gamma$$

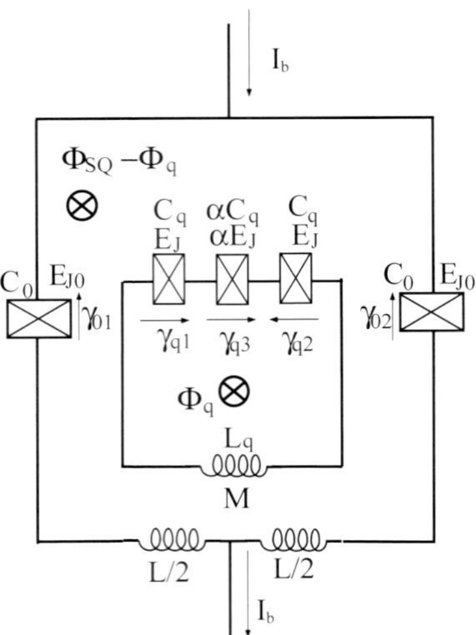



Figure 1.    Measurement of a flux-qubit system with a detector SQUID that has a continuous variable $\gamma = (\gamma_{01} + \gamma_{02})/2$.

$$= (\hbar/2e)^2 \dot{\gamma}^2/(2C) - E_{\rm J}(\Phi_{\rm SQ})(1 - \gamma^2/2 + \gamma^4/24 - \cdots) - (\hbar/2e)I_{\rm b}\gamma,$$

where $C$ is the effective capacitance of the junction and $I_{\rm b}$ is the bias current applied to the SQUID.

A half decade ago, our experimental results, where raw data before averaging behaved, clearly showed that our flux-qubit measurement with a DC-SQUID is not a $\sigma_z$ projection measurement [24,25]. Moreover, we theoretically clarified that the raw data behave in the qubit measurement with a SQUID when we increase the SQUID bias current slowly, from a detailed calculation of the dynamics of the qubit-SQUID coupled system [23]. The measurement projects the coupled system to one of its energy-eigenstates. In the state, the qubit is approximately in an energy-eigenstate of $H_{\rm q}$, and the value (signal) obtained from a single measurement approximately corresponds to $\langle g|\sigma_z|g\rangle$ or $\langle e|\sigma_z|e\rangle$.

When the amplitude of $\gamma$ is large, the oscillator becomes bistable under appropriate operation parameters [26]. One stable state has a small amplitude (low-amplitude state), and the other has a larger amplitude (high-amplitude state).

The critical driving force $f_{\rm c}$ or the critical detuning $\delta_{\rm c}$ for the transition between these two states is very sensitive to small changes in the operational

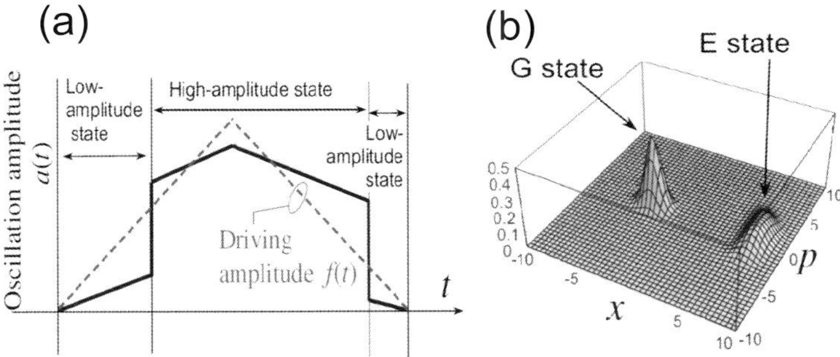



Figure 2. (a) Hysteretic behavior of the oscillation amplitude $a(t)$ of a classical JBA. When the driving force $f(t)$ (broken line) is changed with time $t$, the amplitude (thick line) varies as shown (schematic). (b) Two macroscopic states of JBA (quasi-distribution representation): $p$, $x$ are the momentum and amplitude of the oscillation, respectively.

parameters of the oscillator. For example, when we increase or decrease the driving force continuously, the amplitude of the oscillation behaves hysteretically as shown in Fig. 2. When using a JBA as a qubit state measurement detector, the JBA detects a small change that depends on the qubit state. The bifurcation phenomenon can be discussed only for classical oscillators, and is impossible from the viewpoint of pure quantum mechanics for an isolated system [27].

A quantum mechanical analysis is indispensable if we are to understand the measurement process [13,28]. A JBA can be modeled as an anharmonic oscillator in a rotating frame approximation with a Hamiltonian;

$$H_{\mathrm{J}} = (\Omega - \omega)n_a - \alpha n_a{}^2 - \frac{1}{2}f(a^\dagger + a) \tag{1}$$

where, $a^\dagger(a)$ is the creation (annihilation) operator of the Josephson plasma oscillation. black The normalized amplitude and momentum of an oscillation are expressed by $x = (a^\dagger + a)/2$ and $p = (a - a^\dagger)/(2i)$, respectively. $n_a = a^\dagger a$, and $\Omega$ is the linear resonant frequency of the JBA oscillator. $\omega$ is the driving frequency, which is slightly smaller than $\Omega$ by the detuning $\delta \equiv \Omega - \omega$. $f$ is the driving strength, and $\alpha(> 0)$ is the nonlinearity. In a classical approximation, this model shows the bifurcation in an appropriate parameter region. However, for a quantum-mechanical junction, the transition from $|G\rangle_{\mathrm{J}}$ (low-amplitude state) to $|E\rangle_{\mathrm{J}}$ (high-amplitude state) or, from $|E\rangle_{\mathrm{J}}$ to $|G\rangle_{\mathrm{J}}$ is impossible. In order to reproduce the hysteretic bifurcation behavior, we found that finite decoherence causing stochastic jumps between quasi-enegyeigenstates of the JBA Hamiltonian $H_J$[13].

As an example, hereafter, we introduce linear loss in the oscillator (JBA). The time evolution of the JBA is governed by a Liouville equation [29]:

$$\frac{\mathrm{d}\rho}{\mathrm{d}t} = \frac{1}{i}[H, \rho] + \frac{\Gamma}{2}(2a\rho a^\dagger - a^\dagger a\rho - \rho a^\dagger a),\tag{2}$$

where $\rho$ is the density operator of the system, and $\Gamma$ is the relaxation rate due to the linear loss in the JBA. The qubit-JBA composite system is approximately expressed by the Hamiltonian

$$H = H_J + k\sigma_z n_a + H_q, \qquad H_q = \frac{1}{2}(\varepsilon\sigma_z + \Delta\sigma_x)\tag{3}$$

$k$ is the interaction constant between the qubit and the JBA.

## 3. Information Measures in Quantum Systems

Before we show our calculation results on the dynamics of the qubit-detector (JBA) composite system during the measurement process, we briefly introduce mutual information in classical information theory and the quantum counterpart because later we will discuss the relation between the physical dynamics and the information time-evolution during the measurement process, which is the main topic of this report, in the next section.

When there are two information sources A and B, the correlation between A and B is quantified by the mutual information defined by

$$I_C(A, B) = H(\rho_A) + H(\rho_B) - H(\rho_{AB}),$$

where $\rho_A$ and $\rho_B$ are the probability sets for A, B, respectively. $\rho_{AB}$ is the joint probability set of A and B. $H(\rho)$ is the information (Shannon) entropy for $\rho$. This definition implicitly contains the conditional entropy $K_C(\rho_{B|A}) = \sum_j p_j H(\rho_{B|A=j}) = H(\rho_{AB}) - H(\rho_A)$. The Bayes rule makes the equality satisfied.

On the other hand, the quantum counterparts are given as follows. The density operator of a quantum system $\rho = T^\dagger \begin{pmatrix} p_1 & 0 & 0 & \cdots & 0 \\ 0 & p_2 & 0 & \cdots & 0 \\ 0 & 0 & p_3 & \cdots & 0 \\ \vdots & & & \ddots & \vdots \\ 0 & 0 & 0 & \cdots & p_L \end{pmatrix} T$ is defined, where $T$ is a unitary matrix that diagonalizes $\rho$.

For quantum information, we use von Neumann entropy, $S(\rho) = -\mathrm{Tr}[\rho \log_2 \rho] = -\sum_{l=1}^{L} p_l \log_2 p_l$, instead of Shannon entropy. The quantum mutual information $I_Q$ of a two-body (A and B) system is defined by [30]

$$I_Q(A, B) = S(\rho_A) + S(\rho_B) - S(\rho_{AB}).$$

This agrees with the classical mutual information $I_C$ if there is only a classical correlation between A and B.

$I_Q$ is different from $I_C$ because the Bayes rule is not always valid for quantum events. For quantum events, the conditional probability should be expressed by [31] $K(\rho_{B|A}) = \min_{\Pi_j^A} \sum_j^m p_j S(\rho_{B|\Pi_j^A})$, where $\Pi_j^A$ is an operator that projects the state of A into $j$ and $p_j = \mathrm{Tr}_B[\Pi_j^A \otimes \boldsymbol{I}^B \rho_{AB}]$, $\rho_{B|\Pi_j^A} = \mathrm{Tr}_A[\Pi_j^A \otimes \boldsymbol{I}^B \rho_{AB} \Pi_j^A \otimes \boldsymbol{I}^B]/p_j$ , etc.. Then, we obtain, $K(\rho_{B|A}) \geq S(\rho_{AB}) - S(\rho_A)$ and the equality is not always satisfied. Using the conditional entropy $K$, we can define another type of mutual information, $J_Q(A, B) = S(\rho_B) - K(\rho_{B|A})$, which agrees with $I_Q$ if there is only a classical correlation between A and B. The difference between these two types of mutual information

$$D(\rho_{AB}) = I_Q(A, B) - J_Q(A, B) = K(\rho_{B|A}) - S(\rho_{AB}) + S(\rho_A)$$

is called "Quantum Discord" [31].

Most well known quantum correlation between A and B is entanglement, which is quantified by a measure called "concurrence" [32,33] as

$$C(\rho_{AB}) = \min_{\{p_k | \psi_k\rangle\}} \sum_k p_k C(|\psi_k\rangle),$$

where $|\psi_k\rangle$ and $p_k$ are a pure state and corresponding eigenvalue, with which $\rho_{AB}$ is decomposed as $\rho_{AB} = \sum_k p_k |\psi_k\rangle\langle\psi_k|$, $(0 \leq p_k \leq 1, \sum_k p_k = 1)$, and $C(|\psi_k\rangle) = \sqrt{2(1 - \mathrm{Tr}[\mathrm{Tr}_A[|\psi_k\rangle\langle\psi_k|]^2])}$ is the concurrence for the pure state $|\psi_k\rangle$. Since $\rho_{AB}$ can be decomposed in many ways, $C(\rho_{AB})$ is defined as the realizable minimum value. Hereafter, we adopt an approximate value for $C(\rho_{AB})$ with its lower bound given by K. Chen et al. [33,34], instead of an exact $C(\rho_{AB})$ because the minimization procedure is too difficult. Then,

$$C(\rho_{AB}) \geq \sqrt{2/M(M-1)}\left(\max\{||\rho_{AB}^{T_A}||, ||R(\rho_{AB})||\} - 1\right).$$

where $\rho_{AB}^{T_A}$ is the transposed $\rho_{AB}$ about the indices with A, and $R(\rho_{AB})$ is the transposed $\rho_{AB}$ with both A and B in a cross exchange manner. $||G|| = \mathrm{Tr}[(GG^\dagger)^{\frac{1}{2}}]$ is the trace norm, and we set $M \leq N$.

## 4. Time-evolution during Qubit measurement with a JBA – Physical and Informatics Dynamics –

We carried out numerical calculations [13]. We set $\delta = 0.007\Omega$, blacknonlinearity $\alpha = 8 \times 10^{-5}\Omega$, blackquality factor $Q = 2500$, and blackthe driving force $f$ is operated as $0 \to 0.025\Omega \to 0$. These parameters are similar to those used in actual experiments [12]. To emphasize quantumness, however, $\delta$ and $\Gamma$ are a factor of $10^{-2}$ times smaller than in real cases. More than one hundred basis states are prepared in order to avoid numerical artifacts. Then we can reproduce hysteretic bifurcation phenomena of our JBA.

We show the dynamics during the qubit measurement process in Fig. 3. The initial state is a separable state; $(\frac{1}{\sqrt{2}}|g\rangle_q + \frac{1}{\sqrt{2}}|e\rangle_q) \otimes |G\rangle_J$. Here, $|g\rangle_q$ and $|e\rangle_q$ are the ground and excited states of the qubit, respectively. The qubit parameters are $\varepsilon = 0.2\Omega$, $\Delta/\varepsilon = 1/2$. The coupling between the qubit and the JBA is set at $k = 0.001\Omega$. The driving force $f$ is increased from 0 to $0.012\Omega$ ( slightly larger than the $f_c$ of the JBA) and maintained (see Fig. 4).

The quasi-distribution representations of the JBA state $\mathrm{Tr}_q[\rho]$ are shown in Fig. 3, where $\rho$ is the density operator of the qubit-JBA coupled system, and $\mathrm{Tr}_q[\cdots]$ denotes taking partial trace about qubit degrees of freedom. These peaks constitute an incoherent mixture, so they correspond to two possibilities in the measurement result.

This process can be expressed schematically as

$$\rho(0) = |G\rangle_{JJ}\langle G| \otimes \left(\frac{1}{\sqrt{2}}|g\rangle_q + \frac{1}{\sqrt{2}}|e\rangle_q\right)\left(\frac{1}{\sqrt{2}}{}_q\langle g| + \frac{1}{\sqrt{2}}{}_q\langle e|\right)$$

$$\to \rho(\tau_1) = \frac{1}{2}\left(|G\rangle_J|e\rangle_q + |G'\rangle_J|g\rangle_q\right)\left({}_J\langle G|{}_q\langle e| +{}_J \langle G'|{}_q\langle g|\right)$$

$$\to \rho(\tau_2) = \frac{1}{2}|G\rangle_J|e\rangle_{qJ}\langle G|{}_q\langle e| + \frac{1}{2}|G'\rangle_J|g\rangle_{qJ}\langle G'|{}_q\langle g|$$

$$\to \rho(\tau_3) = \frac{1}{2}|G\rangle_J|e\rangle_{qJ}\langle G|{}_q\langle e| + \frac{1}{2}|E\rangle_J|g\rangle_{qJ}\langle E|{}_q\langle g|. \tag{4}$$

We show in Fig. 5, the time variations of some of the information quantities introduced in the previous section. The horizontal line corresponds to the time from the onset of the driving, normalized by the linear resonant

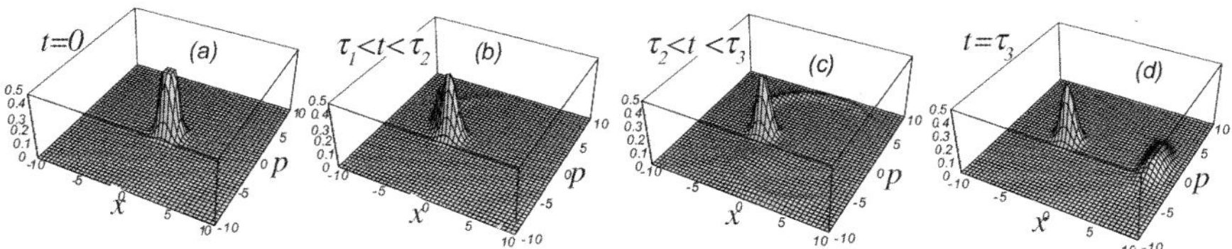

Figure 3. The figures show quasi-distribution representations ($\langle\alpha|\mathrm{Tr}_\mathrm{q}[\rho]|\alpha\rangle$) of the JBA oscillator states, for several times during the measurement process. Here $|\alpha\rangle$ is the coherent state of a complex amplitude $\alpha = x + ip$ (in the rotating frame). (a) Beginning of the measurement. The state of the total system is (schematically) $\rho = |\psi_0\rangle\langle\psi_0|$, with $|\psi_0\rangle = 1/\sqrt{2}(|g\rangle_q + |e\rangle_q)\otimes|G\rangle_\mathrm{J}$. (b) Starting the transition. Entanglement formation and projection are carried out during this period. (c) During the transition. Entanglement has already been destroyed. (d) The entire system has become a mixture of classically correlated states.

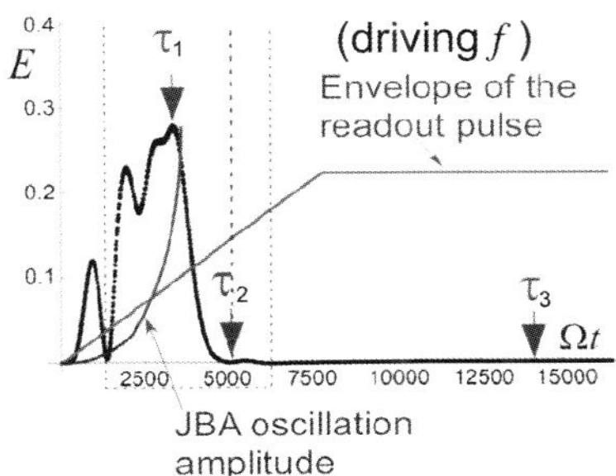


Figure 4. Theoretical calculation of JBA behavior. The horizontal axis is time normalized by the JBA linear resonance frequency $\Omega$. Time $\tau_i (i = 1, 2, 3)$ in the figure are the same as those in Eq. (4). A light line shows the time variation of the driving strength (the amplitude of the readout pulse). Another light line corresponds to the oscillation amplitude of JBA induced by the driving. The black line is the approximate strength of the entanglement formed between the qubit and the JBA. Precise entanglement measure (concurrence) is shown in Fig. 5 (b).

frequency $\Omega$, the vertical line is in bit or qubit units. The concurrence $C(\rho_{AB})$ grows as the driving strength increases. $I_\mathrm{Q}$ and $J_\mathrm{Q}$ behave in the same manner. $J_\mathrm{Q}$ is kept at almost unity after the concurrence $C$ vanishes. If we measure the JBA state classically within this time span, we can obtain maximum information about the qubit state. After $\tau_2$, the quantum discord ($I_\mathrm{Q} - J_\mathrm{Q}$) disappears. That is, after $\tau_2$, a measurement on A or B never destroys the correlation between A and B.

444

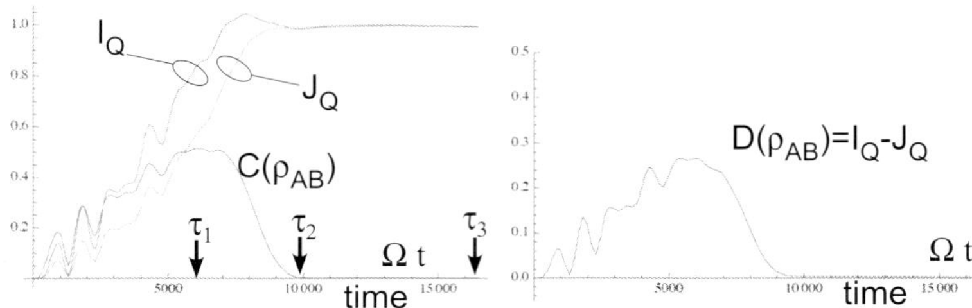


Figure 5. Time evolution of information quantities during the qubit measurement process. $I_Q$: Quantum mutual information. $J_Q$: Classical mututal information. $C$: Concurrence. $D$: Quantum discord

This time evolution behavior in information quantities shows that the correlation initially invoked as a quantum correlation is converted into a classical correlation during the process. Then, then we can obtain information about the projected qubit via a classical measurement on the JBA. Moreover, at time $t = \tau_3$, the required measurement merely involves distinguishing two peaks in Fig. 3 $(t = \tau_3)$. This is a very easy classical measurement. This situation is established via the separation of the two peaks during the time from $\tau_2$ to $\tau_3$. The entanglement has already been destroyed at $\tau_2$. Therefore, this time variation corresponds to classical information amplification.

## 5. Conclusion

A quantum mechanical measurement is information conversion from quantum one to classical one. As an example, we analyzed the Josephson bifurcation amplifier (JBA) measurement process of a superconducting qubit. We theoretically examined the dynamics of the density operator of a detector (a driven nonlinear oscillator) and a qubit coupled system during the measurement process. When the initial state of the qubit is a superposition state, in other words, the qubit contains quantum information initially, we observe the formation of a qubit-detector (JBA) entangled state and it is divided into two separable states at the moment the JBA transition begins. The resulting JBA state shows the measurement result, which is classical information. We also discuss the process from the viewpoint of time variations in information quantities, such as mutual information between the qubit and detector. This clearly shows that the measurement process corresponds to an information conversion from quantum to classical one.

## Acknowledgment

We thank the members of superconducting qubit experiment group in NTT Basic Research Laboratories for fruitful discussions. This work is partially supported by MEXT-KAKENHI, ans JSPS through FIRST program.

## References


1. M. Nielsen and I. Chuang,'*Quantum Computation and Quantum Information*', (Cambridge University Press, Cambridge, 2000).
2. J.Von Neumann, *Mathematical Foundations of Quantum Mechanics* (Princeton University Press, NewYork, 1955).
3. R. Raussendorf, H. J. Briegel, Phys. Rev. Lett. **86,** 5188 (2001).
4. *Quantum Theory and Measurement*, edited by J.A. Wheeler and W. H. Zurek, Princeton Series in Physics (Princeton University Press, Princeton, New Jersey, 1983).
5. W.H.Zurek, Phys. Today 44(10), 36 (1990); revised version : '*Decoherence and the transition from quantum to classical-revisited*', Los Alamos Science, Number 27, 2002 (arxiv : quant-ph/0306072).
6. M. Schlosshauer, Rev. Mod. Phys., **76**, 1267 (2004).
7. '*Decoherence and the appearance of a classical world in quantum theory*', edited by D. Giulini *et al.* (Springer-Verlag, Berlin Heidelberg, 1996).
8. K. Kraus, 'States, Effects, and Operations', Lecture notes in Physics No. 190, (Springer-Verlag, Berlin Heidelberg, 1983).
9. D. F. Walls and G. J. Milburn, 'Quantum Optics' 2nd edition (Springer, Belrin Heigelberg, 2008).
10. H. M. Wiseman and G. J. Milburn, 'Quantum Measurement and Control' (Cambridge Univ. Press, Cambridge, 2010).
11. J. P. Paz and W. H. Zurek, in '*Coherent atomic matter waves*', edited by R. Kaiser, C. Westbrook, and F. David, (Springer-Verlag, Berlin Heidelberg, 2001).
12. I. Siddiqi *et al. Phys. Rev. Lett.* **93** 207002 (2004); I. Siddiqi *et al. ibid.* **94** 027005 (2005); A. Lupascu *et al. ibid.* **96** 127003 (2006); I. Siddiqi *et al. Phys. Rev.* **B 73** 054510 (2006); A. Lupascu *et al. Nature Physics* **3** 119 (2007); Boulant N *et al. Phys. Rev.* **B 76** 014525 (2007).
13. H. Nakano, S. Saito, K. Semba, H. Takayanagi, *Phys. Rev. Lett.* **102** 257003 (2009).
14. J. Mooij *et al. Science* **285**, 1036 (1999 ).
15. L. Tian, S. Lloyd, and T. Orlando, *Phys. Rev.* **B 65** 144516 (2002).
16. C. van der Wal *et al. Science* **290** 773 (2000); C. van der Wal, *Quantum superpositions of persistent Josephson currents* Ph.D. thesis, (Delft: Delft Univ. Press, 2001).
17. I. Chiorescu, Y. Nakamura, C. Harmans, and J. Mooij, *Nature* **431** 159 (2004).
18. I. Chiorescu, P. Bertet, K. Semba, Y. Nakamura, C. Harmans, and J. Mooij, *Science* **299** 1869 (2003).

19. P. Bertet *et al.* cond-mat/0412485 (2004).

20. D. Crankshaw and T. Orlando, *IEEE Trans. Appl. Superconductivity* **11** 1006 (2001); J. You, Y. Nakamura, and F. Nori, *cond- mat/0309491* (2003); A. van Brink, *cond- mat/0310425* (2003).

21. C. van der Wal, F. Wilhelm, C. Harmans, J. E. Mooij, *Eur. Phys. J.* **B 31** 111 (2003).

22. T. Orlando *et al. Phys. Rev.* **B 60** 15398 (1999).

23. H. Nakano and H. Takayanagi, *J. Phys Soc. Jpn.* **72** Suppl. **A** 1 (2003); H. Nakano *et al. arXiv:cond-mat/0406622* (2004); H. Nakano and H. Takayanagi, in *Toward the Controllable Quantum States* (Singapore :World Scientific) p 359 (2003).

24. For review, K. Semba, K. Johansson, K. Kakuyanagi, H. Nakano, S. Saito, H. Tanaka and H. Takayanagi, *Quantum Information Processing*: **8, Issue 2** 199 (2009).

25. H. Tanaka, Y. Sekine, S. Saito, H. Takayanagi, *Physica* **C 368** 300 (2002); H. Takayanagi, H. Tanaka, S. Saito, and H. Nakano, *Physica Scripta* **T102** 95 (2002); H. Tanaka, H. Saito, and H. Takayanagi, *Toward the Controllable Quantum States* (Singapore: World Scientific) p 366 (2003).

26. W. Jordan and P. Smith, *Nonlinear Ordinary Differential Equations* third edition (Oxford: Oxford Univ. Press,1999).

27. M. Dykman and M. Fistul, *Phys. Rev.* **B 71** 140508 (R) (2005); M. Dykman, *Phys. Rev.* **E 75** 011101 (2007); V. Peano and M. Thorwart, *Chem. Phys.* **322** 135 (2006); V. Peano and M. Thorwart, *New J. Phys.* **8** 21 (2006).

28. Y. Makhlin, A. Shnirman, and G. Schön, *Rev. Mod. Phys.* **73** 357 (2001).

29. G. Lindblad, *Commun. Math. Phys.* **48** 119 (1976).

30. B. Schumacher and Westmoreland, *Quantum Processes, Systems & Information* (Cambridge: Cambridge University Press, 2010).

31. H. Ollivier and W. Zurek, 2001 *Phys. Rev. Lett.* **88** 017901 82010).

32. A. Uhlmann, *Phys. Rev.* **A 62** 32307 (2000).

33. R. Horodecki, P. Horodecki, M. Horodecki, and K. Horodecki, *Rev. Mod. Phys.* **81** 865 (2009).

34. K. Chen, S. Albeverio, and S-M. Fei *et al. Phys. Rev. Lett.* **95** 40504 (2005); K. Chen, S. Albeverio, and S-M. Fei *et al. Phys. Rev. Lett.* **95** 210501 (2005).

Quantum Bio-Informatics V

© 2013 World Scientific Publishing Co. Pte. Ltd.

pp. 447–461

# HOW CAN STEGANOGRAPHY BE AN INTERPRETATION OF THE REDUNDANCY IN PRE–MRNA RIBBON?

MASSIMO REGOLI

*Centro Vito Volterra,*
*Università degli Studi di Roma "Tor Vergata", Roma, Italy,*
*E-mail: regoli@uniroma2.it*

In the past years we have developed a new symmetric encryption algorithm based on a new interpretation of the biological phenomenon of the presence of redundant sequences inside pre–mRNA (the introns apparently junk DNA) from a 'science of information' point of view. For the first, we have shown the flow of the algorithm by creating a parallel between the various biological aspects of the phenomenon of redundancy and the corresponding agents in our encryption algorithm. Then we set a strict mathematical terminology identifying spaces and mathematical operators for the correct application and interpretation of the algorithm. Finally, last year, we proved that our algorithm has excellent statistics behavior being able to exceed the standard static tests. This year we will try to add a new operator (agent) that is capable of allowing the introduction of a mechanisms like a steganographic sub message (sub ribbon of mRNA) inside the original message (mRNA ribbon).

## 1. Introduction

The RNA-Crypto System (shortly RCS) is a symmetric key algorithm to cipher data. This algorithm, as shown below, has the peculiarity to expand the message to be encrypted hiding the ciphered message itself within a set of *garbage* and *control information.*

The idea for this new algorithm was originated from the observation of nature. In particular from the observation of RNA behavior and some of its properties.

The RNA sequences include some sections called *Introns* (the name is derived from the term "intragenic regions"). These are non-coding sections of the precursor mRNA (*pre–mRNA*) or other RNAs, that are removed (spliced out of the RNA) before the mature RNA is formed. Once the introns have been spliced out of a *pre–mRNA*, the resulting mRNA sequence is ready to be translated into a protein. The corresponding parts of a gene are known as *introns* as well.

The nature and the role of Introns in the *pre–m*RNA is not clear and is under intensive researches by Biologists. The algorithm described below introduces a mathematical analogue of the Intron sequences in the RNA-Crypto System output as a device to add information which is only apparently chaotic and non coding, but in fact plays an essential role in the decodification of the message.

A general conclusion that can be drawn from the present paper is that, while the use of cryptoanalytic tools in the attempt to decode genetic sequences has a long history, the converse direction, i.e. the exploitation of biological ideas to construct cryptographical algorithms, may produce new fruitful tools in cryptography as well as nontrivial suggestions in the direction of an information theoretical interpretation of biological phenomena such as the *redundancy* in DNA sequences.

This direction will be the object of future investigations.

In the next section we shortly describe the mathematical structure of the new cryptographic algorithm.

The main goal of the present paper is to use the algorithm described below and in some other papers (see [3], [4], [5]) to give a *steganographic* interpretation to the role of introns in the DNA (see below for a definition of steganography).

## 2. The algorithm in brief

### 2.1. *The spaces*

The algorithm uses a multiplicity of sets $X$, which play different roles and which a prior can be different, namely:

- the key alphabet $K$
- the message alphabet $M$
- the *exons* alphabet $A = B_{Ex}$
- the *introns* alphabet $B = B_{Introns}$
- the *control* space

$$S = S_1 \cup S_2$$

where $S_1, S_2 \subset S$ are non–empty subsets satisfying

$$S_1 \cap S_2 = \emptyset \quad ; \quad |S_j| :=< \infty , \ j = 1, 2$$

and, in most concrete cases, we shall choose:

$$K = M = \{0, 1\} \tag{1}$$

Sets of finite ordered sequences with values in one of these spaces will play an important role in the algorithm:

- the space of *Secret Shared Keys* (SSK)

$$\mathcal{K} \subseteq K_0^{\mathbb{N}} := \text{finite ordered sequences of symbols in } K$$

- the space of *Messages*

$$\mathcal{M} \subseteq M_0^{\mathbb{N}} := \text{finite ordered sequences of symbols in } M$$

- the *Exons* space

$$\mathcal{A} = A^{N_A} = \mathcal{B}_{Ex} \subseteq (B_{Ex})_0^{\mathbb{N}} := \text{finite ordered sequences of symbols}$$
$$\text{in } B_{Ex}$$

- the *Introns* space

$$\mathcal{B} = \mathcal{B}_{In} \subseteq (B_{In})_0^{\mathbb{N}} := \text{finite ordered sequences of symbols in } B_{In}$$

- The union of the Exons and Introns spaces defines the output alphabet:

$$B := B_{Ex} \cup B_{In} \text{ (disjoint union)}$$

- the space of *Coded Messages* (or *Output sequences*)

$$\mathcal{C} \subseteq (\mathcal{B}_{Ex} \cup \mathcal{B}_{In})_0^{\mathbb{N}} := \text{finite ordered sequences of symbols in the}$$
$$\text{output alphabet}$$

With the choice (1) the spaces $\mathcal{K}$ and $\mathcal{M}$ become simply finite ordered sequences of standard binary digits and the coded messages will be finite ordered sequences of blocks, representing Exons or Introns.

## 2.2. *Functions*

The algorithm uses several classes of functions.

**Coding functions**

Their role is twofold:

(1) [i)]
(2) to transform a portion of the clear message into a portion of the coded message (*exons*)

(3) to insert some *apparently* redundant information in the coded message (*introns*).

**Operational functions**

We have already said that all functions considered are local. The operational functions specify which blocks of the key, or of other control sequences, have to be used at each step of the algorithm. These functions are parametrized by the output (i.e. coded) sequences

$$\bar{g}_\kappa : \Sigma \in \mathcal{B}_{Ex} \cup \mathcal{B}_{In} \to \bar{g}_\kappa(\Sigma) =: \bar{g}(\kappa, \Sigma) \in \mathcal{K} \tag{2}$$

Intuitively this function indicates which part of the SSK has to be used in the next step of the algorithm as a function of the output in the previous step.

Another interpretation is that the function $\bar{g}$ produces, at each step of the algorithm, the SSK that has to be used in the next step (for this reason some *chaoticity* properties are desirable).

Another operational function used in the algorithm is the characteristic function $\chi_{S_1} : S \to \{0, 1\}$ $(\chi_{S_1}(s) = 1 \Leftrightarrow s \in S_1)$

### 2.3. *Further notations*

Let us emphasize:

(1) [i)]
(2) that the algorithm can produce different coded messages, corresponding to the same message $\sigma$ and the same SSK, simply by using different control sequences $\{\alpha\}$
(3) that it is not necessary, during the decoding phase, to know the sequence $\{\alpha\}$ used in the coding phase. This is because the information about the $\{\alpha\}$ sequence can be deduced from the pair $(\Sigma, \kappa)$ (i.e. coded message and SSK: see formula (2) for a precise formulation of this fact).
(4) the possible role of *introns* to carry important information for the decoding phase, for example the function $\bar{g}(\kappa, \Sigma)$ can use the information in $\Sigma \in \mathcal{B}$ to produces a change in the SSK.

Finally, during the coding phase, there is an expansion of the original messages into the coded message depending on several factors such as the dimensions of the intron sequences and the number of elements of $S_2$ in the control sequence $\{\alpha\}$.

## 2.4. *Biological parallelism*

In the biological interpretation the combined action of the SSK and of the control sequence $\{\alpha\}$ mimic the mechanism of *splicing* through which the *pre–mRNA* is modified by removing certain stretches of non-coding sequences (*introns*), while the coded message is the *pre–mRNA* itself and the clear message is the final *mRNA*.

## 3. The algorithm

### 3.1. *Coding*

The coding operation requires:

- a Pre Shared Key (SSK) $\kappa \in \mathcal{K}$, of length $l_\kappa = N_K$
- a message $\sigma \in \mathcal{M}$ of length $l_\sigma = N_M$
- a pseudo–random $S$–valued sequence $\alpha = (\alpha_i)\ (\alpha_i \in S)$
- two fixed pre shared numbers $n, m \in \mathbb{N}$ such that

$$n \mid l_\sigma \ (n \text{ divides } l_\sigma = N_M)$$

- a finite set of coding functions

$$f_\alpha \ \alpha \in S_1 \cup S_2$$

- a finite set of operational functions

$$g_\alpha \ \alpha \in S_1 \cup S_2$$

Define the following objects:
– the $i$–th bit of the output

$$\Sigma_i := f_{\alpha_i}(\tilde{\sigma}_{l_i}, \tilde{\kappa}_{j_i}) := \begin{cases} \varepsilon_i \in \mathcal{A} & \text{if } \alpha_i \in S_1 \\ \iota_i \in \mathcal{B} & \text{if } \alpha_i \in S_2 \end{cases} \tag{3}$$

where:

- $\tilde{\sigma}_{l_i}$ and $\tilde{\kappa}_{j_i}$ are defined as follows:

  **Definition 1.** For $x \in X_0^{\mathbb{N}}$, $m \in \mathbb{N}$ and $i \in \{1, \ldots, \ell_x\}$ we define the subsequence:

$$\tilde{x}_{i,m} := \left\{ x_i, x_{i+\ell_x 1}, \ldots, x_{i+\ell(m-1)} \right\} \tag{4}$$

or simply:

$$\tilde{x}_i := \left\{ x_i, x_{i+\ell_x 1}, \ldots, x_{i+\ell(m-1)} \right\} \tag{5}$$

if $m$ is fixed.

- $\varepsilon_i$ is the $i$–th bit of the message
- $\iota_i$ is an arbitrary intron
- $l_{i+1}$, i.e. the index of the next bit of the message to be processed, is given by:

$$l_{i+1} = l_i + \chi_{S_1}(\alpha_i)n$$

$$\tilde{\kappa}_{j_{i+1}} = g_{\alpha_i}(\kappa, \Sigma_i)$$

**Definition 2.** The sequence $\Sigma = (\Sigma_1, \ldots, \Sigma_N) \in \mathcal{C}_\mathcal{B}$ is the coded message.

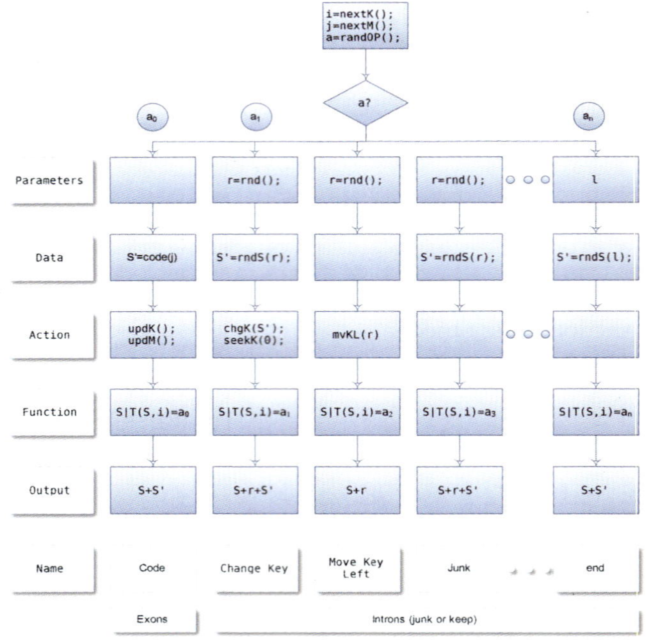


Figure 1.   Coding Schema

## 3.2. *Decoding*

The crucial point for the definition of the function $\bar{f}$, which is at the basis of the decoding procedure, is that, given the output sequence $\Sigma$ (i.e. the entire coded message), which is a sequence of 0's and 1's, $B$ is able to distinguish the single blocks $\Sigma_i$. In particular he can recognize if $\Sigma_i \in \mathcal{A}$ or $\Sigma_i \in \mathcal{B}$.

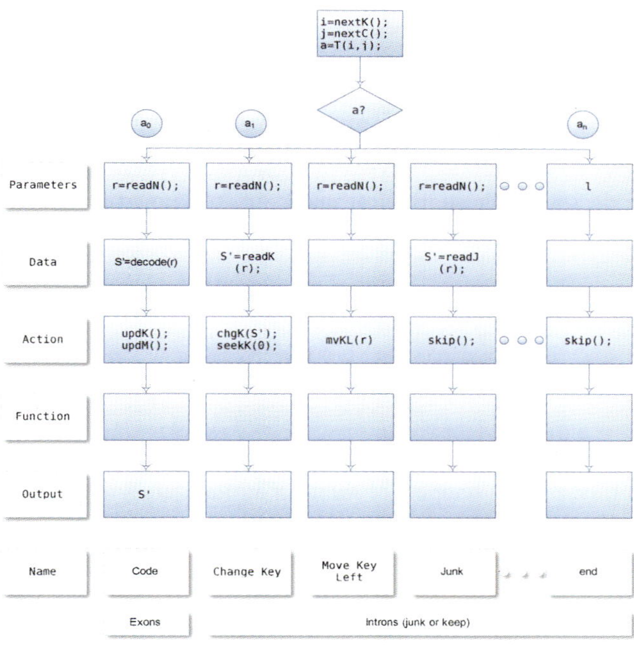


Figure 2.   Decoding Schema

The decoding operation requires:
- an SSK $\kappa \in \mathcal{K}$, of length $l_\kappa = N_K$
- the same fixed pre shared numbers $n, m \in \mathbb{N}$ as in the coding operation.
- a coded message $\Sigma = (\Sigma_1, \ldots, \Sigma_N) \in \mathcal{C}_\mathcal{B}$ of length $l_\Sigma$

Using the above objects and the knowledge of the single blocks $(\Sigma_1, \ldots, \Sigma_N)$ with $\Sigma_i \in \mathcal{A} \cup \mathcal{B}$, the decoding phase computes the $i$–th block of the message $\tilde{\sigma}_i$ and the $j_{i+1}$–th block of the SSK $\tilde{\kappa}_{j_{i+1}}$, to be used in the next step,

by distinguishing two cases:

**Case** $\mathcal{A}$: if $\Sigma_i \in \mathcal{A}$, then the $i$–th block of the message is:

$$\tilde{\sigma}_i = \bar{f}(\Sigma_i, \tilde{\kappa}_{j_i}) \tag{6}$$

and the $j_{i+1}$–th block of the SSK $\tilde{\kappa}_{j_{i+1}}$ is:

$$\tilde{\kappa}_{j_{i+1}} = \bar{g}(\kappa, \Sigma_i)$$

**Case** $\mathcal{B}$: if $\Sigma_i \in \mathcal{B}$, then the $i$–th block of the message is:

$$\bar{f}(\Sigma_i, \tilde{\kappa}_{j_i}) = \emptyset$$

and the $j_{i+1}$–th block of the SSK $\tilde{\kappa}_{j_{i+1}}$ is:

$$\tilde{\kappa}_{j_{i+1}} = \bar{g}(\kappa, \Sigma_i)$$

**Remark.** As mentioned at the end of section (2.2), thanks to the definition of $\bar{g}$, in the decoding phase it is not necessary to know the random sequence $\{\alpha_i\}$.

### 3.3. *More details*

Is is possible to find more details reading [3], [4], [5].

## 4. Steganography

Steganography is the science of writing hidden messages in such a way that no one, apart from the sender and intended recipient, suspects the existence of the message, a form of security through obscurity. The word steganography is of Greek origin and means "concealed writing" from the Greek words steganos ($\sigma\tau\epsilon\gamma\alpha\nu o\varsigma$) meaning "covered or protected", and graphein ($\gamma\rho\alpha\phi\epsilon\iota\nu$) meaning "to write". The first recorded use of the term was in 1499 by Johannes Trithemius in his Steganographia [6], a treatise on cryptography and steganography disguised as a book on magic. Generally, messages will appear to be something else: images, articles, shopping lists, or some other covertext and, classically, the hidden message may be in invisible ink between the visible lines of a private letter.

### 4.1. *Ancient steganographic techniques*

Herodotus tells the story of a Persian nobleman who had cut the hair of a trusted slave, in order to tattoo a message on his skull, and once the hair had grown back, he sent a slave to its destination, the shared secret key is 'cut the hair again'.

## 4.2. *Steganography in nature*

Pre-mRNAs from the D. melanogaster gene dsx contain 6 exons. In males, exons 1,2,3,5 and 6 are joined to form the mRNA, which encodes a transcriptional regulatory protein required for male development. In females, exons 1,2,3 and 4 are joined, and a polyadenylation signal in exon 4 causes cleavage of the mRNA at that point. The resulting mRNA is a transcriptional regulatory protein required for female development.

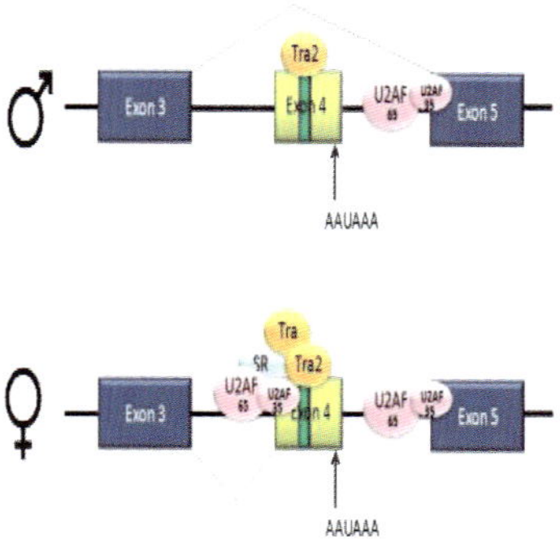


Figure 3. Different interpretation of a portion of Melanogaster pre–mRNA (image from Allen Gathman)

In this example, exon 4 is an hidden (from a steganographic point of view) information for male individuals and exons 5 and 6 are the same for female individuals.

## 4.3. *Insert a steganographic function*

Our idea is to allow that a part of the redundancy, included in the encoded message, can contain a second hidden message to be decoded according to fixed rules or events.

### 4.3.1. *Modification of the Cryptographic protocol*

Here we introduce a new coding function $\varsigma$ (the steganographic function) that will artificially add a hidden message inside the code.

456

### 4.3.2. *Strategies*

The *steganographic* function

**Definition 4.1.** Be $\varsigma : \mathcal{M} \times \mathcal{K} \to \mathcal{C}$ a *coding function*:

$$\varsigma\left(\sigma', \kappa\right) = \Sigma = \iota \tag{7}$$

where $\mathcal{M} \ni \iota$ is the encoded version of $\sigma'$ and $\sigma'$ is a message to hide encoded in the main coded message..

This message, $\sigma'$, can be encoded by continuing the same encryption key $\kappa$ used for the rest of the message or using a different portion of it or using a new different key $\kappa'$ (in this case we must create a selection protocol for the new key).

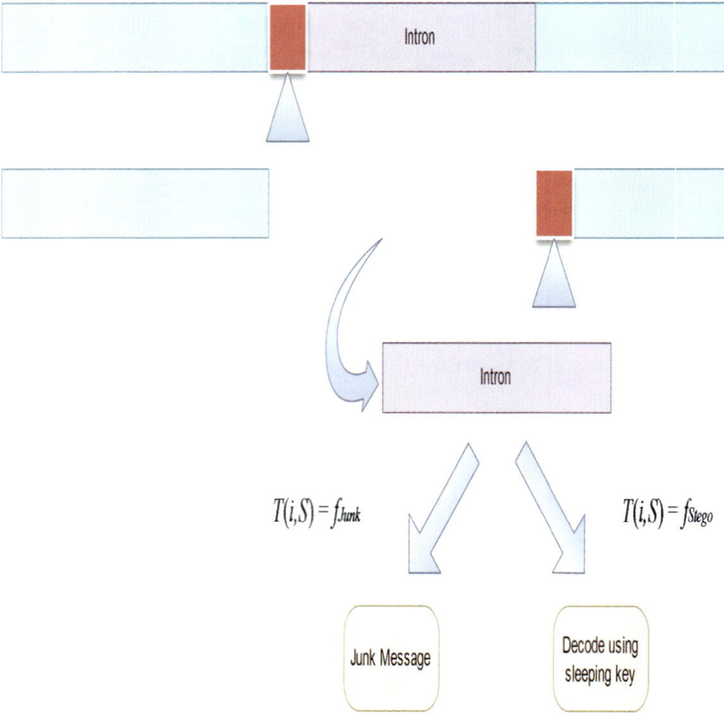


Figure 4.   Different approach to steganograhic message

## 4.4. *Example*

Suppose we want to cipher two messages $\sigma$ and $\sigma'$. For simplicity suppose also to have two shared keys $\kappa$ and $\kappa'$. A possible algorithm to implement the steganographic function, can act as a switch according to the following:

Suppose that at step $i$ during the encoding phase of the message $\sigma$ the system chooses to use the steganographic function and suppose also that the first $n'$ element of the message $\sigma'$ have already been processed previously then:

- If $n'$ is equal to the length of the message $\sigma'$

    - discard the step

- Else

    - Choose a random number $l'$ between 1 and the length of the message $\sigma'$ that remain to encode.
    - write this information on the output sequence
    - Encode the subsequence $\sigma'_{n'+1,l'}$ (here the symbol $*_{n,l}$ stands for a subsequence of $*$ which starts from $n$ and has a length of $l$) using the relative key $\kappa'$
    - write the result on the output sequences

Obviously, the decoding function must:

Assuming that in step $i$, during the decoding phase of the coded message $\Sigma$, the system encounters the steganographic function and assuming also that the first $n'$ element of the message $\sigma'$ have already been processed previously then:

- read from the encoded message the number $l'$.
- decode the rest of the coded message until $l'$ elements are extracted from the message using the alternative key $\kappa'$.
- add the result on the output sequences $\sigma'$

This shows how it is possible to create an easy implementation of the steganographic function inside the RNA-Crypto System algorithm.

It is also possible to create a system in which two different parties react differently deciphering a steganographic message:

- The first one decipher the second message using a *sleeping* key
- The second one discard this part of the message considering it as junk

| | | $S$ | | | | |
|---|---|---|---|---|---|---|
| | | $i_0$ | $i_1$ | $i_2$ | $\cdots$ | $i_n$ |
| $5*i$ | $s_0$ | 0 | $f_1$ | junk | ... | $f_5$ |
| | $s_1$ | 1 | 0 | $f_3$ | ... | $\varsigma$ |
| | $s_2$ | $\varsigma$ | 1 | 0 | ... | $f_4$ |
| | $\vdots$ | | | ... | | |
| | $s_m$ | junk | $\varsigma$ | $f_3$ | ... | $f_1$ |

| | | $S$ | | | | |
|---|---|---|---|---|---|---|
| | | $i_0$ | $i_1$ | $i_2$ | $\cdots$ | $i_n$ |
| $5*i$ | $s_0$ | 0 | $f_1$ | junk | ... | $f_5$ |
| | $s_1$ | 1 | 0 | $f_3$ | ... | *junk* |
| | $s_2$ | *junk* | 1 | 0 | ... | $f_4$ |
| | $\vdots$ | | | ... | | |
| | $s_m$ | junk | *junk* | $f_3$ | ... | $f_1$ |

The differences in tables 1 and 2 (for a full definition of seek table see
[5]) show that in the first case the decoder will treat the next block as a
steganographic message (label $\varsigma$) and in the second case the decoder will
treat the same block as a junk-message (label *junk*).

## 5. Biological interpretation

In recent years, many researches have been born around the phenomenon
of so-called "junk DNA".

Many researchers have put forward various hypotheses about the real
role of *introns* both in the coding phase is and during the production of
amino acids.

In this paper we further advance the hypothesis of a mechanism that
can be activated through a different interpretation of the *introns* trough
the splicing phase, that should be able, therefore, to provide information
(amino acids), when requested, using a different mechanism of splicing, as
an example.

In the cryptographic algorithm this is represented by two tables of cod-
ing/decoding functions:

- in one case completely ignores a series of bits in the message

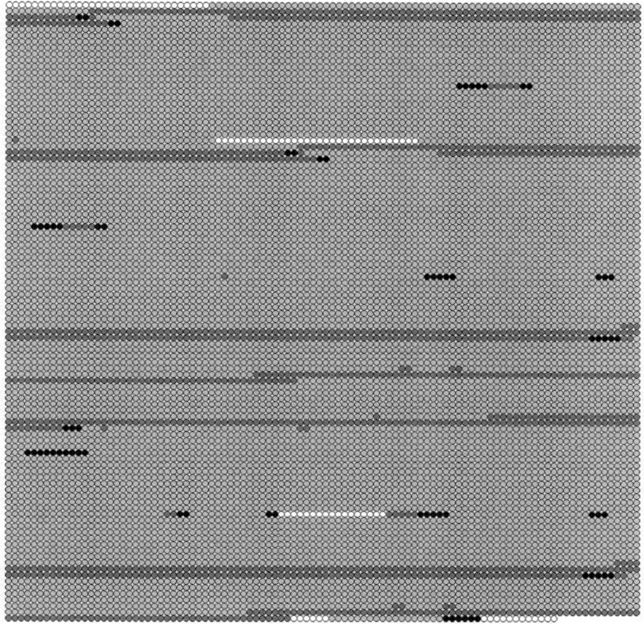

Figure 5. Here the grayscale image for a qualified steganographic decoder (in white the start/stop message, black are Exons and gray levels different kind of Introns (junk or active))

- in other case, the system reads the sequence, and decrypts the message using a special 'sleeping' key

## 6. Conclusions

In the years, evolutionary biologists have told that the bulk of genomes is junk and that this is due to the sloppiness of the evolutionary process.

That is now changing. For instance, Amy Pasquinelli at UCSD, in commenting on long stretches of seemingly non–coding DNA sequences, asks us to "reconsider the contents of such junk DNA sequences in the light of recent reports that a new class of non-coding RNA genes are scattered, perhaps densely, throughout these animal genomes." ("MicroRNAs: Deviants no Longer." Trends in Genetics 18(4) (4 April 2002): 171-3.) ID (Information Detection) theorists should be at the forefront in unpacking the information contained within biological systems. If these systems are designed, we can expect the information to be densely packed and multi-

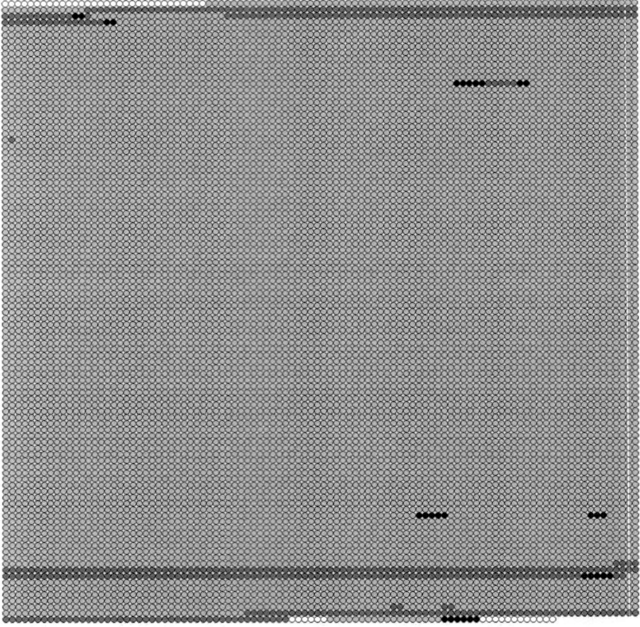

Figure 6. Here the grayscale image for an uqualified steganographic decoder (in white the start/stop message, black are Exons and gray levels different kind of Introns (junk or active))

layered (save where natural forces have attenuated the information). Dense, multi-layered embedding of information is a prediction of ID.

Our approach shows that, also in our cryptographic protocol, it is possible to consider our *junk–DNA* as a *sleeping–DNA* capable of being activated by the need to obtain information or additional or alternative from a coded message.

It is important to underline some facts:

- It is possible to insert inside a RNA-Crypto System  coded message an arbitrary number of sub–messages

  · This in order to confound the spy or to create a hierarchy of messages

- It is NOT possible to detect the amount of sub-messages

  · Of course, it is possible to use an attack looking at the length of the coded message but you cannot affirm if it is one long message or many short messages.

- Furthermore it is NOT possible to locate, in the main coded message, the position of the sub–messages
  - · If you are able to decode the first message (the *dummy* message) you cannot affirm if the 'junk' components of the coded message are real junk or they contain some inner levels of messages

## References


1. Lewin, B.,"Gene VI" *Oxford University Press*, 1997.
2. Menezes, A., van Oorschot, P. and Vanstone, S., "Handbook of Applied Cryptography", CRC Press, 1996
3. Regoli, M., "Bio-Cryptography: A Possible Coding Role for RNA Redundancy", Foundations of Probability and Physics - 5, Vaxjo, Sweden, 24-27 August 2008, Series: AIP Conference Proceedings, Vol. 1101, Accardi, L.; Adenier, G.; Khrennikov, A.Y.; Fuchs, C.; Jaeger, G.; Larsson, J.-A.; Stenholm, S. (Eds.), 2009
4. Regoli, M., "A Simple Symmetric Algorithm Using a Likeness with Introns Behavior in RNA Sequences", QUANTUM BIO-INFORMATICS II From Quantum Information to Bio-Informatics Tokyo University of Science, Japan, 12∼16 March 2008 edited by L. Accardi (Università degli Studi di Roma Vergata, Italy), W. Freudenberg (Brandenburgische Technische Universität Cottbus, Germany), and M. Ohya (Tokyo University of Science, Japan)
5. Regoli, M., "pre-mRNA Introns as a Model for Cryptographic Algorithm: Theory and Experiments", QUANTUM BIO-INFORMATICS III From Quantum Information to Bio-Informatics Tokyo University of Science, Japan, 11∼14 March 2009
6. Trithemius, "Steganographie: Ars per occultam Scripturam animi sui voluntatem absentibus aperiendi certu", 4to, Darmst. 1621. (Written 1500. First printed edition: Frankfurt, 1606.)
7. Amy E. Pasquinelli, MicroRNAs: deviants no longer, TRENDS in Genetics Vol.18 No.4 April 2002

Quantum Bio-Informatics V
© 2013 World Scientific Publishing Co. Pte. Ltd.
pp. 463–472

# COUNTER-FACTUAL PHENOMENON IN QUANTUM MECHANICS

YUTAKA SHIKANO

*Department of Physics, Tokyo Institute of Technology*
*Meguro, Tokyo 152-8551, Japan.*
*Center for Quantum Studies, Schmid College of Science and Technology,*
*Chapman University*
*Orange, CA 92866, USA.*
*E-mail: shikano@th.phys.titech.ac.jp*

Quantum mechanics tells us the wave-particle duality. Therefore, the classical intuitive picture, which is the particle or the wave picture alone cannot be explained. We explain the counter-factual argument as the Hardy paradox. The key point of our explanation is the state-dependent equivalence to be measured by the weak-value measurement.

Keywords: Weak value, Wave-particle duality, Hardy paradox

## 1. Introduction

Quantum mechanics cannot be explained in the classical particle picture only. Also, it cannot be explained in the classical wave picture. From the construction of quantum mechanics from the classical theory, we want to explain non-relativistic quantum mechanics in the classical picture. However, this is prohibited as the wave-particle duality in quantum mechanics. This paradoxical situation is called a "counter-factual" argument. Our ultimate goal is to understand quantum mechanics in the particle picture. As its quantitatively analytical tool, there is a weak value [1,2,3]. The beauty of the weak value is an experimental accessible quantity, see more details in Ref. [4]. While quantum mechanics is based on the probabilistic theory, the explanation by the weak value is focused on the single-shot event due the post-selection. In this paper, we shall explain the quantitative analysis of the counter-factual argument using the Hardy paradox [5].

## 2. Hardy's Paradox

In this section, we analyze the Hardy paradox, which is one of the counterfactual phenomena, by using the weak value.

Let us recall what Hardy's paradox [5] is. The two Mach-Zehnder interferometers for an electron and a positron are combined in such a way that one of the arms, say, $I_e$ of the interferometer for the electron is crossed with one of the arms $I_p$ for the positron. If the electron and the positron meet at the crossing they are supposed to annihilate each other. The setting is illustrated in Fig. 1.

1) Both the electron and positron cannot be in the inner path $I_e$ and $I_p$ at the same time since they would annihilate each other. Therefore, either electron or positron is in the outer path $O_e$ or $O_p$.

2) Let the positron be in the path $O_p$ so that the the electron is not affected by the positron. The Mach-Zehnder interferometer works as usual; the intermediate state is $O_p(I_e + O_e)$ (a shorthand for $|O_p\rangle|\frac{(I_e+O_e)}{\sqrt{2}}\rangle$ which we use in this section for notational simplicity) and therefore the electron registers a click at detector $B_e$, while the positron clicks either the port $B_p$ or $D_p$. Similarly, for the electron in $O_e$, the intermediate state will be $(I_p + O_p)O_e$ and therefore the positron registers a click at detector $B_p$, while the electron clicks either $B_e$ or $D_e$. In either case, only possible cases are: $B_pB_e, D_pB_e, B_pD_e$, but $D_pD_e$ **is not**.

3) However, quantum mechanics tells us that $D_pD_e$ clicks with the probability $1/12$, which has also been verified by experiments. This is because the quantum state after pair annihilation is given by

$$|\psi\rangle = \frac{1}{\sqrt{3}}(|I_pO_e\rangle + |O_pI_e\rangle + |O_pO_e\rangle). \tag{1}$$

Therefore, the quantum state after the second beam splitters,

$$|\psi_2\rangle = \frac{1}{\sqrt{3}}\left(\frac{1}{2}(|B_p\rangle - |D_p\rangle)(|B_e\rangle + |D_e\rangle)\right.$$

$$\left. + \frac{1}{2}(|B_p\rangle + |D_p\rangle)(|B_e\rangle - |D_e\rangle) + \frac{1}{2}(|B_p\rangle + |D_p\rangle)(|B_e\rangle + |D_e\rangle)\right)$$

$$= \frac{1}{2\sqrt{3}}(3|B_pB_e\rangle + |D_pB_e\rangle + |B_pD_e\rangle + |D_pD_e\rangle) \tag{2}$$

to obtain the probability $[1/(2\sqrt{3})]^2 = 1/12$ to simultaneously click $D_pD_e$.

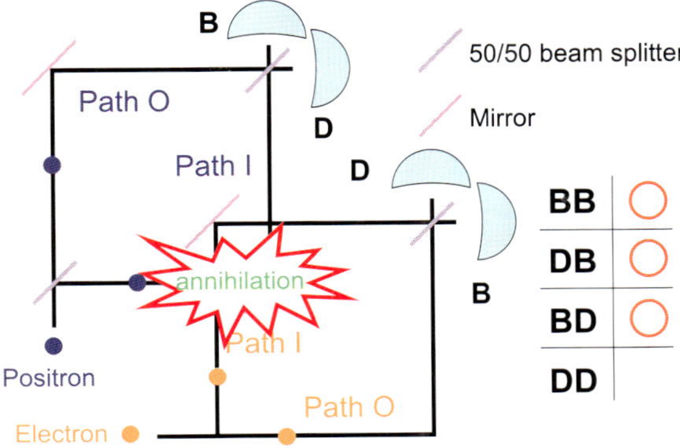



Figure 1.   Hardy's setup: The Hardy paradox is explained in the main text.

The line of reasoning 2) is not a consequence of measurements but an inference by the counter-factual argument [6]. For example, if the positron is found in $O_p$ by measurement, the electron is definitely in $(I_e + O_e)$ ending up with $B_e$, which can be demonstrated as an actual experiment. This is one of definitions on the locality [a] Another different experiment would be the one starting from the fact that the electron found in $O_e$. But one cannot infer anything about the non-measured quantity by combining the two different experiments of non-commuting observables $P[O_p(I_e + O_e)]$ and $P[(I_p + O_p)O_e]$. Here, $P[O_p(I_e + O_e)] := \frac{1}{2}|O_p\rangle(|I_e\rangle + |O_e\rangle)\langle O_p|(\langle I_e| + \langle O_p|)$ is the projection operator. This is based on the Kochen-Specker Theorem [7], that the measured quantity, which corresponds to the eigenvalue for any measured observable, cannot be assigned before the measurement in the case that the dimension of the Hilbert space is more than three. If the theory consists of the commutative observable, this can be assigned *a la* classical intuition. Therefore, this statement comes from quantum measurement for the non-commutative observables. The most obvious way to resolve Hardy's paradox is to dismiss the counter-factual reasoning all together.

---

[a]Rigorously speaking, this is different from the *non-signaling* condition. This is because the definition of the locality is that the operation on a target system is not **simultaneously** affected to the others. However, one of the non-signaling condition is that the operation on the target system is not affected to the others, **which is assumed on the spatially different space to the target system, beyond the light speed.**

## 2.1. *Aharonov et al. Resolution*

Aharonov and his colleagues have long advocated the use of weak values which are experimentally accessible to provide experimental credence to the counter-factual arguments [8] as follows. We set the pre-selected state as Eq. (1) and the post-selected state as

$$|\phi\rangle = |D_p D_e\rangle = V_2 \left[ \frac{1}{2}(|O_p\rangle - |I_p\rangle)(|O_e\rangle - |I_e\rangle) \right], \tag{3}$$

where $V_2$ is the unitary operator on the second beam splitter. We calculate the weak value for the path observables of the electron **or** the positron as

$$\langle P[I_p \otimes Id.]\rangle_w = 1, \tag{4}$$

$$\langle P[Id. \otimes I_e]\rangle_w = 1, \tag{5}$$

$$\langle P[O_p \otimes Id.]\rangle_w = 0, \tag{6}$$

$$\langle P[Id. \otimes O_e]\rangle_w = 0. \tag{7}$$

Furthermore, we calculate the joint-path observables as

$$\langle P[I_p \otimes I_e]\rangle_w = 0, \tag{8}$$

$$\langle P[I_p \otimes O_e]\rangle_w = 1, \tag{9}$$

$$\langle P[O_p \otimes I_e]\rangle_w = 1, \tag{10}$$

$$\langle P[O_p \otimes O_e]\rangle_w = -1. \tag{11}$$

Taking the weak value as the conditional probability in the following equation,

$$|\langle a|\psi\rangle|^2 = \int {}_\phi\langle|a\rangle\langle a|\rangle_\psi^w |\langle\phi|\psi\rangle|^2 d\phi, \tag{12}$$

from Eqs. (4) and (5), the electron or positron goes through the inner path, respectively. However, from Eq. (8), the electron and positron do not *simultaneously* go through the inner path. This is contradiction and is the reason to occur the paradoxical situation. While the weak value outside the range of the spectrum of observable in Eq. (11), from Eqs. (8), (9), (10), and (11), we show that the electron and positron do not simultaneously go through the inner path but otherwise do. Therefore, we obtain the consistent description by the weak value. The crucial reason to occur the paradoxical situation is not to satisfy

$$\langle P[I_p \otimes I_e]\rangle_w = \langle P[I_p \otimes Id.]\rangle_w \langle P[Id. \otimes I_e]\rangle_w \tag{13}$$

except that the pre- and post-selected states are separable states. Therefore, the Hardy paradox tells us the following statement; (i) the detectors

$D_pD_e$ are not correlated, (ii) the second beam splitters are not correlated, or (iii) the pre-selected state is entangled. On the statements (i) and (ii), each equipment is spatially separated and has no interaction. It is difficult to consider these situations. Therefore, from the classical argument, we conclude that the paradoxical situation comes from the pre-selected entangled state. Actually, the joint-path observables have been recently demonstrated in the optical settings [9,10]. It is remarked that the optical version of the Hardy paradox is described by Mølmer [11] to ignore the two-photon detection from the coincidence count between the electron and positron detectors. That is, those experiment is verified that the pre-selected state is entangled.

It is remarked that the reason to take the strange weak value was analyzed in Ref. [12]. It should be noted that the Hardy paradox has also been analyzed in the different contexts, for examples, the consistent histories' approach [13], the ontological approaches [14,15,16], the joint measurability [17], the Feynman-path analysis [18,19,20], the relationship to the Wigner function [21], and the non-locality condition in the sense of the Bell-type argument [22,23,24]. We are going to closely examine the counter-factual step 2) by looking at the corresponding weak values. We will see what is missing in the argument of 2) and how the counter-factual argument works in terms of weak values.

According to Hardy [25], the original motivation to consider the Hardy paradox is to relate the non locality and the Lorenz transformation. However, he failed to derive the Lorenz transformation for his system not to define the concept on the *simultaneous* event. This is still an open problem.

### 2.2. *When the counter-factual situation occurred?*

In this subsection, we justify the standard initial quantum state (1) on the basis of the counter-factual arguments 2) in the original Hardy paradox and then show how to experimentally check whether the initial quantum state is the right hand side of Eq. (1) or not by evaluating the weak values.

The pre-selected state is in general given by

$$|\psi_0\rangle = \eta|I_pI_e\rangle + x|I_pO_e\rangle + y|O_pI_e\rangle + z|O_pO_e\rangle \tag{14}$$

with the coefficients $\eta, x, y, z \in \mathbb{C}$ satisfying the normalization condition, $|\eta|^2 + |x|^2 + |y|^2 + |z|^2 = 1$. We are going to determine the coefficients by the following counter-factual arguments. Since the electron going through the path $O_e$ does not affect the path of the positron, the projection operators $P[O_p(I_e + O_e)]$ and $P[O_p \otimes Id.]$ are equivalent in the initial quantum state $|\psi_0\rangle$. Similarly, $P[(I_p + O_p)O_e]$ and $P[Id. \otimes O_e]$ are equivalent in $|\psi_0\rangle$. Furthermore, since the electron and the positron simultaneously going through the paths $I_p$ and $I_e$ would annihilate each other, $P[I_pO_e]$ and $P[I_p \otimes Id.]$ and $P[O_pI_e]$ and $P[Id. \otimes I_e]$ are also equivalent for $|\psi_0\rangle$. Here, we define the state-dependent equivalence as

$$A \sim_\psi B \overset{def}{\iff} \mathbf{Ex}\left([A - B]\right) = 0, \ \mathbf{Var}\left([A - B]\right) = 0 \tag{15}$$

$$\iff \langle\psi|(A - B)^2|\psi\rangle = \langle\psi|(A - B)|\psi\rangle = 0. \tag{16}$$

This is apparently different from the operator equivalence on the Hilbert space $\mathcal{H}$ as

$$A = B \overset{def}{\iff} (A - B)|\phi\rangle = 0, \ (\forall|\phi\rangle \in \mathcal{H}). \tag{17}$$

The state-dependent equivalence for the operator means that the probability distribution for the quantum state $|\psi\rangle$ is identical since the probability distribution is characterized only by the mean and the variance guaranteed by the central limit theorem. It is noted that the condition of the central limit theorem for the i.i.d. sequence of the quantum state $|\psi\rangle$ is satisfied. From the definition, the above counter-factual arguments demand

$$\langle\psi_0|(P[O_p(I_e + O_e)] - P[O_p \otimes Id.])^2|\psi_0\rangle$$
$$= \langle\psi|P[O_p(I_e + O_e)] - P[O_p \otimes Id.]|\psi_0\rangle = 0 \tag{18}$$
$$\langle\psi_0|(P[(I_p + O_p)O_e] - P[Id. \otimes O_e])^2|\psi_0\rangle$$
$$= \langle\psi_0|(P[(I_p + O_p)O_e] - P[Id. \otimes O_e])|\psi_0\rangle = 0 \tag{19}$$
$$\langle\psi_0|(P[I_pO_e] - P[I_p \otimes Id.])^2|\psi_0\rangle$$
$$= \langle\psi_0|(P[I_pO_e] - P[I_p \otimes Id.])|\psi_0\rangle = 0 \tag{20}$$
$$\langle\psi_0|(P[O_pI_e] - P[Id. \otimes I_e])^2|\psi_0\rangle$$
$$= \langle\psi_0|(P[O_pI_e] - P[Id. \otimes I_e])|\psi_0\rangle = 0 \tag{21}$$

From Eqs. (20) or (21), we obtain $\eta = 0$. From Eqs. (18) and (19), we obtain $x = y = z$. Taking into account the normalization condition, we obtain the standard Hardy initial state (1). The second equations of Eqs. (18), (19), (20) and (21) also imply that the initial quantum state can be experimentally tested by the joint expectation value of the paths for the electron and the positron.

Furthermore, we will show that this initial state can also be verified by the joint weak measurement, that is, the state can be constructed from a set of weak values. The inner product of the pre- and post-selected states are

$$\zeta_{D_p D_e} := \langle D_p D_e | \psi_0 \rangle = \frac{1}{2}(\eta - x - y + z), \tag{22}$$

$$\zeta_{D_p B_e} := \langle D_p B_e | \psi_0 \rangle = \frac{1}{2}(\eta + x - y - z), \tag{23}$$

$$\zeta_{B_p D_e} := \langle B_p D_e | \psi_0 \rangle = \frac{1}{2}(\eta - x + y - z), \tag{24}$$

$$\zeta_{B_p B_e} := \langle B_p B_e | \psi_0 \rangle = \frac{1}{2}(\eta + x + y + z). \tag{25}$$

We obtain the following table of weak values for the pre-selected state $|\psi_0\rangle$ and the post-selected state $|\phi\rangle$:

$$
\begin{array}{c|cccc}
\text{Weight} & \zeta^2_{D_p D_e} & \zeta^2_{D_p B_e} & \zeta^2_{B_p D_e} & \zeta^2_{B_p B_e} \\
\phi & D_p D_e & D_p B_e & B_p D_e & B_p B_e \\
\hline
P[O_p(I_e + O_e)] & 0 & -\dfrac{y+z}{\zeta_{D_p B_e}} & 0 & \dfrac{y+z}{\zeta_{B_p B_e}} \\[2ex]
P[(I_p + O_p)O_e] & 0 & 0 & -\dfrac{x+z}{\zeta_{B_p D_e}} & \dfrac{x+z}{\zeta_{B_p B_e}} \\[2ex]
P[I_p O_e] & -\dfrac{x}{2\zeta_{D_p D_e}} & \dfrac{x}{2\zeta_{D_p B_e}} & -\dfrac{x}{2\zeta_{B_p D_e}} & \dfrac{x}{2\zeta_{B_p B_e}} \\[2ex]
P[O_p I_e] & -\dfrac{y}{2\zeta_{D_p D_e}} & -\dfrac{y}{2\zeta_{D_p B_e}} & \dfrac{y}{2\zeta_{B_p D_e}} & \dfrac{y}{2\zeta_{B_p B_e}} \\[2ex]
P[O_p \otimes Id.] & \dfrac{-y+z}{2\zeta_{D_p D_e}} & \dfrac{-y-z}{2\zeta_{D_p B_e}} & \dfrac{y-z}{2\zeta_{B_p D_e}} & \dfrac{y+z}{2\zeta_{B_p B_e}} \\[2ex]
P[I_p \otimes Id.] & \dfrac{\eta-x}{2\zeta_{D_p D_e}} & \dfrac{\eta+x}{2\zeta_{D_p B_e}} & \dfrac{\eta-x}{2\zeta_{B_p D_e}} & \dfrac{\eta+x}{2\zeta_{B_p B_e}} \\[2ex]
P[Id. \otimes O_e] & \dfrac{-x+z}{2\zeta_{D_p D_e}} & \dfrac{x-z}{2\zeta_{D_p B_e}} & \dfrac{-x-z}{2\zeta_{B_p D_e}} & \dfrac{x+z}{2\zeta_{B_p B_e}} \\[2ex]
P[Id. \otimes I_e] & \dfrac{\eta-y}{2\zeta_{D_p D_e}} & \dfrac{\eta-y}{2\zeta_{D_p B_e}} & \dfrac{\eta+y}{2\zeta_{B_p D_e}} & \dfrac{\eta+y}{2\zeta_{B_p B_e}}
\end{array}, \tag{26}
$$

where the "Weight" is the probability of transition from the pre-selected state $|\psi_0\rangle$ to the post-selected state $|\phi\rangle$. Here, the column of the post-selected state $|\phi\rangle$ shows the weak values for each observable, e.g. $P[O_p(I_e + O_e)]$. Interestingly our table (26) is similar to the one in Feynman's paper [26].

The state-dependent equivalence for the operator can be also represented as

$$A \sim_\psi B \stackrel{def}{\Longleftrightarrow} \mathbf{Ex}\left([A - B]\right) = 0, \ \ \mathbf{Var}\left([A - B]\right) = 0$$

$$\Longleftrightarrow \langle\psi|(A - B)^2|\psi\rangle = \langle\psi|(A - B)|\psi\rangle = 0$$

$$\Longleftrightarrow \sum_{|\phi\rangle} \langle\psi|(A - B)|\phi\rangle\langle\phi|(A - B)|\psi\rangle = 0$$

$$\Longleftrightarrow \sum_{|\phi\rangle} (\langle\phi|(A - B)|\psi\rangle)^2 = 0$$

$$\Longleftrightarrow \sum_{|\phi\rangle} (_\phi\langle(A - B)\rangle_\psi^w)^2 (\langle\phi|\phi\rangle)^2 = 0$$

$$\Longleftrightarrow \ _\phi\langle A\rangle_\psi^w = \ _\phi\langle B\rangle_\psi^w \ (\forall|\phi\rangle). \tag{27}$$

From the same counter-factual arguments as before together with the first equalities of Eqs. (18) – (21) and Eq. (27), we see that

$$InPositron(D_p D_e) := \ _{D_p D_e}\langle P[I_p O_e]\rangle_\psi^w - \ _{D_p D_e}\langle P[I_p \otimes Id.]\rangle_{\psi_0}^w$$
$$= 0, \tag{28}$$

$$InEletron(D_p D_e) := \ _{D_p D_e}\langle P[O_p I_e]\rangle_\psi^w - \ _{D_p D_e}\langle P[Id. \otimes I_e]\rangle_{\psi_0}^w$$
$$= 0, \tag{29}$$

$$OutPositron(D_p D_e) := \ _{D_p D_e}\langle P[O_p(I_e + O_e)]\rangle_\psi^w - \ _{D_p D_e}\langle P[O_p \otimes Id.]\rangle_{\psi_0}^w$$
$$= 0, \tag{30}$$

$$OutEletron(D_p D_e) := \ _{D_p D_e}\langle P[(I_p + O_p)O_e]\rangle_\psi^w - \ _{D_p D_e}\langle P[Id. \otimes O_e]\rangle_{\psi_0}^w$$
$$= 0. \tag{31}$$

Using the results of the above table in Eqs. (28) and (29) yields

$$\eta - x = -x \quad \text{or} \quad \eta - y = -y \tag{32}$$

to obtain that the coefficient of $|I_p I_e\rangle$ must be $\eta = 0$. Similarly from Eqs. (30) and (31), we have

$$-y + z = 0 \quad \text{and} \quad -x + z = 0 \tag{33}$$

to obtain the coefficients $x = y = z = 1/\sqrt{3}$ by the normalization condition.

We therefore conclude that

**Fact 1. (Hosoya and Shikano** [12]**).** The following conditions

(1)    $InPositron(D_p D_e) = InElectron(D_p D_e) = 0,$
(2)    $OutPositron(D_p D_e) = OutElectron(D_p D_e) = 0,$

are satisfied if and only if the pre-selected state is Eq. (1).

Note that we only need a particular post-selected state $|D_p D_e\rangle$ to verify whether a prepared setup gives the standard Hardy's initial state (1) or not.

It is interesting to point out that the above fact is only used in the case that the post-selected state is $D_p D_e$ while Eq. (27) requires the equality for any post-selected state. In the case that the post-selected state is $D_p B_e$, $B_p D_e$, or $B_p B_e$, from the counter-factual argument, the pre-selected state cannot be decided while the condition for the quantum state $|\psi_0\rangle$ that $x = z$ and $\eta = 0$, $y = z$ and $\eta = 0$, or $\eta = 0$ can be restricted, respectively. This means that the roles of the post-selected states are not equal. It is also interesting to point out that the relations between the weak values can be used to evaluate the weak values of the non-local operators like $P[I_p O_e], P[O_p I_e], P[O_p(I_e + O_e)]$, and $P[(I_p + O_p)O_e]$ using those of the local operators $P[I_p \otimes Id.], P[Id. \otimes I_e], P[O_p \otimes Id.]$, and $P[Id. \otimes O_e]$, respectively, provided that the pre-selected state is Eq. (1).

## 3. Conclusion

The counter-factual argument is based on the classical particle picture. In this paper, we explicitly showed that the weak values can be explained on the paradoxical situation of the classical particle picture in the Hardy paradox. However, this is one typical example. We have to show the mathematical relation between the weak value and the counter-factual argument. Also, the state-dependent equivalence is also defined in the context of the perfect correlations [27]. We have to make clear applications to the correlation and the non-classicality.

## Acknowledgement

The author (YS) thanks Akio Hosoya for useful collaboration [12,28]. YS thanks Yakir Aharonov, Lucien Hardy, and Paul Kwiat for useful discussion. YS acknowledges the great hospitality and great support, especially from Margaret Sullivan, at Massachusetts Institute of Technology hosted by Seth Lloyd partially supported by JSPS Excellent Young Researcher Overseas Visit Program. YS is supported by JSPS Research Fellowships for Young Scientists (No. 21008624) and Global Center of Excellence Program "Nanoscience and Quantum Physics" at Tokyo Institute of Technology.

# References


1. Y. Aharonov, D. Albert, A. Casher, and L. Vaidman, Ann. New York Acad. Sciences **480**, 620 (1986).
2. Y. Aharonov, D. Albert, A. Casher, and L. Vaidman, Phys. Lett. A **124**, 199 (1987)
3. Y. Aharonov , D. Albert, and L. Vaidman, Phys. Rev. Lett. **60**, 1351-1354 (1988).
4. Y. Shikano, arXiv:1110.5055.
5. L. Hardy, Phys. Rev. Lett. **68**, 2981 (1992).
6. W. G. Unruh, Phys. Rev. A **59**, 126 (1999).
7. S. Kochen and E. Specker, J. Math. Mech. **17**, 59 (1967).
8. Y. Aharonov, A. Botero, S. Popescu, B. Reznik, and J. Tollaksen, Phys. Lett. A **301**, 130 (2002).
9. J. S. Lundeen and A. M. Steinberg, Phys. Rev. Lett. **102**, 020404 (2009).
10. K. Yokota, T. Yamamoto, M. Koashi, and N. Imoto, New J. Phys. **11**, 033011 (2009).
11. K. Mølmer, Phys. Lett. A **292**, 151 (2001).
12. A. Hosoya and Y. Shikano, J. Phys. A **43**, 385307 (2010).
13. R. E. Kastner, Stud. Hist. Philos. M. P. **35**, 57 (2004).
14. H. P. Stapp, Am. J. Phys. **65**, 300 (1997).
15. N. D. Mermin, Am. J. Phys. **66**, 920 (1998).
16. T. Bigaj, Found. Sci. **12**, 85 (2007).
17. L. Vaidman, Phys. Rev. Lett. **70**, 3369 (1993).
18. R. B. Griffiths, Phys. Rev. A **60**, 5(R) (1999).
19. S. McCall, Found. Phys. Lett. **14**, 95 (2001).
20. D. Sokolovski, I. P. Gimenez, and R. Sala Mayato, Phys. Lett. A **372**, 3784 (2008).
21. M. Tsang, Phys. Rev. A **81**, 013824 (2010).
22. A. Acín, in *Compendium of Quantum Physics: Concepts, Experiments, History and Philosophy* (Springer-Verlag, Berlin Heidelberg, 2009), edited by D. M. Greenberg, K. Hentschel, and F. Weinert, p. 275.
23. S. Mansfield and T. Fritz, arXiv:1105.1819.
24. M. J. Steiner, Proc. SPIE **5436**, 216 (2004).
25. L. Hardy, private communication (2011).
26. R. P. Feynman, Int. J. Theor. Phys. **21**, 467 (1982).
27. M. Ozawa, Ann. Phys. **321**, 744 (2006).
28. Y. Shikano and A. Hosoya, J. Phys. A **43**, 025304 (2010).

Quantum Bio-Informatics V
© 2013 World Scientific Publishing Co. Pte. Ltd.
pp. 473–485

# FROM STRUCTURE AND FUNCTION OF PROTEINS TOWARD *IN SILICO* BIOLOGY

ICHIRO YAMATO

*Department of Biological Science and Technology, Tokyo University of Science, 2641 Yamazaki, Noda, Chiba, 278-8510, Japan*

Researches of biology are targeted on three major flows, materials (or chemicals), energy, and information. I have been mainly concerned with the studies on bioenergy transducing mechanisms. I have studied the mechanism of secondary active transport systems and proposed an affinity change mechanism as a general hypothesis, then tried to confirm that it is applicable to other kinds of bioenergy transducing systems. Choosing $Na^+$-translocating V-type ATPase from *Enterococcus hirae* as target, I hypothesized the affinity change mechanism for the energy transduction of this ATPase. Here I describe several three dimensional structures of parts of the ATPase supporting my hypothesis. From such detailed and extensive researches on protein structure/function relationship, we can proceed toward the *in silico* biology, which I described previously in 2007 ([1] "Toward *in silico* biology").

## 1. Introduction

Life is basically based on the flows of materials, energy, and information. (In other words, living organisms are created on these flows at steady state non-equilibrium state.) Accordingly researches in biology are mainly targeted on such three major flows (Fig. 1) and the biological research fields mainly deal with those respective targets.

*3 kinds of flows*    *"Research field" (examples of researches)*
Flow of matter → "biochemistry" (metabolism, etc)
Flow of energy → "biophysics" (muscle contraction, etc)
Flow of information → "molecular biology" (genetic code,
                                signal transduction, nerve/brain

Fig. 1. Targets of life science (3 major flows)

Around the World War II, the production of (radio)-isotopes by nuclear reactions was established, and researchers used such isotopes as tracers. They found many metabolic pathways, such as Calvin cycle [2] in photosynthesis in plants and Krebs cycle [3]. ATP has been found by Fiske and Subbarow [4] as a high energy substance in biochemical reactions. Muscle contraction occurs depending on the consumption of ATP energy [5] and Huxley and Hanson proposed the sliding mechanism of acto-myosin fibers [6]. In muscle, chemical energy of ATP is converted to mechanical energy of force. There are many kinds of such energy transducing systems, such as photosynthesis, ion-pumping ATPases, and so on. They have been studying such energy transduction mechanisms in biology. Then after proposal of DNA structure by Watson-Crick [7] and owing to recent development of –omic studies (genome, transcriptome, proteome, interactome, etc), information flows or signal transductions in many biological systems have been extensively elucidated, including highly complex information processing systems as cognition and brain. At the first conference of this quantum-bio-informatics research center in 2007, I have described our trials toward *in silico* biology to reproduce living systems in a computer based on the recent –omic data [1]. We know that we have to accumulate more and more biology related data to reproduce living systems in a computer (toward *in silico* biology). We are now on the way. Here I describe our studies on the energy flow using *Entercoccus hirae* V-type sodium-translocating ATPase. By such detailed and extensive studies on protein structure/function relationships, I think we can proceed toward the final goal of this *in silico* biology.

## 2. Energy transducing systems in biology

Living systems are created on the flows of substances and to maintain such flows the energy is the prerequisite. They take up several kinds of energy available in their environment such as light and chemicals and they utilize such energy inputs to exert several kinds of works necessary to maintain the systems such as forces, electrochemical gradients of substances, etc.

Typical bioenergy transducing systems can be classified according to their energy input and output forms (Table 1). Chemical energy is usually derived from the hydrolysis free energy of ATP or oxido-reduction electron transfer through respiratory chain. Electrochemical potential energy means the electrochemical free energy difference of a substance across a barrier (biomembranes). Mechanical energy is the energy generating a force or torque.

Table 1. Bioenergy transduction systems

Light-chemical (photosynthesis)
Light-electrochemical gradient (bacteriorhodopsin)
Chemical (redox)-electrochemical gradient (P-ATPases, etc)
Chemical-mechanical (actomyosin, F- & V-ATPases, etc)
Electrochemical gradient-electrochemical gradient (sympoters, etc)

## 2.1. *Model of bioenergy transduction mechanism for secondary active transport system*

I have been studying secondary active transport systems [8], especially proline/$Na^+$ symporter in *Escherichia coli*. In this system, ion electro-chemical gradient across the cell membrane ($Na^+$ gradient) drives the active accumulation of substrate (proline) into the cell, where a membrane protein (proline/$Na^+$ symporter, PutP) is responsible for this active transport. The substrate binding property to the symporter (PutP) was dependent on the coupling ion ($Na^+$) (Fig. 2): The symporter orienting outside binds first the coupling ion (step1) which causes the conformational change (step2) of the symporter to have the binding ability for substrate (step3), then the symporter translocates to cytoplasmic side (step4) to release the substrate first (step5) and then ion next (step6) [9]. Similar mechanism of the binding reaction was shown to work for several other secondary active transport proteins [10].

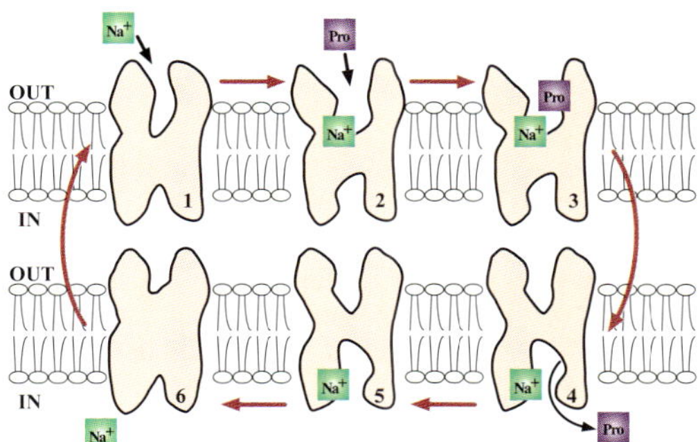


Fig. 2. Reaction mechanism of a secondary active transport (PutP)

Thus, I proposed this affinity change model depending on the binding of coupling ion as the general mechanistic model for secondary active transport [10]. Here we can say that the protein works as a Maxwell's daemon to discriminate the entry or exit of a molecule (substrate) depending on energy input ($Na^+$ binding) at the small hole in the barrier wall of two boxes filled with gas molecules (see Fig. 3). In this case, the protein binds coupling ion easily when the ion concentration is higher at outside, then the ion binding changes the affinity of the protein for substrate, and ternary complex can translocate across the membrane. The protein alone cannot bind substrate at any time, but just after binding of $Na^+$ the protein gains the ability of binding substrate. In this sense, $Na^+$ does not facilitate the translocation movement (speed) of the protein; instead it facilitates obtaining the binding ability (Fig. 2). I named this model as affinity change model of bioenergy transducing mechanism [11]. Recently three dimensional structures of many secondary active transport proteins (very hydrophobic membrane proteins) have been elucidated and within them they have obtained several crystals for inward-orienting, outward-orienting, and occluded forms of Mhp1 corresponding to the transport intermediates in my transport mechanism [12, 13].

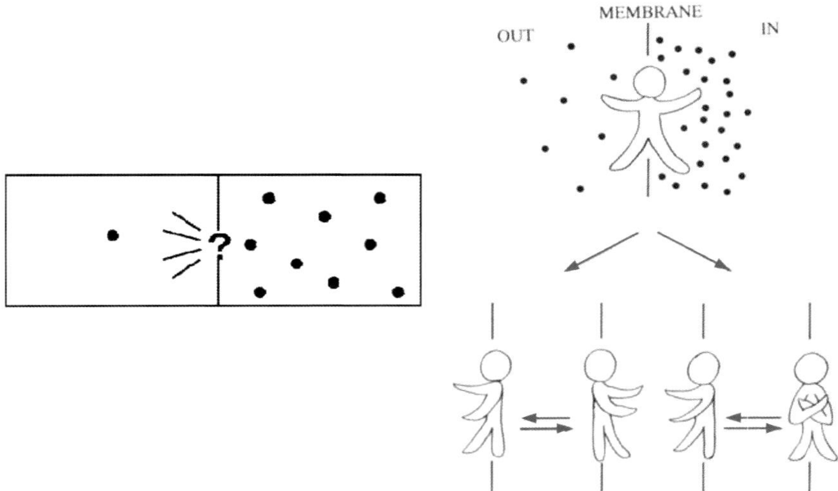


Fig. 3. Two boxes system and a model of a secondary active transporter showing how the Maxell's daemon (transporter) works. Left figure: Two boxes system having a hole with a Maxwell's daemon (marked with ?) controlling the in and out of gas molecules into the boxes. Right figure: Two representative models where a Maxwell's daemon works in an active transport system with energy input by facilitating in either way, changing the binding affinity for substrate as shown by open and closed arms (right case) or changing the speed of translocation of substrate as shown by the different thickness of the translocation arrows (left case) (right figure).

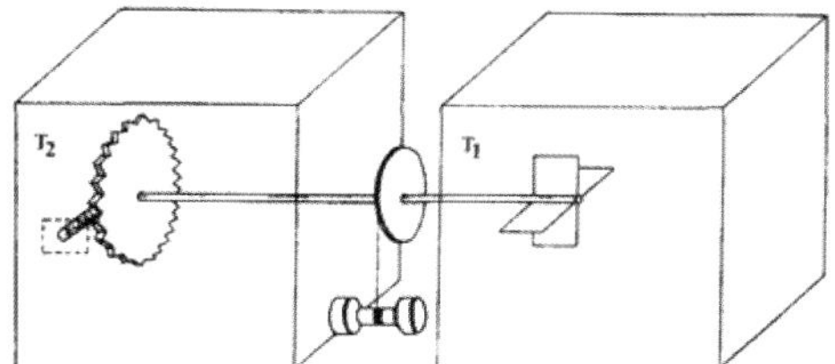

Fig. 4. Feynmann thermal ratchet model from [18]. A ratchet and a pawl isolated at different temperatures make up a motor working by unidirectional motion. The ratchet has a shape of a circular saw with asymmetric teeth, one face of which orienting perpendicular to the circumference and the other creating a shallower angle. A wheel wound with string with a weight is attached to the ratchet and the ratchet and wheel are in turn connected by a rod to a vane. A spring-loaded pawl attached to a fixed plate interdigitates between the ratchet teeth and prevents free rotation of the ratchet. How it works is described in [18].

## 2.2.  *Affinity change model as general bioenergy transduction mechanism --- hypothesis*

In other bioenergy transducing systems, several energy coupling mechanisms are proposed. In P-type ATPase [14, 15], they identified the phosphorylated state of the protein and captured each binding state of substrate (typically $Na^+$ and $K^+$) depending on the phosphorylated and dephosphorylated forms. Therefore, the similar affinity change for substrates was demonstrated depending on the input of energy (ATP) [14]. But in muscle contraction, for a long time, it has been considered that the input of energy (ATP) at the myosin head causes the conformational change of the head to make a rowing motion to generate the force for contraction. Eisenberg and Hill [16] have already demonstrated that the myosin head makes the conformational change to give different affinity for actin filament depending on the input of ATP (ATP form, ADP form, and no nucleotide form). Yanagida et al. [17] observed the single molecule movement of actin filament on myosin using fluorescent probe and recently Vale and Oosawa hypothesized the "Brownian motion of the head" model as the Feynmann thermal ratchet model (Fig. 4) [18]. In this model, we can think that actin filament moves owing to the thermal energy and attaches to myosin head depending on the nucleotide form and also the conformation (direction) of the myosin head. By attaching strongly, even if the actin filament tends to move further toward the backward direction, myosin fiber blocks the movement and if the actin tends to move toward another direction by thermal energy, myosin head changes the nucleotide form to change the conformation to detach from the actin filament, so the actin filament can move freely for some distance. Then if the myosin head changes back to the original position (conformation), it takes another nucleotide binding form to attach the actin filament to block the actin filament movement. In this way, the myosin head by ATP hydrolysis moves one stroke distance to generate the force of muscle contraction.

After my proposal of affinity change model, I became interested whether any micro (bio) molecular energy transduction should follow this model (not by facilitating the speed change, but by causing the affinity change, bioenergy makes the Maxwell's daemon's work --- in this sense, energy input is to give the information of direction whether the protein should bind a substrate (proline or actin filament) or not). There are other several bioenergy transducing systems (Table 1), and I believe they should also follow this affinity change model. Among them, I have been interested in another chemical-mechanical energy transducing system, V-ATPase (Fig. 5) [19]. Since F-ATPase is known as a rotary motor rotating its axis depending on the hydrolysis of ATP [20-23] and V-ATPase is a relative of the F-ATPase [19], many researchers are interested in its mechanism. In F-ATPase, ATP exerts its work by generating the rotation

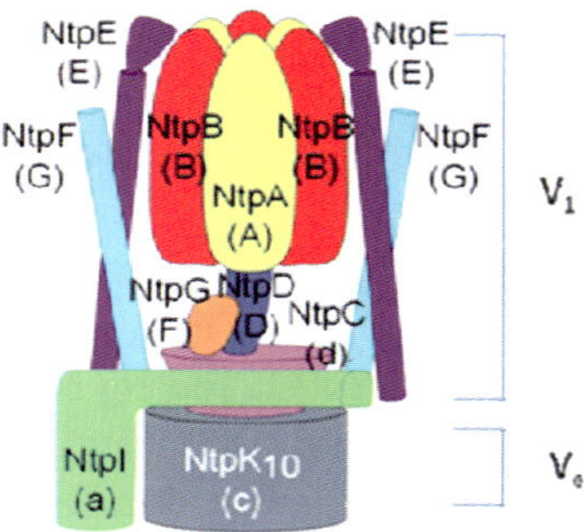


Fig. 5. A model of V-ATPase from *Enterococcus hirae*. V$_1$ indicates catalytic domain (consisting of NtpA$_3$-B$_3$-D-G), V$_o$ indicates membrane domain (consisting of NtpI-K$_{10}$). Peripheral stalk consists of NtpE and NtpF subunits and central stalk consists of NtpC, NtpD, and NtpG subunits (Parentheses show names of corresponding subunits of eukaryotic V-ATPase).

torque directly or like as PutP which can select the direction by giving the information of allowed direction to the axis to rotate from the energy. Still the energy transduction mechanism, whether it works as the force generator changing the velocity of movement or just selecting direction by changing the affinity is not well understood.

## 3. *Enterococcus hirae* Na$^+$-translocating V-type ATPase

Here I describe our studies on the *E. hirae* Na$^+$-translocating V-type ATPase, especially the recent structure/function studies of this protein.

## 3.1.  *Ion-pumping ATPases and V-type ATPase*

Ion pumping ATPases hydrolyze ATP to drive the active transport of ions, such as $H^+$, $Na^+$, $K^+$, $Ca^{2+}$, and so on. These are classified into three types: a) P-type ATPase in cytoplasmic membranes, b) F-type ATPase in mitochondria, chloroplast or cytoplasmic membranes of bacteria which works to synthesize ATP by using electrochemical potential of $H^+$ across membranes and c) V-type ATPase in organellar membranes (Fig. 6).

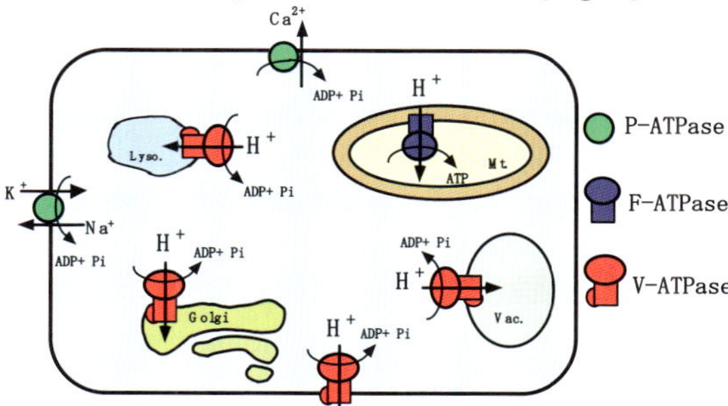


Fig. 6. Location and functions of ion-translocating ATPases in an eukaryotic cell. Green circles, blue and red mushroom like symbols indicate P-type, F-type, and V-type ATPases, respectively. The external rectangle represents an eukaryotic cell membrane. Straight arrows indicate the direction of ion translocation. Mt. represents mitochondria and Vac. represents vacuoles.

V-type $H^+$-translocating ATPases provide the intracellular organelles such as endosomes, Golgi, lysosomes, and so on with acidic environment to facilitate the functions of such organelles.

## 3.2.  E. hirae *Na$^+$-ATPase*

*Enterococcus hirae* is a gram-positive fermentative bacterium. It contains a single cell membrane. *E. hirae* can grow at high salt concentration and/or alkaline pH and $Na^+$-translocating ATPase is responsible for this growth by pumping out $Na^+$ inside of cells [24-26].

*E. hirae* $Na^+$-translocating V-ATPase is a homolog of eukaryotic V-ATPase which physiologically transports $Na^+$ rather than $H^+$ (Fig. 5) [25, 26]. This V-ATPase is composed of a soluble catalytic domain ($V_1$, NtpA$_3$-B$_3$-D-G) and an integral membrane domain (Vo, NtpI-K$_{10}$) [19]. The central stalk of this ATPase is composed of NtpC, NtpD, and NtpG subunits, where NtpD-G is called central axis. And the NtpA$_3$-B$_3$ part hydrolyses the ATP to convert the chemical free energy to the rotation (torque) energy of the axis NtpD-G. It is still not fully understood how the hydrolysis energy is transduced to the rotation energy even after the several studies on the F-ATPases. The rotation of NtpD-G causes the rotation of NtpK rotor ring in membrane embedded $V_0$ part and $Na^+$ is transported through the interface of NtpK and the stator, NtpI. We have solved the 3D structures of our V-ATPase head part NtpA$_3$-B$_3$-D-G, the axis NtpD-G

480

[27], and membrane embedded ion-translocating rotor ring NtpK [28, 29]. Here we describe below the structure/function of the ion-translocating rotor ring.

### 3.3.  *Structure of NtpD-G axis and NtpK rotor ring*

We have obtained the crystal structures of the catalytic part of this ATPase, which converts chemical energy, ATP, to rotation energy of NtpD-G axis. But we have not yet published it (manuscript in preparation), so I cannot describe the d
etails of it. Instead I describe the structures of the axis and the rotor ring, which have been published recently [27-29]. Figure 7A is the 3D structure of the axis. It has coiled-coil structure of α-helix [27]. Figure 7B shows the structure of NtpK rotor ring with Na$^+$ bound at the middle of the transmembrane helix (shown by light blue spheres) [28]. The ring is consisted of 10 NtpK monomers, each of which is composed of 4 transmembrane helices. We modeled another component, NtpC, according to the amino acid sequence homology with the

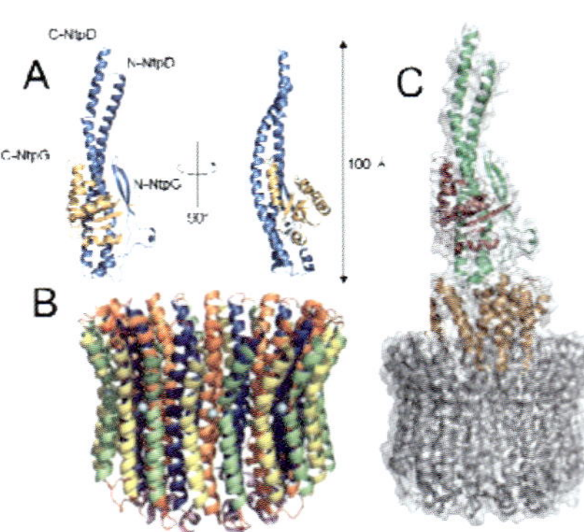

Fig. 7. Structures of the NtpD-G complex and NtpK rotor ring of the *E. hirae* V-ATPase [27, 28]. (A) Model of NtpD-G complex. NtpD and NtpG are shown in blue and orange, respectively. (B) NtpK ring of side view. Helices 0 (residues 1–8), 1 (11–46), 2 (51–79), 3 (85–124), and 4 (127–156), and loops (9–10, 47–50, 80–84, and 125–126) are colored in violet, blue, green, orange, yellow and red in ribbon representations, respectively. At the essential acidic amino acids, Glu, at the middle of transmembrane helices, Na$^+$ shown in light blue spheres were detected bound. (C) The model of the entire rotor complex. NtpC homology model is shown in brownish color.

NtpC homologue from *Thermus thermophilus* [30], since the 3D structure of which is known. By combining them, we modeled the rotor ring connected to the axis (Fig. 7C) [27], which rotates by the driving force of ATP hydrolysis at the catalytic head part of NtpA$_3$-B$_3$.

### 3.4.  *Structure of NtpK rotor ring and the ion-translocation mechanism*

The membrane bound stator, NtpI, has been thought to form channels of ions to transport from inside to outside by the rotation of NtpK ring (Fig. 5 and 8). The Arg (R573) in NtpI is postulated to work for opening and closing of the binding site of Na$^+$ at Glu (E139) in NtpK. The proposed model for an ion transport mechanism [28, 31, 32] is as follows (Fig. 8): Clockwise rotation of the K-ring driven by ATP-hydrolysis energy in $V_1$-ATPase, as viewed from the cytoplasm, brings an occupied $Na^+$-binding site into the K-ring–NtpI interface (Fig. 8; upper left panel). The proximity of the $Na^+$ site to the essential R573 residue of NtpI produces an electrostatic interaction between R573 and E139, disrupting the hydrogen-bond network in the $Na^+$-binding pocket of the K-ring. This results in the release of $Na^+$ into the periplasm via a half-channel in NtpI as indicated by the red arrow in Fig. 8 (upper right panel). Further rotation caused by ATP hydrolysis energy in $V_1$-ATPase, which may disrupt the Arg-Glu interaction, results in the binding of a cytoplasmic $Na^+$ as indicated by the blue arrow in Fig.8 (bottom left panel).

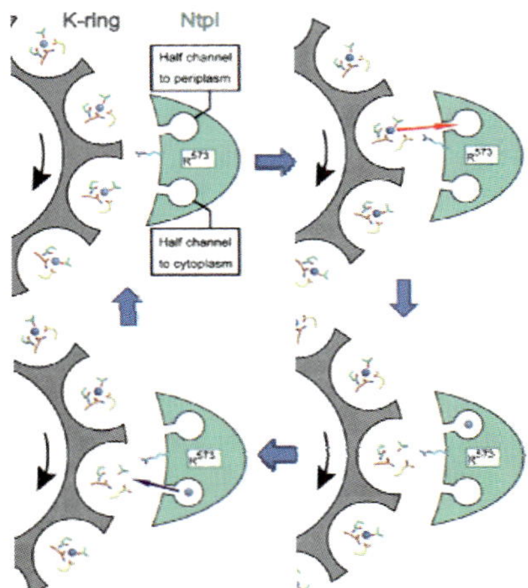


Fig. 8. A model for the ion transport mechanism of $Na^+$-translocating V-ATPase [29]. The model is based on the crystal structures of the K-ring with and without DCCD. The views are from the cytoplasm and show the K-ring–NtpI interface at the level of the $Na^+$ binding sites. Residues involved in $Na^+$ binding are shown in stick representation. NtpI shown in green is in close proximity to the K-ring with the essential R573 (blue stick representation), which has two half-channels connecting to the periplasm and cytoplasm, respectively. The $Na^+$-transport mechanism of V-ATPase is represented by four intermediate states. See text for details.

Mizutani et al. [29] obtained a DCCD (dicyclohexylcarbodiimide)-modified NtpK rotor ring structure, which mimics the low affinity state of the ring. The binding affinity of $Na^+$ to wild-type NtpK was very high at 10 µM order and to the DCCD-modified ring was low at mM order [29]. The modification effectively neutralized the negative charge of the E139 resulting in lower affinity for $Na^+$ at the binding site. We postulated that E139 modified with DCCD takes the "open" conformation mimicking the similar "open" conformation to release the bound $Na^+$ during ion translocation via NtpI channel through formation of an electrostatic interaction with R573.

Thus, we propose an ion transport mechanism of the V-ATPase whereby ion translocation can be accomplished through an affinity change of the $Na^+$ binding sites by neutralization of E139 due to formation of a salt bridge with R573 of the stator. We have not yet obtained the 3D structure of the stator, NtpI, so the above mentioned mechanism is our postulation at present, but at least the different binding affinities of $Na^+$ detected is the strong supporting evidence for the affinity change model. We hope the 3D structure of the whole V-ATPase becomes available in near future, which would give us the evidence for the opening and closing of the binding site according to the movement of NtpK.

### 3.5. *Bioenergy transduction mechanism in general*

I hypothesized that bioenergy transduction mechanism follows the affinity change model [11]. Instead of describing the NtpA$_3$-B$_3$-D-G rotation here, I described the NtpK ion translocation mechanism (mechanical-electrochemical energy transduction) above as an example of such affinity change mechanism.

I hope near future we can establish the affinity change mechanism as a general bioenergy transduction mechanism and reveal the secret of nature why in biological system only affinity change mechanism is allowed to work while velocity change mechanism seems to be able to work equally well.

## 4. Toward *in silico* biology

I think we are now going to be close to the general rule of bioenergy transduction mechanism. Next big target of life science would be on information flow in biosystems. Watson & Crick modeled the DNA structure as double helix in 1953 and explained many facets of genetic rules (information flow). Since then, researches on bioinformation flow appeared and cleared many new aspects. But we have not yet been able to predict the 3D structure formation of proteins from their amino acid sequences (which is called second genetic coding problem). Last year Shaw et al. showed the success of folding simulation of short peptides using special purpose computer [33] and Baker et al. succeeded to predict 3D structures of several proteins on web (Foldit: http://fold.it/portal/user/2861) using public game system by threading method. But still we are at the primitive stage for the final answer toward this second coding problem. I have started our approach toward this goal at the beginning of our QBIC. Although the progress was slow and after development of Brownian dynamics we could not proceed further in this direction, I hope we are progressing to some extent. By studying many kinds of protein structure/function relationships as described above, I hope we can accumulate enough data to construct *in silico* biology in near future.

## Acknowledgements

This work was supported by a Grant-in-Aid for the Academic Frontier Project from MEXT and for the Open Research Center Project from MEXT to I. Y.

**References**


1. I. Yamato, T. Ando, A. Suzuki, K. Harada, S. Itoh, S. Miyazaki, N. Kobayashi and M. Takeda (2008) Toward *in silico* biology (from sequences to systems). Quantum Bio-Informatics (From Quantum Information to Bio-Informatics) p.440-p.455 (eds L. Accardi, W. Freudenberg, & M. Ohya) (Proceedings of the International Symposium of Quantum Bio-Informatics Research Center 2007, Chiba) World Scientific, Singapore.

2. J. Bassham, A. Benson and M. Calvin (1950) The path of carbon in photosynthesis. J. Biol. Chem., 185, 781-787.

3. H. A. Krebs and W. A. Johnson (1937) Metabolism of ketonic acids in animal tissues. Biochem. J., 31, 645-660.

4. C. H. Fiske and Y. Subbarow (1929) Phosphorus compounds of muscle and liver. Science, 70, 381-382.

5. A. Szent-Györgyi and I. Banga (1941) Adenosinetriphosphatase. Science, 93, 158.

6. H. Huxley and J. Hanson (1954) Changes in the cross-striations of muscle during contraction and stretch and their structural interpretation. Nature, 173, 973-976.

7. J. D. Watson and F. H. C. Crick (1953) Molecular structure of nucleic acids: A structure for deoxyribose nucleic acid. Nature, 171, 737-738.

8. I. Yamato and Y. Anraku (1992) $Na^+$/substrate symport in prokaryotes. In Alkali cation transport systems in prokaryotes, p.53-p.76 (ed. by E. P. Bakker) CRC Press, Florida, USA.

9. I. Yamato and Y. Anraku (1990) Mechanism of $Na^+$/proline symport in *Escherichia coli*: Reappraisal of the effect of cation binding to the $Na^+$/proline symport carrier. J. Membrane Biol., 114, 143-151.

10. I. Yamato (1992) Ordered binding model as a general mechanistic mechanism for secondary active transport systems. FEBS Lett., 298, 1-5.

11. I. Yamato (1993) Ordered binding model as a general tight coupling mechanism for bioenergy transduction --- A hypothesis. Proc. Japan Acad., 69, 218-223.

12. S. Weyand, T. Shimamura, S. Yajima, S. Suzuki, O. Mirza, K. Krusong, E. P. Carpenter, N. G. Rutherford, J. M. Hadden, J. O'Reilly, P. Ma, M. Saidijam, S. G. Patching, R. J. Hope, H. T. Norbertczak, P. C. Roach, S. Iwata, P. J. Henderson and A. D. Cameron (2008) Structure and molecular mechanism of a nucleobase-cation-symport-1 family transporter. Science, 322, 709-713.

13. T. Shimamura, S. Weyand, O. Beckstein, N. G. Rutherford, J. M. Hadden, D. Sharples, M. S. Sansom, S. Iwata, P. J. Henderson and A. D. Cameron (2010) Molecular basis of alternating access membrane transport by the sodium-hydantoin transporter Mhp1. Science, 328, 470-473.

14. I. M. Glynn and S. J. D. Karlish (1990) Occluded cations in active transport. Annu. Rev. Biochem., 59, 171-205.

15. J. S. Skou (1957) The influence of some cations on an adenosine triphosphatase from peripheral nerves. Biochim. Biophys. Acta, 23, 394-401.

16. E. Eisenberg and T. L. Hill (1985) Muscle contraction and free energy transduction in biological systems. Science, 227, 999-1006.

17. T. Yanagida, T. Arata and F. Oosawa (1985) Sliding distance of actin filament induced by a myosin crossbridge during one ATP hydrolysis cycle. Nature, 316, 366-369.

18. R. D. Vale and F. Oosawa (1990) Protein motors and Maxwell's demons: Does mechanochemical transduction involve a thermal ratchet? Adv. Biophys., 26, 97-134.

19. T. Murata, I. Yamato and Y. Kakinuma (2005) Structure and mechanism of vacuolar $Na^+$-translocating ATPase from *Enterococcus hirae*. J. Bioenerg. Biomembr., 37, 411-413.

20. P. D. Boyer (1993) The binding change mechanism for ATP synthase - Some probabilities and possibilities. Biochim. Biophys. Acta, 1140, 215-250.

21. J. P. Abrahams, A. G. Leslie, R. Lutter and J. E. Walker (1994) Structure at 2.8 Å resolution of $F_1$-ATPase from bovine heart mitochondria. Nature, 370, 621-628.

22. H. Noji, R. Yasuda, M. Yoshida and K. Kinosita, Jr. (1997) Direct observation of the rotation of F1-ATPase. Nature, 386, 299-302.

23. K. Adachi, K. Oiwa, T. Nishizaka, S. Furuike, H. Noji, H. Itoh, M.Yoshida and K. Kinosita Jr. (2007) Coupling of rotation and catalysis in F1-ATPase revealed by single-molecule imaging and manipulation. Cell, 130, 309-321.

24. D. L. Heefner and F. M. Harold (1982) ATP-driven sodium pump in *Streptococcus faecalis*. Proc. Natl. Acad. Sci. USA, 79, 2798-2802.

25. Y. Kakinuma and K. Igarashi (1990) Release of component of *Streptococcus faecalis* $Na^+$-ATPase from the membranes. FEBS Lett., 271, 102-105.

26. Y. Kakinuma and K. Igarashi (1990) Some features of the *Streptocossu faecalis* $Na^+$-ATPase resemble those of vacuolar-type ATPase. FEBS Lett., 271, 97-101.

27. S. Saijo, S. Arai, K. M. M. Hossain, I. Yamato, K. Suzuki, Y. Kakinuma, Y. Ishizuka-Katsura, N. Ohsawa, T. Terada, M. Shirouzu, S. Yokoyama, S. Iwata and T. Murata (2011) Crystal structure of the central axis DF complex of V-ATPase. Proc. Natl. Acad. Sci. U. S. A., in press.

28. T. Murata, I. Yamato, Y. Kakinuma, A. G. W. Lestlie and J. E. Walker (2005) Structure of the rotor of the V-type $Na^+$-ATPase from *Enterococcus hirae*. Science, 308, 654-659.

29. K. Mizutani, M. Yamamoto, K. Suzuki, I. Yamato, Y. Kakinuma, M. Shirouzu, J. E. Walker, S. Yokoyama, S. Iwata and T. Murata (2011) Structure of the rotor ring modified with N,N'-dicyclohexylcarbodiimide of the $Na^+$-transporting vacuolar ATPase. Proc. Natl. Acad. Sci. U. S. A., 108,

13474-13479.
30. M. Iwata, H. Imamura, E. Stambouli, C. Ikeda, M. Tamakoshi, K. Nagata, H. Makyio, B. Hankamer, J. Barber, M. Yoshida, K. Yokoyama and S. Iwata (2004) Crystal structure of a central stalk subunit C and reversible association/dissociation of vacuole-type ATPase. Proc. Natl. Acad. Sci. USA, 101, 59-64.
31. T. Murata, I. Yamato and Y. Kakinuma (2005) Structure and mechanism of vacuolar $Na^+$-translocating ATPase from *Enterococcus hirae*. J Bioenerg. Biomembr., 37, 411-413.
32. T. Murata, I. Yamato, Y. Kakinuma, M. Shirouzu, J. E. Walker, S. Yokoyama and S. Iwata (2008) Ion binding and selectivity of the rotor ring of the $Na^+$-transporting V-ATPase. Proc. Natl. Acad. Sci. USA, 105, 8607-8612.
33. D. E. Shaw, P. Maragakis, K. Lindorff-Larsen, S. Piana, R. O. Dror, M. P. Eastwood, J. A. Bank, J. M. Jumper, J. K. Salmon, Y. Shan and W. Wriggers (2010) Atomic-level characterization of the structural dynamics of proteins. Science, 330, 341–346.
34. D.Baker (2011) Foldit. http://fold.it/portal/user/2861.

# International Conference QBIC 2011

## March 7-14 2011  Tokyo University of Science  Canal Hall Noda Campus



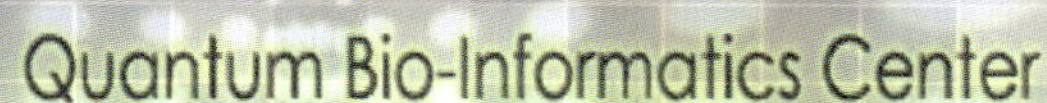



The main aim of QBIC and the conference is to create a new paradigm synthesizing Quantum Information and Bio-Informatics based on efforts by active researchers traversing various fields of Mathematics, Physics, Information and Life Science.

### Speakers in Main Session

| | | | |
|---|---|---|---|
| Luigi Accardi | Denes Petz | Huzihiro Araki | Shigeru Furuichi |
| W. Freudenberg | A. Khrennikov | O. Smolyanov | Hayato Nakano |
| Takeyuki Hida | D. Chruściński | Ludwing Streit | Wonpil Im |
| A. Jamiołkowski | Izumi Ojima | Asao Arai | Alessio Accardi |
| V. P. Belavkin | Si Si | Francesco Fidaleo | Takeshi Obayashi |
| Igor Volovich | M. Michalski | P.v.d. Straten | Irina Basieva |
| K.-H. Fichtner | Massimo Regoli | Un Cig Ji | Roman Belavkin |
| A. Majewski | | Shao-Ming Fei | Tadashi Ando |
| | | F. Mukhamedov | Anton Trushechkin |
| | | J. Jurkowski | |
| | | Dierk Wanke | and some more QBIC members |

### Organizer

M. Ohya, Chief
H. Sakata
N. Watanabe
I. Yamato
/Tokyo University of Science, Japan

### Advisory Committee

L. Accardi /University of Roma II, Italy
W. Freudenberg /Brandenburg Tech. Univ. Cottbus, Germany
T. Hida /Emeritus, Nagoya University, Japan
H. Kamimura /Tokyo University of Science, Japan
I. Ojima /Kyoto University, Japan

### Local Committee

K. Kuchitsu
T. Matsuoka
S. Miyazaki
H. Takayanagi
M. Takeda
Y. Togawa
S. Tomizawa
/Tokyo University of Science, Japan


QBIC project from 2006 to 2011 is sponsored by Japanese Ministry of Education and Research Institute for Science&Technology Tokyo University of Science

Contact Information    http://www.rs.noda.tus.ac.jp/qbic/    +81-4-7124-1501 ext.5249    e-mail: qbic10@rs.noda.tus.ac.jp